GEOMECHANICAL MODELLING IN ENGINEERING PRACTICE

GEOMECHANICAL MODELLING IN ENGINEERING PRACTICE

Edited by

R.DUNGAR
Motor Columbus Consulting Engineers Inc., Baden

J.A.STUDER
GSS Consulting Engineers Inc., Zurich

A.A.BALKEMA / ROTTERDAM / BOSTON / 1986

CIP-DATA KONINKLIJKE BIBLIOTHEEK, DEN HAAG

Geomechanical

Geomechanical modelling in engineering practice / ed. by R. Dungar, J.A. Studer. – Rotterdam [etc.]: Balkema. – Ill.
ISBN 90-6191-518-X bound
SISO 560 UDC 550.34.01
Subject heading: geomechanical modelling; engineering practice.

ISBN 90 6191 518 X

Distributed in USA & Canada by: A.A.Balkema Publishers, P.O.Box 230, Accord, MA 02018

Printed in the Netherlands

Contents

PART 3: SPECIFIC NUMERICAL MODELS AND PARAMETER EVALUATION

Preface

In recent years the finite element method has become an established tool in the static and dynamic analysis of complex earth and rock engineering problems. The successful application of the method depends to a large extent, however, upon the use of appropriate geomechanical models for the given situation, involving the determination of representative material parameters.

For the practicing engineer it is becoming more and more difficult to obtain a clear picture of the progress made in numerical modelling, because of the flood of scientific publications with which he is confronted. The aim of the International Symposium on Numerical Models in Geomechanics* held at the Federal Institute of Technology, Zurich, was to provide a forum for discussion and exchange of views between researchers and practicing engineers. Out of this conference arose the idea of collecting together important representative theories for reference purposes. This is the origin of the present book.

The book has sixteen chapters written on various topics of geomechanical modelling. The first four introductory chapters provide a survey of the foundational principles of soil and rock modelling, the basic theory for the static and dynamic analysis of geotechnical structures with particular emphasis on the Finite Element Method, as well as a survey of laboratory and field methods for determining material parameters. With this basic knowledge, which is related as far as possible to the needs of engineering practice, the reader is in a position either to begin to attack the problem himself or to enlist the aid of specialists in a more objective manner. The next twelve chapters deal with problem-oriented solutions and also details of chosen soil models.

Each chapter has been written by an expert in that particular field. Amongst others, the following topics are covered: design of tunnels near the ground surface, design of earth and rockfill dams under dynamic and static loading, e.g. due to earthquake and possible liquefaction, analysis of sedimentation ponds, as well as several aspects of the modelling of foundation engineering problems. The individual chapters, which are more or less complete in themselves, reflect recent scientific progress, but still attempt to keep in mind the requirements of the practicing engineer, whereby the emphasis is on reliability and thus speculative ideas of a purely research nature have been avoided.

*Dungar, R., G.N. Pande & J.N. Studer (eds.) 1982. *Numerical models in geomechanics, Proceedings of international symposium.* Rotterdam: Balkema.

The editors believe that with this book a contribution has been made to further the successful use of new analytical tools in engineering practice. At the same time it should serve the practicing engineer as an introduction to recent scientific developments, as well as to point the researcher to the needs of practice.

R. Dungar
J.A. Studer

Contributing authors

Y.F. Dafalias, Prof., Department of Civil Engineering, University of California, Davis, California, USA
R. Dungar, Dr, Dams Department, Motor Columbus Cons. Engrs. Inc., Baden, Switzerland
Z. Eisenstein, Prof., Department of Civil Engineering, University of Alberta, Edmonton, Alberta, Canada
J. Ghaboussi, Prof., Department of Civil Engineering, University of Illinois at Urbana-Champaign, Urbana, Illinois, USA
K. Ishihara, Prof., Department of Civil Engineering, University of Tokyo, Bankyo-Ku, Tokyo, Japan
K.J. Kim, Dr, Applied Research Associates, Royalton, Vermont, USA
P.V. Lade, Prof., School of Engineering and Applied Science, University of California, Los Angeles, USA
S. Nemat-Nasser, Prof., Technological Institute, Northwestern University, Evanston, Illinois, USA
V.A. Norris, Dr, Department of Civil Engineering, University College of Swansea, Swansea, UK
G.N. Pande, Dr, Department of Civil Engineering, University College of Swansea, Swansea, UK
S. Pietruszczak, Dr, Department of Civil Engineering, University College of Swansea, UK
E.G. Prater, Dr, Department of Civil Engineering, Federal Institute of Technology, Zurich, Switzerland
J.H. Prevost, Prof., Department of Civil Engineering, Princeton University, Princeton, New Jersey, USA
R. Pyke, Dr, Consulting engineer, Berkeley, California, USA
H.B. Seed, Prof., Department of Civil Engineering, University of California, Berkeley, California, USA
J.A. Studer, Dr, GSS Glauser Studer Stüssi, Cons. Engrs. Inc., Zurich, Switzerland
R.N. Yong, Prof., Geotechnical Research Center, McGill University, Montreal, Canada

Part 1
An introduction to basic concepts

Linear and non-linear modelling of geomechanical media

1

R. DUNGAR

1.1 WHAT ARE GEOMECHANICAL MATERIAL MODELS?

1.1.1 *Introduction*

Simply stated, a material model is a mathematical relationship describing the stress-strain behaviour of a small but finite quantity of material, and is often termed a 'constitutive model'. The formulation of the material model forms an important step in obtaining solutions for practical engineering problems and, once established, it can be employed in a numerical solution code suitable for the analysis of geomechanical structures, with given boundary conditions, material zones and loading. The model should thus be comprehensive enough to represent all important material behaviour occurring within the given structure under the prevailing loading conditions.

A representative formulation for geomechanical media is particularly important because highly nonlinear material behaviour is often exhibited. A whole branch of geomechanical engineering science has thus evolved for the purpose of formulating and calibrating such models, many of which are now of practical usefulness, and some of which are considered in detail in this text.

Certain general modelling concepts are introduced in this chapter, and the main details of specific models are included in later chapters. In order that the uninformed reader may become acquainted with important experimental findings, this chapter also includes an overview of the observed behaviour of geomechanical media. The most popular numerical solution method, the finite element method, is also introduced in Chapter 2, in the context of which a number of different types of geomechanical problems may be solved.

1.1.2 *Introduction to elastic and piecewise-linear modelling*

As illustrated in Figure 1.1, soil and rock, as with other materials, exhibit nonlinear behaviour in the form of either brittle fracture, at a given applied stress or strain, or of plastic behaviour at relatively low stress levels. There is currently much research effort being made to formulate models where a high degree of material nonlinearity is exhibited, but many problems remain for which conditions of linear elastic behaviour may realistically be assumed, at least as a first approximation, or for which a suitable choice of equivalent elastic parameters over the known or estimated stress range will

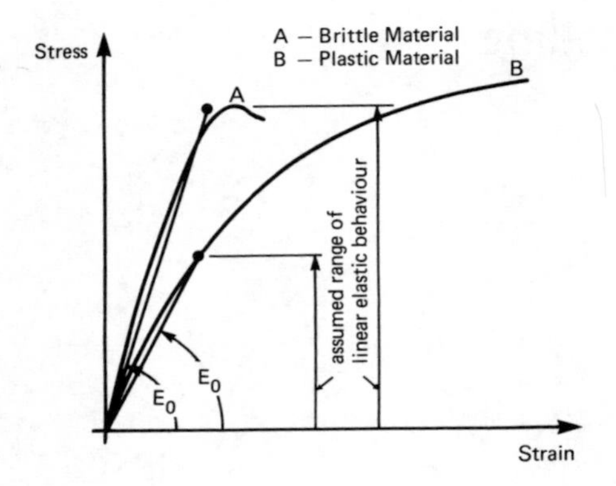

Figure 1.1. Stress-strain characteristics for brittle and plastic material

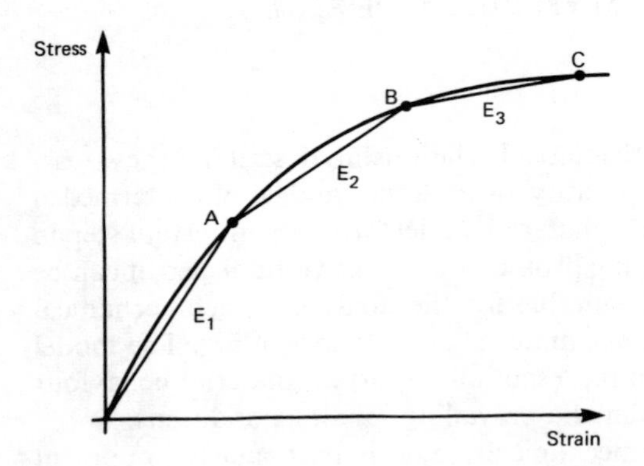

Figure 1.2. Piecewise-elastic idealization of a plastic material

give a solution for both deformation and stress with an accuracy which is entirely acceptable from a practical point of view.

The finite element method of analysis has already been widely used for the solution of practical problems, and many suitable programs now exist for obtaining elastic solutions for a wide range of two- or three-dimensional formulation. With such facilities at hand, historically the next step was to assume elastic material response over a finite range of stress, as illustrated in Figure 1.2, termed 'piecewise linear analysis', and based upon this approach several models have thus been proposed, particularly for soils. The analytical procedure is relatively straightforward, as described in Chapter 3, and basically involves making adjustments to the elastic properties with the load being applied in a finite series of steps. Some of these models will be discussed in detail in this and later chapters.

For the case of dynamic loading, similar piecewise linear procedures may be adopted on a step-by-step basis throughout the time history of the resulting response, in which case a large number of adjustments must be made to the local instantaneous elastic properties, and the problem of modelling the behaviour during reversed loading must also be solved. A more simple approach from the conceptual point of view is to identify 'equivalent elastic parameters' relevant to the prevailing average loading conditions and to conduct the response analysis with these, making iterative adjustments until the assumed conditions match those of the calculated response. This method, known as the equivalent linear method, has found much popularity for the response analysis of soil structures and foundations. This must strictly be classified as a solution method, elastic material parameters being adapted to suit the equivalent conditions, as described in Chapter 3.

1.1.3 *Introduction to elastoplastic models*

Elastoplastic models are based upon the concept that an elastic stress-strain response occurs until a given level of stress, as shown by the line OA of Figure 1.3, followed by a combination of elastic and plastic behaviour, illustrated by ABC. The curve ABC, displayed here in terms of a single component of stress and strain (uniaxial conditions), also defines a simple 'yield curve'. That is, the elastic limit is reached at point A, at a corresponding stress $\sigma_y = \sigma_a$, known as the 'yield stress'. Any further positive increments of stress are associated with increments of irrecoverable plastic strain. Thus, at point B, and corresponding stress $\sigma_y = \sigma_b$, the plastic strain ε^p is retained when unloading occurs, along the line B–O′, whereas the elastic strain, ε, is recovered. On reloading, linear elastic behaviour is assumed until the yield stress, σ_y, is again reached. For this simple example, we may thus write:

$$\sigma_y = f(\varepsilon^p) \tag{1.1}$$

The above relationship is called the 'hardening law' and the material is said to be strain hardening when the yield stress, σ_y, increases with the increment of plastic strain, $d\varepsilon^p$. This is represented in Figure 1.4, together with the case of a 'perfectly plastic' material, for which σ_y remains constant with ε^p. The case of strain softening, for which σ_y decreases as a smooth function of ε^p, and the case of a brittle material for which the value of the yield stress experiences a sudden decrease at a prescribed value of accumulated strain, are also illustrated in Figure 1.4. Geomechanical materials often exhibit combinations of the above behaviour, and thus a material with initial strain hardening may, at a given value of accumulated plastic strain, also undergo a transition to strain softening or even brittle fracture. Strictly speaking, the relationship of Equation 1.1 should be written in terms of stress and strain components in two- or three-dimensional stress space in order that continuum problems may be solved, as discussed in later sections of this chapter.

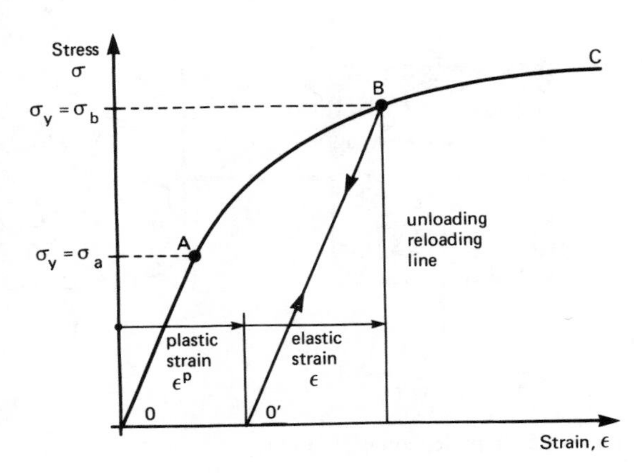

Figure 1.3. Representation of yield stress, plastic strain and unloading-reloading response for a plastic material

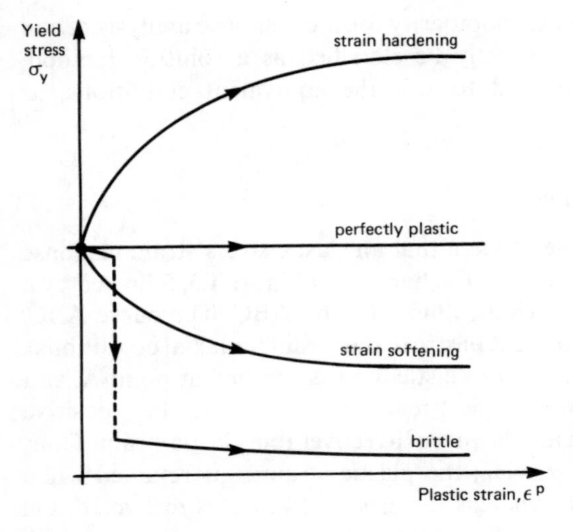

Figure 1.4. Idealised nonlinear stress-strain response of various types of geomechanical material

1.2 EFFECTIVE STRESS, SATURATION AND DRAINAGE

1.2.1 *The principle of effective stress*

The presence of pore fluid and the resulting influence of the pore pressure upon the stress-strain response of geomechanical media is well recognised. Components of stress within a finite volume of material may be separated into contributions from both the solid and fluid phases. The well accepted principle known as the principle of effective stress is illustrated in Figure 1.5, for which it is assumed that the stress acting on the material skeleton is equal to the total stress (stress given by the loads on the boundary of the material element) minus the pore fluid pressure. This results in a matrix relationship which, for the three orthogonal coordinate axes, x, y and z (using

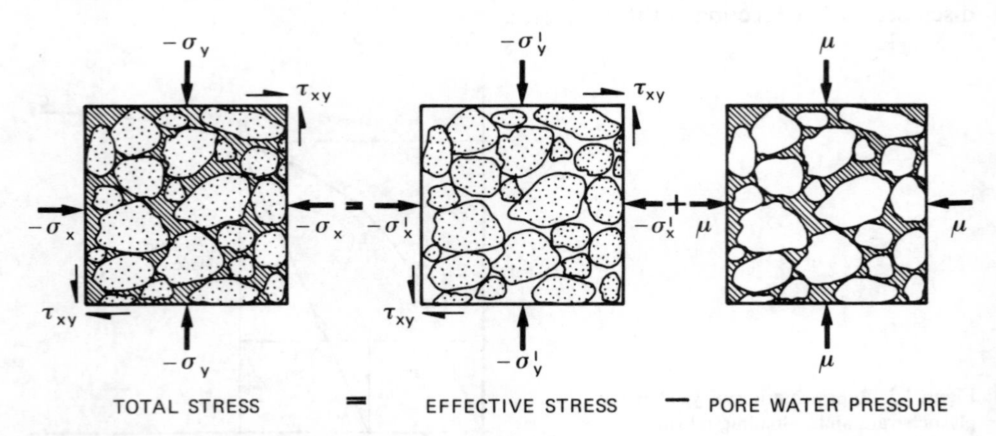

Figure 1.5. Representation of total and effective stress

continuum mechanics notation of positive tensile stress), is given as:

$$\begin{bmatrix}\sigma_x\\ \sigma_y\\ \sigma_z\\ \tau_{xy}\\ \tau_{yz}\\ \tau_{zx}\end{bmatrix} = \begin{bmatrix}\sigma'_x\\ \sigma'_y\\ \sigma'_z\\ \tau_{xy}\\ \tau_{yz}\\ \tau_{zx}\end{bmatrix} - \begin{bmatrix}\mu\\ \mu\\ \mu\\ 0\\ 0\\ 0\end{bmatrix}$$

or:

$$\boldsymbol{\sigma} = \boldsymbol{\sigma}' - \boldsymbol{\mu} \tag{1.2}$$

where $\boldsymbol{\sigma}$ and $\boldsymbol{\sigma}'$ are known as the vectors of total and effective stress, respectively, and $\boldsymbol{\mu}$ the pore pressure vector. The prime (′) is used to denote effective stress.

1.2.2 *Boundary porosity*

The principle of effective stress, as discussed above, is applicable not only to soil but also to rock, provided that sufficient fracturation exists to enable pore pressure to develop throughout the continuum. Thus, when considering the sliding resistance of rock joints, for example, this principle is also applicable because pore pressure may generally develop on the joint plane. It is also common practice to apply the principle in order to calculate the effective stress in situations of only moderate fracturation, for example when calculating the stability of a large dam against foundation sliding, but this may lead to an overconservative estimate of the effective stress.

Consider now the diagram of Figure 1.6. It is observed that the pore fluid pressure provides a decreasing contribution to the total stress with increasing contact area across a given boundary lines, as shown. We may define the effective stresses (see Serafim 1968) from the following relationship:

$$\boldsymbol{\sigma} = \boldsymbol{\sigma}' - \eta\boldsymbol{\mu} \tag{1.3}$$

in which case Equation 1.3 reduces to Equation 1.2 when $\eta = 1$, that is when the area of contact across a given boundary is negligible. Care must be exercised when using the

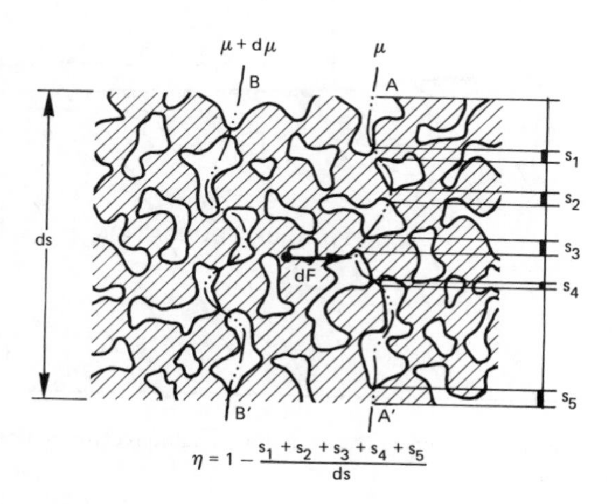

Figure 1.6. Representation of the boundary porosity model

above definition of effective stress, and in passing it should be noted that an additional term must be included in the stress-strain relationship (see Chapter 2), which for isotropic conditions is equivalent to an initial strain (see Section 2.6) given in terms of the three orthogonal components of strain, ε_1, ε_2 and ε_3, by:

$$\varepsilon_1 = \varepsilon_2 = \varepsilon_3 = \mu\left[\frac{(1-2\nu)(1-\eta)}{\mathrm{E}} - \frac{1}{3\mathrm{B_s}}\right] \tag{1.4}$$

where E and ν are the modulus of elasticity and Poisson's ratio of the rock mass without

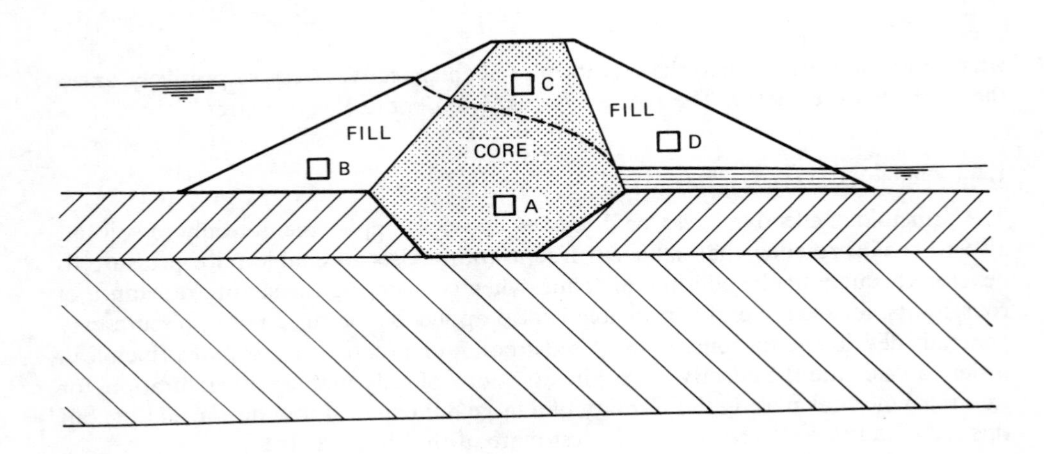

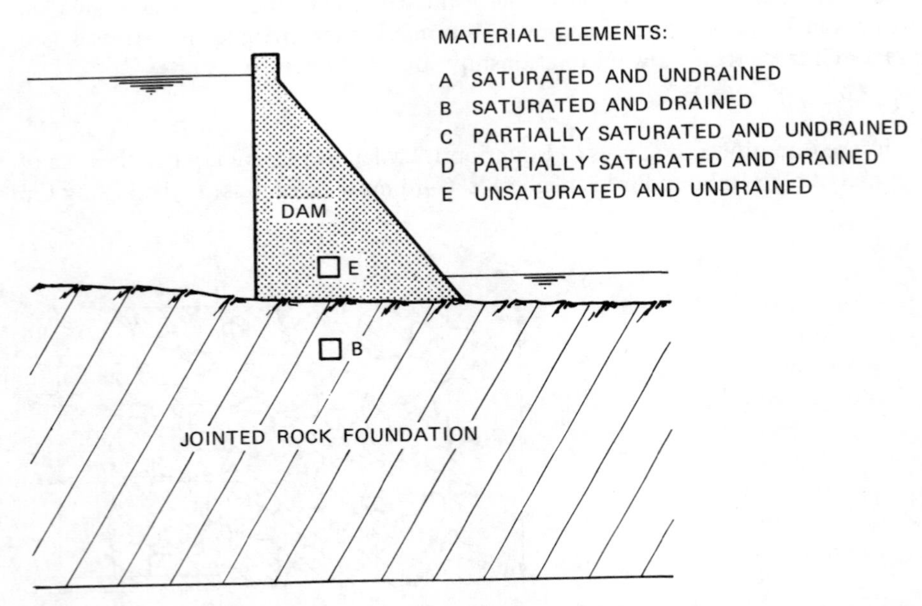

Figure 1.7. Drainage and saturation conditions of materials in various zones of an embankment and gravity dam

any pore fluid, and B_s the bulk modulus of the solid matter (see also Geertsma 1957 for more details).

It is in fact rare that the above boundary porosity equations are used in practice, but it is worth noting that the normal effective stress relationship of Equation 1.2 may lead to an overconservative estimate of the effective stress as mentioned above.

1.2.3 *Saturation and drainage*

The degree of saturation and type of drainage has an important bearing upon the stress-strain response of geomechanical material. A material is fully saturated when the pore voids (soil) or fissures (rock) are completely filled with the pore fluid (usually water). An unsaturated material contains no pore fluid and partial saturation implies the presence of air within the voids.

A saturated material is said to be either drained or undrained according to the time required for pore fluid pressure to dissipate in relation to the time span of the applied load, as will now be illustrated by reference to Figure 1.7.

The core of the embankment, as shown, is of relatively low permeability and may require several years before the pore pressure reaches an equilibrium condition, if at all. Fill material, on the other hand, may reach equilibrium conditions in a matter of months, in the case of a graded soil, or even hours or fractions of a second, in the case of rock fill. The filter zones will in any case be of the relatively undrained category.

The above categorisation must be revised for earthquake loading conditions, the time scale of the applied load being in the order of seconds rather than years. Thus, the upstream fill material, for example, would probably fall into the category of saturated and undrained.

Material models should in general be capable of reflecting the drainage condition unless the model is formulated in terms of effective stress, in which case specialised, and sometimes costly, techniques are required in order to obtain a solution. Further discussion on this topic is given in Section 3.3.1.

1.3 INVARIANTS OF STRESS AND STRAIN

Uniaxial stress-strain behaviour, as considered above, can be used only to describe material stress response in terms of one specific direction of strain-increment. For a material model to be of use in a general numerical solution code it must be capable of representing paths in multi-dimensional stress and strain space.

The simpler numerical models are formulated for isotropic conditions, for which the material fabric, defined by planes of weakness and the orientation of material particles, is ignored. It thus follows that the model may be formulated with any arbitrary orientation of the coordinate system. Hence, the principal components of stress and strain may be employed in the formulation because these are themselves independent of the coordinate directions. The three stress invariants p′, q and θ, illustrated in Figure 1.8, are often used, where:

$$p' = \frac{-1}{3}(\sigma_1' + \sigma_2' + \sigma_3') \qquad (1.5)$$

$$q = \frac{1}{\sqrt{2}}[(\sigma_1' - \sigma_2')^2 + (\sigma_2' - \sigma_3')^2 + (\sigma_3' - \sigma_1')^2]^{1/2} \tag{1.6}$$

and the Lode angle θ (Kirkpatrick 1957) is given as:

$$\tan^{-1}\left[\frac{1}{\sqrt{3}}\frac{(\sigma_1' - 2\sigma_2' + \sigma_3')}{(\sigma_1' - \sigma_3')}\right] \tag{1.7}$$

Figure 1.8 illustrates the principal stress axes together with the hydrostatic axis, $\sigma_1' = \sigma_2' = \sigma_3'$. The plane normal to the hydrostatic axis, called the deviatoric plane or π plane, is also shown, together with the distance from the origin to the π plane, $\sqrt{3}p'$, and the radial distance within the π plane, $\sqrt{2/3}\,q$. The quantities p′ and q, together with the Lode angle, θ, as shown, completely define the stress state σ_1', σ_2', σ_3'.

For the special case of the triaxial test, where $\sigma_2' = \sigma_3'$, it is easily verified that:

$$p' = -\tfrac{1}{3}(\sigma_1' + 2\sigma_3'), \tag{1.8}$$

and:

$$q = \sigma_3 - \sigma_1. \tag{1.9}$$

It will be noted that the above definitions, for p′ and q, are of opposite sign compared with the normal soil mechanics convention. This is because positive stresses are here defined as tensile.

The principal strain increments, $d\varepsilon_1$, $d\varepsilon_2$ and $d\varepsilon_3$, are likewise employed in the formulation of models and these may again be resolved into a spherical (or volumetric) component, dv, and a deviatoric component (or shear component within the octahedral

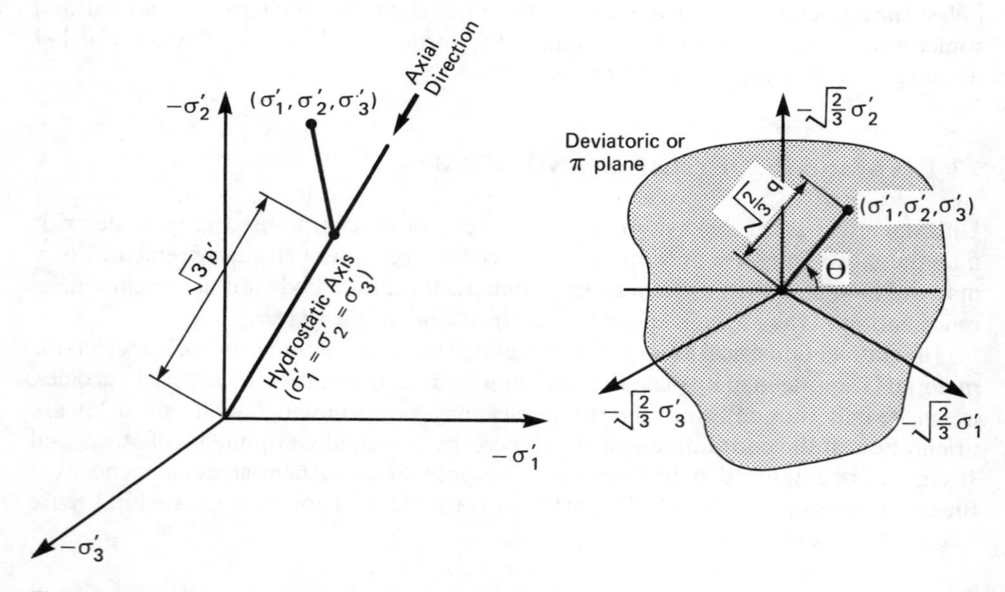

Figure 1.8. Representation of the stress invariants p′, q and θ: the deviatoric or π plane is normal to the hydrostatic axis

or π plane), γ_{oct} (see Valliappan 1981), where:

$$dv = d\varepsilon_1 + d\varepsilon_2 + d\varepsilon_3 \tag{1.10}$$

and:

$$d\gamma_{oct} = \tfrac{2}{3}[(d\varepsilon_1 - d\varepsilon_2)^2 + (d\varepsilon_2 - d\varepsilon_3)^2 + (d\varepsilon_3 - d\varepsilon_1)^2]^{1/2} \tag{1.11}$$

which for triaxial conditions, $d\varepsilon_2 = d\varepsilon_3$, becomes:

$$d\gamma_{oct} = \frac{2\sqrt{2}}{3}(d\varepsilon_3 - d\varepsilon_1) \tag{1.12}$$

The above equations will be employed later in the text.

1.4 AN INTRODUCTION TO OBSERVED GEOMECHANICAL MATERIAL BEHAVIOUR

1.4.1 *Material classification*

Geomechanical materials are here loosely classified as either 'brittle' or 'plastic-granular'. The first category includes different types of jointed, fissured and homogeneous rock as well as certain heavily overconsolidated clays. The second category is related to sands, gravels, rockfill (for dams and similar zoned regions) and also for clays and silts, and are here regarded as having essentially the same characteristic behaviour, provided that similar drainage conditions apply. The expression 'plastic-granular' is used to imply the state of a material in which interparticle flow and grain reorientation freely occurs, as opposed to the brittle state of, for example, a heavily overconsolidated clay.

1.4.2 *Brittle materials*

Brittle materials may be further classified as:

1. materials with broadly homogeneous isotropic conditions;
2. materials with fissures predominantly orientated within certain planes;
3. rocks with fissures and joints which are free from clay and melonite; and
4. rocks with clayey, compressible joints.

Isotropic material properties

In order to quantify both elasticity and strength, it is common practice to conduct both laboratory and in situ testing, usually on small samples. The triaxial laboratory test is commonly employed (see Chapter 4), and besides enabling the value of the elasticity modulus and Poisson's ratio to be obtained from initial elastic conditions, the peak strength attained for various confining pressures allows the Mohr envelope to be constructed, as shown in Figure 1.9. Strain controlled tests enable residual strength conditions to be obtained, as illustrated in Figure 1.10.

Results compiled by Kikuchi et al. (1982) for a wide range of rock type, show a distinct correlation between the elastic modulus and the uniaxial compressive strength,

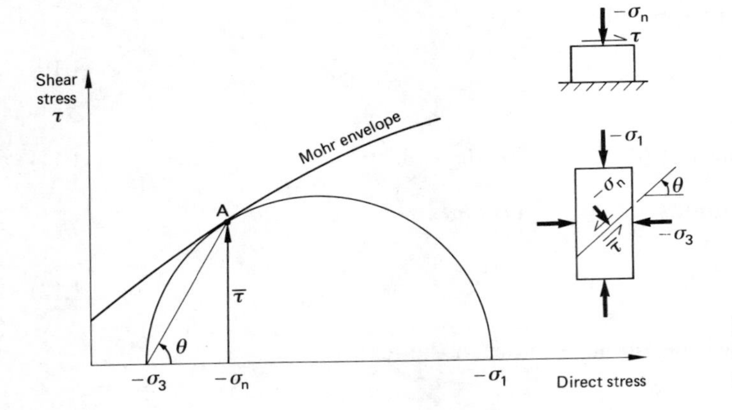

Figure 1.9. Construction of the Mohr envelope from triaxial test results

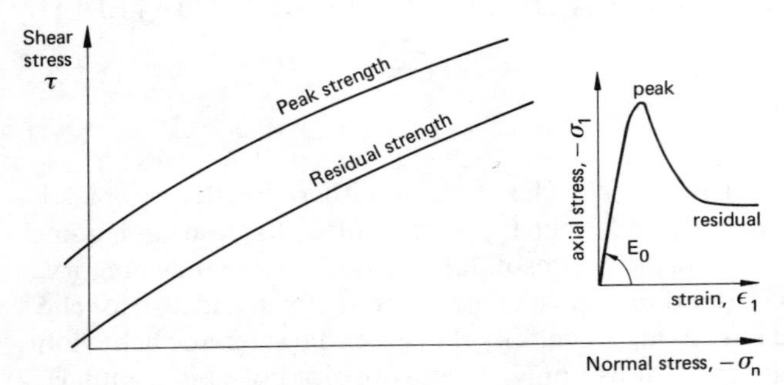

Figure 1.10. Peak and residual strength conditions for rock, as obtained from the triaxial test

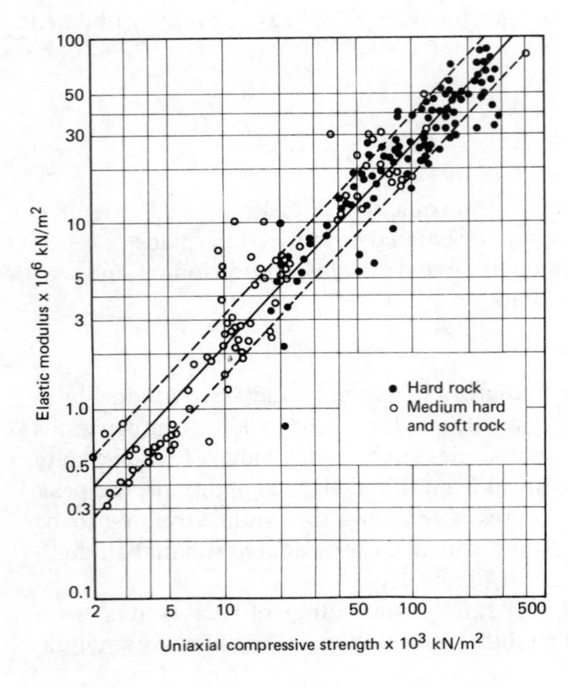

Figure 1.11. Elastic modulus of various rock grades as a function of the uniaxial compressive strength (after Kikuchi et al. 1982)

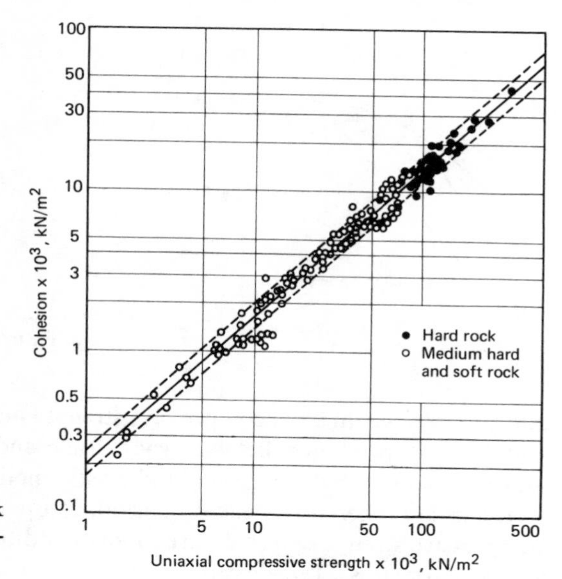

Figure 1.12. Cohesion value of various rock grades as a function of the uniaxial compressive strength (after Kikuchi et al. 1982)

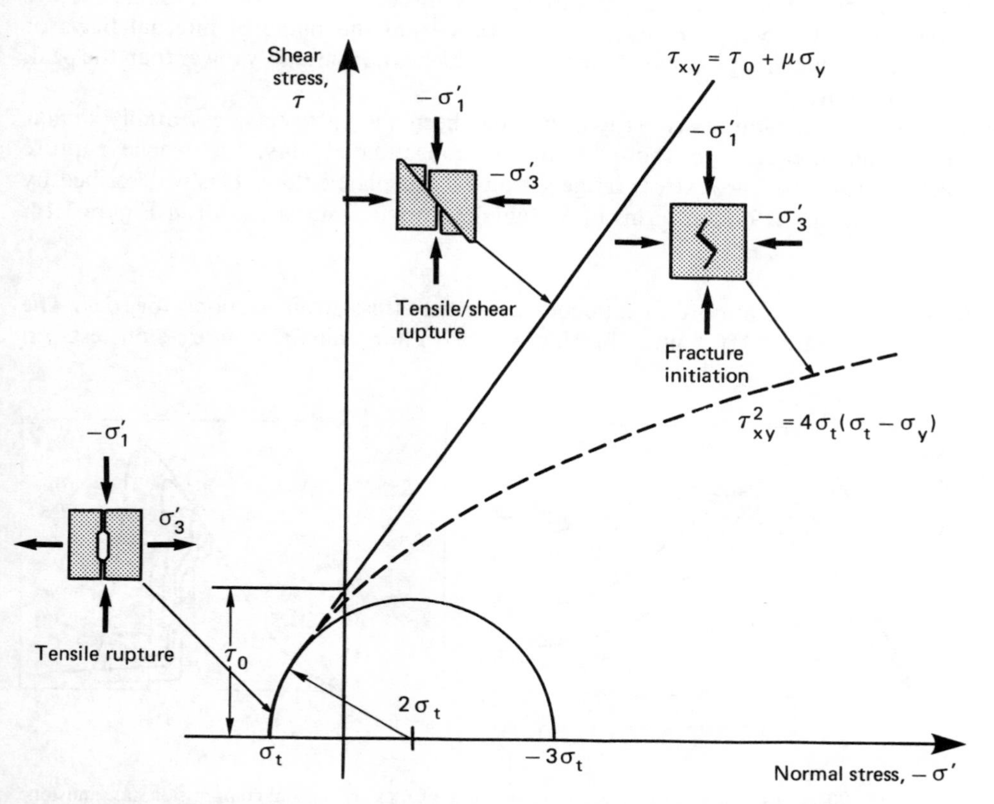

Figure 1.13. A rupture criterion for brittle rocks (showing different rupture mechanisms) (after Hoek 1968)

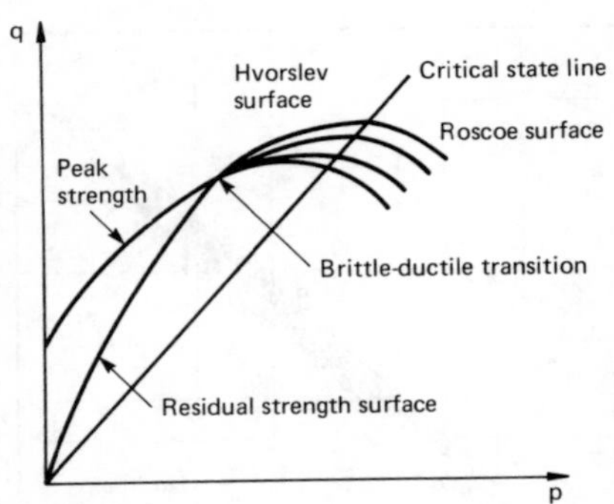

Figure 1.14. A proposed failure mechanism for rocks, (after Price & Farmer 1981)

and between the uniaxial compressive strength and cohesion, as shown in Figures 1.11 and 1.12. A correlation between rock grade and geomechanical properties, such as internal friction angle, cohesion and elastic modulus, was also proposed.

Rocks are inherently weak in tension, compared with the relatively high observed compressive strengths, particularly under conditions of high mean effective stress and low deviator stress. Tensile rupture, Figure 1.13, is experienced in the tensile zone, and in combination with shear rupture under relatively low values of compressive normal stress on potential shear failure surfaces, as discussed by Hoek (1968). Fracture initiation, which will commence propagation from the plane of internal flaws or microcracks, begins at a value of shear stress which is considerably lower than the peak value, as shown.

At very high compressive stress, rock has been shown to have essentially similar failure characteristics to those of an overconsolidated clay, the tensile rupture mechanism of the lower stress range giving way to plastic flow. This is described by Price & Farmer (1981) in terms of a Hvorslev surface, and is shown in Figure 1.14.

Strain rate

Strain rate has an important influence upon the stress-strain response of rock. The results of Figure 1.15(a), after Bieniawski (1970), for uniaxial compression tests on

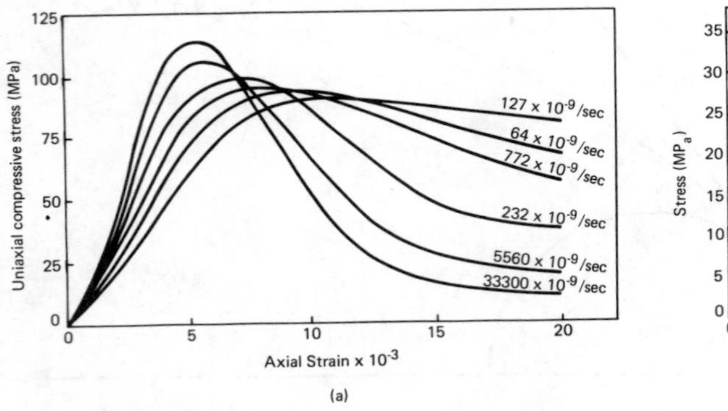

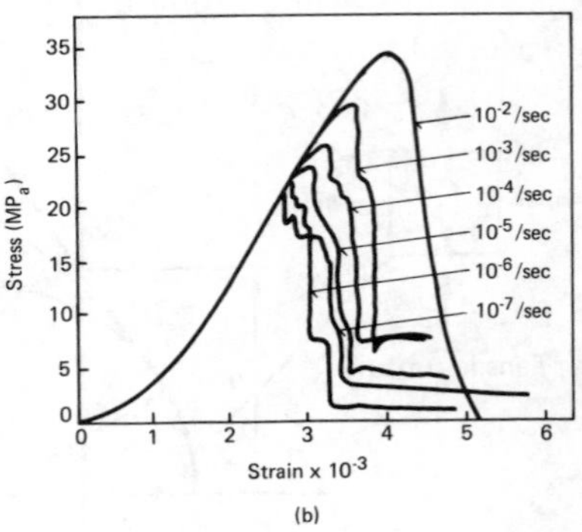

Figure 1.15. Effect of strain rate on stress-strain response of rock in uniaxial compression; (a) sandstone (after Bieniawski 1970); and (b) tuff (after Peng & Podnieks 1972)

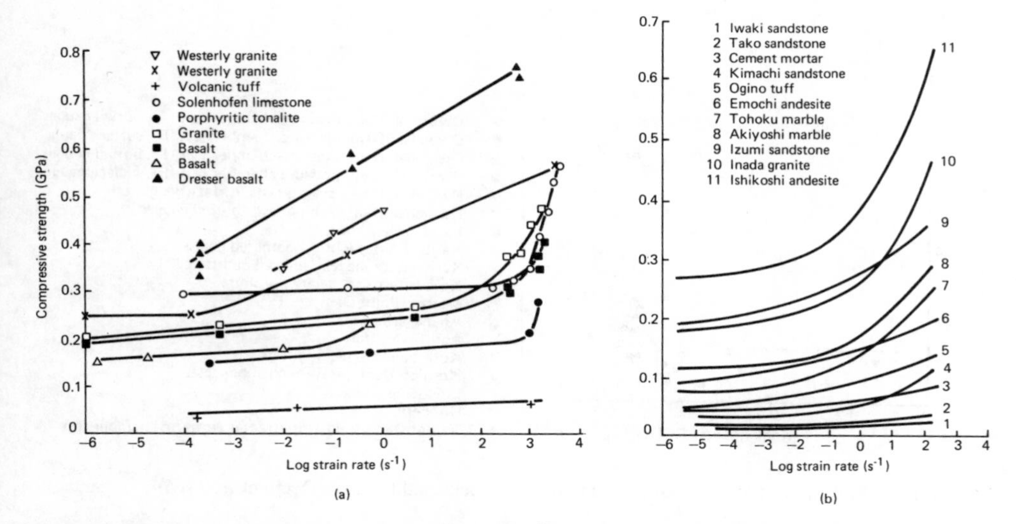

Figure 1.16. Effect of strain rate on compressive strength of rock; (a) after Blanton (1981); and (b) after Kobayaski (1970)

sandstone, conducted at widely varying strain rates, illustrate that lower values of strain rate result in a reduction of the peak stress, but with an increase in the residual stress and also an increase in the strain required to reach the peak stress. The results of Figure 1.15(b), after Peng & Podnieks (1972), for tuff, are also similar to the results from tests on other rocks (Wawersik 1973 and Peng 1973 on marble and sandstone), and also indicate an increase of the peak stress for even higher strain rates. The results of Figure 1.16 also indicate the same trend for even higher strain rates. However, care must be taken when interpreting results obtained for medium to high strain rates (10^{-2} to $10\,\sec^{-1}$), so as to ensure that inertia effects of the test machine are not giving rise to an apparent increase of strength (see Blanton 1981).

Cyclic behaviour

The peak-residual behaviour, obtained from strain controlled triaxial tests under conditions of monotonic straining, is also reproduced when cyclic loading tests are performed, as shown in Figure 1.17, the material showing progressive degradation of both peak strength and elasticity modulus with number of cycles. Indeed, hysteresis

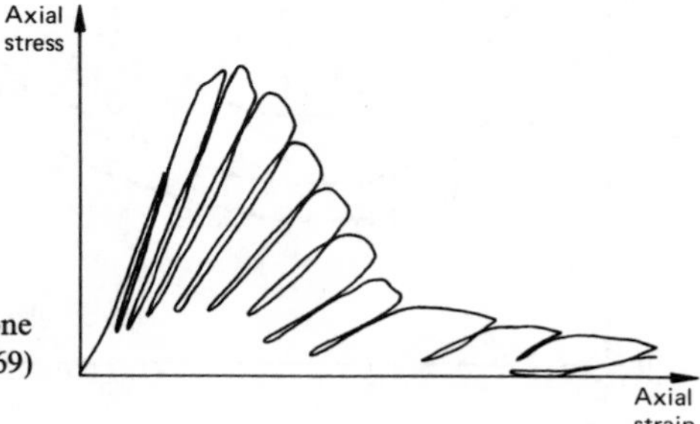

Figure 1.17. Deformational behaviour of sandstone under cyclic uniaxial loading (after Bieniawski 1969)

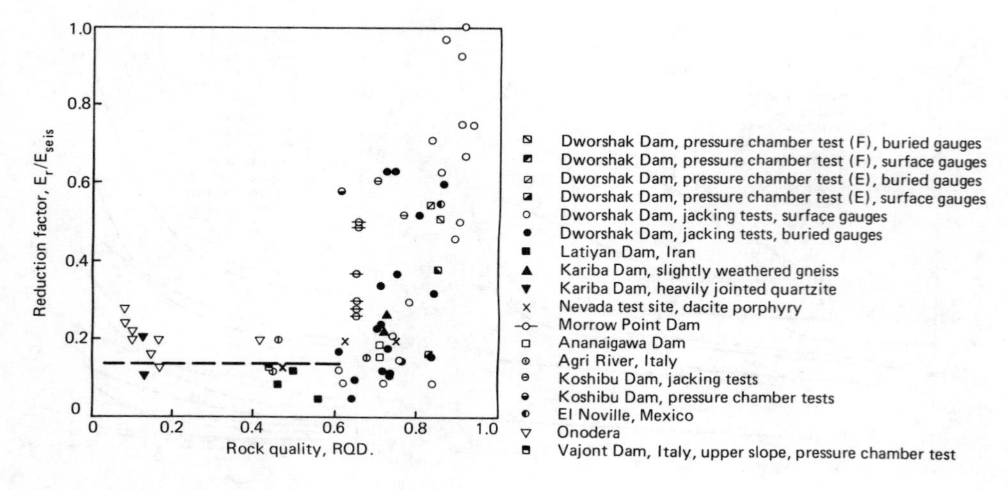

Figure 1.18. Variation of elastic reduction factor with rock quality (after Deere et al. 1966)

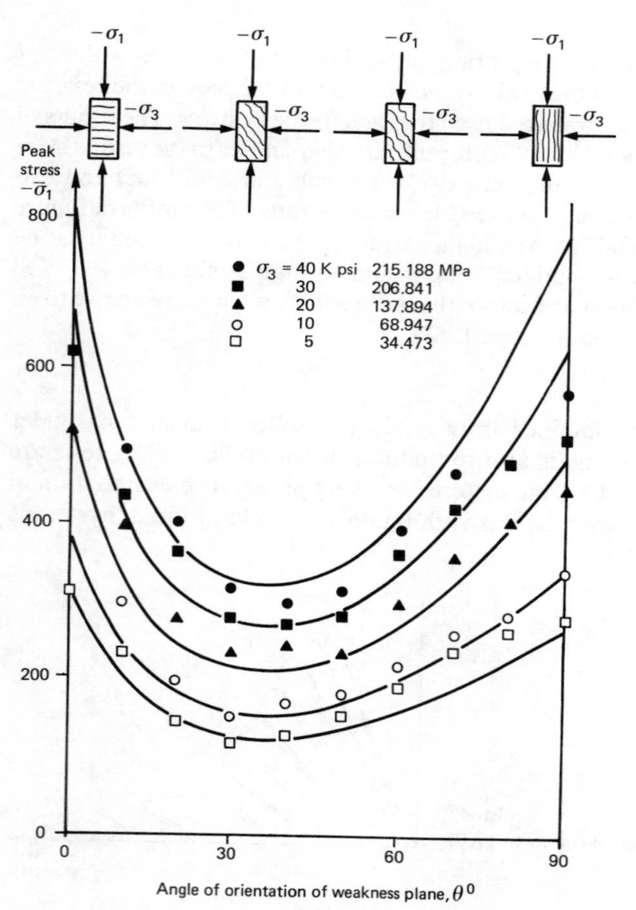

Figure 1.19. Variation of peak strength with orientation of test axis, showing comparison between experimental results on Austin slate and numerical model of Nova (after Nova 1980)

effects due to repeated loading-unloading are commonly reported, for example from instruments situated in large dam foundations, where the load on the dam and foundation due to the reservoir water pressure is periodically applied and removed.

Inherent fracturation and bedding planes
The presence of microfissures has an important influence on the elasticity of the rock matrix, as indicated in Figure 1.18, where the ratio of the laboratory determined elastic modulus and modulus obtained from on site seismic testing (see Chapter 4) is correlated against the rock quality index (RQD).

Metamorphic and sedimentary rocks possess an inherent anisotropy and, apart from exceptional cases, a rock mass will exhibit at least some degree of fracturation leading to weakness planes. Such planes may exist in a number of different directions of orientation, and thus a study of the geomechanical properties, and thus models, for such brittle materials often involves a study of the inherent anisotropy.

Care must be taken when conducting tests on materials with known planes of fracturation, as the peak value of axial stress will vary as some function of the orientation of these planes in relation to the axis of the test apparatus. Triaxial test results performed on Austin slate, for various directions of orientation of the weakness planes, are illustrated in Figure 1.19 in comparison with the model results of Nova (1980).

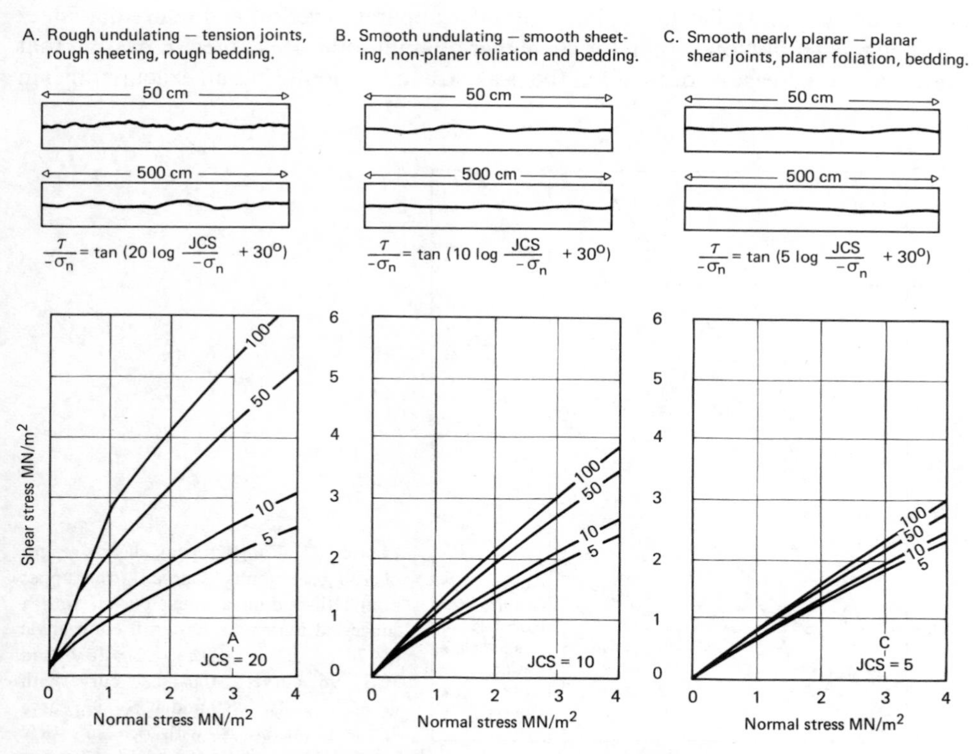

Figure 1.20. Illustration of empirical law of Barton, showing the variation of shear strength of rock joints for various values of JCS (as labelled on each curve) and displayed for three values of JRC, 20, 10 and 5 (after Barton & Choubey 1977)

Joints

Although categorised above as a brittle material, a jointed rock behaves in a similar way to a granular material when observed in relation to the particular joint failure plane. Under the action of continual shear straining, this plane will offer a resistance which is governed by three factors:

- frictional resistance due to simple sliding, quantified by the angle of friction offered by contacting (smooth) surfaces;
- roughness of the given joint plane, whereby the interfering asperities offer a shear resistance associated with dilational straining in a direction normal to the shear direction;
- crushing at asperity contacts, which tends to counteract the above dilational effect resulting in a lowering of the apparent frictional resistance.

The simple empirical relationship of Barton (1973):

$$\bar{\tau} = \tan\left[\text{JRC}\log\left(\frac{\text{JCS}}{-\sigma_n'}\right) + \phi_r\right] \tag{1.13}$$

where $\bar{\tau}$ = peak shear strength, $-\sigma_n'$ = effective normal stress, JRC = Joint Roughness Coefficient, JCS = Joint Wall Compressive Strength, and ϕ_r = residual friction angle, offers a simple means of quantifying the shear resistance of rock joints. The formula is illustrated in Figures 1.20 and 1.21.

Geometrical details and the degree of weathering play an important role in joint behaviour. A smooth flat joint plane will offer apparent friction and cohesion values which are generally lower than for a rougher dilatant joint. It is noted in passing that the apparent increase or decrease in the peak strength of joints has an exact parallel in

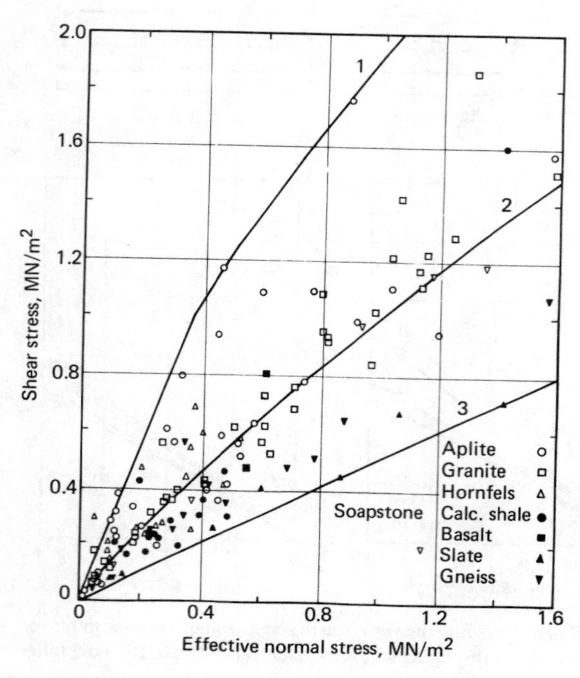

Figure 1.21. Range of peak shear strength of 136 rock joints. Curve 1, the upper bound (including a linear cut-off with a suggested maximum design friction angle of 70°), for $\phi_r = 29°$, JRC = 16.9 and JCS = 96; curve 2, the mean curve with $\phi_r = 27.5°$, JRC = 8.9 and JCS = 92; curve 3, the lower bound with $\phi_r = 26°$, JRC = 0.5, JCS = 50 (after Barton & Choubey 1977)

granular materials where a dense material, which exhibits dilatant behaviour during initial straining, attains a considerably higher peak strength than a loose non-dilatant material.

At high normal stress, and therefore high shear stress, the effects of joint irregularities are diminished because of asperity shearing and crushing.

1.4.3 *Plastic-granular materials*

Behaviour under monotonic loading

All types of plastic-granular materials (clay, sand, gravel) exhibit essentially similar characteristics, according to their void ratio (sand and gravel) or water content (clay).

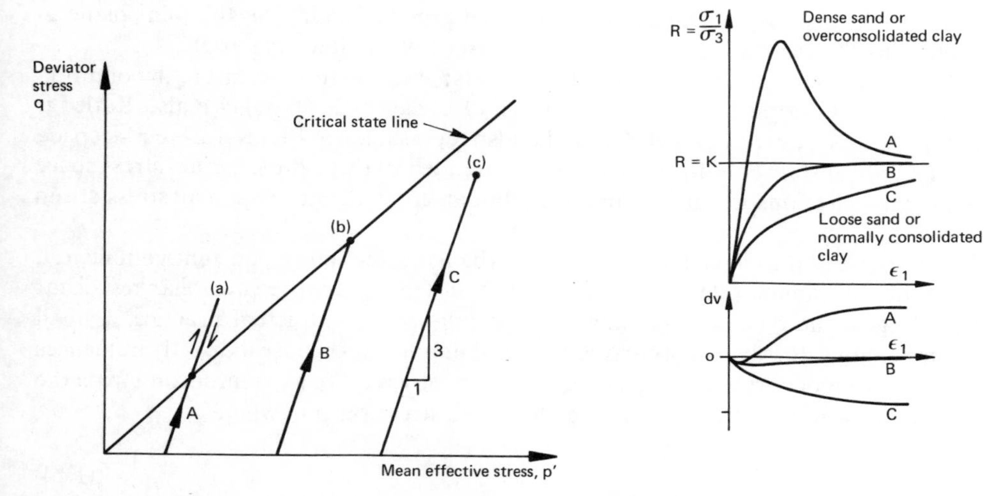

Figure 1.22. Drained stress-strain characteristics for various states of sand and clay under triaxial loading conditions

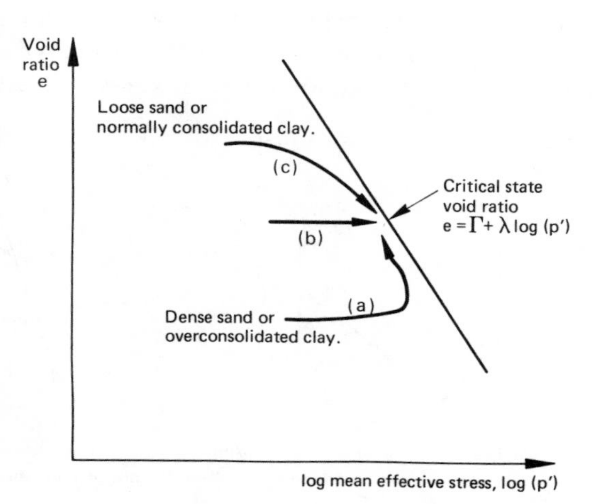

Figure 1.23. Compaction-dilation characteristics for various states of sand and clay under triaxial loading conditions

Shear stress for drained samples of dense sand and heavily overconsolidated clay attains a high peak value followed by 'softening' to a residual strength, as shown in Figure 1.22. The material dilates, as shown in Figure 1.23, until a given void ratio is reached, termed the critical void ratio, at which shearing continues at constant volume. A similar phenomenon occurs for loose granular soils and normally consolidated clays, except that the material now compacts until essentially the same void ratio as for the dense material is attained, following which shearing again continues at constant volume.

It is now well recognised that the critical void ratio may be expressed as a function of the mean effective pressure, the details being given in Section 1.7 in connection with discussion on critical state models. A similar phenomenon was also noted in connection with rock joints, as explained in the previous section, where it is observed that the tendency for a joint to dilate, with an associated increase in the shearing resistance, is also dependent upon the normal pressure acting on the joint plane in relation to the crushing pressure of the asperities of the joint material.

For soils, the residual strength is directly related to the residual strength condition called here the 'critical state' after the work of Casagrande (1936) (see also Rutledge 1947) and Roscoe et al. (1958, 1963), and is also represented in Figure 1.22 in p′-q space.

The critical state condition may also be extended to three-dimensional stress space as shown in Sections 1.6 and 1.7 and has a fundamental influence upon all stress-strain response.

In the case of the triaxial undrained test, which is essentially a constant volume test, the material cannot dilate or compact but instead pore pressure changes occur according to the tendency for dilatation or compaction (in actual practice, a small volume change does in fact occur and is indeed important because it directly influences the generation of pore pressure). Here, the critical stress ratio, M, is important from the point of view of induced strains. The closer the stress ratio η, where

$$\eta = \frac{q}{p'}, \tag{1.14}$$

approaches the critical stress ratio, M, the greater is the induced shear strain component in comparison with the volumetric strain component, and thus large shear strains are developed in the vicinity of the critical state line (see also the discussion of flow rules in Section 1.7). Volumetric strain is compressive below the critical state line and dilative above, and thus as the stress for material type (a) of Figure 1.24 passes this

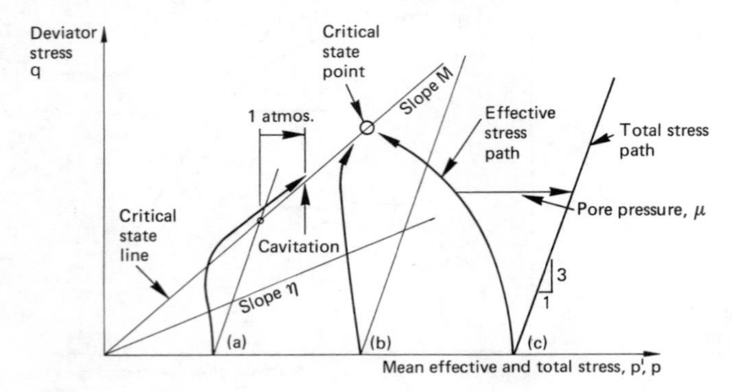

Figure 1.24. Undrained stress-strain characteristics for various states of sand and clay under triaxial loading conditions

line, the pore pressure decreases with increasing shear strain. A point is sometimes reached where cavitation of the pore fluid occurs, as shown. It is noticed that, because the test is at constant volume, the mean effective pressure at the critical state point may be readily determined from such a plot as that of Figure 1.24 and, hence, the maximum deviator stress, $\bar{q}$, calculated.

Behaviour under cyclic loading

Cyclic straining of plastic-granular materials has received much attention in the literature, due to the importance of material behaviour of soil structures and foundations during earthquake and blast (see summary papers in Pande & Zienkiewicz

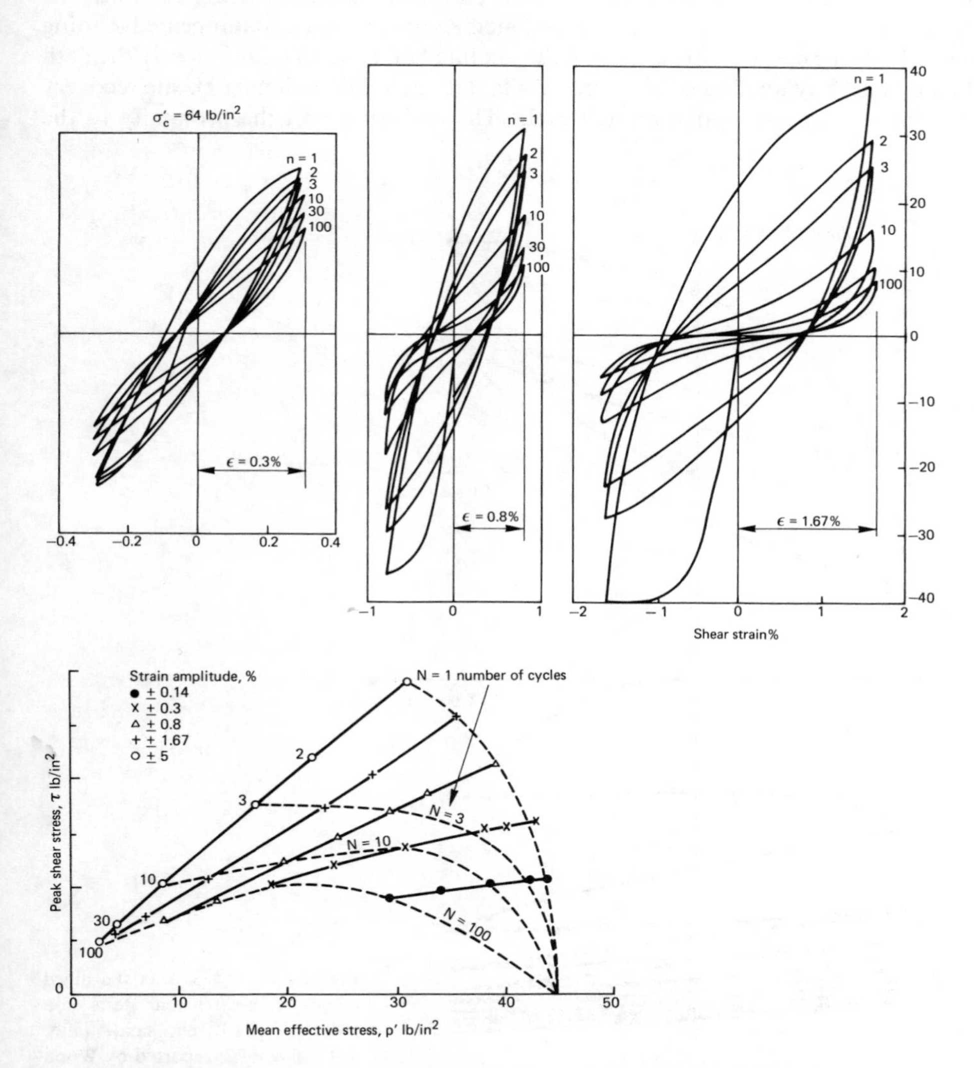

Figure 1.25. Results of cyclic strain controlled triaxial undrained compression-extension tests on normally consolidated halloysite, showing migration of mean effective stress (after Taylor & Bacchus 1969)

(eds.) 1982). Tests are often conducted under cyclic triaxial or cyclic simple shear conditions (see Chapter 4) and either stress-controlled or strain-controlled tests are performed. Both hydrostatically consolidated and K_0 consolidated samples ($\sigma'_1 = K_0\sigma'_3$) are used and one-way or two-way cycling of either stress or strain is employed. Stress-strain results generally indicate hysteresis effects, as shown in Figure 1.25, and cyclic densification of loose drained sands continues until some equilibrium density is reached, as shown in Figure 1.26. The resultant shape of the stress-strain curve is often quantified in terms of equivalent stiffness and damping values, as explained fully in Chapter 10.

Undrained test results show corresponding changes in pore pressure and a general tendency for migration of the effective stress path towards the critical state line. The results of Figure 1.27 for normally consolidated Kaolin, indicate that repeated loading causes the cyclic effective stress path of the loading half cycle to follow closely the path taken by one-way loading of an overconsolidated material, with near elastic recovery occurring during the unloading half cycle. The importance of the proximity to the

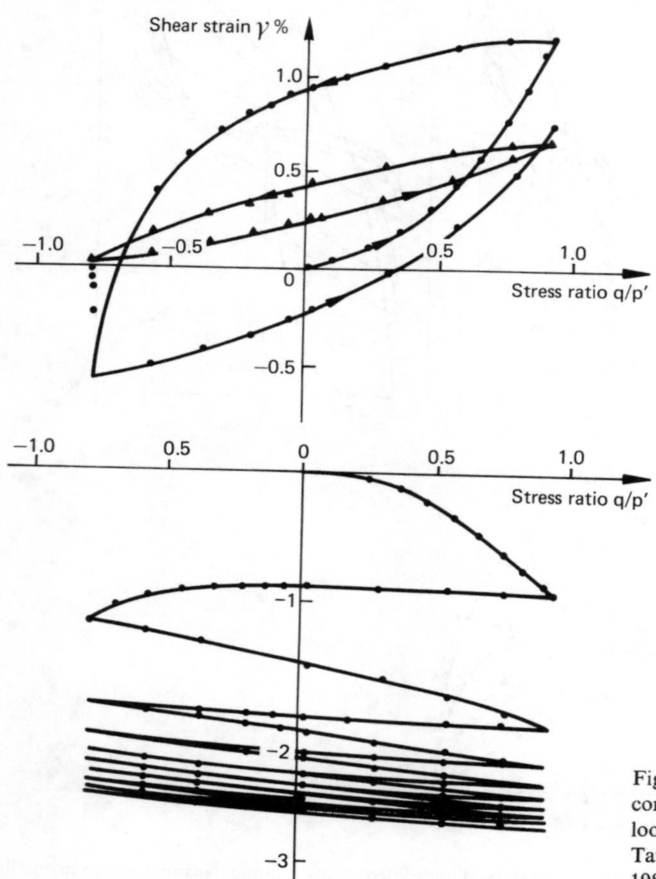

Figure 1.26. Cyclic triaxial drained compression-extension tests on loose Fuji River sand (after Tatsuoka 1972; reported by Wood 1982), with initial void ratio = 0.750 and confining pressure 2 kg/cm²

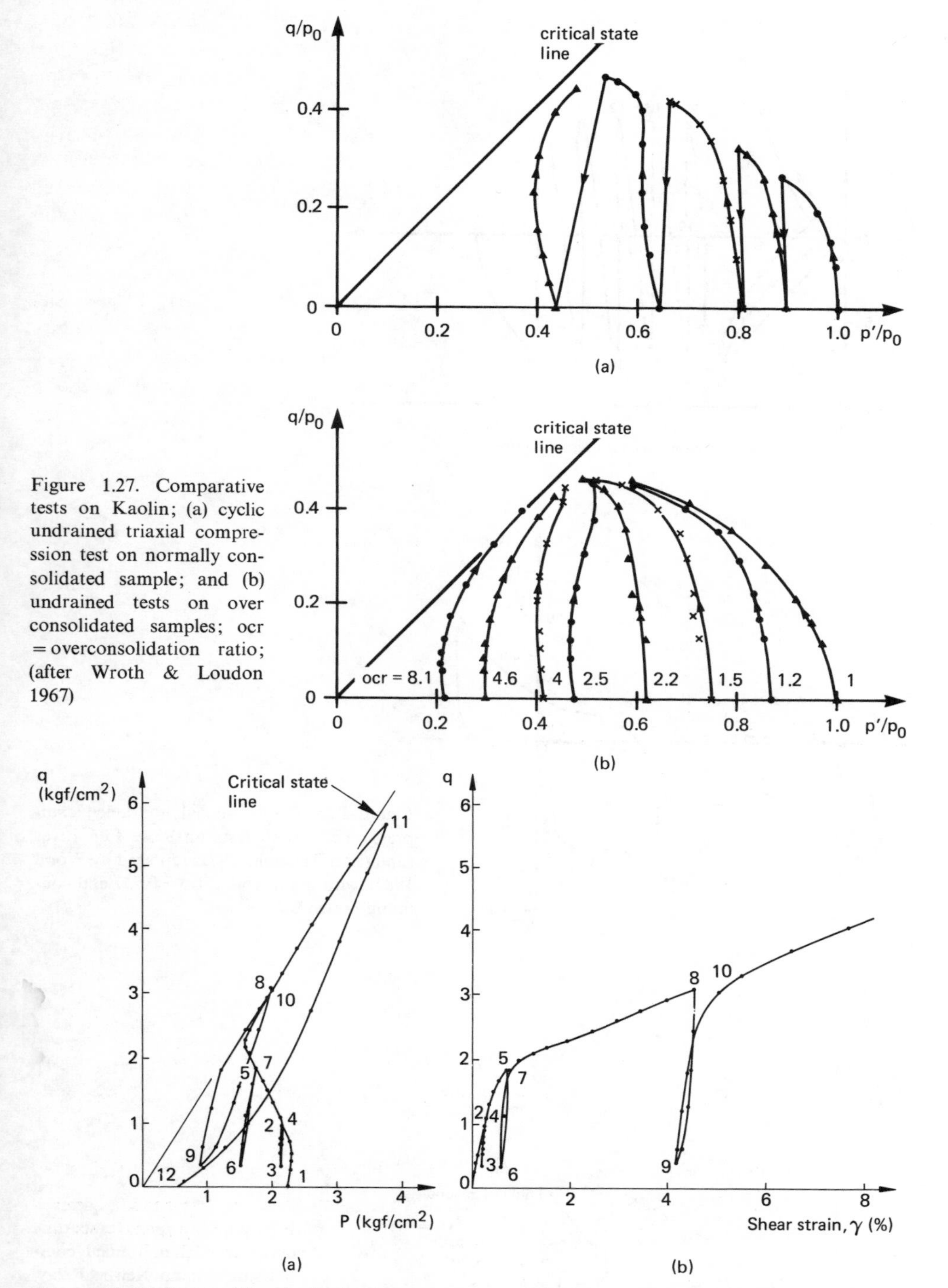

Figure 1.27. Comparative tests on Kaolin; (a) cyclic undrained triaxial compression test on normally consolidated sample; and (b) undrained tests on over consolidated samples; ocr = overconsolidation ratio; (after Wroth & Loudon 1967)

Figure 1.28. Cyclic triaxial undrained compression test on loose Fuji River sand (after Tatsuoka 1972; reported by Wood 1982), with initial void ratio = 0.749 and confining pressure 2.25 kg/cm^2

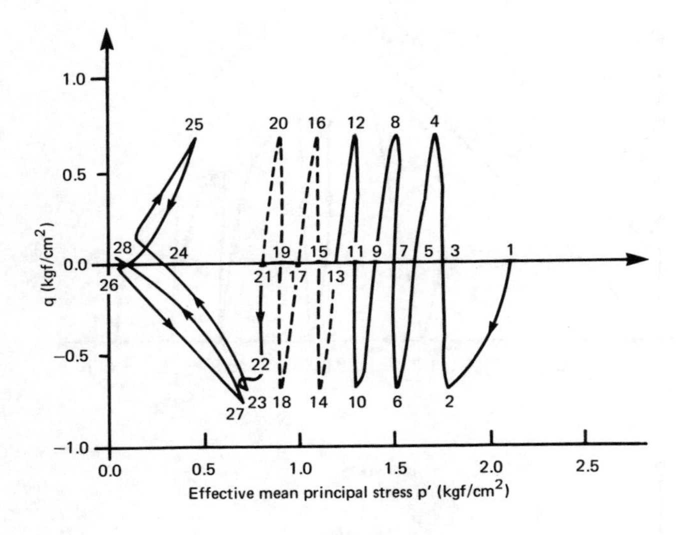

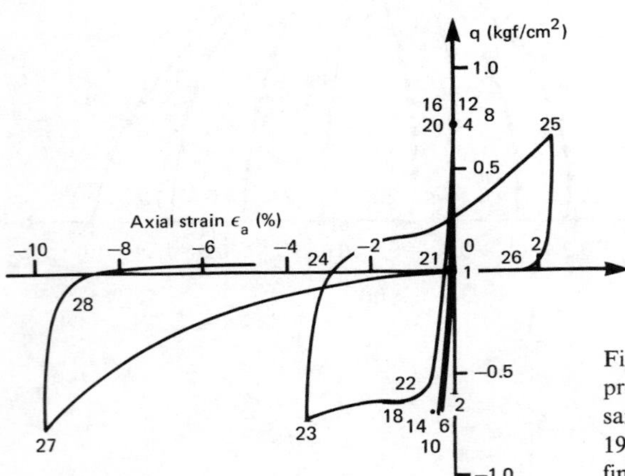

Figure 1.29. Cyclic triaxial undrained compression-extension tests on loose Fuji River sand (after Tatsuoka 1972; reported by Wood 1982), with initial void ratio = 0.737 and confining pressure 2.1 kg/cm²

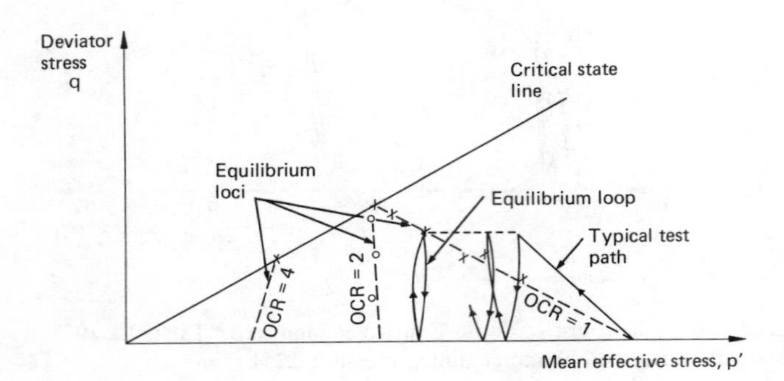

Figure 1.30. Typical test path of cyclic triaxial undrained compression test on Newfield clay, showing equilibrium loci for test series (after Sangrey et al. 1969)

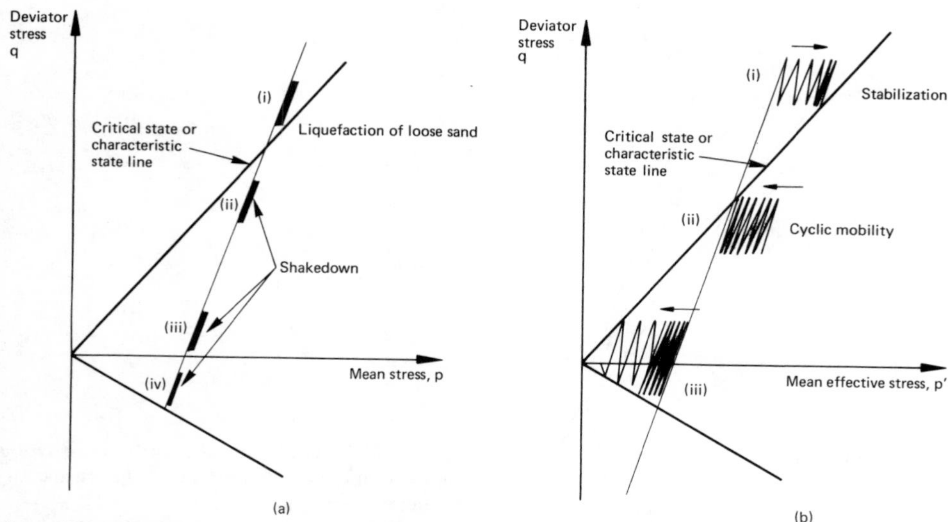

Figure 1.31. Phenomena observed in (a) drained and (b) undrained cyclic loading tests on Fontainebleau sand (after Luong 1980)

critical state line is again evident, as is also shown for the results of Figure 1.28 for random loading of Fuji River sand. In stress controlled undrained tests, a considerable increase in shear strain amplitude is experienced as the material approaches the critical state line, as shown in Figure 1.29. Two-way strain-controlled tests conducted by Taylor & Bacchus (1969) on samples of halloysite also indicated an increase in pore pressure with number of cycles, the effective stress amplitude decreasing with number of cycles.

An important observation regarding pore pressure generation during undrained cyclic tests is that limited generation can occur in stress-controlled tests as illustrated in Figure 1.30, for tests on Newfield clay. Here a typical effective stress path is shown, indicating that a so-called equilibrium loop occurs, for which there is no net change of pore pressure, the mean effective stress, p′, at which this phenomenon occurs being a function of the stress amplitude. The loci of equilibrium points for different initial consolidation ratios are as shown. Again proximity to the critical state line appears to be a governing factor.

The importance of the critical state line* is also demonstrated in the results for Fontainebleau sand (Luong 1980), for which the stress state in proximity to the critical state line again determined the cyclic response. For drained conditions, a material which lay above the critical state line, point (i) of Figure 1.31, experienced incremental collapse due to progressive dilational effects, whilst materials lying under this line,

*Although Luong coined a new phrase 'characteristic state', that is a state at which no compaction or dilation occurs during loading (and presumably unloading) which as he demonstrated occurred at a given stress ratio, $\bar{\eta}$, it is shown in Section 1.7 that this corresponds to the definition of the critical state stress ratio $\bar{\eta} = M$, when the material reaches the critical void ratio. This so-called 'characteristic state' stress ratio is usually assumed to be independent of void ratio, and fabric effects (see Section 1.7) but may indeed vary according to material history, and in particular the difference between the void ratio and the void ratio at the critical state (see Section 1.7).

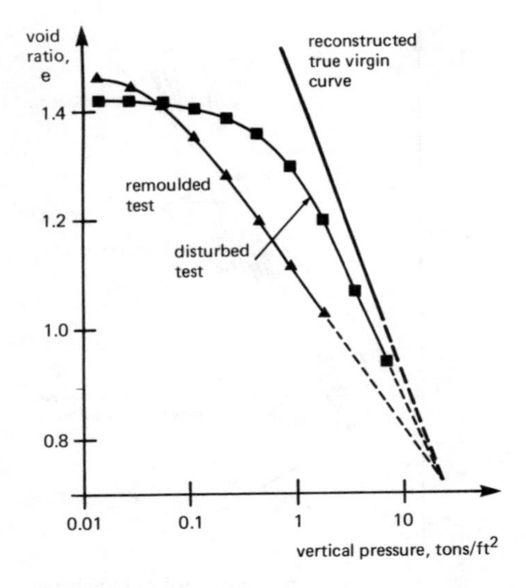

Figure 1.32. Effect of material disturbance on results of normal consolidation tests on marine organic silty clay (after Schmertmann 1955)

points (ii), (iii) and (iv), demonstrated progressive densification and so-called 'shake-down'. For undrained conditions, the tendency for densification at point (i) provided a stabilization due to a corresponding decrease of pore pressure, whilst for materials at point (iii) pore pressure generation occurred, due to the tendency for densification, promoting liquefaction. At intermediate point (ii) pore pressures were again generated but a certain degree of stability was experienced in the region of the critical state line, thus promoting only cyclic mobility (large strain amplitude).

Material disturbance

Materials are often disturbed when sampled from their in-situ state and, although difficult to quantify, effects due to disturbance of material fabric have an important influence upon the resulting geomechanical properties, as often does the remoulding of samples. The results of Figure 1.32 indicate the changes which may occur in the consolidation line, a similar observation being noted by Dungar (1982) with regard to crushing effects in sand samples which were tested at high pressure. Material disturbance during sampling also has an important influence upon the cyclic response.

1.5 HYPERBOLIC MODELS

1.5.1 *Basic models for triaxial and shear conditions*

The stress-strain response of many strain-hardening materials may be approximated as a hyperbolic function, as proposed by Kondner (1963) and as illustrated in Figures 1.33 and 1.34 for triaxial and simple shear test conditions, respectively.

The relevant hyperbolic equations are written in their basic form for the triaxial test as:

$$q = (\sigma_3 - \sigma_1) = -\frac{E_0 \varepsilon_1}{1 + \varepsilon_1/\bar{\varepsilon}_1} \tag{1.15}$$

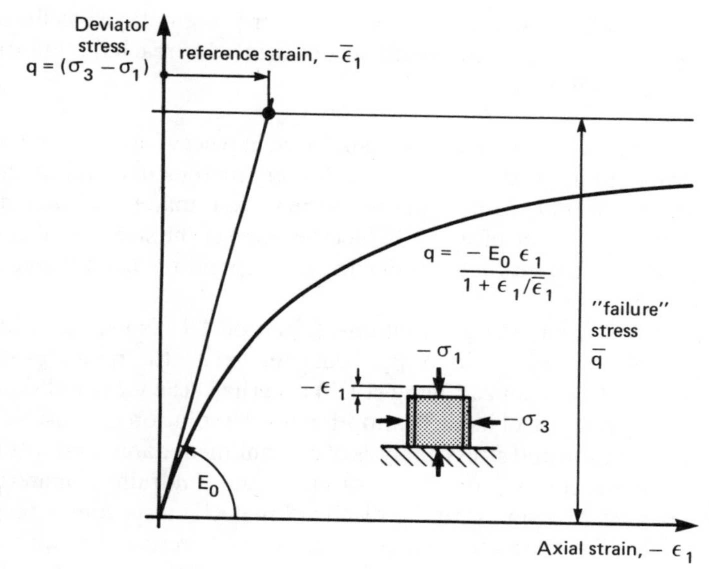

Figure 1.33. A hyperbolic model for triaxial test conditions

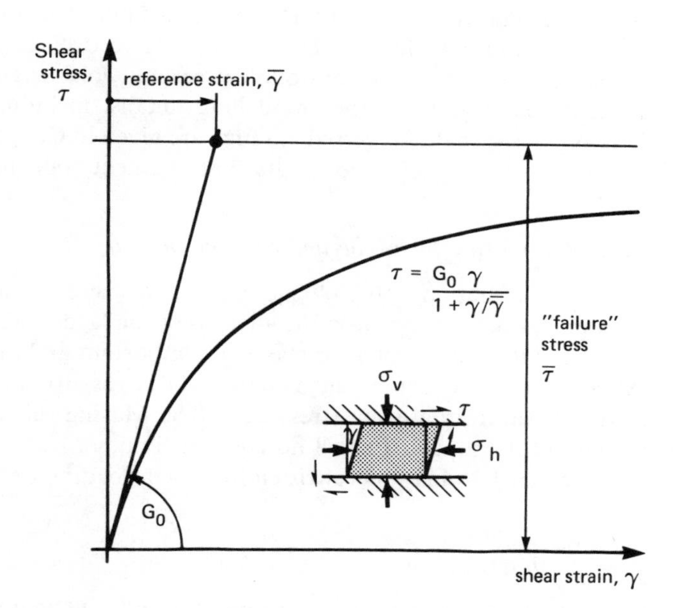

Figure 1.34. A hyperbolic model for simple shear test conditions

where σ_1 and σ_3 are the maximum and minimum principal stress components, $-\varepsilon_1$ the axial strain and E_0 the initial slope. For the simple shear test:

$$\tau = \frac{G_0\gamma}{1 + \gamma/\bar{\gamma}} \tag{1.16}$$

where τ and γ are the shear stress and strain, respectively, and G_0 the initial slope (initial shear modulus).

In both cases, a reference strain, $\bar{\varepsilon}_1$ or $\bar{\gamma}$, is employed as illustrated in the respective figure, and corresponds to the strain required to reach the 'failure stress' when the initial slope is followed.

Both of the above models are employed in piecewise linear analysis, whereby calculated loads are applied in a series of discrete increments, and at the beginning of a given increment the elastic properties of the various material elements are adjusted according to the given state of stress. In fact the slope of the stress-strain curve, called the tangent modulus, is used to determine the corresponding 'instantaneous' elastic parameter (E or G).

For static loading conditions, Equation 1.15 is used as a basis for more advanced models, the initial slope, E_0, being related to the minor principal stress, $-\sigma_3$. The model of Duncan & Chang (1970) is perhaps the most well-known hyperbolic model, mainly because of its early adoption for the solution of constructional and impounding deformation and stress analysis of embankment dams, and subsequent correlation with prototype behaviour. Both drained and undrained material behaviour may be analysed by employing the results of material tests conducted under the appropriate conditions. Care must be exercised when selecting the equivalent Poisson's ratio, ν, which for undrained materials should be set to a value of 0.5 in order to model the condition of no volume charge. However, the finite element method cannot handle this condition, and a slightly lower value of 0.45 to 0.49, depending upon the available numerical accuracy (single or double precision word length) and upon the number of elements employed. Further modelling details, including the use of the variable Poisson's ratio and of reversed loading, are given in Chapter 6, and the methods used for incorporating such models in a finite element code, in Chapter 3.

1.5.2 *Masing rule for cyclic and random loading*

The above Equation 1.16 is often employed for piecewise linear analysis with cyclic or random loads, for example in the analysis of soil layers subjected to earthquake. Here, reversal of load and hysteresis effects are important and the basic model must thus be extended by a procedure which enables the stress-strain response to be redefined for load reversal from a given stress state. The Masing rule is one such procedure and, because of its popularity, will now be briefly described.

Equation 1.16 may be rewritten for virgin loading or unloading as:

$$\tau = \frac{G_0 \gamma}{1 + |\gamma/\bar{\gamma}|} \tag{1.17}$$

where the modulus sign, | |, is introduced to indicate an absolute value. This equation gives the virgin loading curve of Figure 1.35. Reversal of strain from a point A is modelled by simply taking the virgin curve, enlarged in size by a factor of two in both the stress and strain direction, but with the initial condition given by point A (τ_a, γ_a). Thus, Equation 1.17 may be adapted by substituting for the reference strain, $\bar{\gamma}$, a value of $2\bar{\gamma}$ and by taking a charge of coordinates $\tau - \tau_a$ and $\gamma - \gamma_a$, that is:

$$\tau = \tau_a + \frac{G_0(\gamma - \gamma_a)}{1 + |(\gamma - \gamma_a)/2\bar{\gamma}|} \tag{1.18}$$

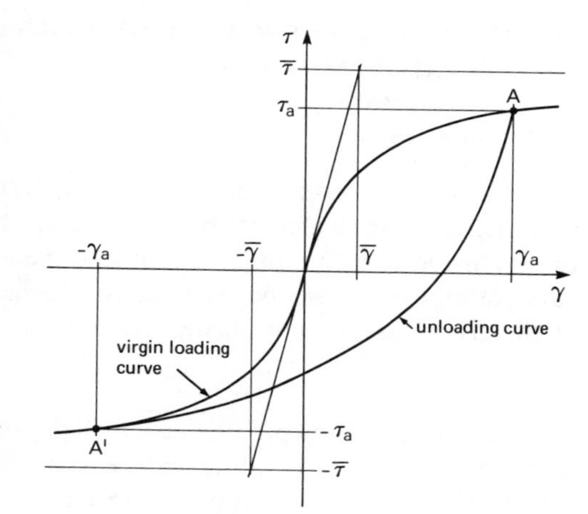

Figure 1.35. Illustration of the Masing rule for modelling unloading-reloading behaviour

It is noted that when $\gamma = -\gamma_a$, the corresponding stress is given as:

$$\tau = \tau_a - \frac{2G_0\gamma_a}{1 + |\gamma_a/\bar{\gamma}|} \tag{1.19}$$

and substituting for τ_a, given from Equation 1.17, we obtain $\tau = -\tau_a$. That is, when the strain is fully reversed to A′ of Figure 1.35, the stress will also be fully reversed. Use of Equation 1.18 enables both small and large strain reversals to be modelled as illustrated in Figure 1.36.

Further discussion of this model and similar models is given in Chapter 10, together with details for the necessary parameter evaluation and accuracy in comparison with cyclic material test results obtained for different strain amplitudes. An introduction to solution methods for continuum problems is also given in Chapter 3.

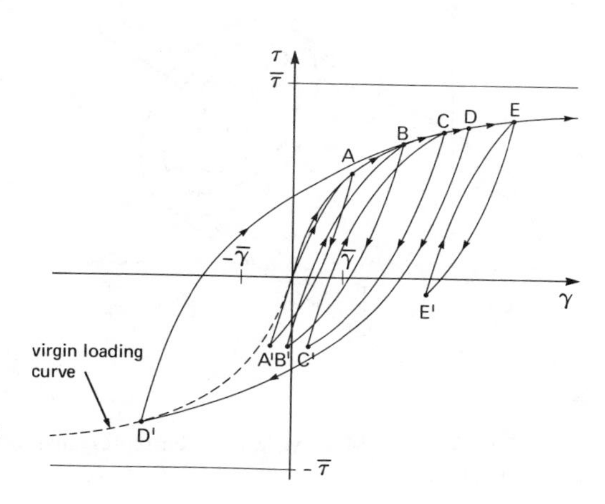

Figure 1.36. Illustration of unloading-reloading paths as given from the Masing rule

1.6 VON MISES, TRESCA, MOHR-COULOMB AND EXTENDED VON MISES MODELS

1.6.1 *Introduction*

We now consider certain details regarding the formulation of elastoplastic models, and in particular three which are based upon well known yield criteria. Firstly we consider the extension of Equation 1.1 to multi-dimensional stress and strain space. The yield curve of Figure 1.3 may be extended to a surface expressed in terms of the components of effective stress $\boldsymbol{\sigma}'$, and plastic strain, $\boldsymbol{\varepsilon}^p$, as:

$$F(\boldsymbol{\sigma}', \kappa) = 0 \tag{1.20}$$

where κ is a 'hardening parameter' which depends upon the plastic strain, $\boldsymbol{\varepsilon}^p$. A family of such functions, corresponding with plastic strains $\boldsymbol{\varepsilon}_1^p, \boldsymbol{\varepsilon}_2^p \ldots \boldsymbol{\varepsilon}_n^p$, are illustrated in Figure 1.37. Thus if we consider a point $p(\boldsymbol{\sigma}')$, the function:

$$F = F(\boldsymbol{\sigma}', \kappa) \tag{1.21}$$

may be regarded as representing a normalised distance from the stress point to the

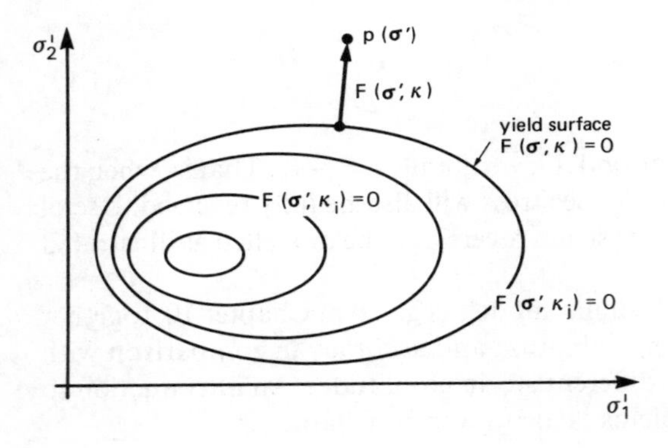

Figure 1.37. The yield surface in two-dimensional stress space

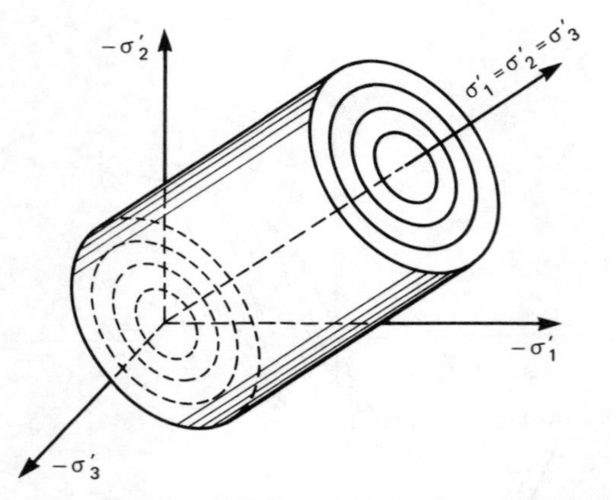

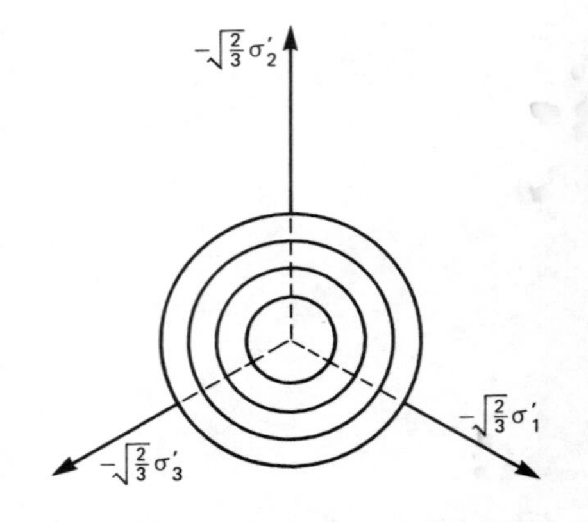

Figure 1.38. The von Mises yield criterion represented in three-dimensional stress space and also on the deviatoric plane

given yield surface, as illustrated. The function F is zero when the stress point lies on the yield surface, given by the corresponding value ε_n^p, positive if lying outside, and negative when lying inside of this surface.

1.6.2 *Von Mises criterion*

As explained in Section 1.3, a simplification may be made for isotropic materials when formulating specific expressions for yield functions, namely that the yield function must be independent of the axes' direction of stress and strain. Thus, the yield function can be expressed in terms of the invariants of stress and plastic strain. Perhaps the simplest yield criterion which may be written for three-dimensional stress space is that due to von Mises.

Here a cylindrical yield surface of radius r is used, with the axis taken as the hydrostatic axis, as illustrated in Figure 1.38. The yield function is simply given as:

$$F = q - r = 0 \tag{1.22}$$

Using the results from triaxial tests, as shown in Figure 1.39, and on substitution into Equation 1.22 for the triaxial test conditions ($\sigma_1' < \sigma_2' = \sigma_3'$), we obtain:

$$F = q - 2c = 0 \tag{1.23}$$

where c is a positive quantity (see Figure 1.39).

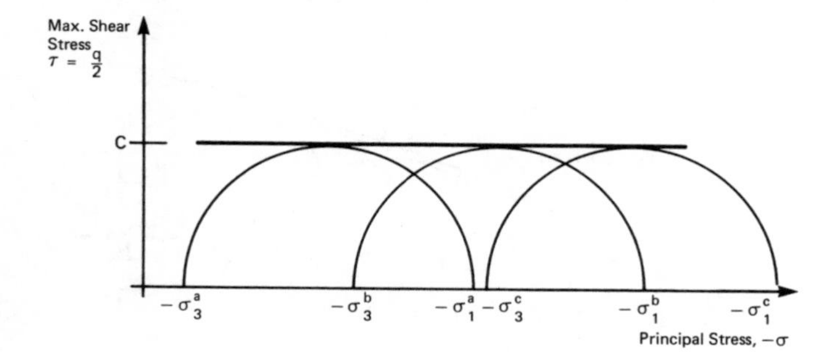

Figure 1.39. Typical triaxial undrained test results for clay, showing the value of the cohesion, c

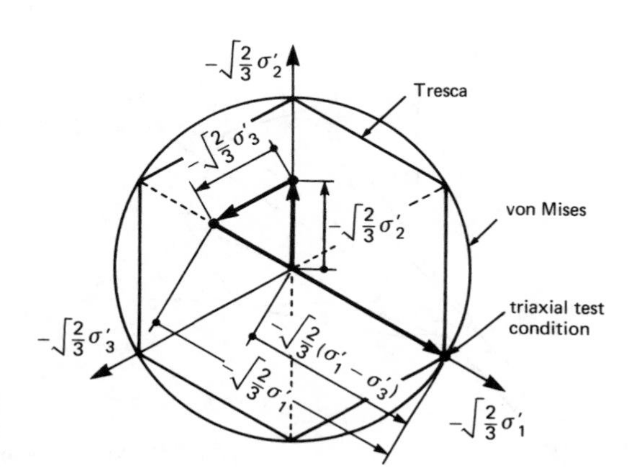

Figure 1.40. Von Mises and Tresca criteria illustrated in the deviatoric or π plane, showing the triaxial compression test condition

1.6.3 *Tresca criterion*

It should be noted that the above expression of Equation 1.23 also corresponds with the maximum shear stress yield criterion of Tresca, but for triaxial compression conditions, only. The general Tresca yield criterion is given as:

$$F = (\sigma_3 - \sigma_1) - 2c = 0 \quad (1.24)$$

and is illustrated in Figure 1.40 for the deviatoric plane, in comparison with the von Mises criterion. The vectorial addition of stress components, and the resultant condition given from Equation 1.24 are also illustrated for triaxial test conditions.

1.6.4 *Mohr-Coulomb criterion*

It is well recognised that drained geomechanical materials exhibit pressure sensitivity, in terms of the maximum shear stress carried by the material. The familiar curve of Figure 1.41, again for triaxial test conditions, illustrates such a failure envelope and one which is often approximated by the Mohr-Coulomb condition:

$$\bar{\tau} = c - \sigma_n \tan \phi \quad (1.25)$$

where $\bar{\tau}$ is the maximum shear stress on the failure plane, $-\sigma_n$ the corresponding normal stress and ϕ a so-called angle of friction. It may be shown that the corresponding failure criteria can be generalised as:

$$F = -\sigma_1'(1 - \sin \phi) + \sigma_3'(1 + \sin \phi) - 2c \cos \phi = 0 \quad (1.26)$$

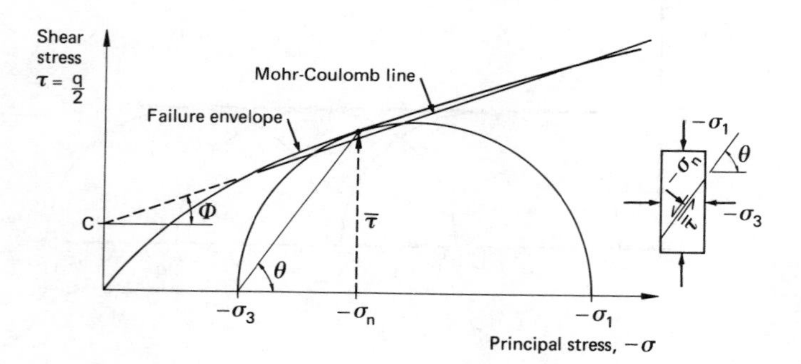

Figure 1.41. Failure envelope obtained from a series of triaxial tests, also showing a constructed Mohr-Coulomb line

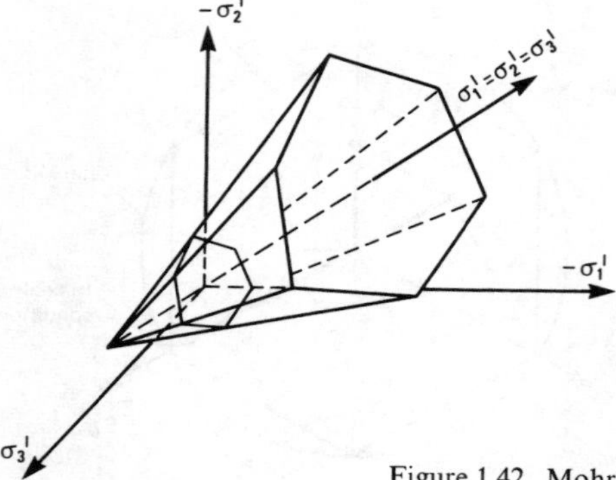

Figure 1.42. Mohr-Coulomb criterion in three-dimensional stress space

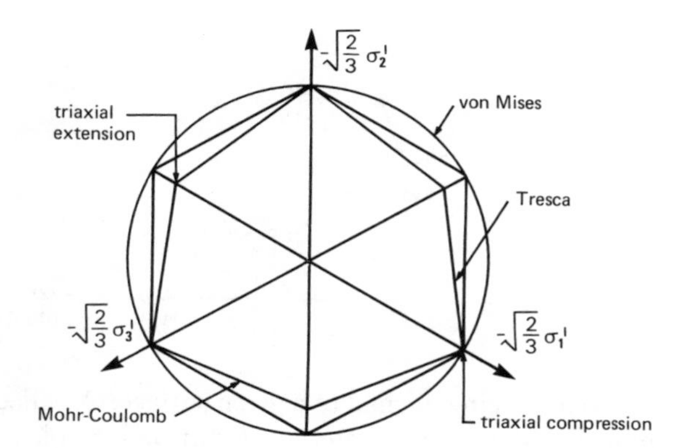

Figure 1.43. Comparison of von Mises, Tresca and Mohr-Coulomb criteria in the deviatoric or π plane

which also reduces to the above Tresca criterion when $\phi = 0$. The yield surface is illustrated in Figure 1.42 in general stress space and in Figure 1.43 on the π plane, in comparison with the Tresca and von Mises criteria.

1.6.5 *Extended von Mises criterion*

The above Mohr–Coulomb criterion, as well as the Tresca criterion, results in corners being encountered in three-dimensional stress space and other approximate formulations have thus been suggested. In particular the conical form of the von Mises criterion is sometimes used with a suitable 'compromise radius' (see Zienkiewicz & Pande 1977 and Figure 3.11) and also a smoothed curve in the π plane, as illustrated in Chapter 12.

The above criteria are often referred to as failure criteria, because they provide an expression for material strength. When such a material element is located within a structure, the term yield criteria is more appropriate, in the sense that when the stress within the given element has reached the given criterion it does not automatically mean that the structure has failed. The material element is still free to undertake excursions along the given yield surface as the redistribution of load, to and from adjacent elements, dictates. In this sense, and with the yield criteria considered above, the material can be regarded as perfectly plastic, because the function, F, is independent of the plastic strain, $\boldsymbol{\varepsilon}^{p}$.

1.6.6 *Plastic strain increment and plastic potential function*

The next step in the formulation is to define the plastic strain increment, $d\boldsymbol{\varepsilon}^{p}$, where the differential here denotes a flow direction which, for a rate dependent plastic material, will correspond to the time rate of the change of the plastic strain, and for a non time-dependent material will simply correspond to the direction of plastic flow for the given stress state. This is illustrated in Figure 1.44 for a given point in stress space, $p(\sigma')$, where the vector, $d\boldsymbol{\varepsilon}^{p}$, for a time-dependent material, given as:

$$d\boldsymbol{\varepsilon}^{p} = \frac{\partial Q}{\partial \boldsymbol{\sigma}'} dt \tag{1.27}$$

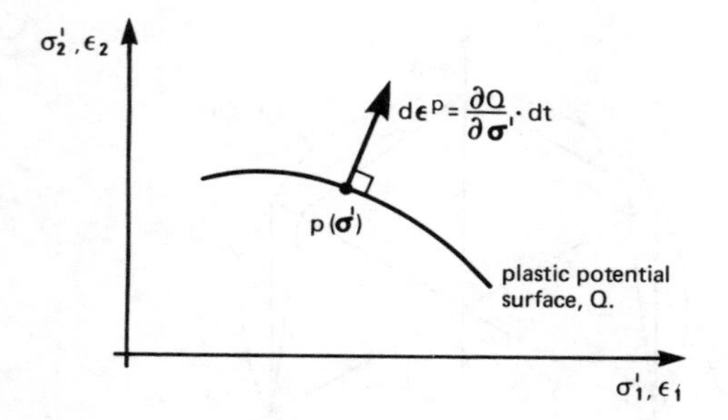

Figure 1.44. Strain-increment vector and plastic potential surface

is taken as being normal to a given surface, Q, called the 'plastic potential surface'. For non time-dependent materials, a normalized plastic strain quantity, λ, is introduced and the increment of plastic strain is given as:

$$d\boldsymbol{\varepsilon}^p = \frac{\partial Q}{\partial \boldsymbol{\sigma}'} d\lambda \tag{1.28}$$

The function Q is often taken as the yield function, F, in which case the model is said to be an 'associated plastic model'. When Q and F are given by different expressions the model is termed 'non-associated'.

In the formulation of an elastoplastic model, as with other models, it is necessary to relate the increments of stress and strain. If we now consider the stress increment, $d\boldsymbol{\sigma}'$, as illustrated in Figure 1.45, the corresponding strain increment, $d\boldsymbol{\varepsilon}$, is given by the sum of the elastic and plastic components, or:

$$d\boldsymbol{\varepsilon} = \mathbf{D}^{-1} d\boldsymbol{\sigma}' + \frac{\partial Q}{\partial \boldsymbol{\sigma}'} d\lambda \tag{1.29}$$

the matrix $\mathbf{D}^{-1}$ relating the increments of effective stress and elastic strain (see Chapter 2). We may now use the additional constraint that, under conditions of plastic straining, the stresses must remain on the yield surface. Thus:

$$dF = \left(\frac{\partial F}{\partial \boldsymbol{\sigma}'}\right)^T d\boldsymbol{\sigma}' + \frac{\partial F}{\partial \kappa} d\kappa = 0 \tag{1.30}$$

the symbol $(\ )^T$ indicating the transpose of the given vector, and where $d\kappa$ is equal to the

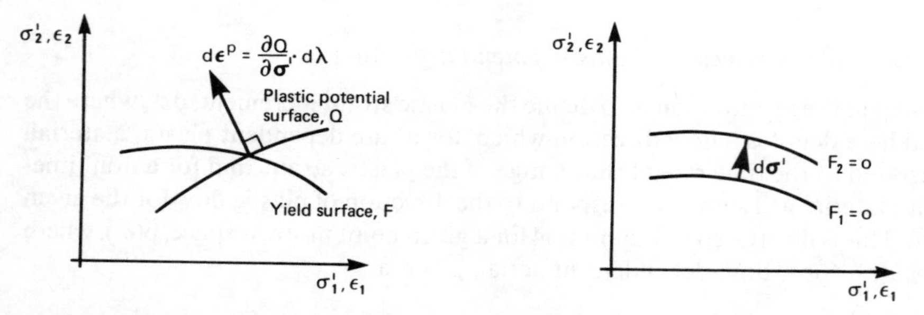

Figure 1.45. Plastic potential and yield surfaces, showing a stress increment under conditions of plastic straining

increment of plastic work:

$$d\kappa = \boldsymbol{\sigma}' d\boldsymbol{\varepsilon}^{p}, \tag{1.31}$$

From Equation 1.28 we obtain:

$$d\kappa = \boldsymbol{\sigma}' \frac{\partial Q}{\partial \boldsymbol{\sigma}'} d\lambda, \tag{1.32}$$

From Equations 1.29, 1.30 and 1.32 it may be shown (see Zienkiewicz 1977) that if:

$$d\boldsymbol{\sigma}' = \mathbf{D}_{ep} d\boldsymbol{\varepsilon} \tag{1.33}$$

where $\mathbf{D}_{\varepsilon p}$ is a matrix relating the increments of effective stress and total strain, then:

$$\mathrm{D}_{ep} = \mathrm{D} - \mathrm{D}\left\{\frac{\partial Q}{\partial \boldsymbol{\sigma}'}\right\}\left\{\frac{\partial F}{\partial \boldsymbol{\sigma}'}\right\}^{T} \mathrm{D}\left[-\frac{\partial F}{\partial \kappa}\boldsymbol{\sigma}^{T}\frac{\partial Q}{\partial \boldsymbol{\sigma}'} + \left\{\frac{\partial F}{\partial \boldsymbol{\sigma}'}\right\}^{T} \mathrm{D}\frac{\partial Q}{\partial \boldsymbol{\sigma}'}\right]^{-1} \tag{1.34}$$

Although at first sight Equation 1.34 appears to be complicated, the vectors $\partial Q/\partial \sigma'$ and $\partial F/\partial \sigma'$ may easily be calculated, once the stress and the functions F and Q are known. The remaining quantity, $\partial F/\partial \kappa$, is defined by the rate of change of the yield function with respect to the plastic work, and is often simply related to a uniaxial stress-plastic strain diagram similar to that illustrated in Figure 1.3.

1.7 INTRODUCTION TO CRITICAL STATE SOIL MODELS

1.7.1 *Stress dilatancy and elliptical flow rules*

The critical state concept plays an important role in the systematic modelling of soils, but before we consider the details, two flow rules will firstly be introduced.

Rowe (1962) introduced what is essentially a flow rule to describe the stress-strain behaviour of drained granular materials in triaxial compression. The stress-dilatancy equation:

$$R = \frac{\sigma_1}{\sigma_3} = \left(1 - \frac{dv}{d\varepsilon_1}\right)K \tag{1.35}$$

or:

$$\frac{dv}{d\varepsilon_1} = 1 - \frac{R}{K} \tag{1.36}$$

essentially relates the ratio of volumetric and axial strain increments to the difference between the ratio of the maximum and minimum principal stress components and a critical stress ratio K. Thus, when $R = K$ there is a zero volumetric strain increment associated with a given axial strain increment. When $R > K$ the material dilates and when $R < K$ it compacts.

The corresponding plastic potential function is illustrated in Figure 1.46(a) where the critical state line of slope M corresponds in p–q space to the critical stress ratio, K. It is noticed that when the stress lies on the p′ axis, the plastic strain increment has both a volumetric and shear component, which implies anisotropic behaviour for a material

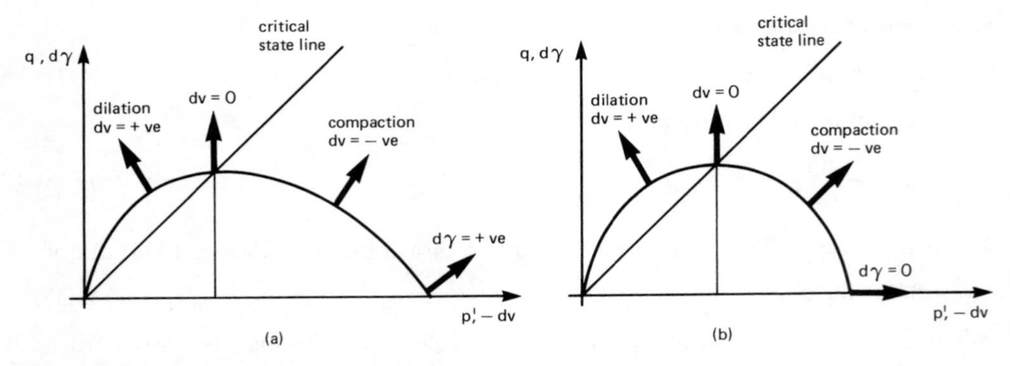

Figure 1.46. Plastic potential functions showing various directions of the plastic strain increment: (a) stress-dilatancy; and (b) elliptical Modified Cam Clay functions

which may, during a subsequent load increment, experience an increase in the mean effective stress, p′.

The elliptical plastic potential function illustrated in Figure 1.46(b) and of the form:

$$F = q^2 - M^2\{p'(p_o - p')\} = 0 \tag{1.37}$$

was adopted by Roscoe & Burland (1968) and overcomes the problem of anisotropy as implied for the above stress-dilatancy equation. It is also noted that the zero volumetric strain condition at the critical stress ratio, $\eta = M$ (and therefore at $R = K$), is again maintained. In both cases, the slope is taken to correspond to the Mohr failure condition. For triaxial compression the slope K is given by:

$$K = \frac{1 + \sin\phi}{1 - \sin\phi} \tag{1.38}$$

and the corresponding slope M, as shown in Figure 1.47, is given for triaxial compression by:

$$M = \frac{6 \sin\phi}{3 - \sin\phi} \tag{1.39}$$

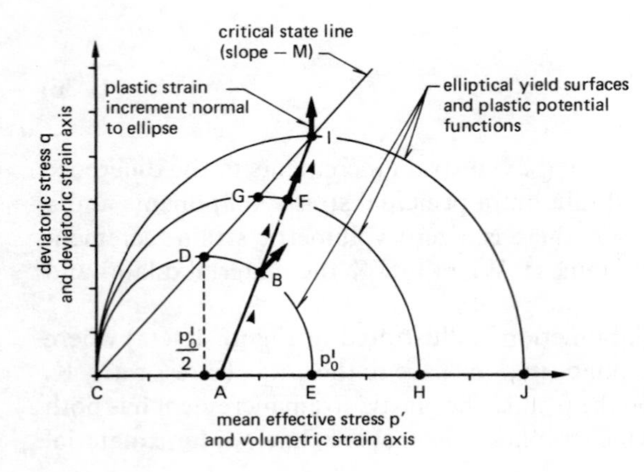

Figure 1.47. Triaxial drained test path, in p′-q space, of an overconsolidated clay or dense sand, showing the elliptical yield surfaces and plastic potential function of the Modified Cam Clay model

and for triaxial extension by:

$$M = \frac{6 \sin \phi}{3 + \sin \phi} \tag{1.40}$$

1.7.2 *The critical state condition*

As suggested by Roscoe et al. (1958), 'a soil element undergoing shear distortion eventually reaches a critical state condition in which it can continue to distort without further change of void ratio, e, and of the components of effective stress, q and p'. This is illustrated in Figure 1.47 for a drained triaxial test on overconsolidated wet clay or loosely compact sand. The material is initially isotropically compressed to point A, followed by triaxial compression (σ_1, increasing) along line ABFI, for which initial elastic behaviour is assumed from A to B until the elliptical yield surface B is reached. Further deformation will cause an expansion of the yield surface with corresponding plastic strain, the direction of which is also defined by the normal to the elliptical plastic potential function, as illustrated. As the yield surface expands, the direction of the plastic strain increment vector is such that an increase in shear strain is experienced in comparison with the volumetric strain until the Mohr or critical state line is reached at point I, at which point the volume change becomes zero and only shear strain occurs.

From the above definition of critical state, the condition of zero volume change can be maintained only under the condition that the void ratio, e, has also reached a critical value. This is important and yet often overlooked. The relation for the void ratio at which the critical state occurs is usually assumed in the form (see for example Schofield & Wroth 1968):

$$e = \Gamma - \lambda \log(p'), \tag{1.41}$$

as illustrated in Figure 1.48. This is termed the critical state line in e-log(p) space. In actual fact, both of the conditions corresponding to the critical state lines of Figures 1.47 and 1.48 must apply, together with the condition that positive shear deformation is occurring, before the material is truly at the critical state.

The virgin compression line is also illustrated in Figure 1.48, where:

$$e = \Gamma' - \lambda \log(p'_0) \tag{1.42}$$

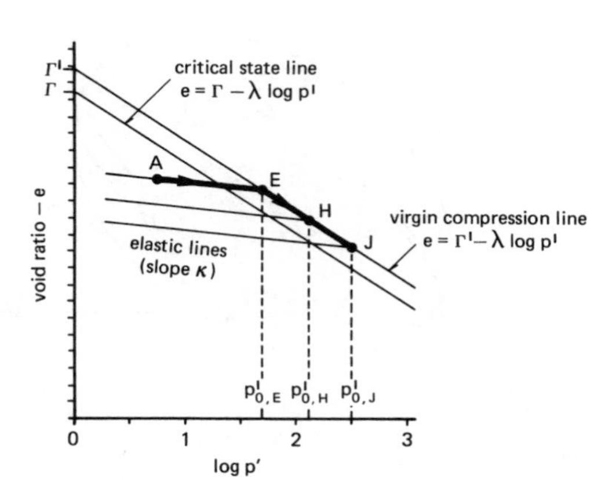

Figure 1.48. Triaxial drained test path, in e – log(p') space, of an overconsolidated clay or dense sand, showing the variation of the pressure p'

and:

$$\Gamma = \Gamma' - (\lambda - \bar{\kappa}) \log(2) \tag{1.43}$$

The value of $\bar{\kappa}$ corresponds to the slope of the elastic swelling line from which, together with the condition:

$$dv = \frac{de}{(1+e)} \tag{1.44}$$

the value of the elastic bulk modulus, K, is given by:

$$K = \frac{(1+e)}{\bar{\kappa}} p' \tag{1.45}$$

It should be noticed that Equation 1.43 is simply given, firstly by the fact that a material at the critical state (as shown by point I of Figure 1.47) which is unloaded elastically in the shear direction and then loaded to the corresponding virgin compression line (point J), will undergo a change in the mean effective stress (I–J) of ratio 2 (because of the elliptical yield function), and secondly by the fact that only elastic volumetric strains occur within the yield surface.

Equation 1.42 is in reality a hardening law for the various elliptical yield surfaces, because it follows that the size of the ellipse in p′–q space is directly controlled by the isotropic consolidation pressure p_0', which in turn is governed by the void ratio, e, as given by Equation (1.42).

The above critical state model, based upon elliptical yield surfaces, is known as the Modified Cam Clay model, and has been used with success in predicting field conditions and notably in connection with the 'MIT Trial Embankment' (Wroth 1977).

The elliptical yield surface has so far been defined by the pressure p_0' and the slope M. It was observed that the value of M is different for triaxial compression and extension and, indeed, if a general model is to be formulated then the full Mohr–Coulomb failure condition must be incorporated. This is easily accomplished by setting M to the

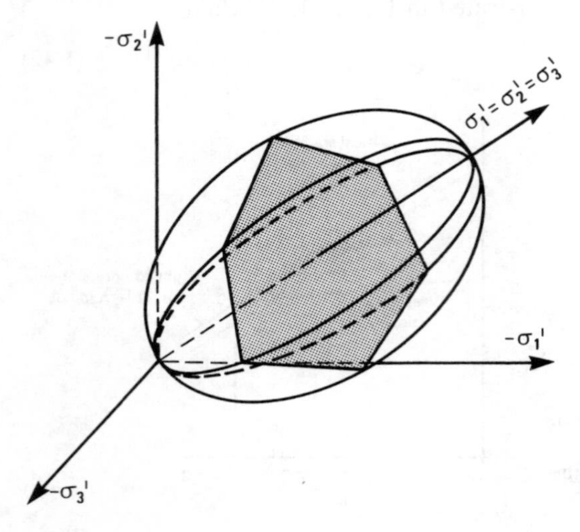

Figure 1.49. Illustration of the Modified Cam Clay ellipse, in combination with the Mohr–Coulomb criterion in three-dimensional stress space

corresponding function of the Lode angle, θ (see Figure 1.8). The resulting yield surface and corresponding plastic potential surface are then as illustrated in Figure 1.49.

1.7.3 *Cyclic loading*

A disadvantage of the above formulation is the inability of the model to perform adequately under cyclic and random loading-unloading. Repeated cyclic loading below the yield surface will produce an elastic response only.

Carter et al. (1982) have attempted to overcome this problem by introducing into the original model a cyclic densification effect, by which the yield surface recedes (softens) during the elastic straining of the unloading-reloading cycle. This modified model gave a limited qualitative correlation with observed laboratory test results for two different saturated clays.

1.7.4 *Endochronic critical state model*

In an attempt to formulate a model for the cyclic response of soil which would also demonstrate an adequate prediction for both the drained and undrained response of overconsolidated clay and dense granular materials, Dungar (1982) employed an endochronic formulation based also upon the critical state model described in Section 1.7.2 above. The main concept is that plastic straining occurs at all conditions of loading and unloading, and the simplest differential equation of the endochronic type which may be used to represent uniaxial material response, with peak stress σ_0, is:

$$d\sigma = E\left(d\varepsilon - \frac{\sigma}{\sigma_0}|d\varepsilon|\right) \tag{1.46}$$

where $d\varepsilon$ is the increment of total strain, E the elastic modulus and the modulus sign, $|\,|$, represents absolute value. This equation was used to obtain the results of Figure 150 with $E = 10$ and $\sigma_0 = 1$ (for illustration only), the strain being reversed at the points a, b, c and d. It is immediately noticeable that both elastic and plastic strains occur at all increments of total strain and that the stress response is well able to qualitatively represent reversals of this strain. The essential difference between a classical plasticity model and an endochronic model is that the yield surface is replaced by a loading surface (plastic straining also occurs on unloading), the stress point always lying on this surface. An elliptical loading surface was again used.

Hardening laws are again required. Here the law based on a differential form of

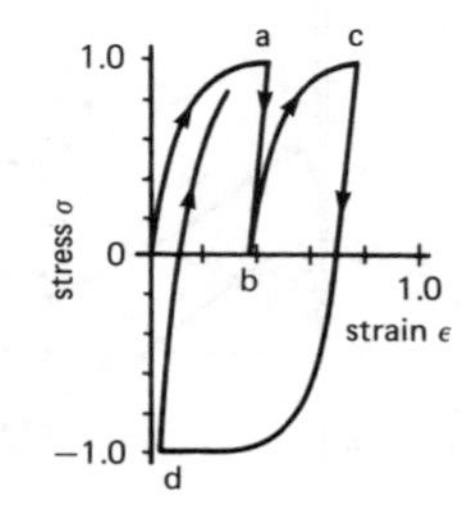

Figure 1.50. Typical stress-strain response showing hysteresis effects when using a simple endochronic model

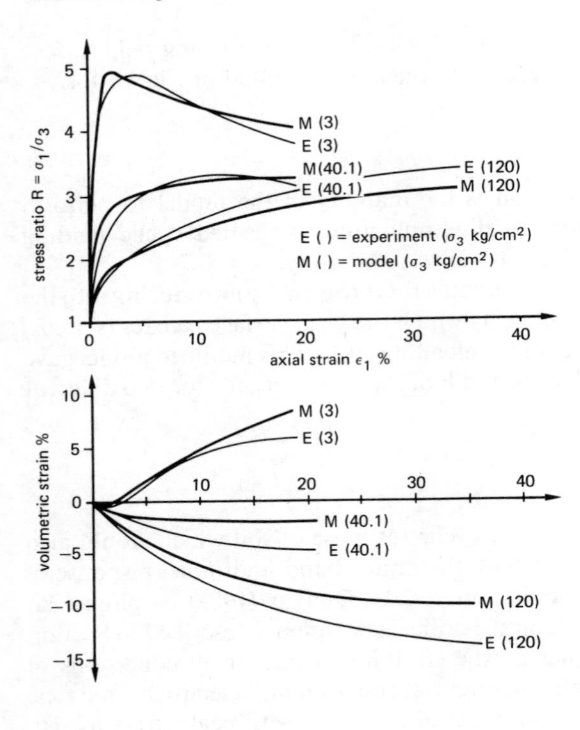

Figure 1.51. Comparison of drained triaxial test results with model predictions (after Dungar 1982)

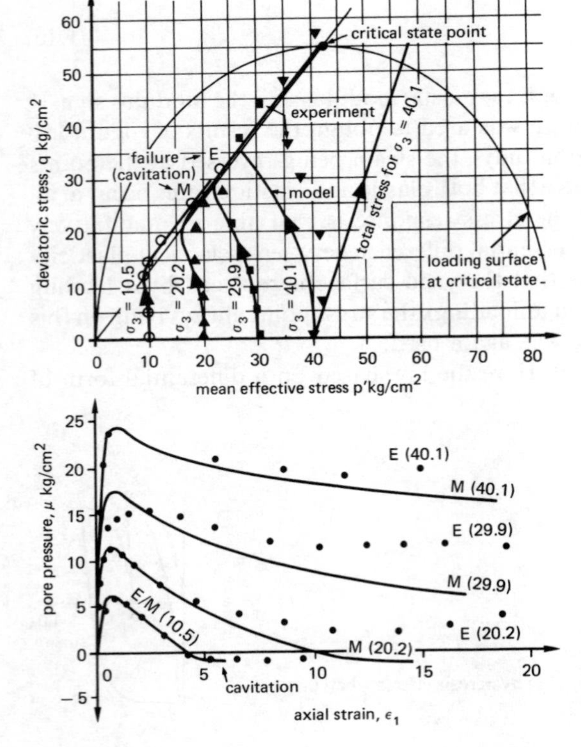

Figure 1.52. Comparison of undrained triaxial test results with model predictions (after Dungar 1982)

Equation 1.42 is need, where:

$$d(\log(p_0')) = E\{[e + e_0 - \Gamma' - 1/((\lambda - \bar{\kappa})E)]d\chi + [e_0 - \lambda \log(p_0')]|d\chi|\} \tag{1.47}$$

and where $d\chi$ is a plastic strain measure:

$$d\chi = f(\alpha)d\gamma^p + dv^p \tag{1.48}$$

and α is related to the difference between the void ratio and the void ratio at the critical state condition corresponding to the given mean effective pressure. E and e_0 are material constants. Results are presented in Figures 1.51 and 1.52 in comparison with drained and undrained tests, respectively, on Sacramento River sand (after Lee & Seed 1967 and Seed & Lee 1967).

It is important to note that the plastic shear strain increment also appears in the hardening law, and this is essential for this type of formulation because it enables the model to 'climb' the critical state line in p–q space until the critical state point (given by the critical void ratio) has been reached.

The above endochronic model also enables qualitative predictions for cyclic and random loading conditions, showing progressive increase in generated pore pressure (see Figure 1.53) and corresponding hysteresis effects. In addition general stress

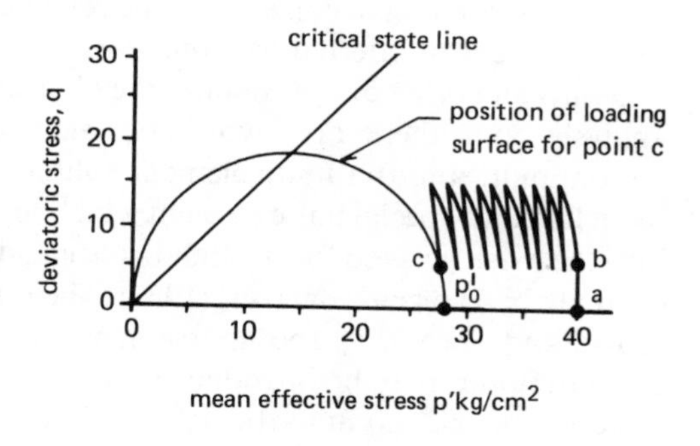

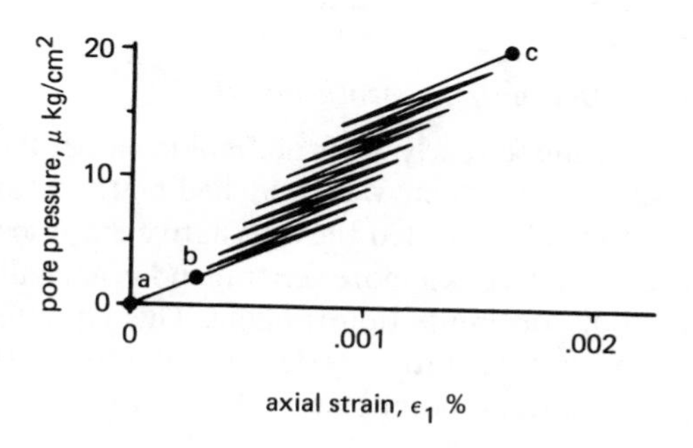

Figure 1.53. Illustration of cyclic pore-pressure generation when using an endochronic model (after Dungar 1982)

histories may be modelled, but precise correlation under all conditions of the strain and stress increment direction has not been made to date.

1.8 ROCK MODELS

1.8.1 *Introduction*

As mentioned in Section 1.4, the study of the geomechanical behaviour of rock and its mathematical formulation is essentially linked with the classification of its macroscopic flaws, microfissures bedding and jointing systems.

The theoretical concept of crack propagation, originally developed by Griffith (1921) for glass, has been applied also to rock. Here, a single void is idealized as an ellipsoid in a plane stress field. Griffith considered a material loaded in tension and assumed a brittle failure mechanism associated with the initiation of cracks at the tips of the ellipsoid. Although the Griffith failure concept may strictly be applied only to tensile stress conditions, several researchers have investigated the conditions around a crack in a compressive stress field. Concentrations of tensile stress are theoretically possible at the tip of a flat crack but if a sufficiently high compressive stress acts normal to the crack axis the crack will close and frictional resistance along this plane influences subsequent deformational behaviour. Thus modifications to the original theory are proposed by Bieniawski (1967). Figure 1.13 summarises the different mechanisms operating for different conditions of normal stress, the parabola (dotted line) indicating the Griffith criteria. Shear rupture is described by the Coulomb-Navier line where τ_0 is the shear stress due to cohesion of the rock matrix and interlocking of asperities, and μ the coefficient of internal friction.

Joints and other discontinuities are of prime importance in practical geomechanical analysis, and there are two basic methods of introducing the effects of such discontinuities into a finite element solution code. Firstly, individual joints may be modelled by a special finite element which represents the joint plane, and is introduced into the mesh so as to form a link between adjacent solid elements, as summarized for example by Heuze & Barvour (1982). The nonlinear equations of joint behaviour are introduced into this special element. Secondly, the effect of multiple, parallel discontinuities may be introduced into the solid finite element itself in the form of a material model, and in particular the overlay model as used by Zienkiewicz & Pande (1977) enables several sets of discontinuities to be analysed in the same element or series of elements. The second method is thus of particular interest here, and will now be discussed in some detail.

1.8.2 *Discontinuous rock model*

Following the early work of Zienkiewicz & Pande in which the peak strength on a given discontinuity plane was modelled by the Mohr–Coulomb failure condition, Pande & Xiong (1982) selected the alternative empirical equation of Barton, Equation 1.13 (see Section 1.4.2), as a more general and practically useful expressions for the peak strength on a discontinuity (joint) plane. This equation is easily incorporated into a standard visco-plastic solution code (see Chapter 3), the various discontinuity systems for the given material point being taken in turn and the resulting plastic strain increment for each plane being added to the corresponding increments for the other planes. This increment is in turn based upon the residual stress (see Chapter 3). Thus the maximum component of shear stress on the given plane, τ_m, together with the associated normal stress, σ_n, evaluated for the given stage in the elastoplastic solution process, is used to obtain the residual stress, τ_r, or:

$$\tau_r = |\tau_m| + \sigma_n \tan\left[\mathrm{JRC}.\log\left(\frac{\mathrm{JCS}}{-\sigma_n}\right) + \phi_r\right] \tag{1.49}$$

where the sign | | means absolute value. The material fluidity parameter (see Chapter 3), if correctly chosen, enables the time-dependent nonlinear deformational and stress response to be calculated. Results from such a model are also given in Chapter 3.

The effects of dilatancy in the discontinuity plane may also be incorporated in the solution routine by employing the usual equation:

$$\tan\psi = \frac{\text{increment of normal strain}}{\text{increment of shear strain}} \tag{1.50}$$

where ψ is known as the dilatancy angle. Thus once the visco-plastic shear strain increment is known, Equation 1.50 enables the corresponding visco-plastic strain increment in the direction normal to the discontinuity plane to be calculated.

Considering the approximate nature of the problem, the empirical formula proposed by Pande & Xiong, based upon observed joint behaviour as reported by Barton & Choubey (1977), appears to be a complicated expression, and Dungar (1985) proposed an alternative expression for ψ, as illustrated in Figure 1.54 and given as:

$$\begin{aligned} \psi &= 1.285\,\mathrm{JRC}\log\left(\frac{\mathrm{JCS}}{-4\sigma_n}\right), & -\sigma_n &\le \frac{\mathrm{JCS}}{4} \\ \psi &= 0, & -\sigma_n &> \frac{\mathrm{JCS}}{4} \end{aligned} \tag{1.51}$$

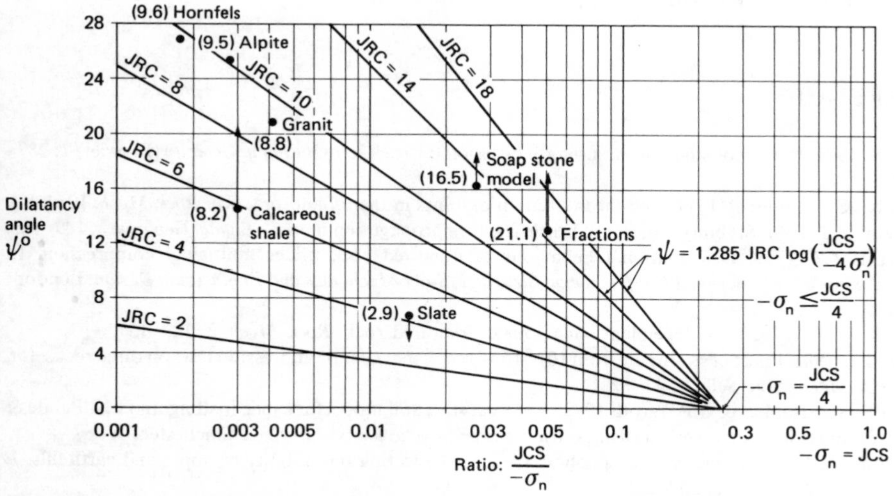

Figure 1.54. Observed dilatancy angle of joints in various rocks, correlated as a function of the normal stress on the joint plane and the joint roughness coefficient

1.8.3 *Pre- and post-peak response of discontinuity systems*

The investigation of nonlinear deformation in the foundation of the Zervreila Arch Dam, as reported by Dungar (1985), has revealed the need for accurate modelling of the pre- and post-peak behaviour on discontinuities when the resulting displacement response is of prime concern. It became clear from an initial series of nonlinear calculations that the use of the above model, together with realistic estimates of the angle of friction, resulted in deformation values in the rock foundation which were far too small compared with the recorded deformation, even though a full three-dimensional solution with a large number of foundation elements was employed (see illustration in Chapter 3, Section 3). This is undoubtedly due to the nature of the model which employs a yield surface with only elastic deformation of the rock matrix occurring before the peak strength is reached.

In relation to a laboratory test, the above condition of elastic behaviour until peak strength corresponds to the joint reaching its peak strength condition at a shear strain corresponding with the elastic shear strain of the constituent rock, a condition which is far from the reality. In practice, a substantial shear strain is observed before the peak strength is reached, being associated with sliding at asperity boundaries, with corresponding dilational behaviour.

Due to the fact that only the correlation of recorded and calculated nonlinear deformation was of interest, Dungar lowered the value of the friction angle on the two sets of discontinuity systems to achieve an acceptable result, but when the stability of the rock mass is of main concern this solution would not be acceptable. A more realistic model must be formulated for general application so that observed stress-strain behaviour, in addition to peak strength, can be fully represented.

Post-peak behaviour will also be important for discontinuity systems which show a peak-residual deformational response, and this may also be accommodated in the model by employing a strain softening relationship which may be taken as a linear function of accumulated plastic shear strain on the given plane (Pande & Xiong 1982) or, alternatively, by taking an exponential function.

REFERENCES

Barton, N. 1973. Review of a new shear strength criterion for rock joints. *Engng. Geology* (Elsevier) 7: 287–332.

Barton, N. & V. Choubey 1977. The shear strength of joints in theory and practice. *Rock Mech.* 10: 1–54.

Bieniawski, Z.T. 1967. Stability concept of brittle fracture propagation in rock. *Engng. Geology* 2: 149–162.

Bieniawski, Z.T. 1969. Deformational behaviour of fractured rock under multiaxial compression. In *Proceedings of International Conference on Structure, Solid Mechanics and Engineering Design*. London: Wiley.

Bieniawski, Z.T. 1970. Time-dependent behaviour of fractured rock. *Rock Mech.* 2: 123–137.

Blanton, T.L. 1981. Effect of strain rate from 10^{-2} to 10 sec^{-1} in triaxial compression tests on three rocks. *Int. J. Rock Mech. Min. Sci.* 18: 47–62.

Carter, J.P., J.R. Booker & C.P. Wroth 1982. A critical state soil model for cyclic loading. In G.N. Pande & O.C. Zienkiewicz (eds.), *Soil mechanics–transient and cyclic loads*: 219–252. Chichester: Wiley.

Casagrande, A. 1936. Characteristics of cohesionless soils affecting the stability of slopes and earth fills. *J. Boston Soc. Civ. Eng.* 1: 257–276.

Deere, D.U., A.J. Hendron, F.D. Patton & E.J. Cording 1966. Design of surface and near-surface construction in rock. In *Proceedings of Symposium on Rock Mechanics* (Minnesota).

Duncan, J.M. & C.Y. Chang 1970. Non-linear analysis of stress and strain in soils. *J. Soil Mech. Fdn. Engng. Div.* (ASCE) 96 (SM 5): 1629–1653.
Dungar, R. 1982. Endochronic critical state models for soils. In *Proceedings of International Symposium on Numerical Models in Geomechanics* (Zurich). Rotterdam: Balkema.
Dungar, R. 1985. Analysis of plastic deformations leading to cracking of grouted contraction joints in Zervreila Arch Dam. In *Proceedings of 15th ICOLD Congress* (Lausanne).
Geertsma, J. 1957. The effect of fluid-pressure decline on volumetric changes of porous rocks. *Trans. AIME* 219: 331.
Griffith, A.A. 1921. The phenomenon of rupture and flow in solids. *Phil. Trans. Roy. Soc. London* A 221: 163–198.
Heuze, F. & T.G. Barbour 1982. New models for rock joints and interfaces. In *Proceedings of ASCE* 108 (GT 5): 757–776.
Hoek, E. 1968. Brittle failure of rock. In *Rock mechanics in engineering practice*: 99–124. London: Wiley.
Kikuchi, K., K. Saito & K. Kusunoki 1982. Geotechnically integrated evaluation on the stability of dam foundation rocks. In *Proceedings of 14th ICOLD Congress* Q 53: 49–75.
Kirkpatrick, W.M. 1957. The condition of failure of sands. In *Proceedings of 4th International Conference on Soil Mechanics and Foundation Engineering* 1: 172–178.
Kobayaski, R. 1970. On mechanical behaviour of rocks under various loading-rates. *Rock mech. Japan* 1: 56.
Kondner, R.L. 1963. Hyperbolic stress-strain response: cohesive soils. *J. Soil Mech. Fdns. Div.* (ASCE) 89 (SM 1): 115–143.
Lee, K.L. & H.B. Seed 1967. Drained characteristics of sand. *J. Soil Mech. Fdns. Div.* (ASCE) 93 (SM 6).
Luong, M.P. 1980. Stress-strain aspects of cohesionless soils under cyclic and transient loading. In *Proceedings of International Symposium on Soils under Cyclic and Transient Loading* (Swansea), 1: 315–324. Rotterdam: Balkema.
Nova, R. 1980. The failure of transversely isotropic rocks in triaxial compression. *Int. J. Rock Mech. Min. Sci. & Geomech. Abstr.* 17: 325–332.
Pande, G.N. & W. Xiong 1982. An improved multilaminate model for jointed rock masses. In *Proceedings of International Symposium on Numerical Models in Geomechanics* (Zurich). Rotterdam: Balkema.
Pande, G.N. & O.C. Zienkiewicz (eds.) 1982. *Soil mechanics–transient and cyclic loads.* Chichester: Wiley.
Peng, S.S. 1973. Time-dependent aspects of rock behaviour as measured by servo-controlled hydraulic testing machine. *Int. J. Rock Mech. Min. Sci.* 10: 235–246.
Peng, S.S. & E.R. Podnieks 1972. Relaxation and the behaviour of failed rock. *Int. J. Rock Mech. Min. Sci.* 9: 699–712.
Price, A.M. & I.W. Farmer 1981. The Hvorslev surface in rock deformation. *Int. J. Rock Mech. Min. Sci. & Geomech. Abstr.* 18: 229–234.
Roscoe, K.H. & J.B. Burland 1968. On the generalized stress-strain behaviour of 'wet' clay. In Heymand & Leckis (eds.), *Engineering plasticity*: 535–609. Cambridge (UK): Cambridge University Press.
Roscoe, K.H., A.N. Schofield & C.P. Wroth 1958. On the yielding of soils. *Geotechnique* 8: 22–53.
Rowe, P.W. 1962. The stress-dilatancy relation for static equilibrium of an assembly of particles in contact. In *Proceedings Royal Soc. London* A 269: 500–527.
Rutledge, P.C. 1947. Progress report on triaxial shear research: 68–104. Waterways Experiment Station, Vicksburg, Miss.
Sangrey, D.A., D.J. Henkel & M.I. Esrig 1969. The effective stress response of a saturated clay soil to repeated loading. *Can. Geotech. J.* 6 (3): 241–252.
Schmertmann, J.H. 1955. The undisturbed consolidation behaviour of clay. *Trans. ASCE* 120: 1201–1233.
Schofield, A.N. & C.P. Wroth 1968. *Critical state soil mechanics.* New York: McGraw-Hill.
Seed, H.B. & K.L. Lee 1967. Undrained strength characteristics of cohesionless soil. *J. Soil Mech. Fdns. Div.* (ASCE) 93 (SM 6).
Serafim, J.L. 1968. Influence of interstitial water on the behaviour of rock masses. In Stagg/Zienkiewicz (eds.), *Rock mechanics in engineering practice.* London: Wiley.
Taylor, P.W. & D.R. Bacchus 1969. Dynamic cyclic strain tests on a clay. In *Proceedings of 7th International Conference on Soil Mechanics and Foundation Engineering* (Mexico City), 1: 401–409.
Tatsuoka, F. 1972. Shear tests in a triaxial apparatus–a fundamental study of deformation of sand (in Japanese). Ph.D. Thesis, University of Tokyo.
Valliappan, S. 1981. *Continuum mechanics fundamentals.* Rotterdam: Balkema.
Wawersik, W.R. 1973. Time-dependent rock behaviour in uniaxial compression. In *Proceedings of 14th Symposium on Rock Mechanics* (Pennsylvania): 85–106.

Wood, D.M. 1982. Laboratory investigations of the behaviour of soils under cyclic loading: a review. In G.N. Pande & O.C. Zienkiewicz (eds.), *Soil mechanics–transient and cyclic loads*: 513–582. Chichester: Wiley.

Wroth, C.P. 1977. The predicted performance of soft clay under a trial embankment loading based on the Cam-clay model. In *Finite elements in geomechanics*. London: Wiley.

Wroth, C.P. & P.A. Loudon 1967. The correlation of strains within a family of triaxial tests on overconsolidated samples of Kaolin. In *Proceedings of Geotechnical Conference* (Oslo), 1: 157–163.

Zienkiewicz, O.C. 1977. *The finite element method*. New York: McGraw-Hill.

Zienkiewicz, O.C. & G.N. Pande 1977. Some useful forms of isotropic yield surfaces for soil and rock mechanics. In *Finite elements in geomechanics*: Chapter 5. London: Wiley.

Zienkiewicz, O.C. & G.N. Pande 1977. Time-dependent multilaminate model for rocks – a numerical study of deformation and failure of rock masses. *Int. J. for Numer. and Analyst. Meth. in Geomech.* 1: 219–247.

Linear analysis of statically and dynamically loaded geomechanical structures 2

R. DUNGAR

2.1 INTRODUCTION

Basic concepts of material modelling for geomechanical media are introduced in the first chapter. This present chapter is concerned with certain methods of incorporating these models into a suitable numerical solution code, such that a realistic idealisation of the structural system may be made, including the various boundary conditions and applied loads. Currently, the most popular solution method is the Finite Element Method (FEM) and is thus the main concern of this and the next chapter.

The principles underlying a linear elastic finite element analysis for geomechanical problems are discussed in this chapter, including the cases of both static and dynamic loading. Methods for nonlinear analysis are introduced in the next chapter.

2.2 INTRODUCTION TO LINEAR STATIC FINITE ELEMENT ANALYSIS

An elastic finite element solution may be used to determine displacement and stress due to given static loads, and will include the following steps:

1. A decision of whether or not a full three-dimensional solution is needed. This will be made on the grounds of need and acceptability of cost, and of the appropriateness of making assumptions which reduce the problem to that of a solution within one or more planes. These assumptions are as follows:
   - zero strain in the direction normal to the specified plane, termed plane strain;
   - zero stress in the direction normal to the specified plane, termed plane stress;
   - a combination of plane stress and plane strain; or
   - the assumption of a symmetrical geometry with respect to a given axis, termed axisymmetric.
2. Selection of appropriate foundation dimensions, in order to obtain a reasonable representation of foundation stiffness.
3. Specification of the different material zones to be analysed.
4. Subdivision of the various material zones into a finite number of units of given geometrical form (triangle, rectangle, cube), termed finite elements. These elements must be selected from those which are available in the 'element library' of the finite element program. This subdivision process is known as 'mesh-generation' and computer programs are often available for the automatic or semi-automatic

performance of this task. The computer program is often referred to as a 'pre-processor', because this work precedes the main work of forming and solving the resulting structural equations.

5. Definition of loading and associated load data (self-weight forces, surface pressures, temperatures, etc.).
6. Main formation process to calculate the structural stiffness properties for each finite element, and to assemble these so as to form the stiffness for the complete structure. The loads for the complete structure are also obtained from the loading of each element.
7. Solution of the stiffness equations and, for each separate load case, the solution of the corresponding displacements.
8. Solution for the stresses corresponding with the above calculated displacements, again for each load case.
9. Plotting of displacements and stresses for each load case, normally by using a 'post-processing program', so-called because this work follows the main solution process of steps 6, 7 and 8 above.

2.3 FINITE ELEMENT STIFFNESS FORMULATION

It is now necessary to introduce some basic concepts and equations related to the elastic (or stiffness) properties of a simple finite element. These equations will also be needed later in the text in relation to nonlinear analysis techniques. Loading conditions will be considered in the next section.

The most popular, and perhaps useful, method of element formulation employs isoparametric finite elements, and discussed here because of its general applicability to many different types of two- and three-dimensional element (as well as beam, plate and shell element). Only the main details are included here, because of space limitations.

An 8-node isoparametric element is illustrated in Figure 2.1, together with its so-called 'mapping space'. This element is of second order because displacements within the element and along the boundary are assumed to be a quadratic function of the nodal displacements. This quadratic function is specified first in terms of the mapping coordinates η and ξ, as:

$$\begin{aligned} u &= N_1 u_1 + N_2 u_2 + N_3 u_3 + N_4 u_4 + N_5 u_5 + N_6 u_6 + N_7 u_7 + N_8 u_8 \\ v &= N_1 v_1 + N_2 v_2 + N_3 v_3 + N_4 v_4 + N_5 v_5 + N_6 v_6 + N_7 v_7 + N_8 v_8 \end{aligned} \tag{2.1}$$

where u and v are displacements of a given point, p(x, y), in the x and y coordinate directions, as shown, and $u_1, u_2 \ldots u_8, v_1, v_2 \ldots v_8$ are the corresponding displacements at the eight element displacement control points, called nodes. The functions N_1, $N_2 \ldots N_8$ are associated with each of the eight element nodes, taken in turn, and are formulated such that when p(x, y) lies at node i the corresponding function N_i has a value of unity and all other functions ($N_j, j \neq i$) are zero. In addition, when p lies on an element boundary the function corresponding to other nodes not on that boundary are also zero. Equation 2.1 is more conveniently written in matrix from as:

$$\begin{bmatrix} u \\ v \end{bmatrix} = \sum_{i=1}^{8} N_i \begin{bmatrix} u_i \\ v_i \end{bmatrix} \tag{2.2}$$

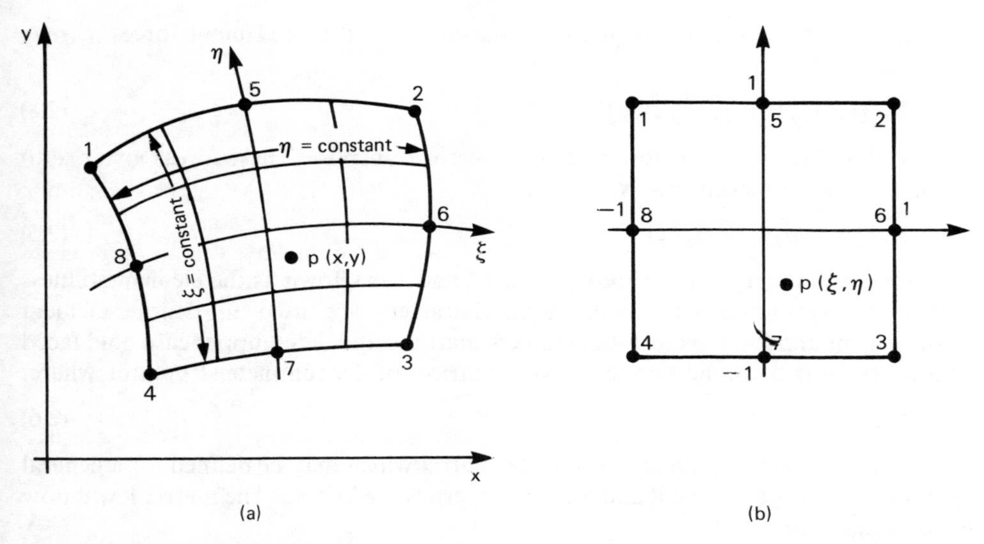

Figure 2.1. An 8-node isoparametric finite element and associated mapping space; (a) 8-node finite element; and (b) mapping space

It is observed from the above that both u and v may be 'mapped' with the same functions, N_i, and these may also be used to map other such quantities which can be specified in terms of given values at the element nodes (such as temperature, pore pressure, etc.). The functions, N_i, are termed 'shape functions'. These shape functions for the illustrated isoparametric element are defined in terms of the mapping coordinates η and ξ as illustrated in Figure 2.1, and are given (see Zienkiewicz 1977) as:

$$N_1 = -\tfrac{1}{4}(1-\xi)(1+\eta)(1+\xi-\eta)$$

$$N_2 = -\tfrac{1}{4}(1+\xi)(1+\eta)(1-\xi-\eta)$$

$$N_3 = -\tfrac{1}{4}(1+\xi)(1-\eta)(1-\xi+\eta)$$

$$N_4 = -\tfrac{1}{4}(1-\xi)(1-\eta)(1+\xi+\eta)$$

$$N_5 = \tfrac{1}{2}(1+\eta)(1-\xi^2)$$

$$N_6 = \tfrac{1}{2}(1+\xi)(1-\eta^2)$$

$$N_7 = \tfrac{1}{2}(1-\eta)(1-\xi^2)$$

and $N_8 = \tfrac{1}{2}(1-\xi)(1-\eta^2)$

or:

$$N_i = N(\xi, \eta) \tag{2.3}$$

where ξ and η are specified in the range $-1 \leq \xi \leq 1$, $-1 \leq \eta \leq 1$.

The stiffness properties for an element of the plane strain type (see Section 2.2 above) will now be formulated.

Firstly, consider a set of forces acting in the x and y directions at the element nodes, $f_{u,1}$,

$f_{u,2} \ldots f_{u,8}, f_{v,1}, f_{v,2} \ldots f_{v,8}$, or, in matrix notation, a vector of element forces, **f**, may be specified as:

$$\mathbf{f} = [f_{u,1} f_{v,1} f_{u_2} f_{v,2} \ldots f_{u,8} f_{v,8}]^T \tag{2.4}$$

The symbol ^{T} is used to denote matrix or vector transpose. These forces give rise to element nodal displacements, **x**, where:

$$\mathbf{x} = [u_1 v_1 u_2 v_2 \ldots u_8 v_8]^T \tag{2.5}$$

The relationship between the above vectors **f** and **x** is known as the element stiffness matrix **k** (here lower case bold faced characters are used to denote element displacement and force vectors and stiffness matrices and, later, upper case bold faced characters for corresponding vectors and matrices of the complete structure), where:

$$\mathbf{f} = \mathbf{kx} \tag{2.6}$$

and **k** is in this case a symmetrical 16 × 16 matrix which may be defined for a general element once the geometrical and elastic properties are known. The matrix **k** will now be formulated in detail.

The strains due to the displacements, **x**, are given, for point p(x, y), as:

$$\varepsilon_x = \partial u/\partial x$$

$$\varepsilon_y = \partial v/\partial y$$

$$\gamma_{xy} = \partial v/\partial x + \partial u/\partial y$$

and, from Equation 2.2:

$$\begin{bmatrix} \varepsilon_x \\ \varepsilon_y \\ \gamma_{xy} \end{bmatrix} = \sum_{i=1}^{n} \begin{bmatrix} \partial N_i/\partial x & 0 \\ 0 & \partial N_i/\partial y \\ \partial N_i/\partial y & \partial N_i/\partial x \end{bmatrix} \begin{bmatrix} u_i \\ v_i \end{bmatrix} \tag{2.7a}$$

or:

$$\boldsymbol{\varepsilon} = \sum_{i=1}^{n} \mathbf{B}_i \mathbf{x} \tag{2.7b}$$

The quantities $\partial N_i/\partial x$, $\partial N_i/\partial y$, etc. are as yet undefined. Equation 2.7b may be expanded to include all components $\mathbf{B}_i$, and may simply be written as:

$$\boldsymbol{\varepsilon} = \mathbf{Bx} \tag{2.7c}$$

Stresses at point p(x, y) are related to the corresponding strains, and for isotropic elastic conditions the following relationships apply:

$$\begin{bmatrix} \varepsilon_x \\ \varepsilon_y \\ \varepsilon_z \\ \gamma_{xy} \end{bmatrix} = \frac{1}{E} \begin{bmatrix} 1 & -\nu & -\nu & 0 \\ -\nu & 1 & -\nu & 0 \\ -\nu & -\nu & 1 & 0 \\ 0 & 0 & 0 & \dfrac{1}{2(1+\nu)} \end{bmatrix} \begin{bmatrix} \sigma_x \\ \sigma_y \\ \sigma_z \\ \tau_{xy} \end{bmatrix} \tag{2.8}$$

where the z direction is normal to the x–y plane, and for plane strain condition, $\varepsilon_z = 0$, the following is obtained:

$$\begin{bmatrix} \sigma_x \\ \sigma_y \\ \tau_{xy} \end{bmatrix} = \frac{E}{(1-\nu^2)} \begin{bmatrix} 1 & \nu & 0 \\ \nu & 1 & 0 \\ 0 & 0 & \dfrac{(1-\nu)}{2} \end{bmatrix} \begin{bmatrix} \varepsilon_x \\ \varepsilon_y \\ \gamma_{xy} \end{bmatrix}$$

or:

$$\boldsymbol{\sigma} = \mathbf{D}\boldsymbol{\varepsilon} \tag{2.9}$$

The matrix **D** is known as the elasticity matrix.* From Equations 2.9 and 2.7 we may write:

$$\boldsymbol{\sigma} = \mathbf{DBx} \tag{2.10}$$

We now use the principle of virtual work to establish an expression for the stiffness matrix, **k**. If the nodes undergo very small (virtual) displacements, $\bar{\mathbf{x}}$, these will produce corresponding strains, $\bar{\boldsymbol{\varepsilon}}$, of magnitude:

$$\bar{\boldsymbol{\varepsilon}} = \mathbf{B}\bar{\mathbf{x}} \tag{2.11}$$

The resulting internal energy (virtual work) per unit volume at p(x, y), Δw_i, is given as:

$$\Delta w_i = \bar{\boldsymbol{\varepsilon}}^T \boldsymbol{\sigma} \tag{2.12}$$

which, on integration over the finite element, results in an expression for the total internal virtual work, w_i, where:

$$w_i = \bar{\mathbf{x}}^T \int_v \mathbf{B}^T\mathbf{DB}\,dv.\mathbf{x} \tag{2.13}$$

An expression for the virtual work may likewise be obtained from consideration of the energy associated with the nodal forces, **f**, which move though the corresponding virtual displacements $\bar{\mathbf{x}}$. Thus the external virtual work, w_e, is given by:

$$w_e = \bar{\mathbf{x}}^T \mathbf{f} \tag{2.14}$$

Equating w_i and w_e, and from Equation 2.6, the stiffness matrix **k** is given by the simple expression:

$$\mathbf{k} = \int_v \mathbf{B}^T\mathbf{DB}\,dv \tag{2.15}$$

Integration of Equation 2.15 is complicated by the fact that the matrix **B** contains terms in ξ and η, and the best method of obtaining this integral is to adopt a numerical integration scheme. The Gauss quadrature formula (Zienkiewicz 1977) enables numerical integration to be performed with minimal computational effort and, here,

*The matrix, **D**, is here a relationship between total stress and total strain, and is formed from the respective elasticity properties of the constituent material. That, is for an undrained material, the effect of the pore fluid is included in the properties E and ν. For a discussion on an effective stress formulation, see Chapter 3, Section 3.

the individual matrix terms of Equation 2.15 are explicitly evaluated for selected positions within the element, the resultant terms being then added, with the inclusion of a 'weighting factor'.

The quantities $\partial N_i/\partial x$, $\partial N_i/\partial y$ etc. and dv are evaluated for the point $p(x, y)$, or $p(\xi, \eta)$, by first noting that the derivative dN_i may be written as:

$$dN_i = \frac{\partial N_i}{\partial x}\cdot dx + \frac{\partial N_i}{\partial y}\cdot dy \tag{2.16}$$

Partial differentiation of Equation 2.16 with respect to ξ and η yields the matrix expression:

$$\begin{bmatrix} \frac{\partial N_i}{\partial \xi} \\ \frac{\partial N_i}{\partial \eta} \end{bmatrix} = \begin{bmatrix} \frac{\partial x}{\partial \xi} & \frac{\partial y}{\partial \xi} \\ \frac{\partial x}{\partial \eta} & \frac{\partial y}{\partial \eta} \end{bmatrix} \begin{bmatrix} \frac{\partial N_i}{\partial x} \\ \frac{\partial N_i}{\partial y} \end{bmatrix} \tag{2.17}$$

where the central matrix, **J**, known as the Jacobian matrix, is composed of terms which may be evaluated for the given point, $p(\xi, \eta)$, by noting that the shape functions N_i may also be used to map the coordinates x and y, or:

$$\begin{bmatrix} x \\ y \end{bmatrix} = \sum_{i=1}^{n} N_i \begin{bmatrix} x_i \\ y_i \end{bmatrix} \tag{2.18}$$

and thus:

$$\mathbf{J} = \begin{bmatrix} \frac{\partial x}{\partial \xi} & \frac{\partial y}{\partial \xi} \\ \frac{\partial x}{\partial \eta} & \frac{\partial y}{\partial \eta} \end{bmatrix} = \sum_{i=1}^{n} \begin{bmatrix} \frac{\partial N_i}{\partial \xi} & \frac{\partial N_i}{\partial \xi} \\ \frac{\partial N_i}{\partial \eta} & \frac{\partial N_i}{\partial \eta} \end{bmatrix} \begin{bmatrix} x_i & \\ & y_i \end{bmatrix} \tag{2.19}$$

Since expressions for $\partial N_i/\partial \xi$ etc. may be derived explicitly from Equation 2.3, and knowing ξ and η, it is a straightforward matter to derive explicitly the matrix **J**. Hence, the terms $\partial N_i/\partial x$, required to form the matrix of Equation 2.7, are given by Equation 2.17 as:

$$\begin{bmatrix} \frac{\partial N_i}{\partial \mathbf{x}} \\ \frac{\partial N_i}{\partial y} \end{bmatrix} = \mathbf{J}^{-1} \begin{bmatrix} \frac{\partial N_i}{\partial \xi} \\ \frac{\partial N_i}{\partial \eta} \end{bmatrix} \tag{2.20}$$

2.4 STIFFNESS PROPERTIES FOR THE COMPLETE STRUCTURE

The stiffness properties for individual finite elements, **k**, as defined in the above section (see Equation 2.15) are added to give the stiffness properties for the complete structure, **K**. Thus, if the structural displacement and force vectors are denoted by **X** and **F** respectively, the structural stiffness is given by a matrix expression similar to Equation 2.15, namely:

$$\mathbf{F} = \mathbf{K}\mathbf{X} \tag{2.21}$$

Care must be exercised when forming the matrix **K**, so that the element stiffness properties, **k**, are added, element by element, into the appropriate rows and columns of **K**. The exact location of a given element of **k** within **K** is defined by the position of the corresponding element nodal displacement and force component in the vector **X** or **F**. In locating this position it is usual to eliminate those displacements (degrees of freedom) which have a known fixity or zero displacement condition. Thus the vectors **X** and **F** have components associated with non-zero displacements only.

The forces for the complete structure, **F**, are similarly formed by adding the forces associated with each element taken in turn (see next section).

Solution of Equation 2.21, for a given force vector (load vector), **F**, enables the corresponding displacements, **X**, to be obtained and from this the element displacement vector, **x**, may be extracted. Knowing **x**, the stresses for the given element are simply given by Equation 2.10.

2.5 STATIC LOADING CONSIDERATIONS AND CONSISTENT LOAD FORMULATION

The most common loading conditions are due to the self-weight of the constituent material and superstructure, surface loading due to hydrostatic fluid forces acting on impermeable interfaces, forces acting within geomechanical regions due to changes of fluid pressure (seepage pressure), loads generated by temperature changes and prescribed movements of walls or other structures supporting soil or rock masses. Self-weight and hydrostatic surface and seepage forces are commonly encountered and will be given detailed attention in this section. Prescribed displacements and temperature loads, although important in geomechanical analysis, are beyond the scope of this introduction.

Particular attention must be given to the method of calculating the element force vector, **f**, and, although the simple method of 'lumping' element loads to corresponding nodes is often used the accurate method of load calculation is to employ a so-called 'consistent load' formulation (see Section 2.5.2).

2.5.1 *Loading sequence, surface and body forces*

For linear problems, the principle of superposition enables the effects from different load cases to be added, but for nonlinear problems the sequence in which loads are applied is important. Consider now the particular case of the analysis of a gravity dam, for which both stress and displacement are to be calculated. The results of a direct analysis for self-weight loading of the dam and foundation are illustrated in Figure 2.2. This method, also known as 'gravity turn-on', simply consists of applying the total load in one step. Thus, as illustrated, finite displacements will be calculated for all regions of the foundation and will thus be unrepresentative of the actual situation, even though the calculated stresses will be reasonably well estimated. In reality, displacements under the dam will be much smaller than those indicated, the displacements at the top of the dam being zero at the end of construction (the dam being constructed to its design height and location). The correct solution for this type of problem clearly depends upon the correct sequence of the applied load and the so-called 'stage construction' method, as described in Section 2.7, should be used.

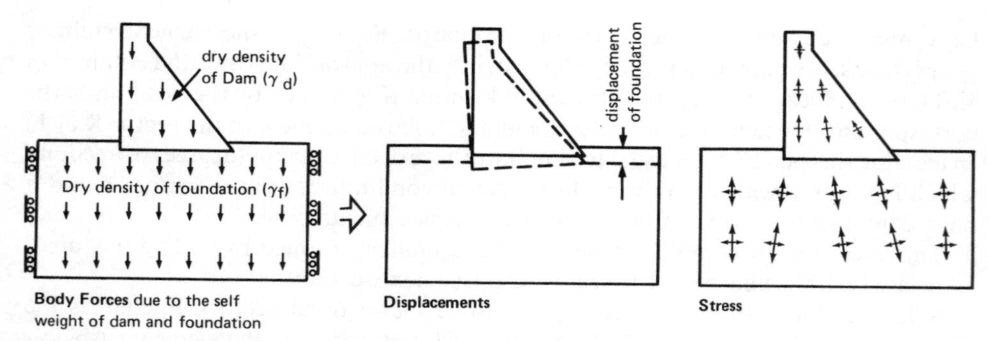

Figure 2.2. Displacement and stress calculation of a gravity dam, due to the self-weight of the dam and foundation, using the direct or gravity turn-on method

In the above example hydrostatic surface and seepage forces will also be required. Grout and drainage curtains are thus illustrated in Figure 2.3 and known pressure at the surface of the finite element mesh may be employed in two ways. Firstly, the foundation region will usually be taken as a drained zone, corresponding with a free flow of ground water, and the surface pressures are then used to calculate the seepage pressures in the foundation. Following this calculation, the pressures at the dam foundation interface are also known. Secondly, the known pressures on the undrained dam zone, together with the calculated interface pressures, enable the corresponding surface forces, acting on this undrained zone, to be calculated.

The total force on the dam and foundation is given by the addition of the seepage and surface forces, as illustrated, enabling the displacement and stress to be calculated. In practice, the above solution process may be automated so that, within one run of the finite element program, the seepage calculation can be made using the same finite element mesh as is used for the structure (see Cathie & Dungar 1975).

Two types of force have been mentioned above, namely those which act within the mass or body of the structure or foundation (self-weight, seepage), termed 'body forces', and those which act at the surface of the structure or at the interface of two zones, termed 'surface tractions'. Surface tractions as discussed above are due to pressure acting normal to the given surface, but may also include forces which act in a tangential direction, as for example due to force transferred to a foundation from wind acting on a building which, unless the whole building is also modelled, is applied as a tangential force to the foundation interface.

2.5.2 *Consistent load formulation*

The element load vector, **f** of Equation 2.4, resulting from body forces (see section above) may be calculated on the basis of the principle of virtual work. Virtual nodal displacements, $\bar{\mathbf{x}}$, produce external virtual work as given by Equation 2.14.

The body force acting on a small increment of the element at p(x, y) may be resolved into the coordinate directions (x and y for a two-dimensional problem) and expressing this as a force per unit volume, **b**, where:

$$\mathbf{b} = \begin{bmatrix} b_x \\ b_y \end{bmatrix} \tag{2.22}$$

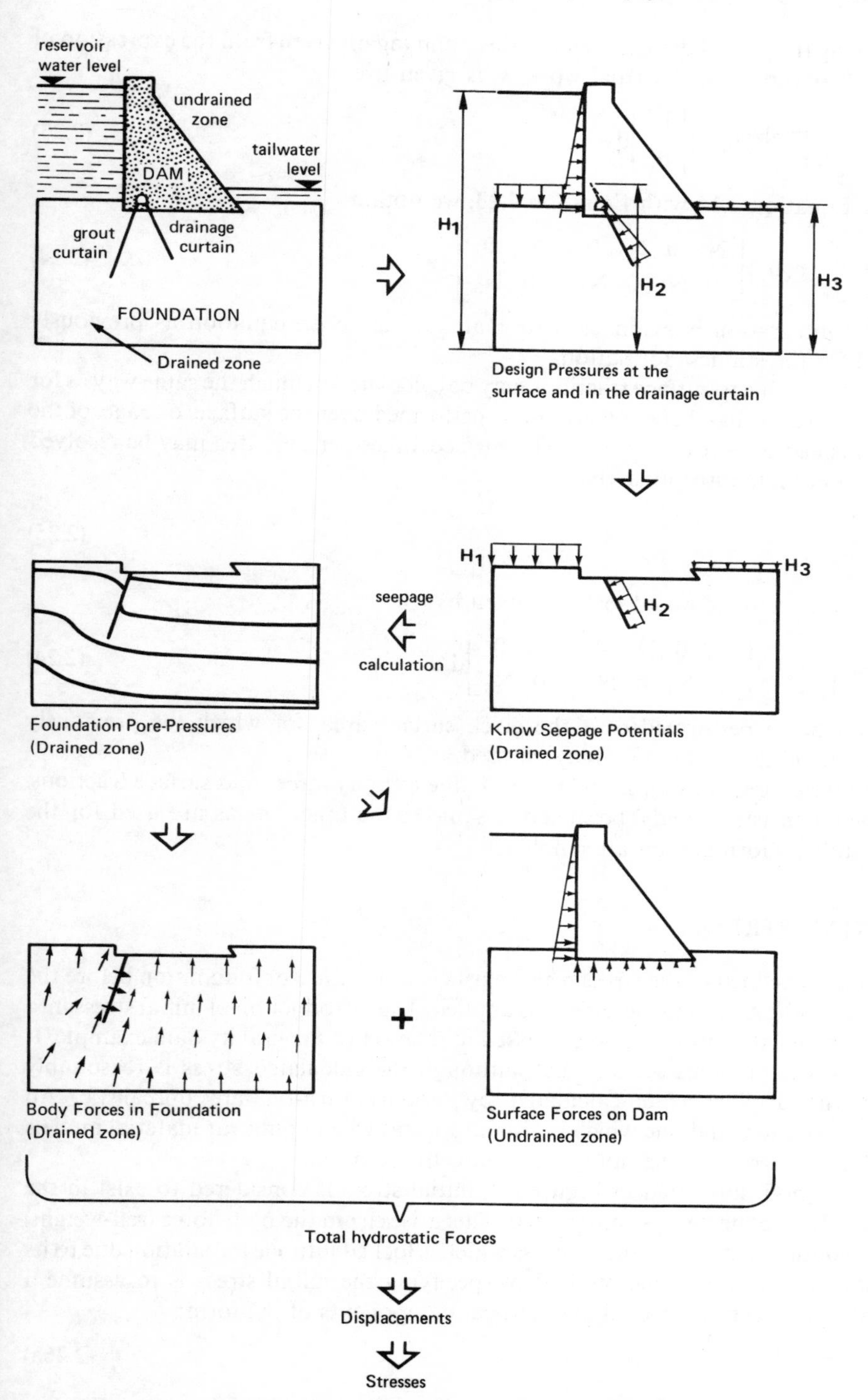

Figure 2.3. Hydrostatic surface force and coupled seepage-body force analysis of a gravity dam and its foundation

and, knowing the virtual displacement at this point (again given from the expression of Equation 2.2), the internal virtual work w_i is given by:

$$w_i = \int_v \sum_{i=1}^{8} N_i[\bar{u}_i \bar{v}_i] \begin{bmatrix} b_x \\ b_y \end{bmatrix} dv \qquad (2.23)$$

Equating Equation 2.14 with Equation 2.23, we obtain:

$$\mathbf{f}^T = \int_v [b_x b_y] \begin{bmatrix} N_1 & 0 & N_2 & 0 & \ldots N_8 & 0 \\ 0 & N_1 & 0 & N_2 \ldots & 0 & N_8 \end{bmatrix} dv \qquad (2.24)$$

Numerical integration is again used to evaluate the above equation as previously explained for the stiffness formation.

Load vectors due to surface tractions may be calculated in much the same way as for body forces except that, here, integration is performed over the surface, or edge, of the element, instead of over the volume. The surface forces per unit area may be resolved into the coordinate directions, that is:

$$\mathbf{s} = \begin{bmatrix} s_x \\ s_y \end{bmatrix} \qquad (2.25)$$

resulting in a vector of nodal loads, **f**, given by:

$$\mathbf{f}^T = \int_v [s_x s_y] \begin{bmatrix} N_1 & 0 & N_2 & 0 & \ldots N_8 & 0 \\ 0 & N_1 & 0 & N_2 \ldots & 0 & N_8 \end{bmatrix} dA \qquad (2.24)$$

integration being performed over the given surface area, for which the previously described numerical techniques may be used.

The above expressions for nodal forces, **f**, due to body forces and surface tractions, are termed 'consistent loads' because the same shape functions as are used for the element stiffness formulation are employed.

2.6 INITIAL STRESS

Initial stress is defined as the stress which exists in a structure or foundation before the considered loading, or load increment, is applied. The introduction of initial stress into a finite element calculation is also discussed in relation to the gravity dam example. It was seen in the previous section that, although the calculated stress is reasonably accurate, the displacements calculated by 'gravity turn-on' are unrealistic. An alternative solution, and one which is also important when nonlinear material models are used, is to introduce the initial stress into the solution.

In the method illustrated in Figure 2.4, initial stress is considered to exist in the foundation before the dam is constructed. This arises from the body force (self-weight) of the foundation and is also due to stress which is locked into the foundation due to its prior geological history. One method of specifying the initial stress is to assume a relationship between horizontal and vertical components of the form:

$$\sigma_x = \sigma_h = K_0 \sigma_v \qquad (2.26a)$$

and:

$$\sigma_y = \sigma_v = \rho H \qquad (2.26b)$$

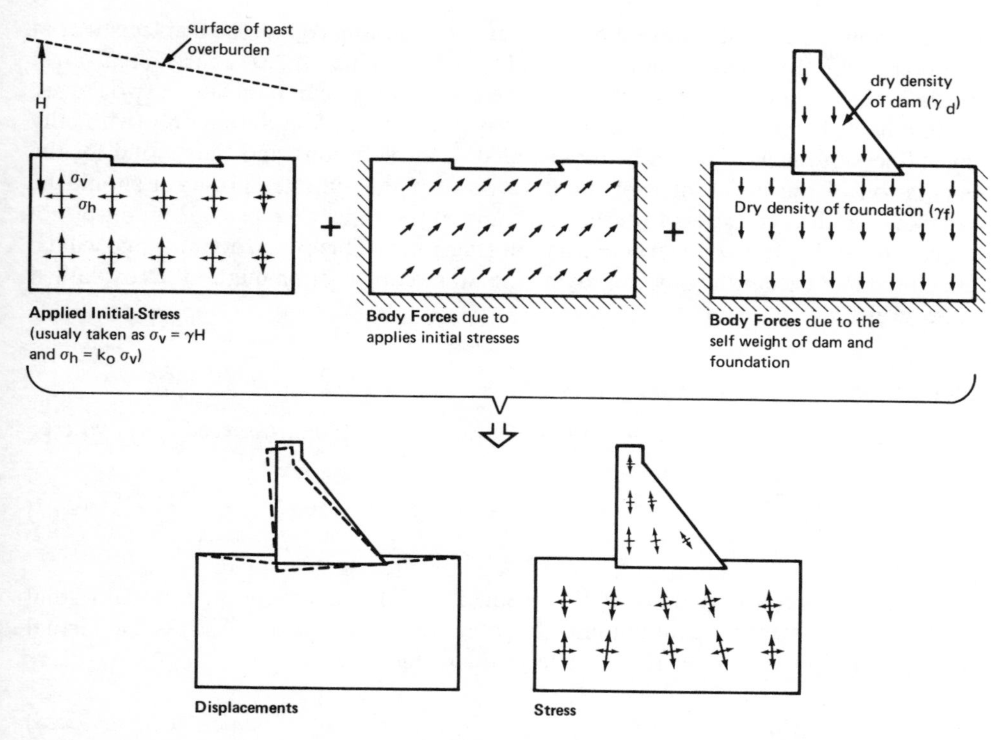

Figure 2.4. Displacement and stress calculation of a gravity dam and foundation, employing initial stresses within the foundation

where ρ is the material density, H a height to some assumed datum, which may be taken as:

1. the present surface level, or
2. the surface level of an estimated or assumed level of past overburden, in which case the value of K_0 will also be taken as some value which could have reasonably applied at the given time.

The initial stress so defined is used in two steps. Firstly, a vector of accumulated stress is set to the value of the initial stress. Secondly, the initial stress is used to define an equivalent nodal force (body force), which is then added to the force due to the material self-weight, to give the total body force. The resulting displacement and stress are then calculated in the usual way, the calculated stress being added to the accumulated stress.

It will be noted that if the distance H (see Figure 2.4) is taken as the present ground surface (option 1 above) the total body force will be zero, the system being in equilibrium with the given initial stress. If option 2 above is adopted, some net change of displacement will occur, corresponding with the heave expected from the past removal of material.

It is clear that if the body force due to the weight of the dam is included in the calculation (in addition to the stiffness of the dam and foundation), the displacements so calculated will correspond to changes occurring during a period from the time of past overburden up until the end of the dam construction. Again, the displacement at the crest will not correspond with reality (zero value of 'as constructed' crest displacement). To remedy this situation, the computation can be made in two stages, the first being for the calculation of 'total body force' for the foundation only (with only the stiffness of the foundation being included in the solution), and the second for the calculation of the body force of the dam body. The computer program should be capable of adding the stress for each stage and at the same time provide an output of displacement due to the given construction stage only. This construction stage will be calculated most accurately by employing the 'stage construction method', as explained in Section 2.7.

2.6.1 *Nodal loads due to initial stress*

It now remains to define the nodal loads due to the initial stress vector, $\boldsymbol{\sigma}_0$, where:

$$\boldsymbol{\sigma}_0 = \begin{bmatrix} \sigma_x \\ \sigma_y \\ \tau_{xy} \end{bmatrix}_{\text{initial}} \tag{2.27}$$

Again we consider a virtual nodal displacement, $\bar{\mathbf{x}}$. The corresponding internal virtual work for the element is again obtained by integrating the product of stress and virtual strain (Equation 2.11) over the volume of the element, or:

$$w_i = \bar{\mathbf{x}} \int_v \mathbf{B}^T \boldsymbol{\sigma}_0 dv \tag{2.28}$$

and by comparison with Equation 2.14, the corresponding nodal forces, **f**, are given as:

$$\mathbf{f} = \int_v \mathbf{B}^T \boldsymbol{\sigma}_0 dv \tag{2.29}$$

Numerical integration is again used to evaluate the above expression.

2.7 STAGE CONSTRUCTION, STAGE EXCAVATION AND STAGE LOADING

It was observed in the above gravity dam example that, even when the initial stress formulation is used to obtain a realistic estimate for the foundation stress, the deformation due to constructional load cannot be accurately predicted for the upper part of the dam unless the analysis considers the construction as a distinct number of 'lifts' or 'stages', the final lift being placed so that the crest level is at the design elevation which, neglecting creep effects, results in zero deformation at the dam crest. This problem is accentuated for embankment dams, where larger settlements are expected in the central zones during construction and these are often recorded, together with impounding deformations, as a control on performance. It is often important to perform a corresponding analysis for constructional deformations by simulating the constructional sequence of a finite number of layers.

Each step of the analysis in which a new layer of material is considered, is called a 'stage' and the term 'stage construction analysis' is used. Likewise, if layers are 'removed' during a stage, the term 'stage excavation' is used. Both types of analysis are 'stage loading', but this term also includes other cases where the geometry does not change but the load is simply considered to be applied in a finite number of steps (or stages), as in the case of the impounding of an embankment dam.

The analysis proceeds in an exactly similar way for all types of stage loading, the procedure described above under 'initial stress' being used. Thus stress from a previous solution stage is applied as initial stress for the current stage. The solution technique is equally applicable to problems of stage excavation. Thus, for example, excavation of soil from in front of a retaining wall, or soil or rock from a tunnel, may be analysed with essentially the same computer code, the accumulated stress calculated from a previous stage of excavating being applied as 'initial stress' for the next stage. It should be noted that when loads due to initial stress are combined with loads due to the material self-weight and other applied loads, the resultant load vector is non-zero and this vector represents the effect of 'stress release' which occurs as the material is excavated.

A stage construction or excavation analysis is of particular importance when nonlinear models are employed because, here, the stiffness for each element changes with increase (or decrease) of load.

2.8 INTRODUCTION TO THE DYNAMIC ANALYSIS OF LINEAR NON-RADIATING SYSTEMS

2.8.1 *Introduction*

The principles described in the above sections are easily extended to the calculation of the dynamic response of linear non-radiating systems. The term 'non-radiating' means that energy dissipation takes place via damping within the considered finite material domain but not via wave propagation to and within other material domains. Thus the foundation of the gravity dam example, illustrated above, is considered to be completely disconnected from adjacent material, by making certain assumptions regarding the boundary conditions for the boundary of the finite element mesh.

From a practical point of view, such an analysis employing a finite pre-defined foundation region may be unrealistic for certain problems where the effects of wave propagation are significant (often termed 'radiation damping', as opposed to internal energy dissipation due to material damping) but, when used with a certain degree of engineering judgement, satisfactory solutions for a large number of problems may be obtained (see also the comments in Section 2.8.4 below).

The equations of motion which govern a non-radiating system may be written in matrix form as:

$$\mathbf{M}\ddot{\mathbf{X}} + \mathbf{C}\dot{\mathbf{X}} + \mathbf{K}\mathbf{X} = \mathbf{F}(t) \tag{2.30}$$

where **K**, **C** and **M** are the stiffness, damping and mass matrices, written in terms of the n degrees of freedom for the complete system. The vectors **X** and **F** are the corresponding displacement and force vectors, respectively, and the dot denotes differentiation with respect to the time variable, t. As in the case of the stiffness matrix, the mass matrix may easily be compiled from corresponding element matrices and the formulation details for an isoparametric finite element are given below. The damping matrix, **C**, is more difficult to specify but one of two assumptions are commonly employed:

1. The damping matrix for each element is related, linearly, to the element stiffness and mass.
2. The normal mode theory is employed, by which, at any given instant of time, t, the response $\mathbf{X}(t)$ may be expressed as the weighted sum of the normal mode shapes (mode shapes of the undamped system), $\mathbf{X}_1, \mathbf{X}_2 \ldots \mathbf{X}_n$.

Assumption 1 is often made in the solution of earth structures with seismic loads, for which the the equivalent linear solution method is used (see Chapter 3 for details). The use of the second assumption enables a large number of problems to be solved, and will thus be considered here in detail, but firstly the details of the element mass matrix formulation will be given.

2.8.2 *Element mass formulation*

The shape functions used for forming the element stiffness may also be used when forming the element mass matrix, to give a so-called 'consistent mass matrix' (in contrast to simply 'lumping' the total element mass to the various element node points). The mass matrix, **m**, defines the relationship between element nodal forces, **f**, and nodal accelerations, $\ddot{\mathbf{x}}$, in the form:

$$\mathbf{f} = \mathbf{m}\ddot{\mathbf{x}} \tag{2.31}$$

and by differentiation of Equation 2.2, the acceleration components of a point p(x, y) are given for the considered 8-node isoparametric element as:

$$\begin{bmatrix} \ddot{u} \\ \ddot{v} \end{bmatrix} = \sum_{i=1}^{8} N_i \begin{bmatrix} \ddot{u}_i \\ \ddot{v}_i \end{bmatrix} \tag{2.32}$$

These accelerations give rise to body forces (see Section 2.4) in the form:

$$\begin{bmatrix} b_x \\ b_y \end{bmatrix} = \rho \begin{bmatrix} \ddot{u} \\ \ddot{v} \end{bmatrix} \tag{2.33}$$

where ρ is the mass density of the material. Virtual work is again used and we thus consider a set of virtual displacements, $\bar{\mathbf{x}}$, for which the external virtual work, w_e, is given by Equation 2.14. The internal virtual work is given simply from the product of the body force and the corresponding virtual displacement, and on integration over the volume of the element, and from Equations 2.23, 2.32 and 2.33 we obtain:

$$w_i = \int_v \rho \sum_{i=1}^{8} N_i [\bar{u}_i \bar{v}_i] \cdot \sum_{i=1}^{8} N_i \begin{bmatrix} \ddot{u}_i \\ \ddot{v}_i \end{bmatrix} dv \tag{2.34}$$

By suitably rearranging and equating with Equations 2.14 and 2.31 we obtain for the given 8-node element:

$$\mathbf{m} = \int_v \rho \begin{bmatrix} N_1 & 0 \\ 0 & N_1 \\ N_2 & 0 \\ \cdot & \cdot \\ \cdot & \cdot \\ 0 & N_8 \end{bmatrix} \begin{bmatrix} N_1 & 0 & N_2 & 0 & \ldots & 0 \\ 0 & N_1 & 0 & N_2 & \ldots & N_8 \end{bmatrix} dv \tag{2.35}$$

Numerical integration may be used as described for the stiffness matrix formulation. In practice, the mass matrix for a given finite element is usually derived at the same time as the corresponding stiffness matrix, and the element mass matrices may be added to form the mass matrix, **M**, for the complete system (see the description for the stiffness formulation, Section 2.3).

2.8.3 *System damping*

When using the normal mode theory (see above) the damping matrix does not have to be defined explicitly. If we now consider vibration in one of the normal modes, $\mathbf{X}_i$, under steady state sinusoidal excitation with:

$$\mathbf{F}(t) = \mathbf{F}\sin(\omega t), \tag{2.36}$$

and a corresponding steady state response:

$$\left.\begin{aligned} \mathbf{X}(t) &= \mathbf{X}_i z(t) = \mathbf{X}_i z_i \sin(\omega t + \theta) \\ \dot{\mathbf{X}}(t) &= \mathbf{X}_i \dot{z}(t) = \mathbf{X}_i \omega z_i \cos(\omega t + \theta) \\ \ddot{\mathbf{X}}(t) &= \mathbf{X}_i \ddot{z}(t) = -\mathbf{X}_i \omega^2 z_i \sin(\omega t + \theta) \end{aligned}\right\} \tag{2.37}$$

where z_i is the response amplitude for mode $\mathbf{X}_i$ and θ is the response phase angle. When $\theta = 90$ we may write two equations of the form:

$$\omega_i^2 \mathbf{C}\mathbf{X}_i = \mathbf{F}_i \tag{2.38}$$

and:

$$\mathbf{K}\mathbf{X}_i - \omega^2 \mathbf{M}\mathbf{X}_i = 0 \tag{2.39}$$

where ω_i is known as the resonant frequency of the i'th mode. The use of Equation 2.39 enables the frequency for the i'th mode, and thus for all modes, to be calculated, and several standard methods now exist for obtaining these modal properties, ω_i and $\mathbf{X}_i$, of which the most appropriate method for linear elastic geomechanical system is, perhaps the method of 'subspace iteration', which may easily be programmed by a suitable adaption of a code which performs 'inverse iteration', which is itself formulated by a modification of the static solution routine.

From the above Equation 2.38, the 'force shape' is given for mode i, namely $\mathbf{F}_i$. This is in reality a set of force components which will excite the i'th mode and only that mode. This means that the energy given by this force in any other mode of vibration. $\mathbf{X}_j$, must be zero for all frequencies of excitation. Thus, on substituting for vibration in the i'th mode into Equation 2.30, and from Equation 2.37 we obtain:

$$\mathbf{X}_j^T \mathbf{M}\mathbf{X}_i \ddot{z}_i + \mathbf{X}_j^T \mathbf{C}\mathbf{X}_i \dot{z}_i + \mathbf{X}_j^T \mathbf{K}\mathbf{X}_i z_i = \mathbf{X}_j^T \mathbf{F}_i(t) = 0 \tag{2.40}$$

which must hold for all excitation frequencies, ω, and for which the necessary and sufficient conditions are:

$$\left.\begin{aligned} \mathbf{X}_j^T \mathbf{M}\mathbf{X}_i &= 0 \\ \mathbf{X}_j^T \mathbf{C}\,\mathbf{X}_i &= 0 \\ \mathbf{X}_j^T \mathbf{K}\,\mathbf{X}_i &= 0 \end{aligned}\right\} \tag{2.41}$$

These are known as the orthogonality conditions, and will be used later.

We may now express the response, $\mathbf{X}$, as a sum of the response in each of the n modes of vibration, or:

$$\mathbf{X} = [\mathbf{X}_1\mathbf{X}_2\mathbf{X}_3 \ldots \mathbf{X}_n]\mathbf{z}$$

or:

$$\mathbf{X} = \mathbf{A}\mathbf{z} \tag{2.42}$$

where the matrix $\mathbf{A}$ is composed of columns which are the normal modes, $\mathbf{X}_1, \mathbf{X}_2 \ldots \mathbf{X}_n$, and from Equation 2.30 and by premultiplying by $\mathbf{A}^T$ (the transpose of A) we obtain:

$$\mathbf{A}^T\mathbf{M}\mathbf{A}\ddot{\mathbf{z}} + \mathbf{A}^T\mathbf{C}\mathbf{A}\dot{\mathbf{z}} + \mathbf{A}^T\mathbf{K}\mathbf{A}\mathbf{z} = \mathbf{A}^T\mathbf{F}(t) \tag{2.43}$$

From the orthogonality conditions of Equations 2.41, the above equation may be expressed as n independent equations, and if:

$$\mathbf{X}_i^T\mathbf{M}\mathbf{X}_i = \beta_i \tag{2.44a}$$

$$\mathbf{X}_i^T\mathbf{C}\mathbf{X}_i = 2c_i\omega_i\beta_i \tag{2.44b}$$

and

$$\mathbf{X}_i^T\mathbf{K}\mathbf{X}_i = \omega_i^2\beta_i \tag{2.44c}$$

we obtain an expression for vibration in the i'th mode, in the form:

$$\ddot{z}_i + 2c_i\omega_i\dot{z}_i + \omega_i^2 z_i = \frac{\mathbf{X}_i^T\mathbf{F}(t)}{\beta_i} \tag{2.45}$$

The modal quantity β_i is known as the generalised mass for the i'th mode. It is now seen that the damping in the i'th mode is specified by means of a single coefficient, c_i, known as the 'critical damping factor'. Equation 2.45 may be solved by standard means for the given forcing function $\mathbf{F}(t)$, to give the required response in the i'th mode, $z_i(t)$.

In practice, only a certain number of modes are considered to contribute to the response of the system, which are usually the first s modes, and having obtained the response $z_i(t)$ for each of these, the response, $\mathbf{X}$, is simply given from Equation 2.42. Stresses are likewise calculated. The stress for each mode shape is first calculated and then, for time t, the total stress given by the summation of the corresponding modal contributions.

One common form of dynamic loading is that due to earthquake, for which the applied force is given as:

$$\mathbf{F}(t) = -\sum_{k=1}^{3} \mathbf{M}\boldsymbol{\delta}_k a_k(t) \tag{2.46}$$

where the 'direction vector', $\boldsymbol{\delta}_k$, is a vector of 0's and 1's which specify those degrees of freedom which are orientated in the direction of the k'th component of the earthquake acceleration-time history $a_k(t)$. For a three-component ground motion, the three components are considered in turn, but for a single component motion, for example due to horizontal ground shaking only, one value of k only need be considered. On substitution into Equation 2.45, we obtain:

$$\ddot{z}_i + 2c_i\omega_i\dot{z}_i + \omega_i^2 z_i = -\sum_{k=1}^{3} \frac{\alpha_{i,k}}{\beta_i} a_k(t) \tag{2.47}$$

where:

$$\alpha_{i,k} = \mathbf{X}_i^T \mathbf{M} \delta_k \tag{2.48}$$

Equation 2.47 may be solved for the given modal constants and for each component of the given earthquake acceleration-time history. Once z_i is obtained, the displacement and stress response of the complete structure is given as described above.

2.8.4 *Radiating or non-radiating systems: a practical suggestion*

A practical suggestion should be made here, especially when geomechanical structures are considered. If we again consider the case of the gravity dam, illustrated in Figure 2.2, it is perfectly possible to follow the procedure described above to obtain frequencies and mode shapes for the first few modes of vibration (say, 10). The resultant frequencies will however depend upon the considered foundation size; the greater the foundation size the lower will be the frequency for the first and higher modes. In the limit, for a large foundation with a relatively small dam, the frequency for the first mode will approach the frequency of vertically propagating waves for a body with a free surface.

In practice, there will be no such 'contained response', the energy associated with vibration of the dam structure being 'radiated away' through the foundation. One practical solution is to consider the mass of the structure, only, in the mass matrix, M, in which case the calculated response will correspond with the 'change of response' caused by the presence of the structure. That is, the change in response from so-called 'free field conditions' (no structure) to conditions with the structure.

In many cases, especially when the elasticity modulus of the foundation increases with depth, either by stiffening of a particular material zone or by transition to a stiff bed-rock layer, a modal solution considering the mass of the structure only, will provide a reasonably acceptable and cost effective solution.

REFERENCES

Cathie, D.N. & R. Dungar 1975. The influence of the pressure-permeability relationship on the stability of a rock-filled dam. In *Proceedings of International Symposium on Criteria and Assumptions for Numerical Analysis of Dams* (Swansea).

Zienkiewicz, O.C. 1977. *The finite element method.* New York: McGraw-Hill.

Piecewise linear, equivalent linear and elastoplastic solution methods 3

R. DUNGAR

3.1 INTRODUCTION

The principles presented in Chapter 2 for the elastic analysis of geomechanical systems are extended in this chapter to the solution of nonlinear problems. Both static and dynamic loading conditions are discussed. Solution methods which attempt to model the material behaviour by taking a full elastoplastic model or by more simple linearisation techniques are considered.

In the case of static loading, two basic solution methods are here considered, namely the 'piecewise linearisation' method, which uses a model based upon 'tangent stiffness' properties, and residual force methods for the solution of elastoplastic problems.

Dynamic loading is discussed in relation to three different methods. The 'equivalent linear method', which is often used in the analysis of embankment structures and important foundation, is described together with a method which allows for a more detailed piecewise linearisation of the material stress-strain relationship. Finally, cyclic and random loading response analysis techniques, employing elastoplastic material models, are discussed.

3.2 A DIRECT METHOD OF PIECEWISE LINEAR ANALYSIS FOR STATIC LOADING

An introduction to piecewise linear (variable-elastic) total stress models is given in Chapter 1. Here, we consider the steps required for the incorporation of such models into a suitable finite element code. If such a code includes the ability to analyse stage construction problems, as described in Section 2.7, the additional work required to implement a variable elastic formulation is then relatively small. The necessary steps that a computer code must follow to obtain a solution by the method which is here termed the direct method are illustrated in Figure 3.1, the details being as follows:

1. 'Initial stresses' are set for each element, as calculated for a new material layer (see Section 2.6) or as obtained from the accumulated stress vector of the previous stage (construction stage or stage of applied load as explained in Section 2.7).
2. Loads due to body forces for elements to be added during the given stage (construction stage) are included in the load increment vector.
3. Loads due to those surface tractions and point loads which are to be added during

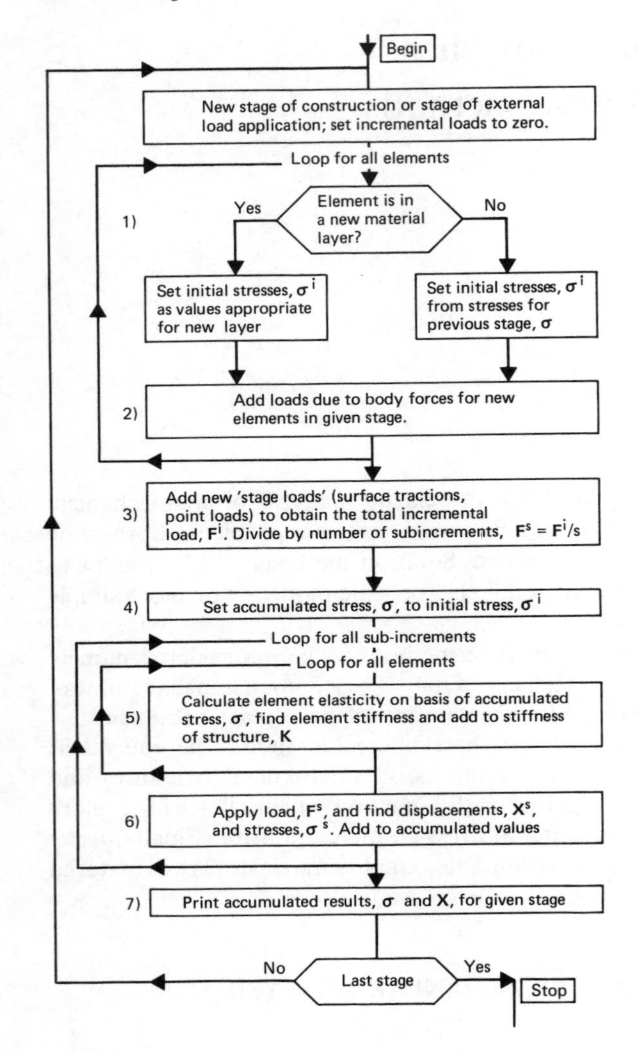

Figure 3.1. Flow diagram for the direct method of piecewise linear analysis

the given stage are included in the load increment vector. The load increment vector is divided into sub-increments, to give the load to be applied as a number of small steps.

4. The accumulated stress is set to the initial stress, given by step 1 above.
5. The elastic properties for each element are determined on the basis of the accumulated stress and these are then used to obtain the element stiffness properties (tangent stiffness matrix), and thus the element stiffness matrix, $\mathbf{k}$, for the given sub-increment. Matrix $\mathbf{k}$ is then added to the matrix for the complete structure, $\mathbf{K}$.
6. The load for sub-increment 's' is applied, and the resulting incremental displacement, $\mathbf{X}^s$, and stress, $\boldsymbol{\sigma}^s$, obtained for a normal elastic solution. This incremental displacement and stress is then added to the corresponding accumulated vector.
7. After repeating the above steps 5 and 6, for all sub-increments, the results for the

given stage are printed, and the above process, 1–6, repeated until all construction and additional loading stages are complete.

The above method essentially consists of obtaining the load increment for the given stage, and applying this as a series of sub-increments. For each sub-increment, the stiffness of the structure is updated to correspond with the stress which has accumulated until that point in the solution. This method is relatively simple to program, but has the disadvantage that a stage excavation solution requires additional steps in order to account for stress which is 'released' by elements which are 'excavated' in the given stage.

The influence of the above 'released stress' may be accounted for by applying equivalent surface tractions to the adjacent elements of the remaining mesh, as illustrated in Figure 3.2. The effect of unloading of a material element may be incorporated by employing an unloading tangential stiffness for the given sub-increment. However, before an appropriate unloading modulus may be chosen the program must know whether unloading is expected during the given sub-increment. This decision is often made on the basis of the stress increments calculated for the previous sub-increment. It must be noted however that a load applied to the whole structure may produce loading in certain material zones but unloading in others. For example, hydrostatic loading due to impounding of a clay core embankment will produce loading on the initially undrained core but unloading of the drained upstream shell.

Certain other problems associated with the selection of appropriate values for the tangent modulus exist, and particularly the problem of obtaining a solution which is consistent with material test results (used in the first instance to derive the model parameters). As explained in Chapter 6, in the early work of Duncan & Chang (1970), two of the pioneers of this type of analysis, tangent stiffness properties were based entirely upon the values of tangent modulus of data obtained from triaxial tests conducted at constant confining pressure only. In general, a material element will follow a stress path with changing minimum principal stress and, in effect, the solution process must follow not one but a series of triaxial test paths. The use of the above direct

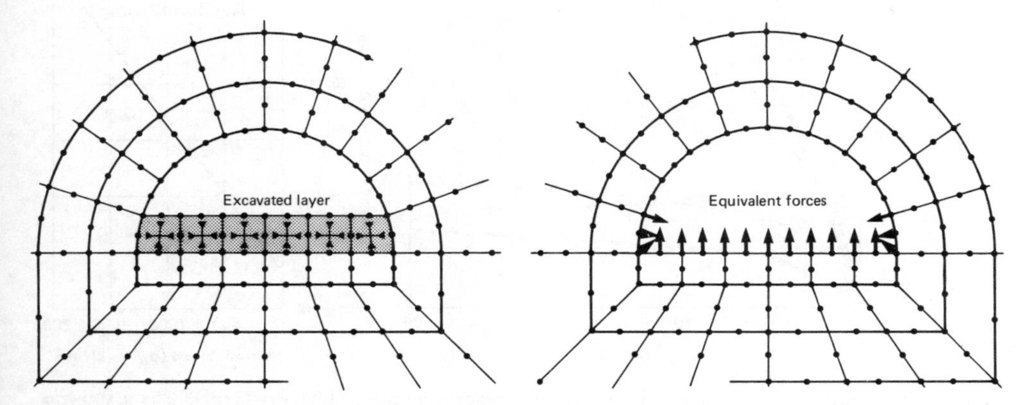

Figure 3.2. Representation of 'excavated elements' indicating the equivalent surface forces which must be applied, when using the direct method, in order to compensate for 'locked in' stress

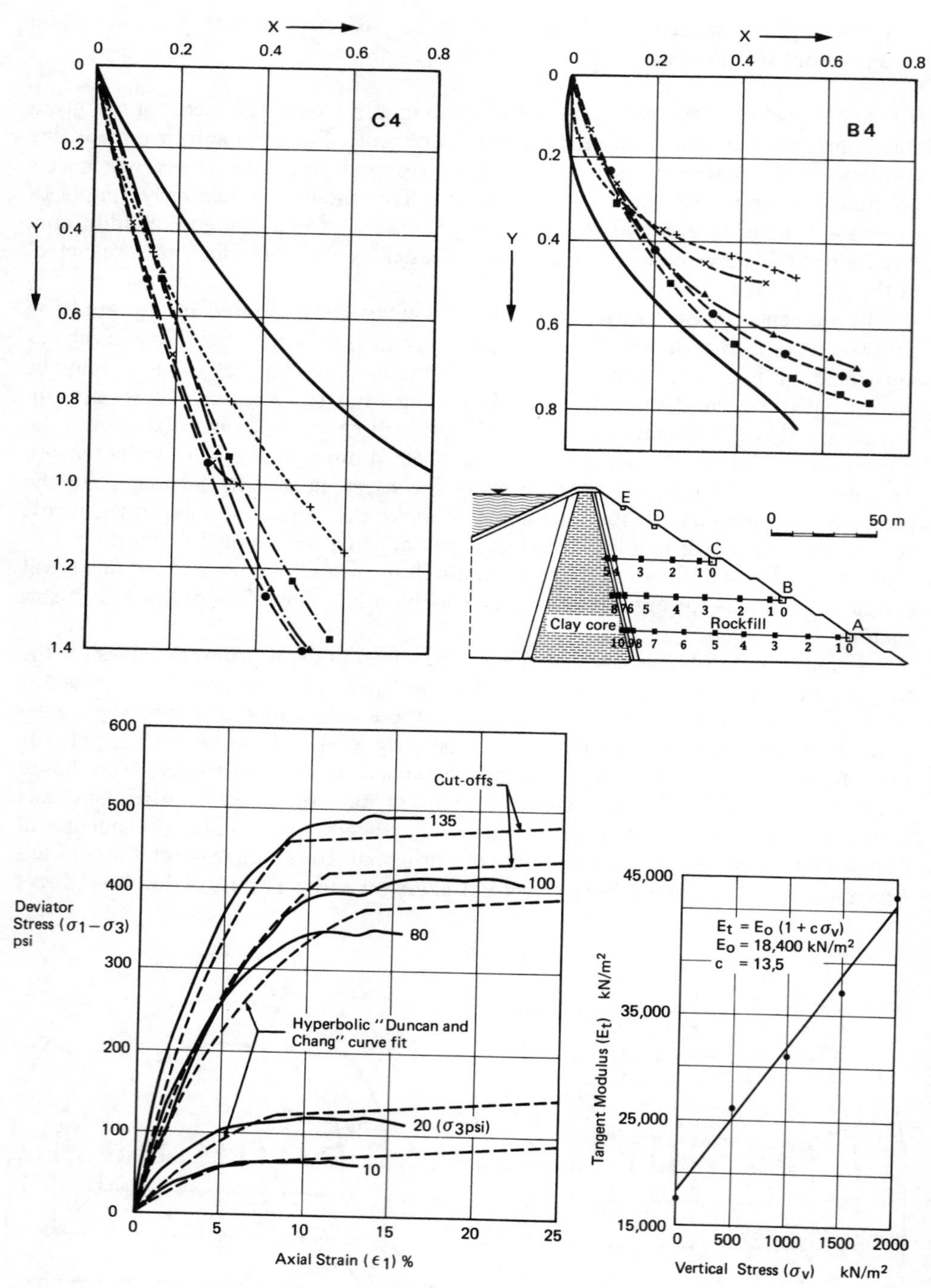

Figure 3.3. Comparison of calculated and measured displacements for the Llyn Briani rockfill dam, showing the hyperbolic stress-strain curve-fit and a fit of tangent modulus as a linear function of vertical stress. Two- and three-dimensional solutions are compared (after Cathie & Dungar 1978)

ROCKFILL PARAMETERS FOR THE VARIOUS ANALYSES				
Description of Analysis	Code	Modulus kN/m²	Poissons Ratio	Other Parameters kN/m²
Elastic 2-D	E27	27,600	0.3	
	E40	40,000	0.3	
Elastic 3-D	E27D	27,600	0.3	
Non-Linear elastic				
(a) E a function of σ_v	EFS	E_t	0.3	$E_o = 18{,}400$, $c = 13.5$
(b) 'Duncan and Chang'	DAC	E_t	0.3	$K' = 561$, $n = 0.71$, $R_f = 0.54$, $c' = 138$, $\phi' = 38^{\circ}$

Figure 3.3. (Continued)

method will often show that, as the solution process progresses, the calculated strain does not correspond with that expected by simple reference to the original triaxial test data, for the prevailing minimum principal stress. This may have significant influence on the calculated displacement, such as the typical results shown in Figure 3.3 (after Cathie & Dungar 1978). This problem is as much a problem of material modelling as of the solution process, and later attempts to produce a more satisfactory solution have included control of both the shear and bulk moduli (see Chapter 6 and Naylor et al. 1981).

3.3 EFFECTIVE STRESS AND RESIDUAL FORCE METHODS

3.3.1 *Effective stress*

The use of piecewise linear models is described in the previous section. In practice, stress-strain data are obtained from laboratory tests (usually triaxial tests) for either drained or undrained conditions, depending upon the particular problem. In either case the analysis is performed as a total stress analysis, that is pore pressures are not explicitly calculated. However, it is an easy task to modify a total stress computer code such that both total and effective stresses are calculated. The necessary steps will now be considered.

The well known principle of effective stress is given by Equation 1.2 as:

$$\boldsymbol{\sigma} = \boldsymbol{\sigma}' - \boldsymbol{\mu} \tag{3.1}$$

the prime (′) denoting effective stress and $\boldsymbol{\mu}$ being the pore fluid pressure vector. This principle is illustrated in Figure 1.5 and implies that the stress acting on the soil skeleton, termed the effective stress, acts in parallel with the pore fluid pressure. That is, a change in the total stress, $\mathrm{d}\boldsymbol{\sigma}$, will in general produce a change of pore pressure, $\mathrm{d}\boldsymbol{\mu}$, as well as a change in the effective stress, $\mathrm{d}\boldsymbol{\sigma}$, or:

$$\mathrm{d}\boldsymbol{\sigma} = \mathrm{d}\boldsymbol{\sigma}' - \mathrm{d}\boldsymbol{\mu} \tag{3.2}$$

The above change of fluid pressure will be governed by both the drainage condition and degree of saturation. Here we are only concerned with fully saturated undrained behaviour; the problem of fluid pressure dissipation, being a time-dependent problem, is outside of the scope of this introduction. The change of fluid pressure is related to the change of volume of the combined solid-fluid system (see Equation 1.10):

$$dv = d\varepsilon_1 + d\varepsilon_2 + d\varepsilon_3 \tag{3.3}$$

by a bulk modulus K_f, which corresponds to the compressibility of both the pore fluid and material grains. The term μ of matrix $\boldsymbol{\mu}$ (see Equation 1.2) is thus given in differential from by:

$$d\mu = -K_f dv \tag{3.4}$$

If we now introduce a notation similar to that of Equation 2.9, but in terms of the effective stress, $\boldsymbol{\sigma}'$, we may write:

$$\boldsymbol{\sigma}' = \mathbf{D}'\boldsymbol{\varepsilon} \qquad 3.5)$$

where the matrix $\mathbf{D}'$ defines the relationship between the effective stress and total strain. From Equations 3.5, 3.3 and 3.4 we obtain:

$$\boldsymbol{\sigma} = [\mathbf{D}' + \mathbf{D}_f]\boldsymbol{\varepsilon} \tag{3.6}$$

where:

$$\mathbf{D}_f = \begin{bmatrix} K_f & K_f & K_f & 0 & 0 & 0 \\ K_f & K_f & K_f & 0 & 0 & 0 \\ K_f & K_f & K_f & 0 & 0 & 0 \\ 0 & 0 & 0 & 0 & 0 & 0 \\ 0 & 0 & 0 & 0 & 0 & 0 \\ 0 & 0 & 0 & 0 & 0 & 0 \end{bmatrix} \tag{3.7}$$

From comparison with Equation 2.9 we obtain:

$$\mathbf{D} = \mathbf{D}' + \mathbf{D}_f \tag{3.8}$$

Equation 3.6 is now a relationship expressing the total stress in terms of the total strain (elastic and plastic strain components are later considered). By comparing the above equations with those of the stiffness formulation given in Chapter 2 it should be clear that any computer code written to solve problems in terms of total stress (the normal analyses as described in previous sections) may easily be modified to enable stress-strain relationships to be used which are formulated in terms of the elastic properties of the material skeleton, by simply adding the appropriate terms for the fluid bulk modulus, K_f. In practice the value of K_f is often several orders of magnitude greater than that of the bulk modulus of the material skeleton and the resultant stiffness is mainly governed, in compression, by the stiffness of the fluid, but in shear the fluid has no influence.

Equation 3.6 is formulated in terms of the elastic properties of the material skeleton and we may thus introduce the notation:

$$\boldsymbol{\sigma}^e = \mathbf{D}\boldsymbol{\varepsilon} \tag{3.9}$$

where $\boldsymbol{\sigma}^e$ indicates a stress due to these elastic properties.

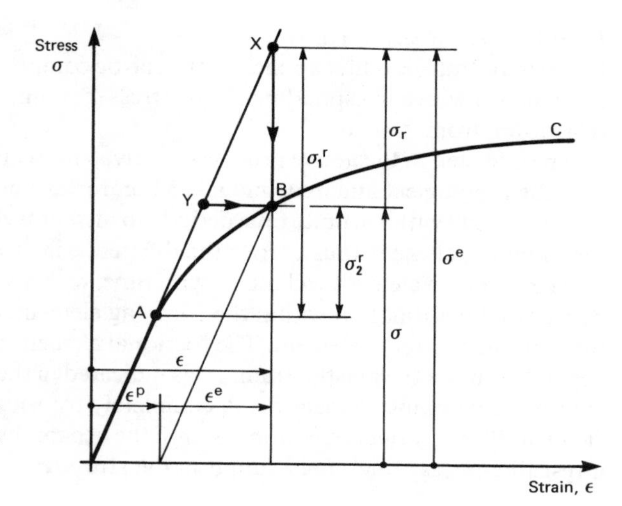

Figure 3.4. Idealised stress-strain curve showing two paths: Paths: 1 A-X-B; Path 2 A-Y-B

3.3.2 *A rheological analogue of material stress-strain response*

Three residual force methods of nonlinear analysis will be considered in the following sections, but before turning to the details a simple rheological analogue is described.

Considered the material stress-strain diagram of Figure 3.4 which in general consists of a combination of both elastic and plastic behaviour. That is, at point **B** the total strain, ε, is given as the sum of the elastic and plastic components, ε^e and ε^p, or:

$$\varepsilon = \varepsilon^e + \varepsilon^p \tag{3.10}$$

In order to arrive at point **B** on the stress-strain diagram, two distinct paths will be considered, as follows:

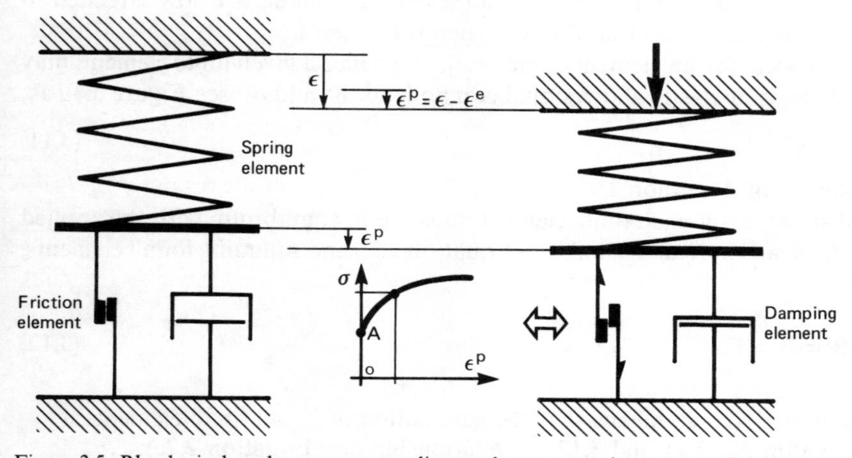

Figure 3.5. Rheological analogue corresponding to the stress-strain response illustrated in Figure 3.4

Path 1: *Applied strain path*
If the total strain, ε, is first applied to the elastic components of the system, so that point X is reached with corresponding elastic stress σ^e, point B may then be reached by stress relaxation, from X to B.

The first step in the above process is equivalent to applying an instantaneous strain, ε, to the rheological analogue model of Figure 3.5, and because the damping element can move only within a finite time period, no instantaneous plastic strain is allowed and the damping element thus carries the difference in stress, σ_1^r, as shown in Figure 3.4.

The damping element will move, with time, with a velocity proportional to the load carried and deformation within the damping element will allow stress transfer to take place to the frictional element. The frictional element is able to carry a stress which is dependent upon the plastic strain, ε^p (as indicated in the small 'σ-ε^p' diagram), the stress within the damping element being completely transferred, with time, to the frictional element. Plastic strain, ε^p, continues until the accumulated value is sufficient to provide a resisting stress which fully supports the stress, σ.

Path 2: *Applied stress path*
Here, the stress, σ, is first applied to the elastic system so that point Y of Figure 3.4 is reached. This is then followed by strain relaxation from Y to B.

Again, from the analogue of Figure 3.5, an instantaneous stress, σ, results in an instantaneous elastic strain, ε^e. The initial stress difference (which in this case is σ_2^r) is carried again by the damping element. Time dependent plastic strain again occurs, and stress is transferred from the damping element until the point is again reached where the frictional element is able to carry the applied stress.

Several solution methods exist which are based upon the concepts introduced in the above simple rheological model, and three popular methods will now be described. These are all termed residual force methods.

3.3.3 *Residal force methods*

The equilibrium conditions for the above simple analogue model are now extended so as to define the equilibrium of nonlinear structural systems.

The total stress, $\boldsymbol{\sigma}$, for an element of material, and hence a given finite element, may be written in terms of elastic and residual components. $\boldsymbol{\sigma}^e$ and $\boldsymbol{\sigma}^r$ (see Figure 3.4), as:

$$\boldsymbol{\sigma} = \boldsymbol{\sigma}^e - \boldsymbol{\sigma}^r \tag{3.11}$$

where $\boldsymbol{\sigma}^e$ is given by Equation 3.9.

The total stress, $\boldsymbol{\sigma}$, for each finite element, must be in equilibrium with the applied load, **f**, and, from an expression similar to Equation 2.29 and summing for all elements, we thus obtain:

$$\sum_e \int_v \mathbf{B}^T \boldsymbol{\sigma} \, dv = \mathbf{F} \tag{3.12}$$

where the summation sign means that the summation is for all finite elements.

From Equation 3.9, 3.11 and 3.12 the relationship (see Equation 2.7c):

$$\boldsymbol{\varepsilon} = \mathbf{B}\mathbf{x} \tag{3.13}$$

together with the notation:

$$\mathbf{F}^{r} = \sum_{e} \int_{v} \mathbf{B}^{T} \boldsymbol{\sigma}^{r} dv \tag{3.14}$$

where $\mathbf{F}^r$ is termed the residual force, we obtain the expression:

$$\sum_{e} \int_{v} \mathbf{B}^{T}\mathbf{D}\mathbf{B} dv.\mathbf{x} = \mathbf{F} + \mathbf{F}^{r} \tag{3.15}$$

or in a more familiar form:

$$\mathbf{KX} = \mathbf{F} + \mathbf{F}^{r} \tag{3.16}$$

where $\mathbf{K}$ is the elastic stiffness matrix for the structure and $\mathbf{X}$ the vector of nodal displacements.

It thus follows that, if we can obtain a suitable expression for the above residual stress, $\boldsymbol{\sigma}^r$, we can apply the corresponding residual force, $\mathbf{F}^r$, defined by Equation 3.14, together with the applied force, $\mathbf{F}$, to the elastic system, in order to obtain the required displacements, X. The residual stress must also be subtracted from the elastic stress, $\boldsymbol{\sigma}^e$, to give the total stress, $\boldsymbol{\sigma}$, as shown by Equation 3.11. The residual stress is thus treated in a similar way to 'initial stress' described in Chapter 2, and one such solution method is in fact called the initial stress method. The details will now be given.

The initial stress method

The computational steps taken when using the initial stress method are illustrated in Figure 3.6, for one stress point in a finite element, the stress state being represented by one stress component only.

The solution process proceeds in a similar way to that described in Section 3.3.2 above, except that the elastic properties and structural stiffness are formed only once

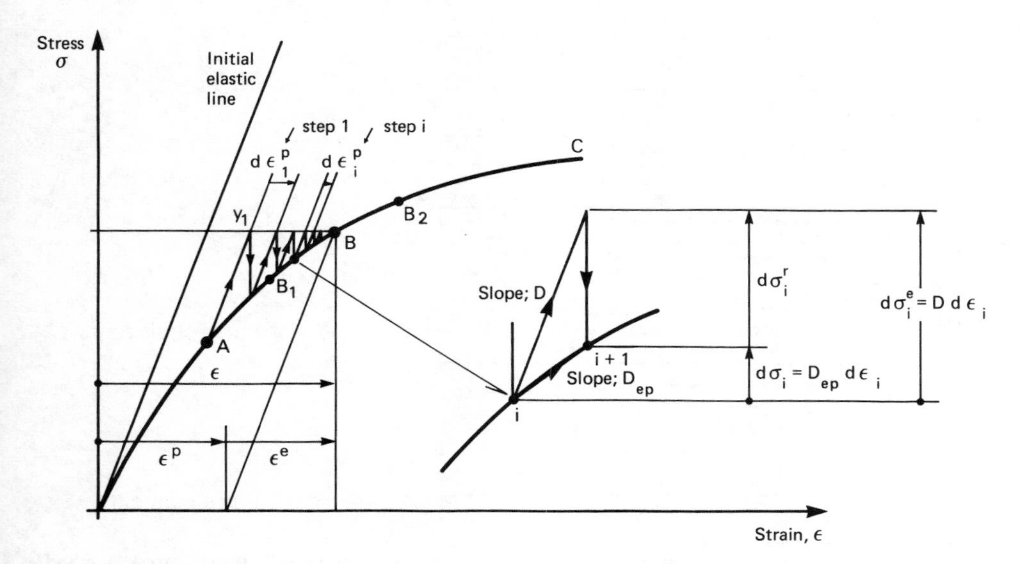

Figure 3.6. Initial stress method: idealised stress-strain history of one finite element stress point

for a given geometrical configuration, or construction stage, and for an effective stress solution (see above) the fluid elasticity, $\mathbf{D}_f$, is also included. The flow diagram of Figure 3.7 is seen to be a modified version of Figure 3.1. The accumulated stress is set to the initial stress, for the given stage, and the force corresponding to this initial stress is combined with the total applied stage loads (from other body forces, surface tractions and point loads) to give the incremental force, **F** (see steps 2–4 of Figure 3.7). It is noted that the applied loads are in equilibrium with the initial stress when the incremental load, **F**, is zero (see Sections 2.6 and 2.7).

The considered element of material, initially lying at point A of Figure 3.6, will reach

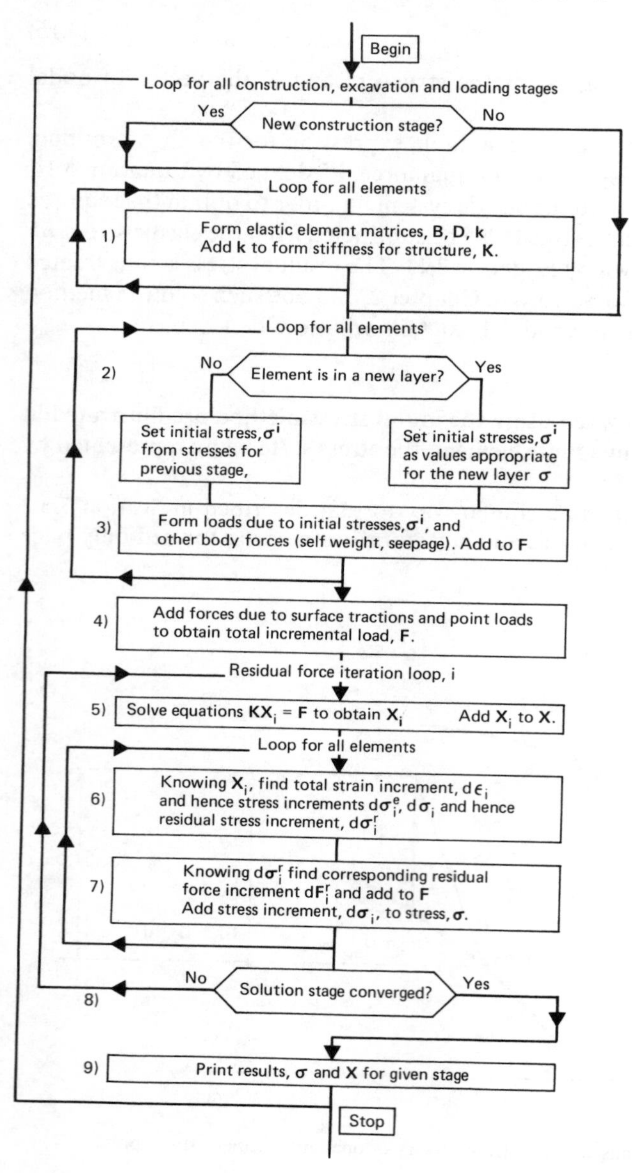

Figure 3.7. Initial stress method: flow diagram

point Y_1 when the incremental load, **F**, is applied to the elastic system. A series of steps are taken from this point, as illustrated, until point B is reached. These steps are made in the residual force iteration loop of Figure 3.7 and consist of the following details:

– On application of force **F** to the elastic system (step 5) the displacements for the i'th iteration, $\mathbf{X}_i$, are known and hence the increment of total strain, $d\boldsymbol{\varepsilon}_i$, may be calculated for each stress point.

– For each stress point, the increment of elastic stress, $d\boldsymbol{\sigma}_i^e$, is given from the increment of total strain, $d\boldsymbol{\varepsilon}_i$, and the elastic modulus matrix, **D** (see Figure 3.6), using the differential form of Equation 3.9.

– The increment of total strain is also used to calculate the increment of total stress, $d\boldsymbol{\sigma}_i$, required for the material to remain on the given stress-strain curve. Here a so-called elastoplastic modulus matrix, $\mathbf{D}_{ep}$, is employed (see Figure 3.6 and Chapter 1), or:

$$d\boldsymbol{\sigma}_i = \mathbf{D}_{ep} d\boldsymbol{\varepsilon}_i \tag{3.17}$$

The stress difference, $d\boldsymbol{\sigma}^r$, as shown in Figure 3.6, where:

$$d\boldsymbol{\sigma}_i^r = d\boldsymbol{\sigma}_i^e - d\boldsymbol{\sigma}_i \tag{3.18}$$

is used to calculate the residual force increment, $\mathbf{F}_i^r$, using Equation 3.14. This is added to the force, **F**, following Equation 3.16, and the iteration repeated until convergence is obtained. The stress increment, $d\boldsymbol{\sigma}_i$, is added to the accumulated stress, $\boldsymbol{\sigma}$.

It should be noted that convergence to point B (Figure 3.6) occurs along line Y–B as shown for the simple uniaxial condition considered. A general material element will experience load transfer to or from adjacent elements during the residual force iteration sequence and the final stress state could be as shown by point $\mathbf{B}_1$ or $\mathbf{B}_2$.

The initial strain method

The so-called initial strain method is similar to the above solution method, except that

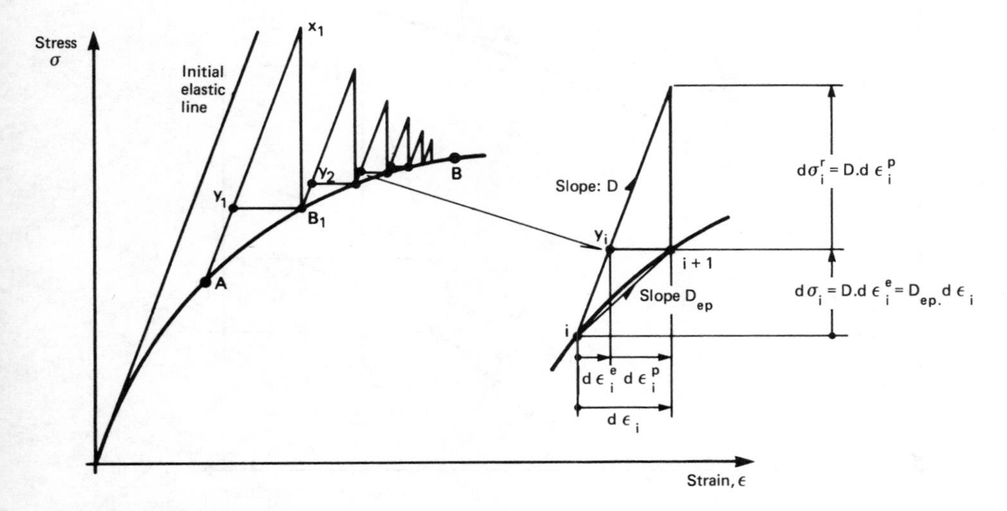

Figure 3.8. Initial strain method: idealised stress-strain history of one finite element stress point

having applied the loading for a given step, i, the stress $d\boldsymbol{\sigma}_i^e$ is again calculated but is used to determine the increment of total strain, $d\boldsymbol{\varepsilon}$, required to arrive at the material yield surface, $A–B_1–B$, as shown in Figure 3.8. The procedure for incorporating this method in a finite element code is exactly similar to that described for the above initial stress method, except that the increment of residual stress, $d\boldsymbol{\sigma}_i^r$, is based upon the plastic strain increment, $d\boldsymbol{\varepsilon}_i^p$, where:

$$d\boldsymbol{\sigma}_i^r = \mathbf{D} d\boldsymbol{\varepsilon}_i^p \tag{3.19}$$

and $d\boldsymbol{\varepsilon}_i^p$ being obtained from the relationship as illustrated (in one dimension only) in Figure 3.8. Hence the name 'initial strain'.

Convergence would take place in one step, to B, if no load transfer were to take place between adjacent elements. In general, however, a given point will follow several iteration cycles as shown in Figure 3.8, until convergence to point B is reached.

The method is known to give numerical instability problems for certain types of solution, but it forms a stepping stone to the following method.

The visco-plastic method

The visco-plastic method is a variation of the above initial strain method whereby the plastic strains are considered to be 'time-dependent' as illustrated by the damping element of the rheological model (Figure 3.5), or:

$$d\boldsymbol{\varepsilon}_i^p \equiv \Delta t \cdot \dot{\boldsymbol{\varepsilon}}_i^p \tag{3.20}$$

where Δt is the time increment and $\dot{\boldsymbol{\varepsilon}}_i^p$ the plastic strain rate for the i'th iteration.

The terms strain rate and time increment are often unrelated to real time, in which case they are used as an artifice to obtain a non-time-dependent solution. However, real

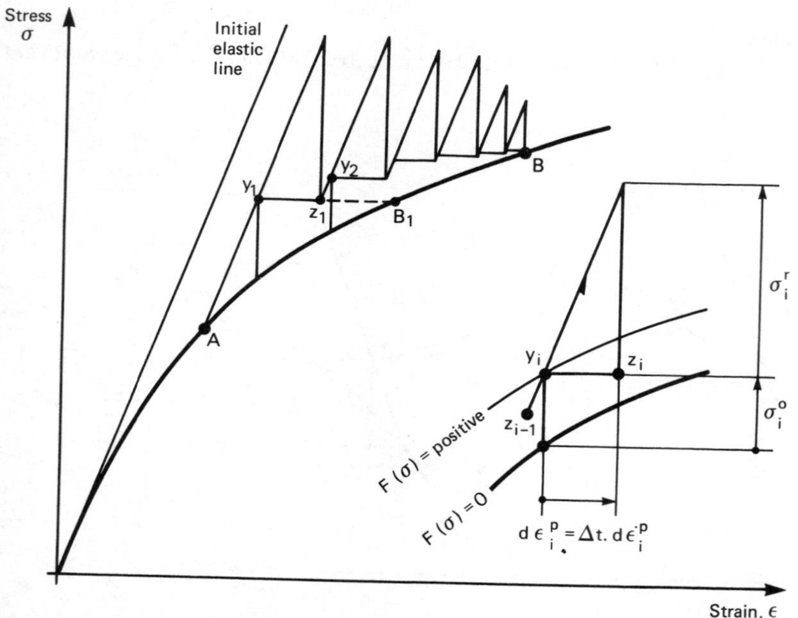

Figure 3.9. Visco-plastic method: idealised stress-strain history of one finite element stress point

time parameters may be employed, in which case a problem in real time is then able to be solved.

Consider now a point, A, lying on the yield surface as shown in Figure 3.9. Application of a load increment to the elastic system will result in point Y_1, and for the i'th increment point Y_i, being reached. In general, there will be an overstress at this point, as shown in Figure 3.9, given by the difference in the stress at Y_i and the stress on the yield surface. The yield surface may in general be expressed by a function of the six stress components, $\boldsymbol{\sigma}'$, by an expression known as the yield function (see Chapter 1, Section 1.6), in the form $F(\boldsymbol{\sigma}', \kappa)$ where (see Equation 1.21):

$F(\boldsymbol{\sigma}', \kappa) = 0$ when the material lies on yield surface,

$F(\boldsymbol{\sigma}', \kappa) > 0$ when the material is in the plastic range, and

$F(\boldsymbol{\sigma}', \kappa) < 0$ when the material is in the elastic range,

κ being a hardening parameter specified as a function of the accumulated plastic strain $\kappa(\boldsymbol{\varepsilon}^p)$.

Similarly, the direction of straining is defined by a function $Q(\boldsymbol{\sigma}', \kappa_2)$, the plastic potential. The strain rate, $\dot{\varepsilon}_i^p$, may thus be written:

$$\dot{\boldsymbol{\varepsilon}}^p = \langle \Lambda(F) \rangle \frac{\partial Q}{\partial \boldsymbol{\sigma}}, \tag{3.21}$$

where the sign $\langle\rangle$ implies a zero quantity for $F < 0$, and $\Lambda(F)$ is known as the fluidity parameter. The condition $Q = F$ is known as associated plasticity (see Chapter 1), for which the direction of the strain increment vector, $\dot{\boldsymbol{\varepsilon}}^p$, is directly associated with the yield function F. In the uniaxial example of Figure 3.9, the function F is simply given by the yield curve $\sigma'(\varepsilon^p)$. The strain increment direction $\partial Q/\partial \boldsymbol{\sigma}'$ is in this simple case not required because the direction of the plastic strain increment is also in the given axis direction. The above quantities F, Q and $\dot{\boldsymbol{\varepsilon}}^p$, are also illustrated in Figure 3.10 in terms of a two-component stress space.

The procedure for obtaining an engineering solution when using the visco-plastic method is exactly similar to that previously described for the initial strain and initial stress solution methods, the same basic routines being employed. The distinct

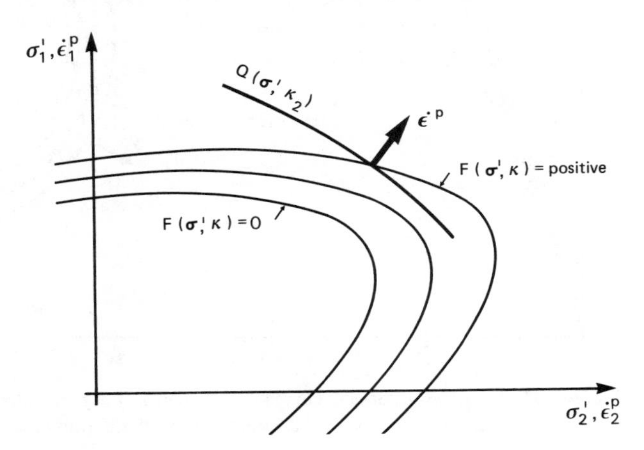

Figure 3.10. Yield and plastic potential functions in a two-dimensional stress space, also showing the strain increment direction

advantage here, besides the ability to solve for time-dependent effects, is that rapid convergence may be obtained provided that the correct choice of the time increment, Δt, is made. Care must be exercised when choosing this value because there exists a critical value, Δt_{crit}, which is an upper bound and above which convergence cannot be guaranteed. This value depends upon the material yield function and upon the expression for the fluidity Λ(F). For the case when Λ(F) is a constant, λ, Cormeau (1976) has obtained the following:

$$\text{Tresca:} \qquad \Delta t_{crit} = \frac{2}{\lambda G} \tag{3.22a}$$

$$\text{Von Mises:} \qquad \Delta t_{crit} = \frac{2}{3\lambda G} \tag{3.22b}$$

$$\text{Mohr-Coulomb:} \qquad \Delta t_{crit} = \frac{2(1-2\nu)}{\lambda G(1-2\nu+\sin\phi)} \tag{3.22c}$$

where G is the shear modulus, ν the Poisson's ratio and φ the effective angle of friction.

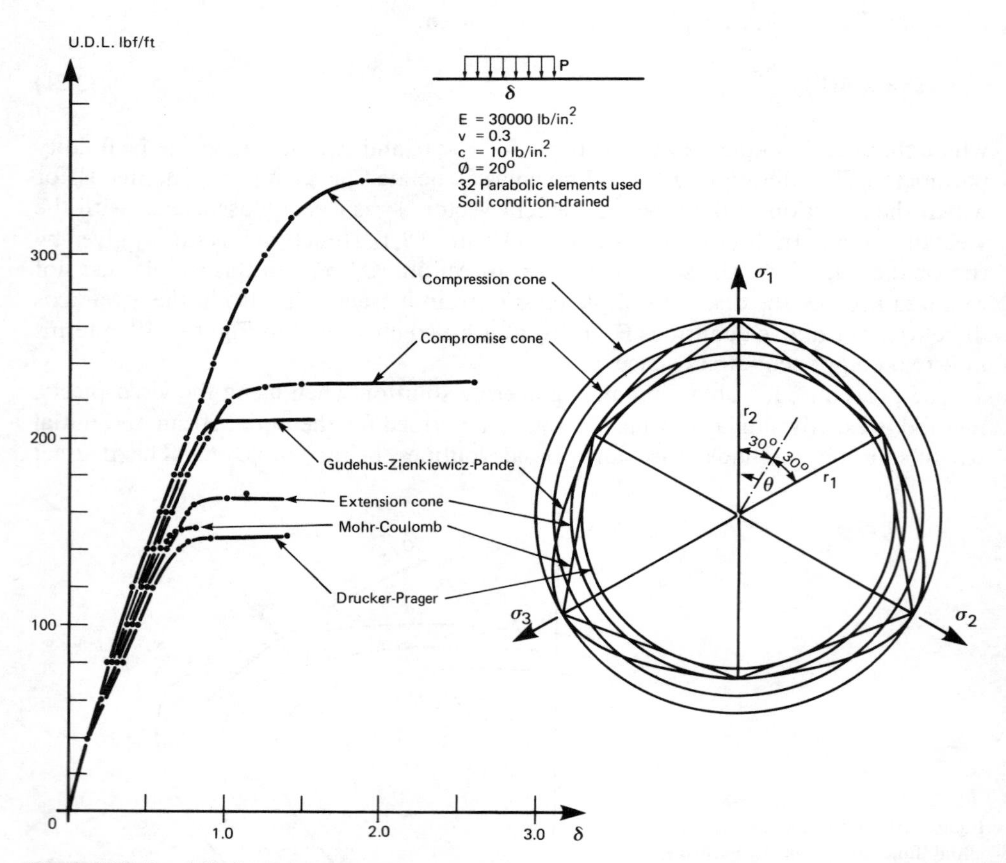

Figure 3.11. Load-displacement curves for a strip footing, calculated on the basis of ideal associated plasticity for various forms of Mohr-Coulomb approximation (after Zienkiewicz et al. 1977)

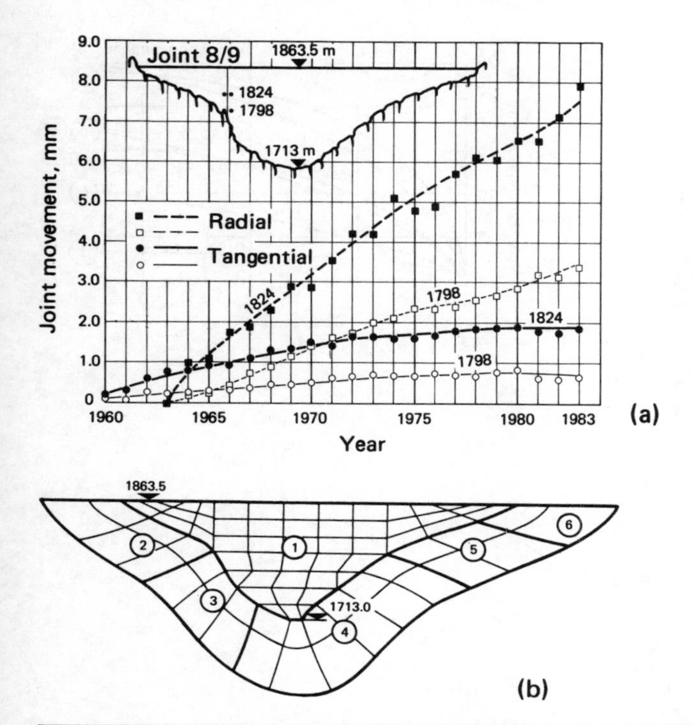

Solution number	Young's modulus (10^6 kN/m^2), Material zone number					
	1	2	3	4	5	6
1	22.00	27.00	27.00	20.00	16.00	15.00
2	19.30	10.00	10.00	6.00	3.75	3.75
5	23.16	12.00	35.40	4.50	2.75	7.00

(c)

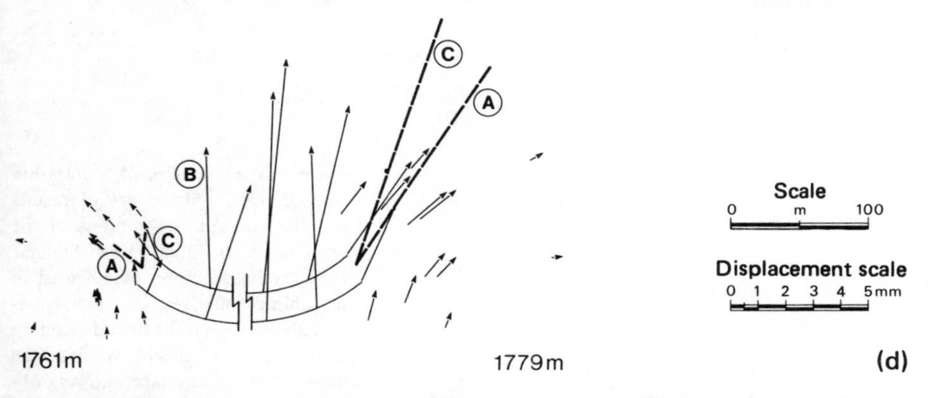

Figure 3.12. Observed contraction joint movement and extracts from the elastic correlation of the Zervreila Arch Dam (after Dungar 1985); (a) observed movement in joint 8/9 at two different elevations, showing continuing radial joint movement with time; (b) material zones for the 3D, F.E., elastic solutions; (c) table of elastic modulus values for various trial solutions; and (d) observed and calculated displacements in two rockmeter groups: Ⓐ-observed, Ⓑ-calculated vectors, Ⓒ-calculated displacement corresponding with rockmeter reading

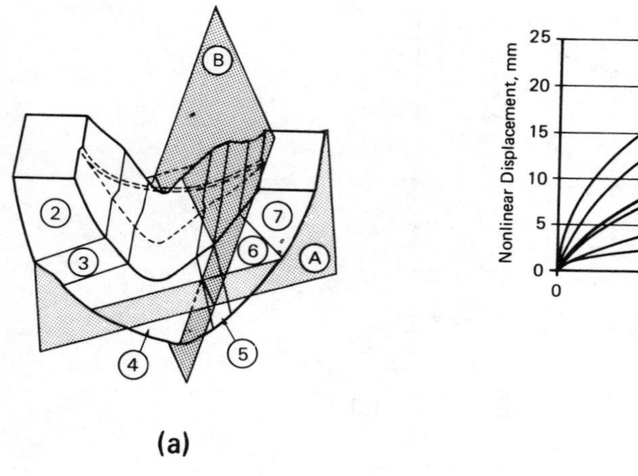

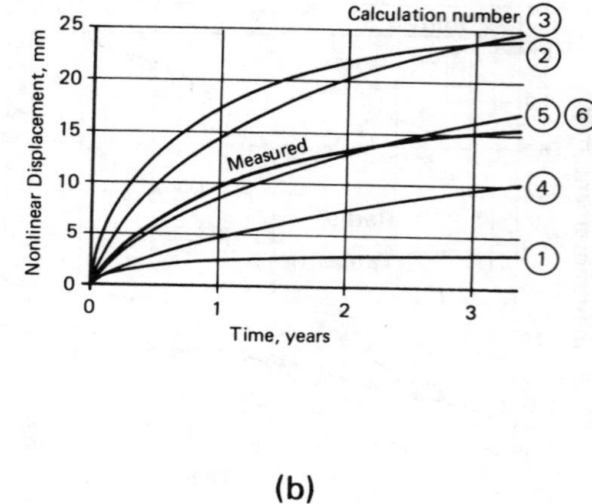

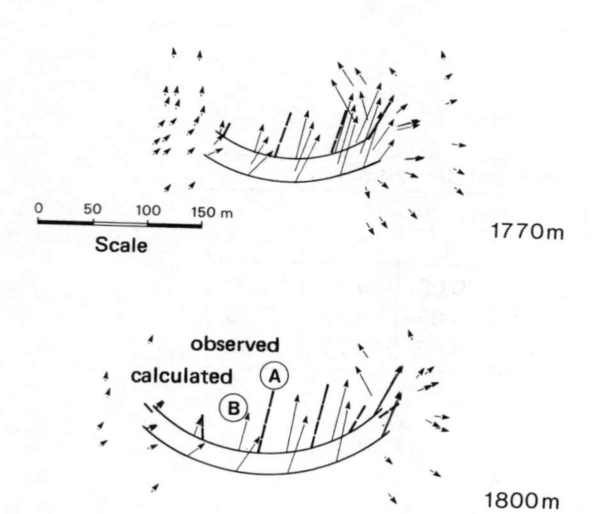

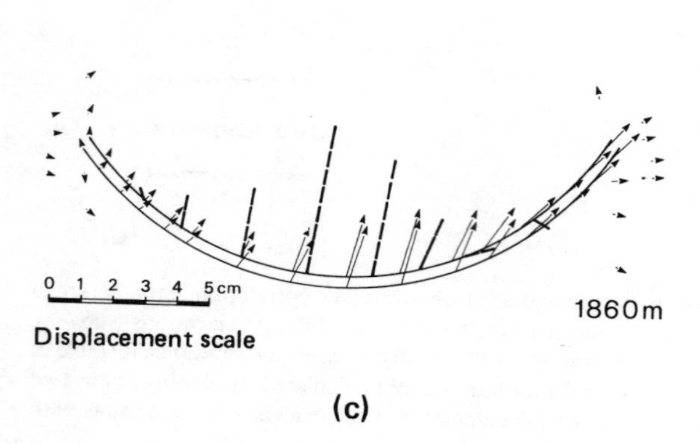

Figure 3.13. Directions of foundation discontinuities and extract of results from the visco-plastic analysis of the Zervreila Arch Dam (after Dungar 1985); (a) discontinuity planes used in the nonlinear analysis Ⓐ–Ⓑ and material zones ②–⑦; (b) time-dependent deformation at a typical location on the dam-foundation interface, showing observed and calculated values; and (c) observed nonlinear deformation of dam body showing comparison with calculation for nonlinear foundation deformation (solution 6)

The results of Figure 3.11 are here presented to demonstrate the effects of assuming different yield criteria, as shown, for the case of a strip footing on a clay foundation.

The results of Figure 3.12 are given to illustrate the accuracy that may be obtained when using a three-dimensional finite element mesh for the analysis of deformation in an arch dam and its associated foundation. The reason for performing such a detailed analysis of this particular dam (Zervreila, Switzerland) was that permanent movement in both the radial and tangential direction of contraction joint 8/9 has occurred since its construction and period of first impoundment (1957), as shown. Particular attention was paid to the effects of foundation deformation because it was clear from the outset that significant permanent deformation had occurred, mainly on the lower right flank. Both flanks consist of gneiss but the geomechanical properties in the various zones show significant variations, mainly due to the direction of the bedding plane (plane A as shown in Figure 3.13). The elastic correlation was performed mainly on the basis of comparing calculated and observed deformation in the four groups of rock meters, as shown, and finally by controlling the elastic modulus of the dam (see Dungar 1985 for more details).

The results of Figure 3.13 illustrate the accuracy that was achieved for the correlation of nonlinear deformation in the foundation, and the reader is also directed to the comments in Chapter 1, Section 1.9 concerning the details of the model. The use of the visco-plastic solution method enabled the time-dependent deformation history to be checked against the corresponding observed history. The influence of this nonlinear foundation deformation upon the deformation of the dam body is also illustrated. From this study it became clear that the dam itself is also experiencing a time-dependent deformation which is independent of that due to the deformation of the foundation (not shown). From an assessment of the induced stress and strain under reservoir draw-down, it becomes clear that the joint opening is to be expected under such conditions. The current safety of the dam was also investigated.

The use of the visco-plastic solution method and corresponding numerical material model has thus proved to be a valuable aid in the engineering assessment of Zervreila dam.

3.4 EQUIVALENT LINEAR TOTAL STRESS ANALYSIS FOR DYNAMIC LOADING

3.4.1 *Introduction*

The response of geomechanical structures to earthquake, blast and other dynamic loads, usually involves a highly nonlinear material response, as illustrated by the stress-strain curve of Figure 3.14(a). As stated in Chapter 8, two methods of solution are open to the engineer for the analysis of this type of problem, one of which, the equivalent linear method, is described below and involves solving the structural response by using appropriate equivalent elastic properties. Other methods attempt to follow the full time-dependent stress-strain history of the material, and these will be briefly discussed in Sections 3.5 and 3.6.

The method of equivalent linear analysis, as proposed by Seed (1966) and as extensively used for the analysis of a slope failure of the Lower San Fernando Dam

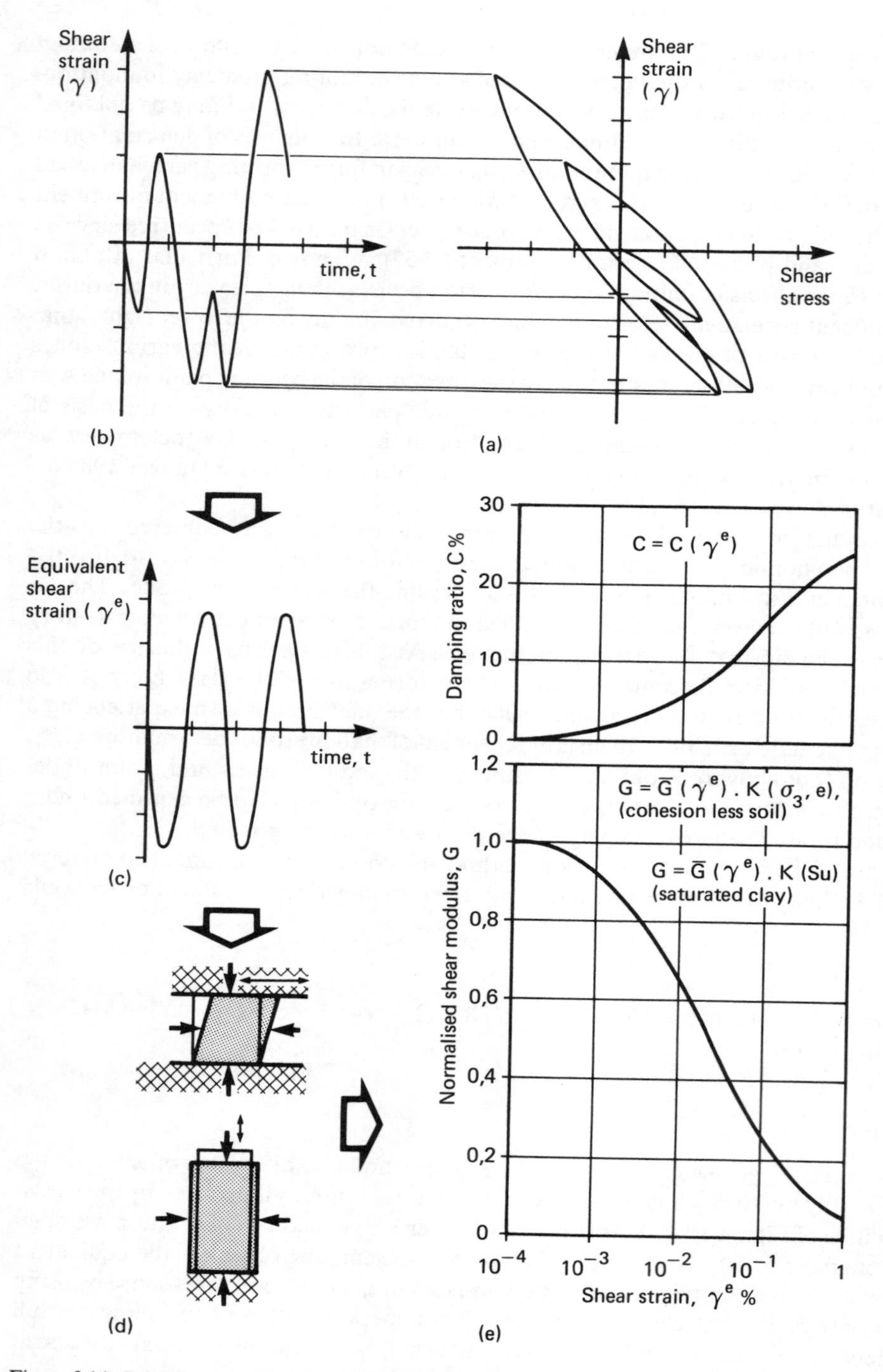

Figure 3.14. Representation of the analytical process used for obtaining equivalent shear modulus and damping ratio from a material stress-strain history; (a) material stress-strain history; (b) shear strain history; (c) sinusoidal history; (d) sinusoidal laboratory tests; and (e) equivalent shear modulus and damping ratio

during the San Fernando earthquake of 1971 (Seed 1979) has since been widely adopted for the design analysis of a number of large embankments. Although strictly a 'method of analysis', an implied material model is used which gives an approximation to the actual material behaviour, a common feature of all material models. The basic steps will now be introduced followed by certain details and variations of the method.

3.4.2 *The basic steps*

The basic steps are as follows:

1. A finite element mesh (two- or three-dimensional) is defined and the static stress (total or effective) existing prior to the dynamic load is estimated or, more accurately, calculated using a static finite element solution (see previous sections of this chapter).
2. The above static stress is assumed to exist as an average stress throughout the period of the dynamic load, and is used to determine the appropriate shear modulus for the given material element.

   For cohesionless soil, the equivalent sinusoidal shear modulus, G, may be taken as

   $$G = \bar{G}(\gamma^e)\cdot K(\sigma_3, e) \tag{3.23}$$

   where e is the void ratio, and γ^e an equivalent sinusoidal shear strain amplitude, given as an estimated value or obtained from a previously calculated shear strain-time history. The constant $K(\sigma_3, e)$ is defined from cyclic material tests as a function of the static stress, σ_3, and void ratio, e (see Chapter 10 for more details). A typical variation of parameter G with respect to sinusoidal shear strain amplitude, γ^e, is illustrated in Figure 3.14(e). As an alternative, if the in-situ shear wave velocity, V_s, is available from on-site testing, this may be used to estimate the value of K (given by G_{max} when γ^e is small) from the expression:

   $$K = V_s^2 \tag{3.24}$$

   where ρ is the mass density (density/gravitational acceleration).

   For an element of saturated clay, the shear modus, G, may be obtained from:

   $$G = \bar{G}(\gamma^e)\cdot K(S_u) \tag{3.25}$$

   where S_u is the undrained shear strength (see Chapter 10 for details).
3. The critical damping value for the soil element is given, again as a function of the equivalent sinusoidal shear strain amplitude, γ^e, from cyclic test results as:

   $$C = C(\gamma^e) \tag{3.26}$$

   as is also illustrated in Figure 3.14(e) (see also Chapter 10).
4. The above shear modulus, G, is employed to obtain the stiffness properties for each finite element taken in turn, the value of the bulk modulus, K, being maintained constant (given from the static value), and the stiffness for the complete structure is again established. In the first instance, estimated properties are used and these are updated in subsequent iteration steps (see below).
5. Damping may be introduced by forming the element damping matrix, **c**, from the combination of the element stiffness and mass, **k** and **m**, where:

   $$\mathbf{c} = C(\omega_1\mathbf{m} + \mathbf{k}/\omega_1), \tag{3.27}$$

and C is the critical damping coefficient, as given from step 3 above and ω, the frequency of the first vibrational mode. It will be observed that if C is constant for the complete structure (uniform cyclic shear strain over the structure), the structural damping matrix, **C**, is such that Equation 2.43 reduces to Equation 2.44 for the first mode of vibration (this assumption was first made by Rayleigh 1945), but for geomechanical structures it is infrequent that such a condition of uniform cyclic shear strain occurs in practice.

6a. Equation 2.30 is solved directly by a time-marching scheme (step-by-step method). The method of Wilson & Clough (1962) is an example of such a solution, the basis of which is that a linear variation of acceleration is assumed over the time increment, Δt, and the unknown displacements for time t may thus be expressed in terms of values at time t-Δt by:

$$\mathbf{X}(t) = \left[\mathbf{K} + \frac{6}{\Delta t^2}\mathbf{M} + \frac{3\mathbf{C}}{\Delta t}\right]^{-1} \bar{\mathbf{F}}(t) \tag{3.28a}$$

$$\bar{\mathbf{F}}(t) = \mathbf{F}(t) + \mathbf{A}^T(t)\cdot\mathbf{M} + \mathbf{B}^T(t)\cdot\mathbf{C} \tag{3.28b}$$

$$\mathbf{A}(t) = \frac{6}{\Delta t^2}\mathbf{X}(t-\Delta t) + \frac{6}{\Delta t}\dot{\mathbf{X}}(t-\Delta t) + 2\ddot{\mathbf{X}}(t-\Delta t) \tag{3.28c}$$

$$\mathbf{B}(t) = \frac{3}{\Delta t}\mathbf{X}(t-\Delta t) + 2\dot{\mathbf{X}}(t-\Delta t) + \frac{\Delta t}{2}\ddot{\mathbf{X}}(t-\Delta t) \tag{3.28d}$$

$$\dot{\mathbf{X}}(t) = \frac{3}{\Delta t}\mathbf{X}(t) - \mathbf{B}(t) \tag{3.28e}$$

$$\ddot{\mathbf{X}}(t) = \frac{6}{\Delta t^2}\mathbf{X}(t) - \mathbf{A}(t) \tag{3.28f}$$

Knowing **X**(t), the strain and stress for each element may be calculated in the usual way. This method is employed in the 'QUAD4' computer code (Idriss et al. 1973).

6b. As an alternative, Equation 2.30 is not solved directly, but the normal mode theory is 'imposed', that is, although the resultant damping matrix, **C**, does not obey the orthogonality conditions of Equation 2.41, nevertheless the first 's' modes X_1, $X_2 \ldots X_5$ are assumed as being 'uncoupled'. The corresponding critical damping coefficient for the i'th mode, c_i, is thus given from Equation 2.44b as:

$$c_i = \frac{1}{2\omega_i\beta_i}\mathbf{X}_i^T\mathbf{C}\mathbf{X}_i \tag{3.29}$$

The response of the complete structure may then be calculated using the above 's' modes, by using Equation 2.45. This method is used by the author.

7. From the results of the dynamic (non-uniform) shear strain amplitude, the equivalent sinusoidal shear strain amplitude is estimated for each stress point (see the illustration of Figure 3.14).
8. Steps 2–7 above are repeated until a suitable convergence is obtained.

The above equivalent linear method forms a corner stone in the methodology for the assessment of seismic stability of dams, as proposed by Seed and as summarised in

Chapter 8. The method also forms the basis for procedures written for foundation analysis as are often employed for the important problem of the assessment of nuclear power plant safety.

In the author's opinion, the approximations inherent in the above equivalent linear calculation process justify the additional assumption inherent in the method of option 6b, above. This method has the advantage that, for earthquake loading in particular, spectral displacement curves may be directly used to solve the system response. The reader is also directed to the comments given in Chapter 8 regarding the required solution accuracy.

3.4.3 *Pore pressure generation*

From a practical point of view, the above basic steps used for the response calculation form a forerunner to the calculation of dynamically induced pore pressure and permanent deformation. The basic method is reported by Serff et al. (1976) and Seed (1979) and from this several variations are possible. Details as used by the present author will now be discussed.

Results taken from stress-controlled cyclic triaxial tests on undrained samples are as illustrated in Figure 3.15. Following the proposal of Prater (1980) (see also Bossoney & Dungar 1980), the maximum generated pore pressure is taken to correspond with the value $\bar{\mu}$, as illustrated. That is, for the initial conditions of point A, where:

$$K_0 = \left(\frac{\sigma_1}{\sigma_3}\right)_{\text{static}} \tag{3.30}$$

and the assumption that the initial static total stress is maintained throughout the duration of the earthquake, together with the assumption that maximum cyclic pore pressure generation corresponds with the displayed critical state line (see also Chapter 1, where the importance of the critical state condition is discussed), we obtain for

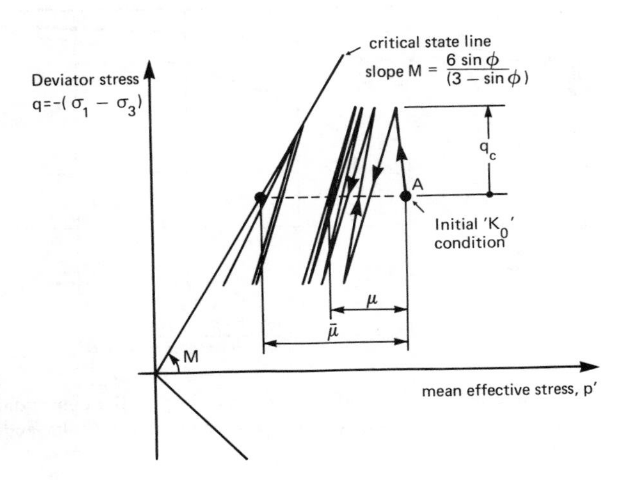

Figure 3.15. Representation of the stress history of an undrained material subjected to stress-controlled cyclic loading

triaxial conditions:

$$\bar{\mu} = -\frac{\sigma_3}{3}\left\{K_0 + 2 - \frac{(K_0 - 1)(3 - \sin\phi)}{2\sin\phi}\right\} \tag{3.31}$$

Cyclic triaxial test results, such as those of Figure 3.16,* are used to determine the number of uniform cycles, $\bar{N}$, required to produce the maximum generated pore pressure, $\bar{\mu}$. Knowing the cyclic deviator stress, q_c, the initial consolidation ratio, K_0, the value of $\bar{N}$ may easily be obtained from such a figure. In the EFESYS program (Dungar 1979) an interpolation scheme is used, the material curves being stored in the data-base as a set of spline functions.

The results of Figure 3.17 illustrate typical pore pressure generation curves for different values of K_0. Knowing the actual number of equivalent uniform cycles, estimated from the above response calculation (for details see Seed 1979) or as estimated from a knowledge of the earthquake duration and period of structural vibration, the ratio:

$$\frac{N}{\bar{N}} = \frac{\text{Number of uniform cycles}}{\text{Number of cycles for maximum pore pressure}} \tag{3.32}$$

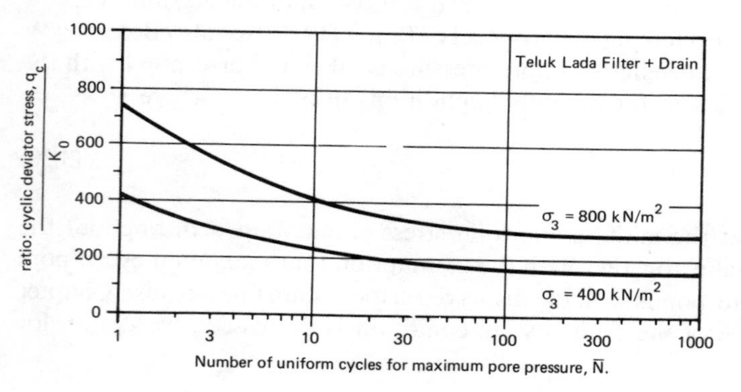

Figure 3.16. Typical undrained cyclic triaxial test results showing the influence of the cyclic deviator stress on the number of cycles required to reach the maximum generated pore pressure

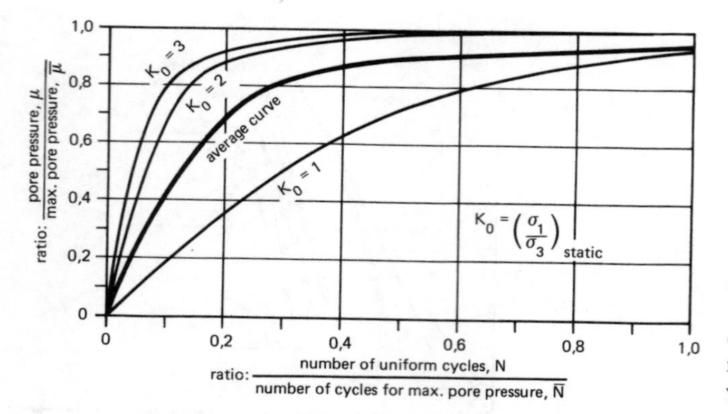

Figure 3.17. Typical undrained cyclic triaxial test results showing the influence of the number of uniform cycles on the generated pore pressure, for different values of K_0

*The value of q_c is here normalised according to the consolidation ratio, K_0, but the validity of this normalisation process will depend upon the given test results, and this is not mandatory for following the given process.

may be calculated. This ratio, together with the corresponding value of K_0, enables the ratio:

$$\frac{\mu}{\bar{\mu}} = \frac{\text{Generated pore pressure}}{\text{Maximum generated pore pressure}} \tag{3.33}$$

to be obtained (interpolation routines may again be used). From this ratio, and knowing $\bar{\mu}$ (see Equation 3.31), the generated pore pressure, μ, is thus obtained.

The above process is reasonably straightforward and suitable post-processing routines enable the pore-pressure to be automatically calculated with the aid of interpolation functions. The material test data may also be stored as a series of splines.

3.4.4 *Cyclic mobility and permanent displacement*

Permanent displacement is somewhat more complicated to calculate than generated pore pressure. The steps are as follows:

1. The above calculated or estimated number of equivalent uniform cycles, N, together with the calculated cyclic deviator stress, q_c, minimum principal stress, σ_3, and consolidation ratio, K_0, enables the equivalent axial permanent strain, ε_1^p (corresponding with cyclic triaxial test conditions), to be established. Typical results from undrained cyclic triaxial tests are displayed in Figure 3.18.
2. Axial permanent strain (often termed strain potential), as calculated above, is used to estimate the permanent displacement induced by the earthquake. A review of four solution methods investigated for the Upper San Fernando Dam (earthquake of 1971), is given by Serff et al. (1976). The strain potential used in this study is also shown in Figure 3.19 together with static shear stress as calculated for the horizontal

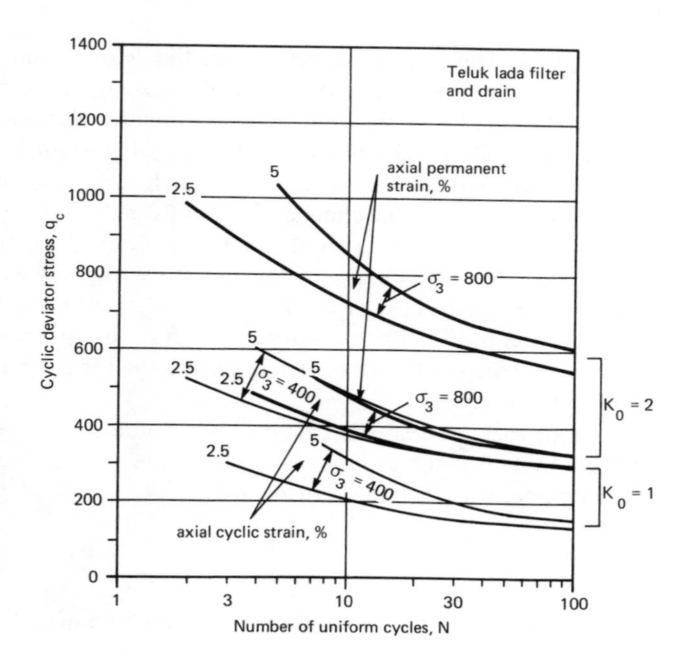

Figure 3.18. Typical undrained cyclic triaxial test results showing the influence of the number of uniform cycles and the cyclic deviator stress on the axial permanent strain

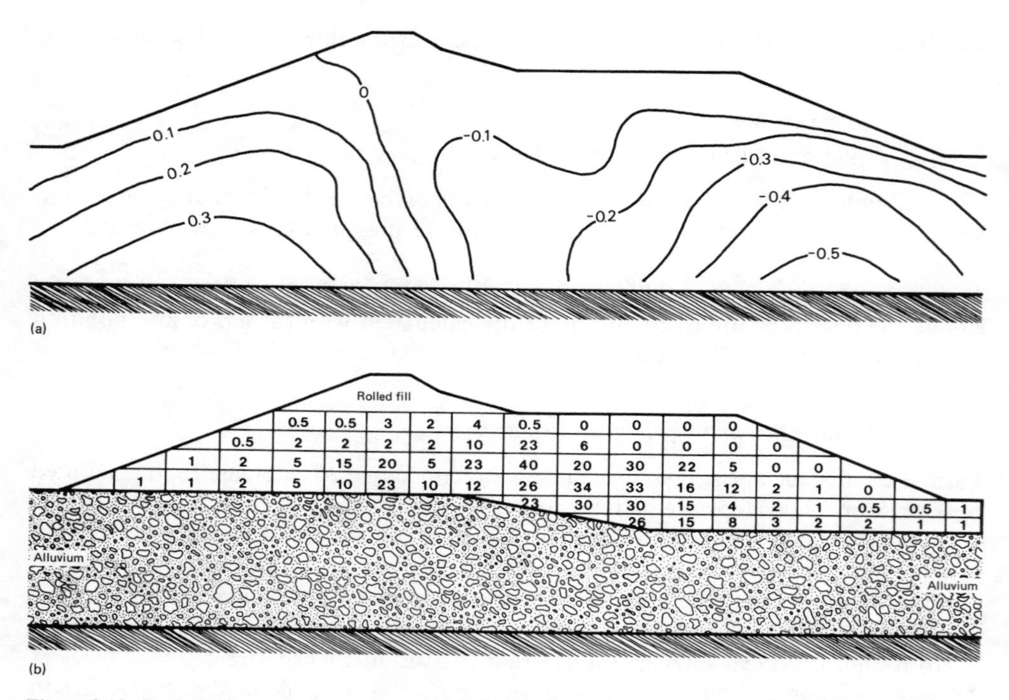

Figure 3.19. Static horizontal shear stress and calculated strain potential as used by Serff et al. (1976) for the calculation of the permanent displacement of the Upper San Fernando Dam due to the earthquake of 1971; (a) static shear stress on xy plane, τ_{xy} tons/ft^2; and (b) finite element mesh showing calculated strain potential

(xy) plane. The calculated 'best fit' for the deformation of the dam, in comparison with the observed behaviour, was obtained by a method employing 'equivalent nodal forces'. These were calculated from material stress-strain curves as illustrated in Figure 3.20, on the basis that a material element located at A (drained curve) before the earthquake, is after the earthquake assumed to lie on curve OBC (undrained curve). The change of strain from A to B was neglected, being assumed small in comparison with the induced strain potential, ε^p. Nodal forces were calculated from the change of deviator stress, Δq, and were applied to the structural stiffness properties as derived from the undrained material curve, OBC, a piecewise linear method being employed, for this 'quasi-static' loading. The resulting structural displacement was then taken to be the permanent displacement induced by the earthquake.

A further assumption that is often made is that the horizontal plane is the predominant plane of 'induced dynamic shear stress' and the shear stress acting on this plane, $\Delta\tau_{xy}$, is related to Δq by:

$$\Delta\tau_{xy} = \frac{\Delta q}{2} \tag{3.34}$$

the direction of $\Delta\tau_{xy}$ being taken as the direction of the pre-earthquake shear stress.

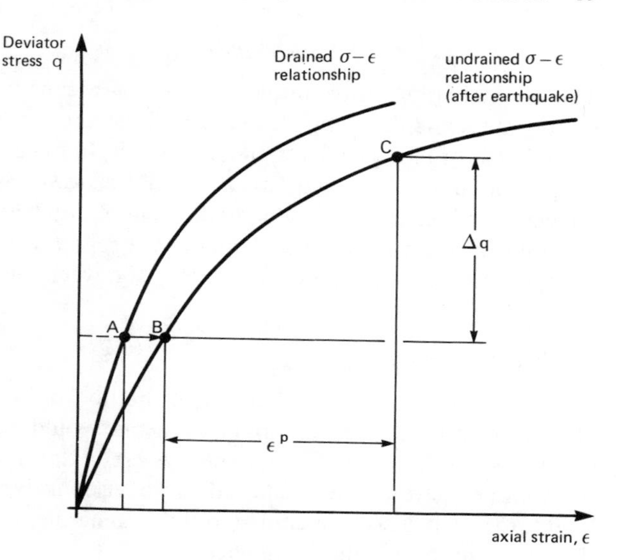

Figure 3.20. Typical stress-strain curves obtained from drained and undrained triaxial tests showing the change of deviator stress, Δq, corresponding with the earthquake-induced strain potential, ε^p, as proposed by Serff et al. (1976)

It should be noted that the calculation of permanent displacement as adopted in step 2 above in fact follows the initial stress solution method, as described in Chapter 2, post-earthquake undrained material properties being used to define the nonlinear quasi-static stress-strain law (see Section 3.2 above).

3.5 PIECEWISE LINEAR TOTAL AND EFFECTIVE STRESS ANALYSIS FOR DYNAMIC LOADS

3.5.1 *Introduction*

The equivalent linear method of response analysis, described in the previous section, has the distinct advantage that standard routines and techniques used for linear dynamic response calculations (see Chapter 2) may easily be adapted for the purpose of estimating the response of nonlinear structures. The main disadvantages are that material changes which progressively occur during the response, and in particular pore pressure changes, are not taken into account and also certain assumptions must be made when using the given equivalent linear properties. For certain classes of problem, both of these disadvantages may be overcome by adopting a piecewise linear solution in which a piecewise linear stress-strain law, similar to that described in Section 3.2 for static loading, is adopted.

The methodology has the same advantages and disadvantages as for static loading, namely related to the difficulty of extending the material model, as formulated for one-dimensional behaviour, to a two- or three-dimensional continuum problem. For static loading, it is normal practice to use models based upon triaxial test data, and for dynamic analysis upon simple shear data. Solutions given to date are only for the one-dimensional case of a vertically propagating shear wave in level ground.

3.5.2 *Numerical solution techniques*

The normal solution technique consists of idealising the layered site as a series of spring, mass and damping elements, as illustrated in Figure 3.21, where the mass for a given layer, or part of a layer, of height h_i, is 'lumped' to the degrees of freedom x_i and x_{i+1}. The mass term, m_i, at a given freedom may thus be regarded as being concentrated on an infinitesimally thin 'plate', and between such plates the stiffness and damping of the various soil elements are represented as equivalent springs, k_i, and damping elements, μ_i, the shear strain, γ_i, between the plates being constant, and given as:

$$\gamma_i = (x_1 - x_{i+1})\hat{}h_i \tag{3.35}$$

When employing an equivalent linear method of analysis (see previous section), the illustrated stiffness and damping elements would be replaced by strain-dependent constants, but here the stiffness and hysteretic damping is, in fact, a representation of the nonlinear stress-strain material curve (viscous damping due to velocity-dependent fluid forces may also be included, but this generally has little influence compared with the above hysteretic damping effects).

Having established the stress-strain law (see Chapter 1), the instantaneous force acting on the soil layer, f_i (which, for a column of unit area, is equal to the shear stress), may be expressed in terms of the strain (and hence the displacement). Figure 3.22 illustrates typical stress changes within a layer in which force f_0, corresponding with stress τ_0, exists at time t. The variation over time step Δt is given in terms of the instantaneous shear modulus G_0 as:

$$\tau = \tau_0 + G_0(\gamma - \gamma_0) \tag{3.36}$$

The equations of motion at time $t + \Delta t$ may thus be written as:

$$\mathbf{M}\ddot{\mathbf{X}}(t + \Delta t) + \mathbf{C}\dot{\mathbf{X}}(t + \Delta t) + \mathbf{K}_0\{\mathbf{X}(t + \Delta t) - \mathbf{X}(t)\} = \mathbf{F}(t + \Delta t) - \mathbf{F}_0(t) \tag{3.37}$$

where the matrix $\mathbf{C}$ is the viscous damping matrix (usually taken as a null matrix), $\mathbf{K}_0$ is the stiffness matrix corresponding to G_0, being due to the change of displacement over

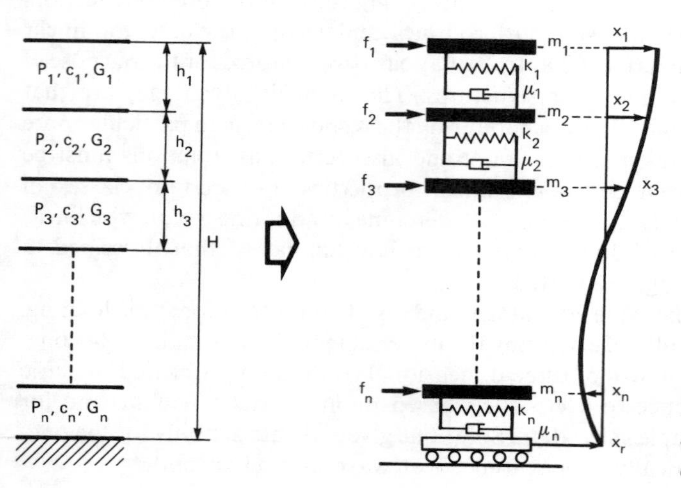

Figure 3.21. Numerical analogue of a layered soil, showing the equivalent mass, spring and damping elements

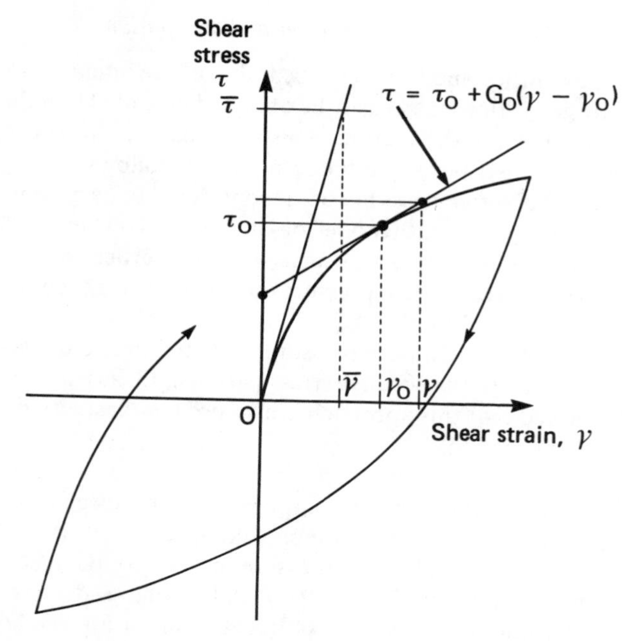

Figure 3.22. Instantaneous shear modulus, G_o, used for the time-step length Δt in a piecewise linear analysis

the time step Δt and $\mathbf{F}(t + \Delta t)$ is the applied force which, for earthquake motion, is given as:

$$\mathbf{F}(t + \Delta t) = -\mathbf{M}\ddot{x}_r(t + \Delta t) \tag{3.38}$$

the nodal displacements $\mathbf{x}$ being, here, measured with respect to the rigid base displacement, x_r, as shown.

Equation 3.37 is of the same form as Equation 2.30 and here a solution method which allows a step-by-step solution in the time domain is needed, an example of which is that of Wilson & Clough given in Equation 3.28 above. Details of stress-strain relationships used in this type of analysis are given in Chapters 1 and 10 and are based upon 'initial loading' shear stress histories taken as a hyperbolic function of shear strain.

Results obtained from the use of such piecewise linear models will in general be different to those from the equivalent linear method when non-cyclic loads are under investigation. Finn (1982) has made a comparison of results obtained from two computer codes (CHARSOIL and DESRA) which are based upon the above piecewise linear method, in comparison with results from an equivalent linear solution (SHAKE). These results obtained on the basis of a total stress analysis indicated a '50% increase in the maximum amplification of the pseudo-acceleration' when the equivalent linear solution was used, in comparison with the piecewise linear solution.

In addition to the above differences due to the 'model', the effects of pore pressure generation and dissipation will often have an important influence on the resulting response.

3.5.3 *Pore pressure generation and dissipation*

The solution method described above is formulated in terms of total stress. Effective stress solutions may easily be obtained provided that the time-dependent pore pressure generation or dissipation is also calculated, and two approaches may be used when employing a piecewise linear model, as follows:

1. The effective stress history is considered to be dependent upon past history effects, as in the 'stress path model' of Ishihara & Towhata (1980), whereby shear stress on the horizontal plane, and corresponding vertical stress is modelled in such a way that the generated pore pressure for a given stress excursion is dependent upon some past 'yield condition'.
2. The generated pore pressure is related to a 'damage parameter', κ, which in turn is related to the effective stress path length, as used by Finn (1980, 1982). It should be noted that this approach is also used in plasticity based models, as explained in the next section.

The first method above is applicable to the specific conditions for which the model is designed and is rather complex compared with the second approach.

Dissipation of pore pressure may also be incorporated into the analysis, as demonstrated by Finn (1982) and Ishihara & Towhata (1980), thus delaying or eliminating the condition of liquefaction of a given soil layer.

3.6 CYCLIC AND DYNAMIC RESPONSE ANALYSIS EMPLOYING ELASTOPLASTIC MODELS

3.6.1 *Introduction*

In principle, an elastoplastic model is formulated such that both loading and unloading histories may be accommodated, unloading generally being modelled by recovery of elastic increments of strain, a yield surface also being encountered in the corresponding unloading direction. Such a formulation is easily handled by normal elastoplastic computer codes (see Section 3.3 for details). The same solution process may, with some change, be adapted to solve full dynamic problems, as shown in a later subsection. Firstly, however, it is appropriate to consider the somewhat simpler topic of cyclic loading, for which inertia effects are not considered.

3.6.2 *Cyclic analysis*

Problems arise when a normal elastoplastic code is extended to solve for cyclic loading, which may be done by using standard solution routines.

The simpler isotropic hardening models may show only elastic response after the first one, or two, cycles of load. The reason is illustrated in Figure 3.23, where the stress history from the origin 0 is such that during the first loading cycle elastic deformation occurs until the yield surface A is reached. On further loading, both elastic and plastic deformation occur (with associated hardening), the yield surface expanding from B to C. On reversal of the load direction, elastic deformation only will occur until the expanded yield surface is again reached at point D, after which plastic deformation, and

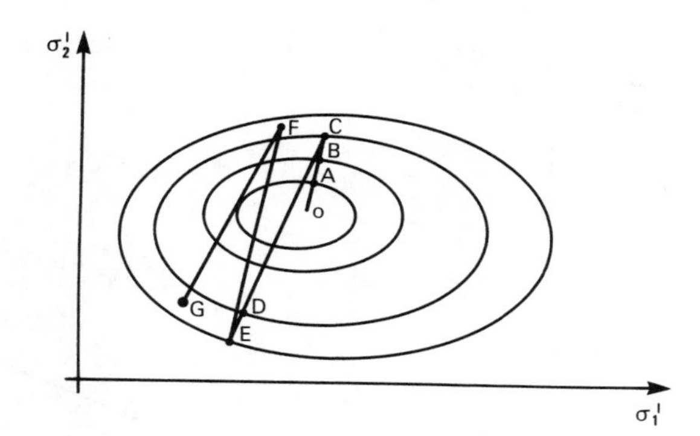

Figure 3.23. Expansion of the yield surface of an isotropic hardening model during cyclic straining

a further expansion of the yield surface from D to E, may again take place. A condition is thus reached where further cycling will occur within this expanded yield surface, as shown by path E–F–G.

The consequence of this elastic behaviour may be important and if we consider, for example, the cyclic behaviour of saturated undrained sand, results of cyclic laboratory tests generally show a continued nonlinear response for all cycles. Cyclical pore pressure generation occurs at a relatively uniform rate as also does the accumulated permanent strain. It is thus clear that normal isotropic strain hardening models are not suitable for modelling cyclic material response. There are basically two ways of overcoming this difficulty, as follows:

1. A so-called 'kinematic hardening' plasticity model may be employed for which the yield surface does not simply expand, isotropically, as illustrated in Figure 3.23, but rather 'floats' in the direction of the applied stress increment, as illustrated in Figure 3.24 (see also Chapter 9). This option is employed by many authors.
2. A primary elastoplastic model for 'virgin' loading may be employed, together with a secondary model to account for deviations from the elastic response which occurs when straining within the primary yield surface takes place. Two options have been used in the past for this secondary model:
   a. The so-called 'paraelastic' stress-strain formulation of Hueckel & Nova (1979) is an attempt at formulating a secondary model in terms of a full hysteretic stress-strain response.

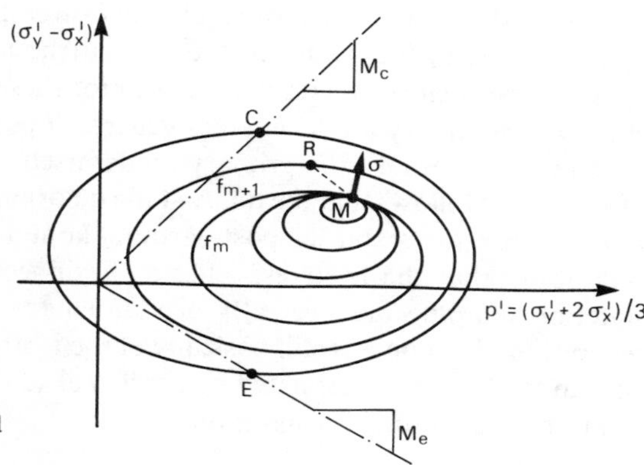

Figure 3.24. Movement of the multiple yield surfaces of a kinematic hardening model

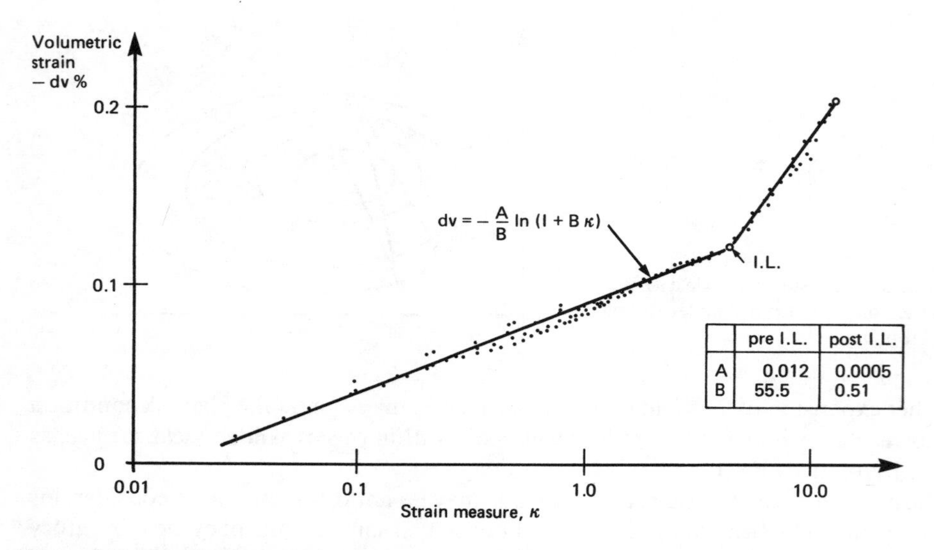

Figure 3.25. Densification model: volumetric strain displayed as a function of the strain measure, κ (after Zienkiewicz et al. 1981)

b. The so-called 'densification' model of Zienkiewicz et al. (1981) accounts for the secondary effects of progressive densification of a drained granular material as illustrated in Figure 3.25 or, alternatively, progressive pore pressure increase of an undrained material. These changes are taken as a function of an effective strain path length, κ, which has a direct analogy with the 'intrinsic time' parameter used in an endochronic model (see Chapter 1).

Kinematic hardening models, although mathematically complicated, appear to offer the promise of providing a general solution to problems in geomechanical engineering practice (see Chapters 14 and 15) and future developments in this direction should be carefully followed. The 'paraelastic' formulation has not as yet been applied to practical problems and is still in the modelling stage.

The 'densification model' has been applied to the dynamic response analysis of level ground and a back calculation of the Lower San Fernando Dam which suffered significant sliding failure during the 1971 earthquake (see also Chapter 8 for comments on the Upper Dam). The results are clear for these problems so far considered. The use of such a secondary model enables generated pore pressure or volume changes to be modelled but because the primary model itself does not account for volume change effects, it also follows that the quasi-static deformation modelled by this primary model will not allow for 'stabilised post-earthquake sliding' due to dilation effects. Further, it is doubtful that the hysteresis effects of stiffness degradation and strain-dependent material damping are correctly accounted for because, apart from pore pressure generation, the model is still basically of the isotropic hardening type. The Upper San Fernando Dam, as investigated by Seriff et al. (1976), would form a better check on the ability of such a dual formulation.

3.6.3 *Dynamic analysis*

The basic difference between cyclic and dynamic conditions lies in the inertia terms to be included in the equations of motion. Although the change of volume, and thus pore fluid content, should strictly be carried into the inertia terms, it is usual practice to neglect such effects, particularly when the formulation is made with small strain theory. It is also usual practice to uncouple the effects of the pore pressure change due to dissipation within a time interval, Δt, from the corresponding change in the body force, and hence the response within the given time step. In the case of undrained materials, this problem does not arise because dissipation does not take place and we will thus consider here the solution of either an undrained or fully drained material.

Equation 3.37 may again be employed, the tangential stiffness matrix being established for each time step, t, together with the force $\mathbf{F}_0(t)$, and the equation solved on a step-by-step basis. However, it is more convenient when using an elastoplastic model to take Equation 2.30 and to replace the force term due to the structural stiffness by that given in Equation 3.12, or:

$$\mathbf{M}\ddot{\mathbf{X}} + \mathbf{C}\dot{\mathbf{X}} + \sum_{e} \int_{v} \mathbf{B}^{T}\boldsymbol{\sigma} dv = \mathbf{F}(t) \tag{3.39}$$

where $\mathbf{F}(t)$ includes forces due to both static and dynamic loads. For example, in the case of earthquake, the load is given by Equation 2.46. The static load, due to the given static body forces and surface tractions, may be introduced as an initial load at time $t=0$ and the solution, $\mathbf{X}(t=0)$ and $\boldsymbol{\sigma}(t=0)$, obtained in the usual way (see Section 3.3), or alternatively corresponding initial stresses may be introduced from a previous static calculation.

Bicanic (1978) proposed a method whereby, with the omission of the viscous damping matrix, $\mathbf{C}$, Equation 3.39 may be solved using a central difference time stepping scheme, where $\mathbf{X}(t)$ is written in the form:

$$\mathbf{X}(t) = \frac{1}{\Delta t^2}\{\mathbf{X}(t+\Delta t) - 2\mathbf{X}(t) + \mathbf{X}(t-\Delta t)\} \tag{3.40}$$

from which we obtain:

$$\mathbf{X}(t+\Delta t) = \Delta t^2 \mathbf{M}^{-1}\left\{\mathbf{F}(t) - \sum_{e}\int_{v} \mathbf{B}^{T}\boldsymbol{\sigma}(t) dv\right\} + 2\mathbf{X}(t) - \mathbf{X}(t-\Delta t) \tag{3.41}$$

and Δt being chosen according to:

$$\Delta t \leq h/V_c \tag{3.42}$$

h being the size of the smallest finite element and V_c the compression wave velocity ($V_c = \sqrt{K/\rho}$, K = bulk modulus and ρ = the mass density).

The total stress vector $\boldsymbol{\sigma}(t)$ of Equation 3.41 may be replaced as shown, by Equation 3.11, and from Equations 3.9 and 3.19 we obtain:

$$\boldsymbol{\sigma}(t) = \mathbf{D}(\boldsymbol{\varepsilon} - \boldsymbol{\varepsilon}^{p}) \tag{3.43}$$

where $\boldsymbol{\varepsilon}$ is given from $\mathbf{X}(t)$ by Equation 2.7, and $\boldsymbol{\varepsilon}^{p}$ is given by the summation of the plastic strain increments, $\Delta\varepsilon^p$, for all time steps, or:

$$\varepsilon^p(t) = \sum_{i=1}^{t} \Delta\varepsilon_i^p \tag{3.44}$$

Knowing $\mathbf{X}(t)$, the procedure for calculating $\Delta\varepsilon^p(t)$ for the given time step, Δt, is exactly as given by one of the residual force methods, the initial strain solution method being recommended for non strain rate-dependent materials, and the visco-plastic method for rate-dependent materials (together with a fluidity parameter λ given by the corresponding material parameter and with only one time increment being employed, namely the 'real time' increment Δt).

The above solution method has proved to be successful for brittle materials and also for soil structures with drained or undrained conditions. However, the time step Δt must be chosen with care, and when modelling undrained behaviour, the time step must necessarily be short because of the high stiffness given from the pore fluid (see Section 3.3 above).

In view of problems encountered when using the above method, related to necessarily small values of the time step, Dungar (1982) proposed a solution whereby the same basic principles may be used, but with the additional assumption that the first few modes of vibration of the elastic system are used, together with the quasi-static response. Known as 'the imposed force summation method', this method may also be adapted from any computer program which is designed to solve elastoplastic static problems using one of the residual force methods, and which also includes the facility for calculating the resonant modes of the elastic system.

REFERENCES

Bicanic, N. 1978. Non-linear finite element transient response of concrete structures. Ph.D. thesis, University of Wales, Swansea.

Bossoney, C. & R. Dungar 1980. Past experience and future demands related to dynamic analysis of soil and rockfill structures and foundations. In *Proceedings of International Symposium on Soils under Cyclic and Transient Loading* (Swansea), 2: 809–820. Rotterdam: Balkema.

Cathie, D.N. & R. Dungar 1978. Evaluation of finite element predictions for constructional behaviour of a rockfill dam. In *Proceedings Instn. Civ. Engrs.* 2: 65, 551–568.

Cormeau, I.C. 1975. Numerical stability in quasi-static elasto-viscoplasticity. *Int. J. for Numer. Meth. in Engng.* 9: 109–127.

Duncan, J.M. & C.Y. Chang 1970. Non-linear analysis of stress and strain in soils. *J. Soil Mech. Fdn. Engng. Div.* (ASCE) 96 (SM 5): 1629–1653.

Dungar, R. 1979. EFESYS–An engineering finite element system. *Advances in Engng. Software* 1 (3). Southampton: CML Pub.

Dungar, R. 1982. An imposed force summation method for non-linear dynamic analysis. *Earthquake Engng. and Struct. Dynamics* 10: 165–170.

Dungar, R. 1985. Analysis of plastic deformations leading to cracking of grouted contraction joints in Zervreila Arch Dam. In *Proceedings of 15th ICOLD Congress* (Lausanne).

Finn, W.D.L. 1980. Endochronic theory of sand liquefaction. In *Proceedings of 7th World Conference on Earthquake Engineering* (Istanbul).

Finn, W.D.L. 1982. Dynamic response analysis of saturated sand. In G.N. Pande & O.C. Zienkiewicz (eds.), *Soil mechanics–transient and cyclic loads.* Chichester: Wiley.

Hueckel, R. & T. Nova 1979. On paraelastic hysteresis of soils and rock. *Bull. Acad. Pol. des Sciences, Sec. Sc. Techn.* 27 (1): 49–55.

Idriss, I.M., J. Lysmer, R. Hwang & H.B. Seed 1973. QUAD4–A computer program for evaluating the seismic response of soil structures by variable damping finite element procedures. Report EERC 73–16, University of California, Berkeley.

Ishihara, K. & I. Towhata 1980. Effective stress method in one-dimensional soil response. In *Proceedings of 7th World Conference on Earthquake Engineering* (Istanbul).

Kondner, R.L. 1963. Hyperbolic stress-strain response: cohesive soils. *J. Soil Mech. Fdns. Div.* (ASCE) 89 (SM 1): 115–143.

Naylor, D.J., G.N. Pande, B. Simpson & R. Tabb 1981. *Finite elements in geotechnical engineering*. Swansea: Pineridge Press.

Prater, E.G. 1980. On the interpretation of cyclic triaxial test data with application to seismic behaviour of fill dams. In *Proceedings of International Symposium on Soils under Cyclic and Transient Loading* (Swansea), A.A. Balkema Rotterdam: Balkema.

Seed, H.B. 1966. Method of earthquake resistant design of earth dams. *J. Soil Mech. Fdns. Div.* (ASCE) 92 (SM 1): 13–41.

Seed, H.B. 1979. Considerations in the earthquake-resistant design of earth and rockfill dams. *Geotechnique*, 29 (3): 215–263.

Serff, N., H.B. Seed, F.I. Makdisi & C.Y. Chang 1976. Earthquake induced deformations of earth dams. Report EERC 76–4, University of California, Berkeley.

Wilson, E.L. & R.W. Clough 1962. Dynamic response by step-by-step matrix analysis. In *Proceedings of Symposium on Use of Computers in Civil Engineering* (Lisbon)

Zienkiewicz, O.C. 1977. *The finite element method*. New York: McGraw-Hill.

Zienkiewicz, O.C., K.H. Leung, E. Hinton & C.T. Chang 1981. Earthdam analysis for earthquakes: numerical solutions and constitutive relations for non-linear (damage) analysis. International Conference on Dams and Earthquake, ICE (London).

Zienkiewicz, O.C., V.A. Norris, L.A. Winnicki, D.J. Naylor & R.W. Lewis 1977. A unified approach to soil mechanics problems of offshore foundations. International Symposium on Numerical Methods in Offshore Engineering (Swansea).

Material testing procedures and equipment 4

E.G. PRATER & J.A. STUDER

4.1 INTRODUCTION

In this chapter the aim is to briefly outline and review laboratory and in-situ testing in geotechnology. In particular, an attempt is made to relate new developments in testing technique to questions of material modelling.

After presenting the basic principles for the investigation of soil and rock properties in Section 4.2 a survey of soil testing equipment is given in Section 4.3. The subsequent Sections 4.4 to 4.6 indicate briefly the sources of error and the most important difficulties encountered in conducting the tests, while Section 4.7 gives a short introduction to cyclic testing, an area of increasing importance. The first part of the chapter, which is concerned mainly with soil testing procedures, is concluded in Section 4.8, in which the choice of a suitable test or combination of tests for providing the data for material models is discussed.

Sections 4.9 and 4.10 deal with laboratory investigations of rock properties, whereby Section 4.10 attempts to survey and evaluate the more important test equipment. For dynamic and transient loading problems appropriate parameters for the material models are required. In such applications the range of strain encountered is of great importance. Some of the available low to medium strain testing devices are briefly mentioned in Section 4.11.

Field testing may also be valuable in the investigation of material properties. Section 4.12 describes some of the more important field tests.

Soils and rocks exhibit nonlinear stress-strain behaviour over a wide range of strain. For technical reasons it is clearly not possible to investigate the whole range of strain with one type of apparatus. Table 4.1 gives an overview of the relevant strain ranges in a selection of geotechnical problems with the associated field and laboratory testing equipment. The table shows that for the very low strain range $10^{-5} - 10^{-4}\%$ a different type of test must be adopted than for strains approaching failure, i.e. around $10^{-1} - 10\%$.

In recent years there has continued to be much development in apparatus and technique with a trend towards more sophistication in in-situ testing, especially for major engineering structures. Centrifuge testing is also playing an increasing role in the study of soil and soil-structure behaviour.

The scope of this chapter is to survey the field of material testing and try to relate developments to the question of material modelling. It is shown that even in the case of

Table 4.1. Field and laboratory testing methods with applications

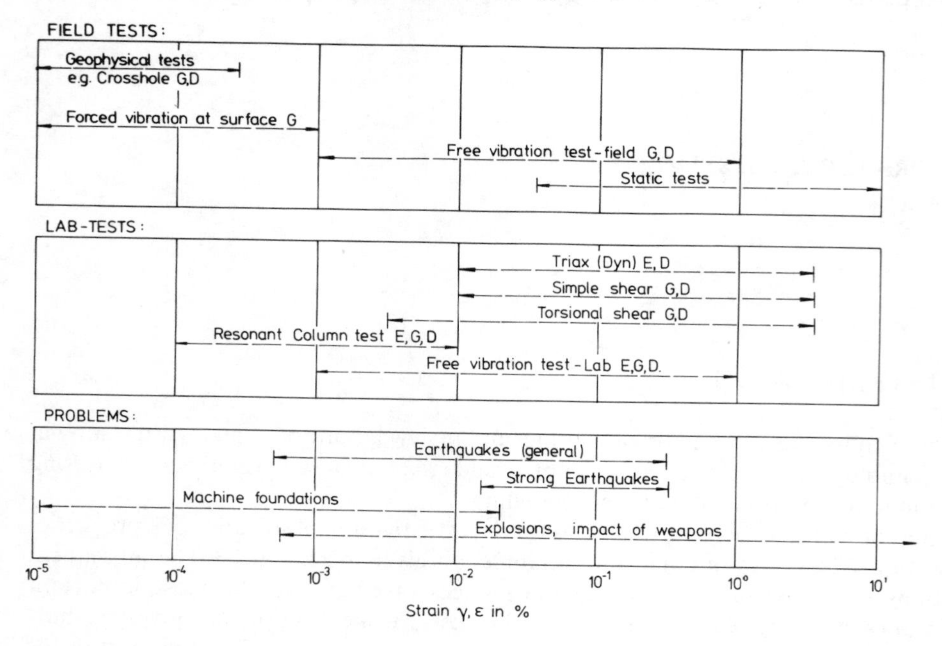

sophisticated testing there is still the need for 'interpretation' of the results as it is almost impossible to eradicate imperfections of apparatus and measurement.

In future there is a greater need to perform more case history studies to obtain soil properties from field behaviour. Advances in field measuring techniques is an important factor here. Back-analysis with the simultaneous adjustment of various parameters is not an easy matter. However, the basic computational tools are now available and the time has come to use these to test various geomechanical models. Only in this way can the value of the big effort put into material modelling in the last decade or so be appraised.

4.2 PRINCIPLES OF GEOMECHANICAL MATERIAL TESTING

Both field and laboratory tests can be used to investigate the mechanical behaviour of soils and rocks. The two types of test complement each other. They also have their specific advantages and disadvantages. Thus a suitable testing programme has to be devised to meet the specific aims and the local conditions encountered.

The properties of soils and rocks in-situ can be measured by various means, such as plate loading tests, borehole pressure-meter tests. These tests have a number of advantages in that the material is usually less disturbed by the test set up and the zone of influence may include natural features such as fissuring, jointing, layering etc. However, there are also certain disadvantages, including a more complex and usually unknown state of stress and the danger of extrapolating from the actual zone of influence to obtain more global property values. The best in-situ test, of course, involves monitoring the behaviour of full-scale structures, e.g. the deformations of dams or the walls and base of an excavation. Unless the necessary back-analysis tools are available, however, the predictive power of such field observations is still somewhat limited. It should not be overlooked too that there are certain pitfalls that can affect back-analysis (Leroneil & Tavenas 1981).

Laboratory tests, on the other hand, have the principal advantage of well-controlled testing conditions. The basic requirement of phenomenological testing, namely that the element of material must be subjected to as homogeneous a stress-state as possible, can also be better adhered to. The main disadvantages are that it is often difficult or even impossible in some cases to obtain undisturbed samples, and, especially in the case of rocks, the properties of an intact sample may have little relevance to in-situ behaviour. The problem of sample disturbance can be quite serious in certain types of soil or field condition. Under these circumstances the quasi-elastic stiffness determination is particularly prone to error.

In soil mechanics the stress path method has been devised to take into account more realistically the given field conditions. Take for example the heave of the bottom of an excavation, for which the soil stiffness properties depend very much on the stress path imposed. With this method the stress conditions applied to a laboratory sample simulate as closely as possible the in-situ conditions including the sequence of loading. Measured deformations are then related directly to the field performance of the structure (Marr & Hoeg 1979).

The classical, more elegant approach, however, is to describe the stress-strain properties of the soil in terms of material laws. To match modern refinements in modelling more attention has to be given to testing principles and technique than is the case for routine, commercially-oriented testing. With the latter the parameters determined sometimes have more value as classification indices to be correlated with experience than they possess as fundamental mechanical properties. For advanced material description, on the other hand, a precise knowledge of the state of stress in the sample is desirable. In effect, the test represents a boundary value problem in its own right. Further, since the aim is to formulate a material model valid in three-dimensional stress space with possible variation in stress path, e.g. cyclic loading states, the test type and conditions should be able to supply the required information. This has necessitated, in part, new and refined testing equipment and procedures.

4.3 SURVEY OF SOIL TESTING EQUIPMENT

Soil mechanics had its beginnings two centuries ago in Coulomb's investigations of failure and the concept of planar sliding. As a result, the historical development of soil testing procedures is characterized by a primary interest in strength testing and, in fact, the earliest devices were designed to produce failure in a plane through a block of soil.

The evaluation of failure loads has tended to overshadow that of deformation under working loads. With the development of the consolidation theory and the oedometer device a basis was provided for investigating deformations due to settlement. The emergence of elastoplastic analysis has had a unifying effect, strength and deformation concepts being handled within one material framework.

Various laboratory devices for the determination of the strength and deformational

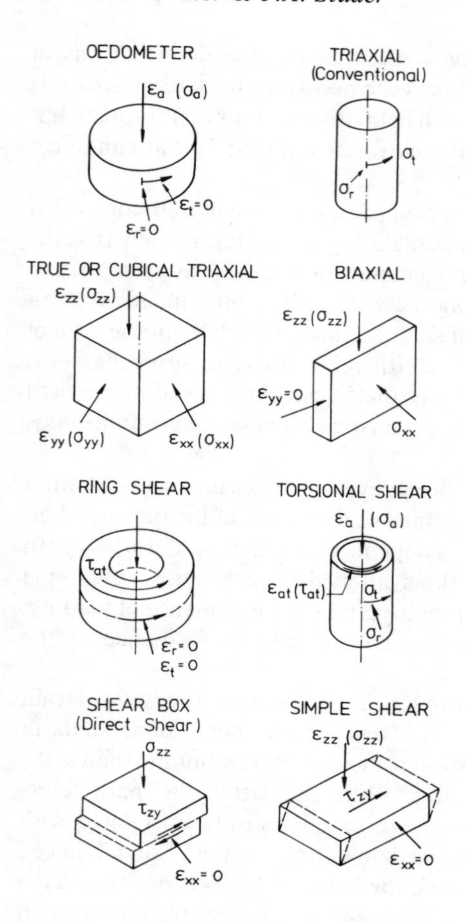

Figure 4.1. Overview of soil testing devices

characteristics of soils are shown in Figure 4.1. Table 4.2 provides a summary and evaluation of these devices, some of which are described briefly in the following.

4.3.1 *Direct and simple shear devices*

The direct shear test derives originally from Coulomb. In this test a sample of soil is contained in a split rigid box of square of circular cross-section. Under a given normal load failure is induced by shearing across the plane defined by the split.

The direct shear test is not suitable for accurate shear strain measurements and can provide unreliable results. Already in the 1930's Hvorslev (1937) observed that the strain conditions are not uniform exhibiting concentrations near the edges, which favours progressive failure. Although this behaviour resembles that observed in slope failure it renders the test unsuitable for material modelling. The latter requires simple, uniform stress states. A further disadvantage is that as deformations become large the area of contact between the two halves of the specimen reduces non-negligibly. To overcome this effect the ring shear test has been devised. It also allows determination of

Table 4.2. Overview of laboratory soil testing devices

Device	Capability	Limitations
Direct shear	τ_{zy}, σ_{zz} in plane to determine strength	Complex strain pattern developed with concentrations at edges, shear strain discontinuity
Simple shear	Apply τ_{zy}, σ_{zz}, measure γ_{zy}, ε_{zz} for strength/stress-strain properties	Complementary shear stress not always applied, σ_{yy} difficult to determine, non-uniform strains usually result
Cylindrical triaxial	Apply ε_z (σ_z) and σ_r, measure, ε_{volume}, strength/stress-strain properties	Non-uniform strains due to end restraint on rigid rough platens, membrane effects, and σ_t always $= \sigma_r$, no rotation of principal stress
Cubical triaxial	Apply 3 principal stresses (strains) independently	Sophisticated test, time-consuming sample installation, no rotation of principal stresses
Plane strain	Apply σ_{zz} (ε_{zz}) σ_{xx}, $\varepsilon_{yy} = 0$ measure ε_{volume}, strength etc.	Similar to triaxial tests
Directional shear cell (Arthur et al.)	As plane strain with applied shear, principal stress rotation	Low stress range, complicated sample installation and test execution
Ring shear	Improved direct shear test	No information on shear strains (discontinuity)
Hollow cylinder	Short sample: like simple shear long sample: general stress path control	Non-uniform strains can be great, sample installation more complicated
Oedometer	Volume-change K_0 characteristics for proportional stress loading ($\varepsilon_r = 0$)	Mantle friction excessive for cohesionless soils, no failure properties determined
Shaking table	Similar to simple shear on long specimens to simulate seismic loading for liquefaction properties	Membrane penetration effect if normal pressure applied via membrane, limited stress state
Centrifuge	Simulation of self-weight stresses in models	Expensive hardware, laborious model preparation

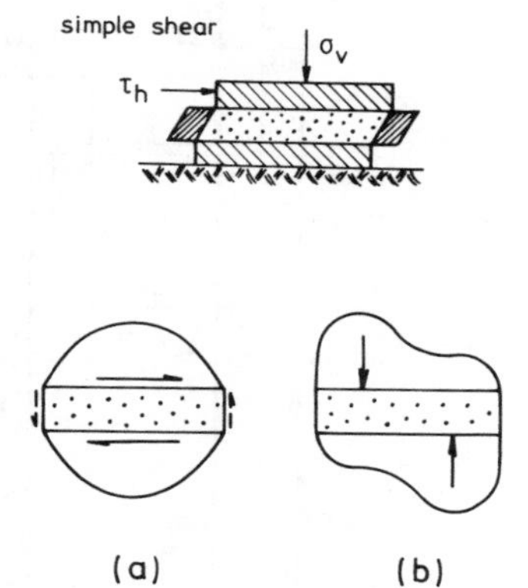

Figure 4.2. Non-uniform stress conditions in simple shear apparatus (after Wood 1982); (a) shear stress; and (b) normal stress

the residual strength of clays. Bishop et al. (1971) survey the various types of direct shear apparatus in existence.

In order to obtain a more uniform state of stress in the soil element as a whole several types of simple shear device have been designed. This test produces zonal shear in contrast to the localized planar shear in the direct shear test. To achieve a state of simple shear, complementary shear stresses are necessary. These, however, are sometimes lacking as in the Geonor device, in which the lateral boundary consists of a reinforced rubber membrane preventing lateral but not vertical expansion. Thus even with some simple shear devices non-homogeneous stress and strain states may result as illustrated in Figure 4.2 after Wood (1982).

4.3.2 *Torsional shear devices*

There are various devices for applying a torsional shear stress to either hollow or solid samples. Samples may be short, i.e. disc-shaped, or long.

A novel device intended to produce a uniform state of shear is due to Ishibashi & Sherif (1974), and a comparable piece of equipment is that of Yoshimi & Oh-oka (1975). They consist of a ring whose thickness varies with radius to give theoretically a uniform torsional shear strain (Figure 4.3). The disadvantage of this device is that it suffers like

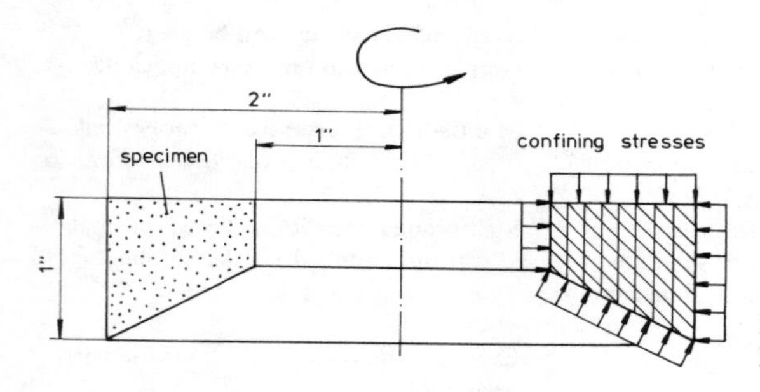

Figure 4.3. Ring shear apparatus of Ishibashi & Sherif (1974)

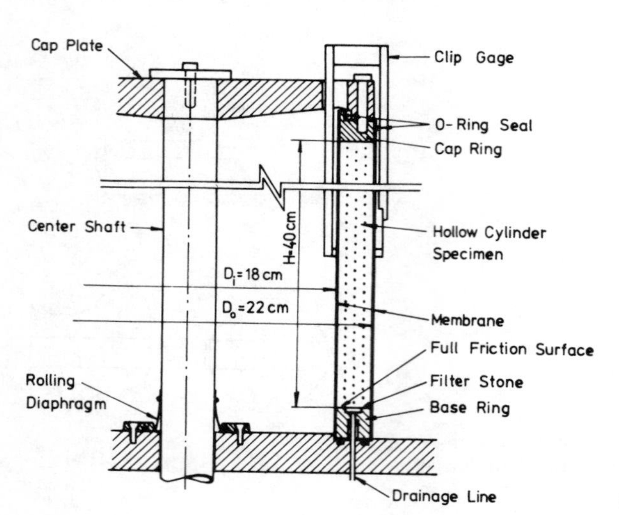

Figure 4.4. Design of Lade's torsion shear device on hollow specimens

all those for short samples by boundary restraint if the sample tends to dilate. Non-uniform strains may also be induced in the consolidation phase of a test.

One way to overcome the above problems is to adopt long samples (L/D > 2.5) as in the hollow cylinder torsional test. In this test the sample is enclosed in a rubber membrane between end plates and a confining pressure is applied both internally and externally. To transmit torque 'rough ends' are necessary, but the restraint effect is much more localized. Most researchers apply equal external and internal confining pressure, as otherwise the stress normal to the wall of the sample varies (Saada & Townsend 1981), but Hight et al. (1983) opt for unequal pressures to facilitate greater generality in the stress path. Figure 4.4 shows the apparatus due to Lade (1981). Apart from the above references, other useful papers with results for clays are Broms & Casbarian (1965) and Saada & Ou (1973).

Whereas the direct and simple shear devices provide information only on shear stress-strain behaviour, the hollow cylinder device allows control of the principal stresses with principal stress rotation and versatility with respect to stress paths. For this reason it is a genuine alternative to the cubical triaxial device described below.

4.3.3 *Plane strain and cubical shear devices*

In the plane strain device stresses are applied to a prismatic specimen, one pair of opposite faces being restrained with respect to lateral movement. Since two of the principal stress are controlled the test is sometimes called the biaxial test. Some devices are essentially modifications of the triaxial cell with the sample enclosed in a rubber membrane such that the cell pressure acts on one pair of faces. Al-Husseini (1971) has reviewed the various types of plane strain apparatus available. More recently, by means of a novel design, Hambly (1972) applied the loads to four moveable rigid platens (Figure 4.5). Based on this principle Kuntsche (1982) developed the apparatus sketched in Figure 4.6.

The plane strain device suffers from the same disadvantages as the triaxial test (to be described later), but compared to the latter it gives a better representation of many field situations. From the results of various independent investigations it appears that the angle of internal friction under plane strain conditions is of the order of 4°–7° greater than for the triaxial compression of medium dense samples (e.g. Cornforth 1964, Lade & Duncan 1973, Marachi et al. 1981). The variation throughout the full range of intermediate principal stress is shown in Figure 4.7. At high confining pressures (> 1 MN/m^2) the differences become less pronounced. In some devices the stress can

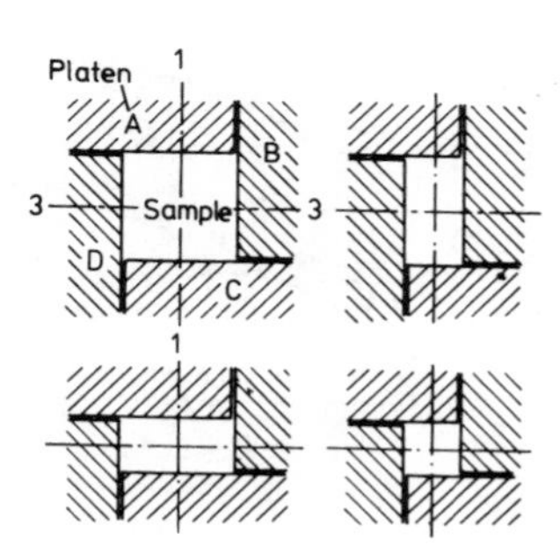

Figure 4.5 Hambly's principle of operation for biaxial testing device (plane strain)

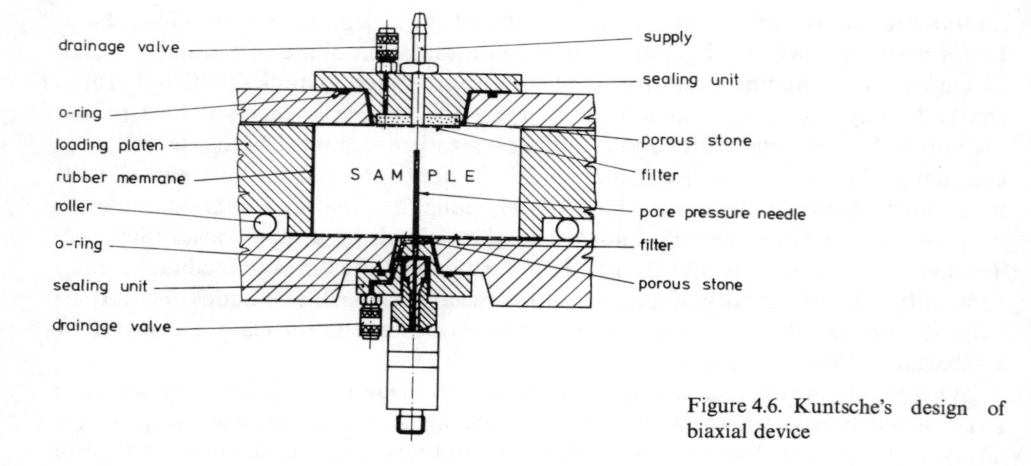

Figure 4.6. Kuntsche's design of biaxial device

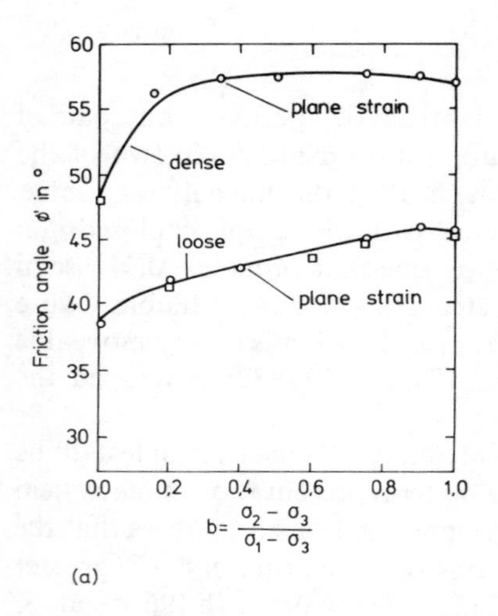

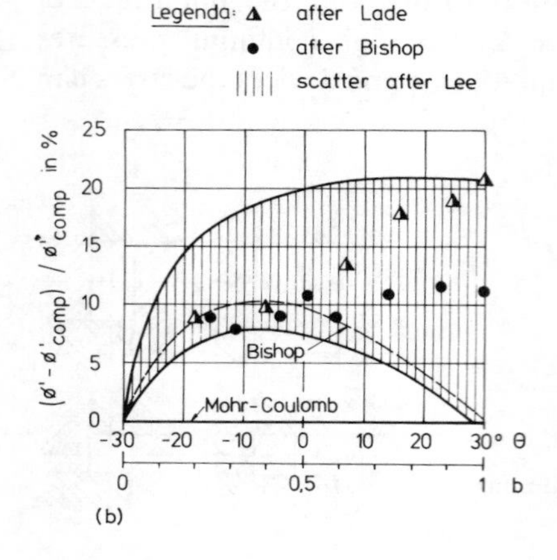

Figure 4.7. Influence of intermediate principal stress on friction angle; (a) after Lade & Duncan (1973)

also be measured in the direction of zero principal strain using a strain-gauged diaphragm, i.e. plane strain K_0 tests are also possible.

In the cubical or 'true triaxial' shear device there is control of all three principal stresses and the corresponding strains are measured. The sample is either enclosed in a rubber membrane (Ko & Scott (1967) or between rigid platens (Pearce 1971). In the case of the former it is difficult to avoid distortion at the edges, while for the latter care must be taken to avoid frictional restraint at the boundaries. This may be achieved by applying silicone grease to the platens with a rubber membrane separating the soil from the grease.

Apart from the relatively complicated sample installation, the cubical triaxial apparatus has the disadvantage of requiring a complicated loading system. A further limitation is that the principal stress directions are fixed.

The bi- and triaxial devices described above do not have the facility of applying boundary shear forces. A special device has been constructed by Arthur et al. (1980) for testing cubical sand samples. The device is a modified plane strain test with the conditions $\varepsilon_2 = 0$, σ_x, σ_y, $\tau_{xy} \neq 0$. An MIT evaluation of the device has also been carried out (Arthur et al. 1981). Apart from allowing principal stress rotation the test gives information about stress-induced anisotropy. Its main limitation is the low range of stress that can be applied.

4.3.4 *Consolidation testing devices*

The state of stress induced in a soil is not always such that failure is reached. Sometimes it is desired to investigate deformational behaviour at stresses well below failure. This type of problem is traditionally termed consolidation. The basic device, called oedometer or consolidometer, consists normally of a cell confining a disc-shaped sample under conditions of no lateral strain, i.e. the walls of the cell are rigid. In the conventional oedometer it is not possible to measure the lateral stresses developed. In some new, more sophisticated devices the sides of the cell are constructed as membranes either allowing lateral stress measurement using strain gauges or all lateral deformation is prevented by fluid pressure compensation which gives directly the lateral pressure.

The one-dimensional consolidation test is in principle a special case of a constant stress ratio test path. The same test can be performed in the triaxial apparatus if the apparatus is modified to prevent lateral strains in the sample. In this case the test is called a K_0 test.

A few types of consolidation cell are shown in Figure 4.8. The main defect of the oedometer test is the presence of friction on the inside of the ring enclosing the sample. This may be reduced by having a smooth surface with a non-corroding ring material and the use of silicone grease helps too. Another method of overcoming the frictional effect is to employ a floating ring set-up (Figure 4.8(b)) or a compressiometer (Figure 4.8(c)) in which the sample is enclosed in a rubber membrane reinforced by a number of thin rings offering no resistance to vertical load but acting rigidly in the radial direction. In the Rowe consolidation cell, which is generally used for larger size samples, the load is applied by means of a flexible rubber diaphragm under the action of pressure, whilst the settlement is measured at the axis of the sample. This apparatus, however, may not work so well for soft, highly compressible soils.

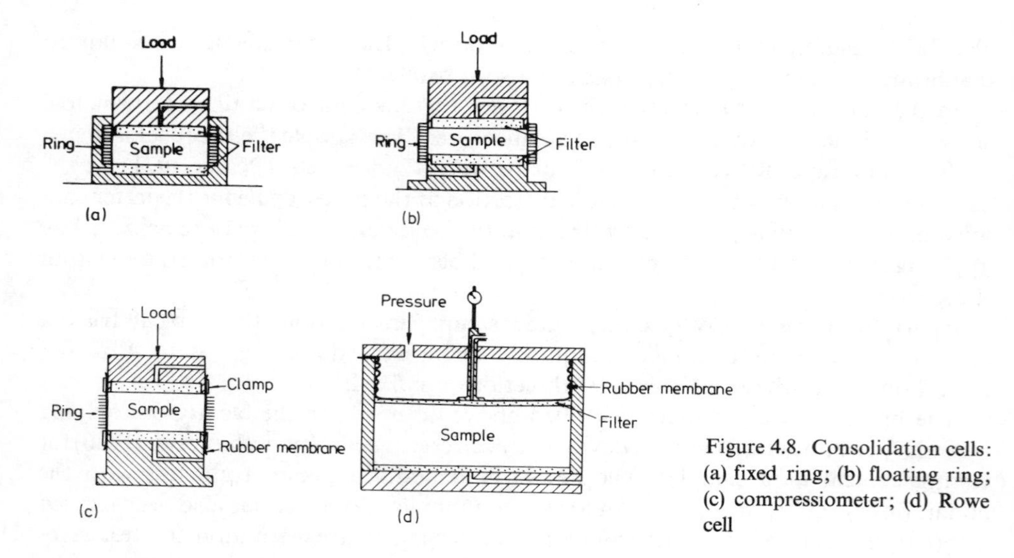

Figure 4.8. Consolidation cells: (a) fixed ring; (b) floating ring; (c) compressiometer; (d) Rowe cell

For coarse grained soils lateral friction effects may be substantial. In this case there may be no alternative to accepting that the stress distribution is non-uniform within the sample and to allow for it by measuring the load transferred to the base of the sample and calculating an average value of stress.

Further modifications to the testing equipment and procedures include measurement of pore water pressure at the base of the sample, drainage being permitted at the top only, and for highly stratified soils, e.g. varved clays, by providing radial drainage at the periphery and in a central core of the sample. The Rowe consolidometer may be easily modified for such purposes.

Apart from the above developments in apparatus and technique a notable advance in recent years has been to move away from the incremental testing procedure, whereby load increments are applied sequentially, usually in 24 hours intervals. Instead, various forms of continuous load application have been introduced. The advantages of this are that the test may be run automatically, it may be speeded up and the pre-consolidation pressure may be identified due to the continuous increase of effective pressure. The pre-consolidation pressure itself, however, may be affected by the loading procedure and, in constant rate of strain tests, by the strain rate. The tests may be run by controlling either the rate of strain (CRS-test) or the pore water pressure at the base of the sample. In either case a pore pressure transducer is required at the bottom end and drainage is one way only. The most sophisticated test (Janbu et al. 1981) maintains a constant ratio of induced pore pressure to applied stress $\Delta u_b/\Delta\sigma$, equal to a value close to 0.1, by automatically adjusting the strain rate. With the CRS-test the strain rate must be set to a value which will give a maximum value of $\Delta u_b/\Delta\sigma < 0.15$. The standard rate will be of the order 0.002 mm/min for medium to high plasticity clays. According to Smith & Wahls (1969) the average effective stress in the sample is then approximately $\sigma' = \sigma - \frac{2}{3}u_b$.

It is assumed here that the pore pressure distribution is parabolic, which may not be a good assumption for an interval of load in which the pre-consolidation pressure is passed, there being a higher value of oedometer modulus E_{oed} in the lower part of the sample which has not yet reached the pre-consolidation pressure.

4.3.5 *The conventional cylindrical triaxial test*

Triaxial is really a misnomer for this test as the state of stress is axisymmetric with the intermediate circumferential stress always equalling theoretically the radial stress. When the latter is the major principal stress a state of extension is produced.

In this test a solid cylindrical soil sample is enclosed in a rubber membrane between rigid platens for axial loading. The fluid pressure in the cell confines the sample. A sketch of the testing principle is found in Figure 4.1.

It is not the aim here to supply a detailed description of the conventional triaxial testing equipment and procedures. For that the reader is referred to the standard text by Bishop & Henkel (1962). However, certain testing effects relevant to the triaxial test are discussed in depth in Sections 4.5 to 4.7, where an evaluation of the test is also given. As soil engineers have come to rely heavily on this test and, in addition, it plays an important part in material modelling it is given special consideration here.

4.4 EFFECTIVE STRESS CONSIDERATION AND PORE PRESSURE MEASUREMENT

Pore water pressures are very important in soil and rock mechanics. In many cases, however, it is extremely difficult from the point of view of measuring technique to obtain accurate measurements especially in-situ but also in the laboratory.

In practice, there are two extreme conditions in which it is possible to avoid pore pressure measurements, namely long-term and short-term strengths for stability analysis. In the first case a fully-drained test is carried out. However, for clays such a slow rate of testing may be required to ensure that only negligible pore pressures are generated, that the testing becomes prohibitive. Reducing the sample size and thus the time to failure improves matters but it may be inconvenient working with small size samples. In the second case the so-called $\phi = 0$ condition is investigated to find the apparent cohesion s_u under conditions of no drainage. As a result, for clays undrained tests are usually carried out.

If the clay soil is unsaturated the test is not strictly undrained. However, in temperate climates natural clayey soils are generally in a state close to complete saturation, $S_r \approx 1$. For partially saturated soils, e.g. clays compacted dry of the optimum water content, it would be necessary to measure, using special techniques, both pore water and pore air pressures (Bishop et al. 1960).

In most cases, however, samples are fully saturated by means of back-pressure (Lowe & Karafiath 1960). For this purpose it is better to have a constant sample volume with increasing water content. For an initial value of $S_r = 0.8$ at least 10 bars of back-pressure will be required for full saturation. Such a pre-stress effect would not, of course, be present in-situ except at great depths (ca. 100 m) below the groundwater table.

Even if the samples themselves are fully saturated and tested undrained, great care must be taken to remove air from the connections to the pore pressure measuring system in order to prevent errors and a time-lag, i.e. a sluggish response. Expansion of tubings, fittings and valves also interferes with the measurement of pore water pressures and is a further source of what is nowadays called system compliance.

Sand samples are seldom tested in the saturated, undrained condition. The main reason is that the dilatancy developed during shearing causes a considerable reduction in pore pressures for medium dense and dense samples. In other words, if back-pressure is applied to remove all air at the beginning of the test it may be so depleted during the test that full saturation is no longer guaranteed. Dense samples may even induce a state of cavitation.

Nevertheless, it is not uncommon to investigate the behaviour of medium dense sands with measurement of pore water pressure under conditions of cyclic loading. During the unloading phase of a cycle, especially under two-way stress reversal loading, positive pore pressures are built up even for dense samples.

In recent years it has been recognized that a very important compliance effect can be present, namely that of membrane penetration, as discussed in the next section.

4.5 SOURCES OF TESTING ERROR

Soil element testing involves the application of stress or kinematic boundary conditions with measurement of pore fluid pressures. The main errors involved are described with respect to the conventional triaxial test.

The following sources of error: piston friction; migration of water and air through the membrane; pore pressure measuring equipment and techniques; and non-uniform distribution of stresses and water content are the most important (Casagrande & Wilson 1960).

The problem of piston friction was formerly solved using either rotating pistons or bushings. Nowadays, for special tests, it is common practice to measure the load actually reaching the sample by means of a load transducer placed within the triaxial cell.

Migration of water or air through the rubber membrane is a problem affecting long-term tests on clays, which may be remedied by the use of two membranes separated by a film of silicone oil.

The last two topics concerning pore pressure and boundary friction restraint are much more problematical and are treated below in greater depth. The importance of compliance in the pore water pressure measuring system due to membrane penetration effects has only been really appreciated in the last two decades.

4.5.1 *Membrane penetration effects*

The most obvious effect of membrane penetration arises in isotropic compression tests on sand, in which only part of the volumetric compression measured is due to the compression of the soil skeleton. If the latex membrane thickness is 0.3 mm (typical for samples of 56.4 mm diameter) the penetration effect is not negligible for average grain sizes above 0.15 mm and may be as great as 25% for medium-coarse grained sands. This error affects directly the measured bulk modulus.

For saturated sand confined in a thin rubber membrane changes in pore water pressure can also cause the amount of penetration of the rubber into the pores to change positively or negatively, so that the test is no longer a true undrained test. The magnitude of the error introduced depends on various factors: thickness of membrane, average grain size, elastic modulus of membrane, and the ratio of the surface area of membrane to volume of sample.

The compliance errors present can be quite considerable with the difference in measured pore pressure being as high as a factor 2 (Lade & Hernandez 1977, Kiekbusch & Schuppener 1977). The interesting fact is that the effective strength parameters (c′, ϕ') are largely insensitive to membrane penetration, though the effective

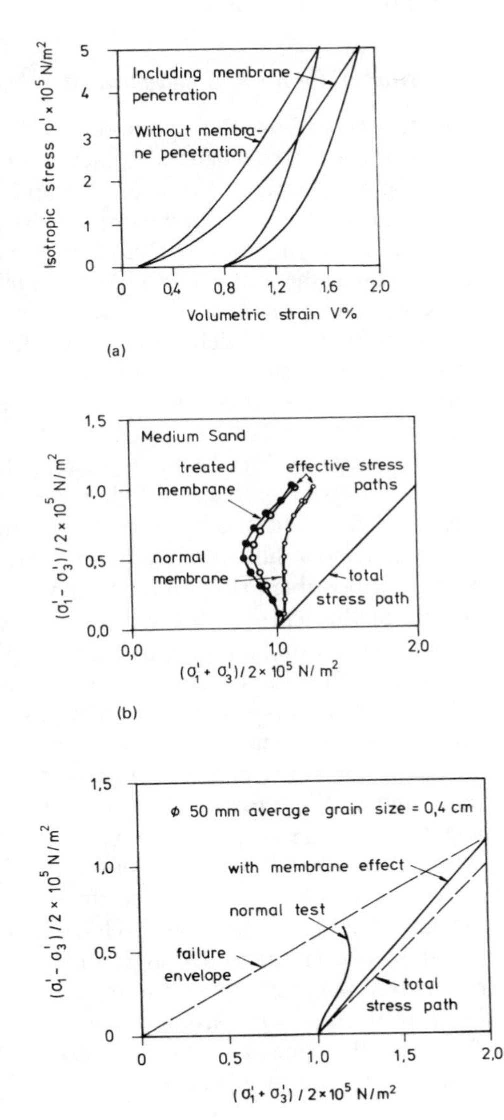

Figure 4.9. Influence of membrane penetration on: (a) isotropic compression curve for sand; (b) and (c) stress path in p′–q′ diagrams, both experimentally (Kiekbusch & Schuppener 1977) and theoretically (Flavigny & Darve 1977)

stress path in a p′-q′ diagram and the undrained strength (s_u) can be greatly affected (Figure 4.9). The effect is quite critical in cyclic loading liquefaction tests as unconservative results are obtained for the more endangered loose sands, while over-conservativeness is present for the dense sands (Martin et al. 1978), i.e. true liquefaction effects may be masked and dilatancy effects suppressed.

Experimentally, membrane penetration effects are rather difficult to correct for, whereas theoretically the characteristics of the constitutive model, i.e. that which is the object of the investigation, have to be known beforehand. When possible, therefore, larger size samples with thicker rubber membranes (say 1 mm for 150 mm diameter size samples) should be adopted to minimize this source of error, and preferably drained tests should be conducted.

4.5.2 *Non-uniform stress and strain development*

Transfer of load to the specimen may lead to problems not only with flexible membranes, but also when rigid platens are used. It is common practice to place rough porous plates between the end caps and the sample for facilitating pore pressure measurement. Thus at the loading surface friction is developed restraining lateral deformations of the sample. This leads to paraboidal cone-shaped zones which do not undergo the same amount of plastic straining as the material in the central bulging zones. For this reason one often speaks of 'dead zones'.

The mechanics of deformation is affected by frictional shear stresses varying between zero at the circumference and at the axis of symmetry, which create non-uniform stress and strain conditions as confirmed by the results of X-ray investigations (Deman 1975).

For sufficiently long samples there is little influence of end effects on strength determinations, i.e. with a length/diameter ratio of about 2.5, depending on the angle of internal friction. The middle third of the sample is largely unaffected by end restraint so that if axial and volumetric strains and pore pressures are measured in this part of the sample it is possible to determine the parameters for soil models. To measure the pore pressure at the centre of the sample probes may be inserted penetrating the rubber membrane enclosing the sample. However, there will still be an unavoidable tendency for pore pressure gradients to develop leading to moisture variations within the sample.

An alternative approach is to attempt to eliminate frictional effects on the loading platens. The most successful technique is that of Rowe & Barden (1964), in which silicone grease is smeared on the platens and separated from the soil by thin latex membranes. To prevent lateral slip the sample aspect ratio is reduced (height/diameter ≈ 1). To measure pore water pressure either small porous discs or a perforated needle filled with wool fibre are placed in the middle of the platens.

One difference observed when 'free ends' are adopted is that there is a strong tendency to suppress the single preferential failure plane obtained with dense samples of sand or clay. The measured undrained strength is usually a little higher (Lee 1978). It appears that the premature development of a failure plane and the characteristic drop in post-peak stress when using rough platens is partly associated with the problem of bifurcation (Kolymbas 1981). The effect of 'free ends' is like that of increasing the confining pressure in this case.

4.6 STRESS PATH CONSIDERATIONS IN CONVENTIONAL TRIAXIAL TEST

A restriction of a theoretical nature concerns the stress path the sample is subjected to in a conventional triaxial test. Due to axial symmetry $\sigma_2 = \sigma_3$ and the stress state is limited to a plane in stress space as shown in Figure 4.10. Thus it is only possible to investigate a section through the failure or yield surface. Nevertheless, within this plane (called the Rendulic plane) some variation of stress path is possible, as indicated in Figure 4.10. Thus the primary (consolidation) stresses may first be established. Usually these correspond to an isotropic stress state (point b). Then paths in compression or extension for either constant cell pressure bb′, constant axial stress bb″ or constant mean principal stress bb‴ may be applied, whereby for the latter some form of electronic control will be necessary for cyclic and repeated loading. For undrained tests the control of the applied stresses has little influence on the effective stress path to failure, which is determined by the pore pressure induced, e.g. paths generally of the form aa′, aa″ are obtained for normally consolidated clays.

The conventional triaxial test allows for relative changes in the magnitudes of the principal stresses but no change in direction except the sudden reversal of principal stresses in cyclic extension-compression tests, which is a very extreme loading condition.

Due to the limitations of (i) the Rendulic stress plane and (ii) no possibility of gradual rotation of principal stress direction, researchers in soil constitutive laws have turned to other testing devices to investiage their models. It is still normal practice to calibrate the

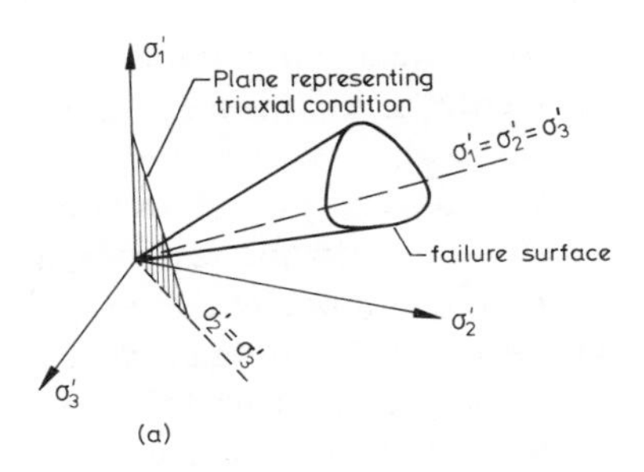

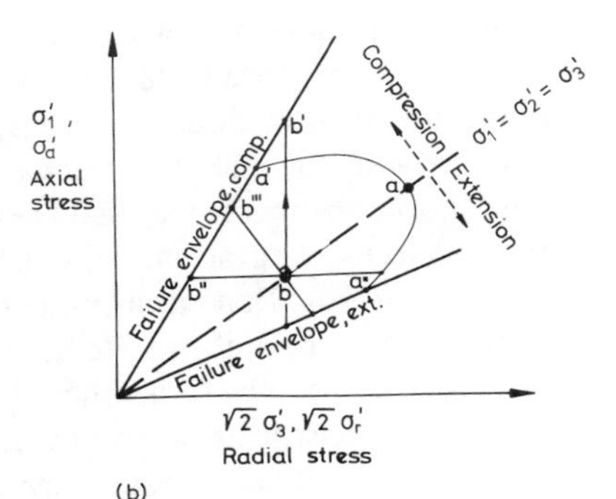

Figure 4.10. (a) Stress space showing failure surface and 'triaxial' plane; and (b) stress paths in triaxial (Rendulic) plane

model with the results of 'cylindrical' triaxial tests. A model's validation, however, requires more general stress path testing. In other words, most models are constructed in such a way that their qualitative response to stress probes outside the Rendulic plane is based on the author's interpretation of published data from special tests, while the quantifiable parameters of the model are given solely by triaxial test data. Whether the non-calibrated features of the model apply to the particular material under investigation would thus have to be checked.

4.7 CYCLIC LOADING TESTS

In recent years much attention has been given to cyclic loading tests. Most work has been conducted using cylindrical triaxial and simple shear devices, while Yamada & Ishihara (1979) have investigated the liquefaction characteristics of sands in the true triaxial apparatus.

For repeated loading in compression an accumulation of compressive strain is obtained. If the load changes symmetrically from compression to extension the tendency is for net extensional strains to develop. The reason for this is that in relation to the shear stresses causing failure according to the Mohr-Coulomb hypothesis the level of applied stress is higher in extension. For drained tests a volumetric strain results, whereas for undrained conditions the soil develops pore water pressures. The nature of the pore pressure, i.e. positive or negative increments, depends upon the initial density of packing, showing either a tendency to densify or to loosen.

One particular problem associated with triaxial cyclic loading is that there is a sudden jump in the directions of the principal stresses during stress reversal, which tends to create a state of anisotropy within the specimen. From the point of view of experimental technique it appears that the problem of end restraint is also important, as cyclic loading accentuates the non-homogeneous behaviour. In particular, the striction that usually develops is of an unstable character in a dense material. The many liquefaction investigations carried out with standard equipment, therefore, have to be treated with caution, as these localization effects can have a dominant influence.

One surprising finding of cyclic testing is that even dense sand can develop positive pore water pressures under two-way cyclic loading (cf. Figure 4.11). The sample undergoes a tendency both to contract (densify) and dilate (loosen) in the course of a cycle of large stress magnitude. At the close of the test there is a net pore pressure increase. Although some of this behaviour is no doubt due to apparatus effects (i.e. end restraint) causing pore water migration to and loosening of the central striction zone, part of the effect is due to wearing down of points of contact and slight particle rearrangement. In the case of dense clays this phenomenon is not apparent.

The simple shear test has also been much used for cyclic testing of soils. Although it is possible to rotate gradually the principal stresses and apply equal amounts of stress reversal it seems that non-uniform stress conditions can also be very serious (Wood 1982), as bad, if not worse, than with the triaxial test (cf. Figure 4.2).

A further point to be remembered in the case of cyclic triaxial tests is the rate of load application. Since the usual application is for earthquake problems there is a preference to apply the load at a frequency corresponding to the predominant earthquake frequency, generally in the order of one Hertz. For undrained tests this can lead to

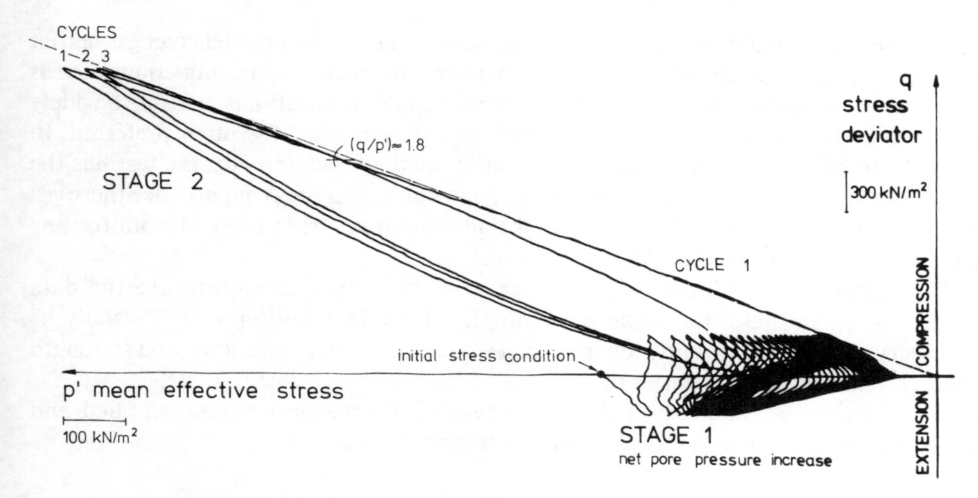

Figure 4.11. Cyclic pore pressures produced in dense sand (ETH-Z test data): undrained test on saturated sample with constant mean total stress, with two-way (compression-extension) followed by one-way (compression) loading. Characteristic pore pressure build-up under stress-reversals and dilatancy behaviour at large deviator stresses
(Note: direct XY-plot of axial force and pore pressure is equivalent to p′–q′ diagram)

problems for clays and silts (Sangrey et al. 1978). The difficulty is that a thorough investigation of the behaviour is hampered by the inability to make reliable pore pressure measurements in the central part of the sample. If end restraint is present, as in most commercial testing, there is undoubtedly a substantial difference in pore pressure at the centre and the ends of the sample, with increments of opposite sign being possible. For a clay tested for say 20 cycles at 1 Hertz the sample remains physically homogeneous, i.e. no pore water migration, but it is problematical to determine a respresentative pore water pressure.

The use of free ends seems to be the only solution to this problem, since there is much evidence to show that if the speed is reduced to allow equalization of pore pressure (1 cycle lasting perhaps 8 hours) the increase of water content in zones of dilatancy may be unacceptable, unless water content distribution is measured after the test and an appropriate correction applied.

Finally, strength increase due to rapid strain rates is reviewed by Morgan & Moore (1968), and is not discussed here, except to remark that the pore pressure gradients mentioned above in connection with clays are probably a significant factor that is often overlooked.

4.8 SELECTION OF TEST TYPE FOR USE WITH SOIL MODELS

Faced with the possibility of using sophisticated material models in computer programs the question arises as to a suitable test to obtain data for calibrating the model i.e. parameter identification. Most models proposed to date can, as mentioned previously, be calibrated using the 'cylindrical' triaxial test, despite the stress path limitations discussed in Section 4.6. The authors of the various models recognize that their model has little chance of acceptance unless the necessary testing equipment is universally available. However, although the triaxial apparatus is the most widely available testing equipment this is not the only reason why it is often preferred. In addition to relative ease of sample preparation, load application etc., the test has the merit of considerable versatility. Further, in geotechnical engineering practice there is a wealth of data correlating field behaviour and the parameters obtained from routine triaxial testing. This in itself is of great value.

Nevertheless, if available, other devices could be employed to improve the data input, e.g. biaxial tests for plane strain applications. In addition, it will usually be necessary to obtain volume change characteristics (consolidation) and strength characteristics for full material description.

For the sake of generality Table 4.2 provides information on data supplied and difficulties encountered with the main soil testing devices.

4.9 STRENGTH AND DEFORMATIONAL BEHAVIOUR OF ROCKS

As a background to rock testing procedures and equipment a brief summary of the mechanical behaviour of rocks is provided in this section. The behaviour of rocks covers a wide field ranging into the domain of structural geology with the description of faults and joints. Highly weathered and fragmented rocks have much in common with soils. Normally, however, the fabric properties of rock are different from those of soil, even in the case of non-igneous rock, as a result of the processes of deformation and metamorphism with joint patterns etc.

The basic rock material consists essentially of a bonded crystalline agglomerate whose structure or fabric exhibits on the microscale randomly oriented flaws (pores and microfissures at grain boundaries). Rock strength behaviour still has affinities with that of soil with Mohr representation of failure stresses. However, brittle fracture phenomena, usually described by means of Griffith's crack theory, play an important role (Jaeger & Cook 1976).

The strength criterion that results from the Griffith's failure concept may be usefully applied for rocks only in the tensile region of a biaxial stress plot, whereby the uniaxial compressive stress forms roughly the boundary between tensile fracture and brittle compressive failure (cf. Chapter 1, Figure 1.13). At high confining pressures there is a transition to ductile behaviour (Franklin 1971).

Due to the inherent difficulties associated with the application of the Griffith's fracture theory in the compressive quadrant of stress space some authors (Franklin 1971, Goodman 1980) prefer description of the failure criteria using purely empirical models, based on the results of tension, uniaxial and triaxial compression tests. The current trend, though, is to describe rock deformational behaviour in terms of elastoplastic models.

Time-dependent (creep) deformational behaviour can be important for certain rock types in tunneling and mining applications (Attewell & Farmer 1976). As a result, rate of loading is a factor requiring control and standardisation in rock testing, even more so than is the case for soils.

Although not generally appreciated, pore pressure effects are also of considerable

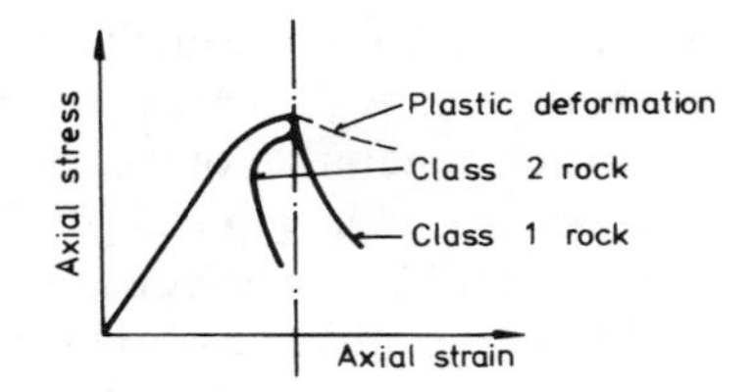

Figure 4.12. Post-failure deformation criteria

importance in rocks. In-situ rocks have moisture contents ranging between 1% and as much as 35% (in sandstone), and joints and fractures may be filled with water. If care has not been exercised to prevent drying out unrepresentative results may be obtained, especially in swelling rocks like gypsum. It is often the practice to test samples in an air-dried condition, but it is better to simulate field conditions closely, and, as wetting tends to weaken rocks, testing in a saturated state is advisable in doubtful circumstances (Hoek 1968).

One of the features of the mechanical testing of rocks which has received much attention in recent years is post-failure deformation behaviour. Broadly speaking, for uniaxial compression two classes can be distinguished (Wawersik & Fairhurst 1970), depending on the form of the post-peak unloading of the curve (see Figure 4.12). Class 1 rocks (generally of weaker type) store further energy and unload in a stable manner, whereas Class 2 rocks exhibit self-sustaining fracture as the stored energy is released with reduction of strain. At high confining pressures fracture and dilatancy tendencies are less pronounced and plastic deformation, possibly with post-peak reduction of strength to a residual value, may be observed.

4.10 SURVEY OF ROCK TESTING APPARATUS

The principles of mechanical testing of rock and many of the types of apparatus in use are similar to those for soil testing. Apparatus effects, for example friction at loading platens, are also the same, though in some cases the influence is less pronounced, in others more pronounced, than it is for soils. The main difference is the greater importance attached to tensile strength in rock which is in the order of one tenth the compressive strength.

4.10.1 *Uniaxial strength testing*

For uniaxial testing the specimen is unjacketed. Both uniaxial, unconfined compression and extension tests are carried out. These are described in standard tests, e.g. Roberts (1977), Goodman (1980).

Uniaxial compression

The main problem associated with tests on cylinders (both uniaxial and triaxial) is the effect of end restraint. Various techniques have been devised to overcome this problem. The use of long specimens with an aspect ratio > 2 does not necessarily provide better results as stress concentrations developing at the circumference of the specimen in contact with the platen can affect behaviour. The use of lubricants (e.g. graphite) is also problematical as it is usually observed that longitudinal tensile splitting occurs. Other methods include the use of conical loading platens, matching end-pieces whose lateral expansion equals that of the specimen and shaping the specimen to a necked form with a central gauge length (Jaeger & Cook 1976; cf. Figure 4.16(b)). Obviously all three methods are very time-consuming and unsuitable for routine testing.

A further problem associated with the ends of the specimen is that they must be flat to provide a uniform contact with the platens. This can be achieved by surface grinding. Parallelness is also very important. It is common practice to provide a swivel bearing at the base of the specimen. This has two functions: accommodation of lack of parallelness between the specimen's ends and preventing induced moments when cleavage of the specimen occurs (cf. Figure 4.16(a) on p. 120). Care must be taken to align the spherical head and in centering the specimen accurately as substantial errors may otherwise be introduced (Obert & Duvall 1967).

Uniaxial extension

In tensile strength tests the earlier practice was to cement the specimen into grips. Gripping the specimen can produce lateral stress with an effect analogous to that of end

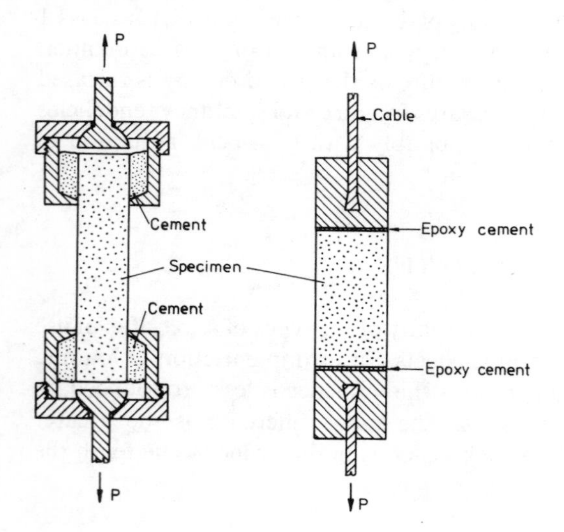

Figure 4.13. Simple extension testing devices

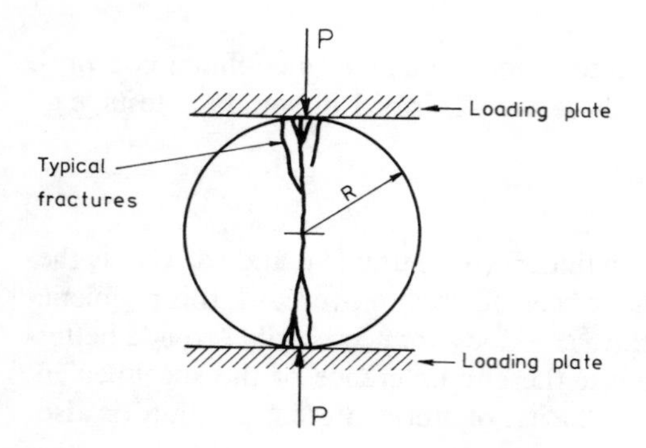

Figure 4.14. Compression of rock cylinder across a diameter (so-called Brazilian test)

restraint. An improved method involves fixing the specimen's ends to steel loading caps using Epoxy resin and applying the tensile force by means of flexible cables (Figure 4.13). Some investigators carefully shape specimens to produce a central gauge length. Flaws and surface scratches can have a big influence on the tensile strength measured, so that specimens need to be well-polished. This is one of the reasons why there is such a great variability in measured tensile strengths. Due to the scatter problem simpler indirect tests are usually preferred.

4.10.2 *Brazilian and point load tests*

Despite the fact that the state of stress in a cylinder compressed across its diameter is highly non-uniform (which is theoretically not the case in uniaxial extension) this type of test (called the Brazilian test) has established itself as a convenient and viable alternative method of determining tensile strength (Figure 4.14). Based on elastic theory or photoelastic studies a good estimate of the lateral tensile stresses at fracture under the given line loading is possible. In practice a pattern of tensile cracks is obtained. However, for less brittle materials localized crushing tends to occur under the loading platens. To overcome this effect a small hole is sometimes drilled at the centre of the cylinder to perform the so-called ring test. This technique is supposed to initiate tensile failure at the holes before crushing takes place; it suffers, however, from size effects.

Some investigators are sceptical about the worth of the Brazilian test. Its main value is that of a classification test and can be compared as a test to the point load index test, which involves loading a rock specimen between hardened steel cones. This test can be performed in the field and gives remarkably good results (reproducibility, correlation with compressive strength) even on irregular specimens (Attewell & Farmer 1976).

4.10.3 *Triaxial testing systems*

The main testing equipment for obtaining the stress-strain properties of rock is the triaxial apparatus. Many laboratories today are equipped with electric servo-hydraulic systems. One of the basic requirements is that the loading frame is very stiff ('hard' machine) as soft machines store elastic energy, which is released as the specimen fractures preventing a post peak investigation of the deformational behaviour. Also for non-fracturing rocks in the post peak stress region a stiff testing machine is indispensable, cf. the load-displacement curves sketched in Figure 4.15.

Various triaxial cells have been designed. A popular one is that of Franklin & Hoek (1970) which is sketched in Figure 4.16(c). The specimen is fitted up with strain gauges

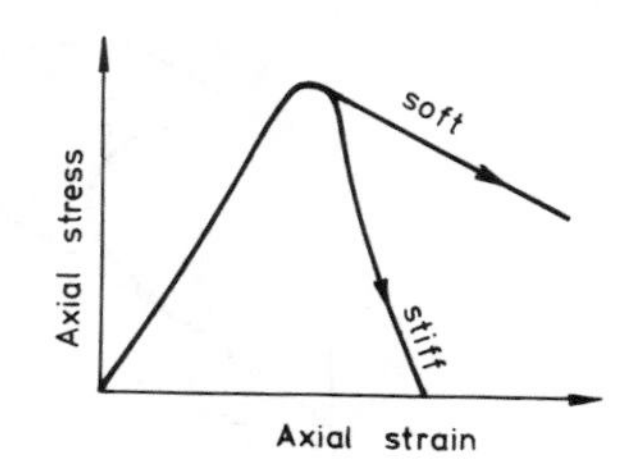

Figure 4.15. Effect of machine stiffness on measured stress-strain curve

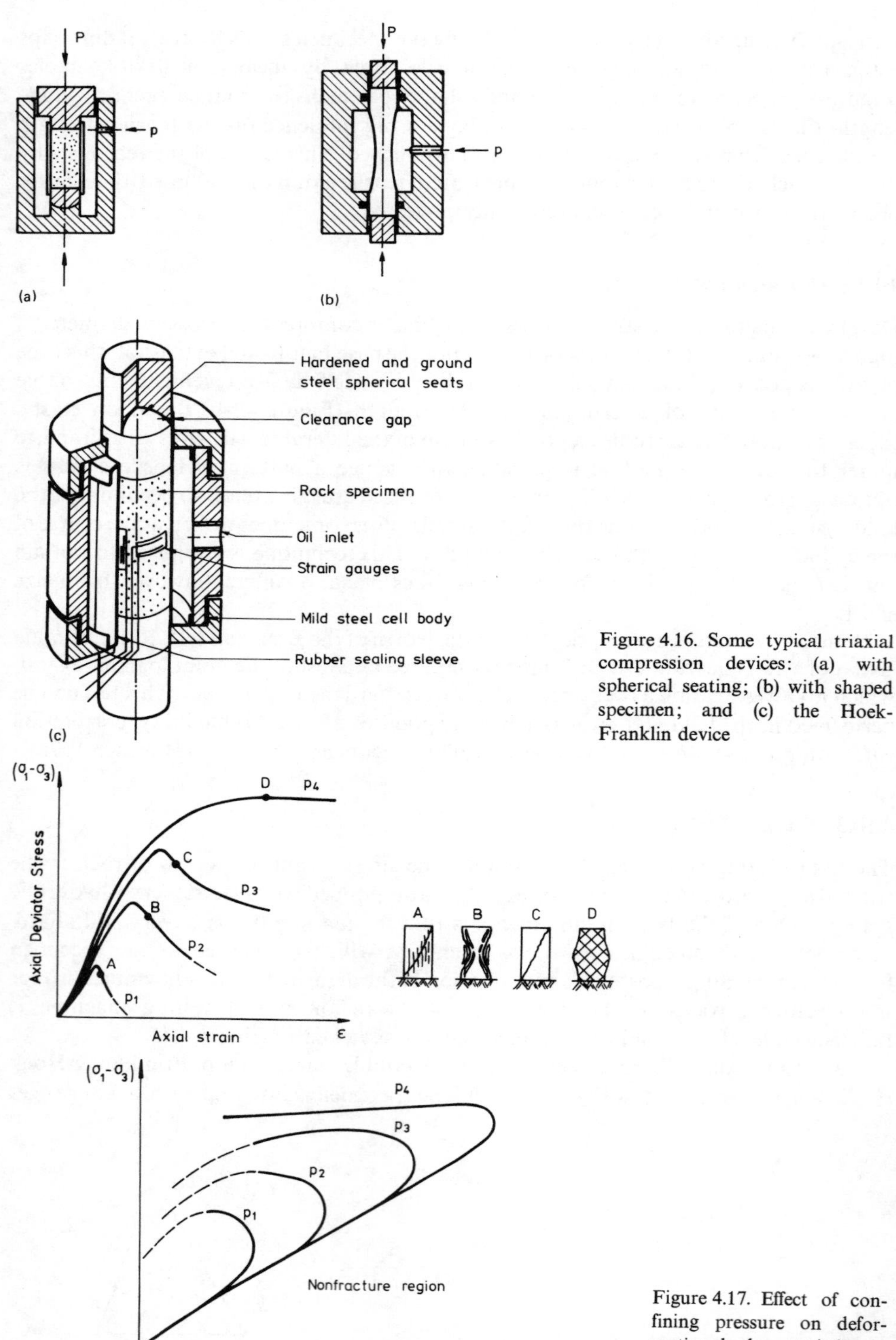

Figure 4.16. Some typical triaxial compression devices: (a) with spherical seating; (b) with shaped specimen; and (c) the Hoek-Franklin device

Figure 4.17. Effect of confining pressure on deformational characteristics in the triaxial test

to measure axial and lateral strains. It is enclosed in a rubber sleeve between hardened spherically seated steel caps of the same diameter as the specimen. Cell pressure is applied via oil up to 70 N/mm^2. With this fairly simple cell only undrained tests can be carried out. Other types allowing drainage or measurement of pore pressure are available.

The influence of confining pressure on the stress-strain characteristics as well as the mechanism of failure are illustrated in Figure 4.17.

4.10.4 *Shear and bending test devices*

Various forms of shear device have been applied in rock testing. The most versatile of these is probably the hollow cylinder torsional test, in which axial load and fluid confining pressure can be controlled.

Other test arrangements in which the specimen is not confined include the punch shear test on rock discs and the double shear test on rock beams (Figure 4.18(a)). A test similar to the latter is the bending test to determine the modulus of rupture, which is dependent on the tensile strength of the extreme bottom fibre for a specimen loaded in a

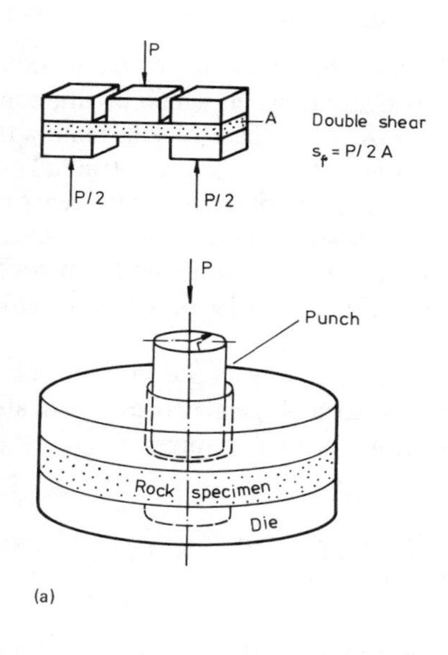

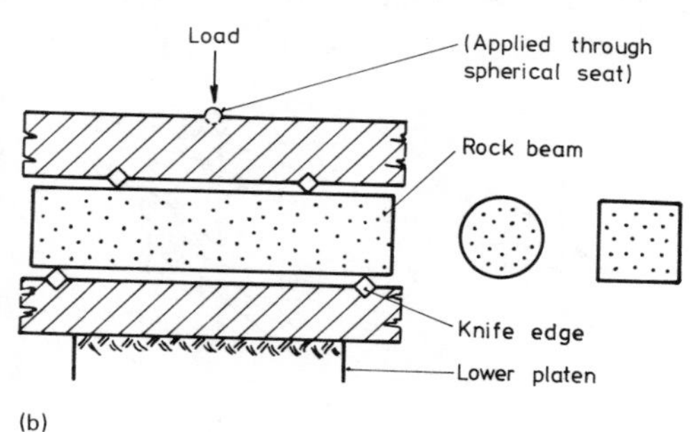

Figure 4.18. (a) Punching shear tests; and (b) bending test on rock beam

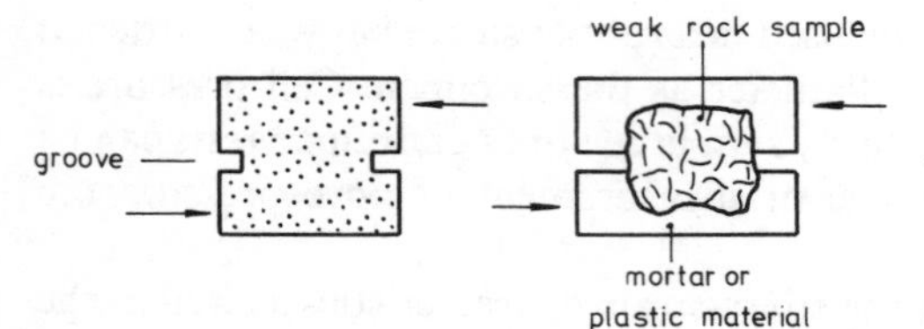

Figure 4.19. Various procedures for direct shear tests

four-point arrangement (Figure 4.18(b)). The tensile strength found in this manner is somewhat higher than that obtained by the direct uniaxial test.

The direct shear test is also of great importance in rock testing since it is often necessary to investigate behaviour along a pre-selected failure plane. The sample is prepared either by cutting a groove in it, or in the case of a weak rock sample the specimen is encastred in mortar on two opposite sides (Figure 4.19). The test is then conducted in the classical manner with measurement of the relative displacement of the two halves of the box. The shearing properties of rock joints, in which dilatancy effects predominate, may also be investigated with the direct shear apparatus.

4.10.5 *Concluding remarks on static laboratory testing of rocks*

Many of the tests described above are strictly speaking more useful for classification purposes, especially those in which in-situ confining pressures cannot be simulated in the laboratory. As in the case of soils testing the tests providing most information are the hollow-cylinder torsion and the triaxial compression test. The latter in particular has been considerably refined in recent years both in terms of equipment features with servo-hydraulic systems permitting the investigation of repeated and cyclic stress-strain behaviour and of an improved control at failure following a multi-stage testing technique (Kovari & Tisa 1975), which has also been recently applied to direct shear testing.

Not discussed above was the problem of inherent anisotropy in rocks. Preferred orientations due to geological processes will affect test results. The problem, however, is beyond the scope of this chapter.

4.11 LABORATORY DETERMINATION OF DYNAMIC ROCK AND SOIL PROPERTIES

In certain geotechnical applications the dynamic elastic and damping properties of soils and rocks are required. Usually the behaviour at low strains is of interest, as occur

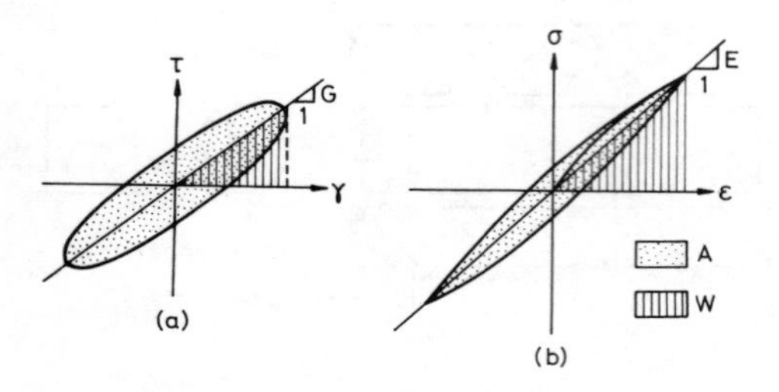

Figure 4.20. Equivalent linear modulus and damping determination; (a) ellipse: ideal viscous damping (shear); (b) hysteretic soil material (triaxial test) where A: loop area, and W: stored strain energy; where damping given by $\left(\frac{A}{4\pi W}\right)$

in vibrations due to blasting, seismic inspection and moderate earthquake motions. High intensity shaking, on the other hand, can produce plastification effects in soils and weak rocks and use of the cyclic triaxial test may be made to obtain the corresponding nonlinear or 'equivalent linear' material properties (Figure 4.20). For the intermediate range of strain the free vibration torsional shear apparatus finds some application (Taylor & Parton 1973). For the small stress range a growing number of laboratories are equipped today with the resonant column apparatus (Drnevich et al., 1978), while pulse propagation and impact tests are also in use.

In the resonant column test a cylindrical specimen with particular end conditions (fixed or free with added mass) is vibrated axially and torsionally at varying frequencies until resonance is obtained. The strains induced in the specimen are less than about

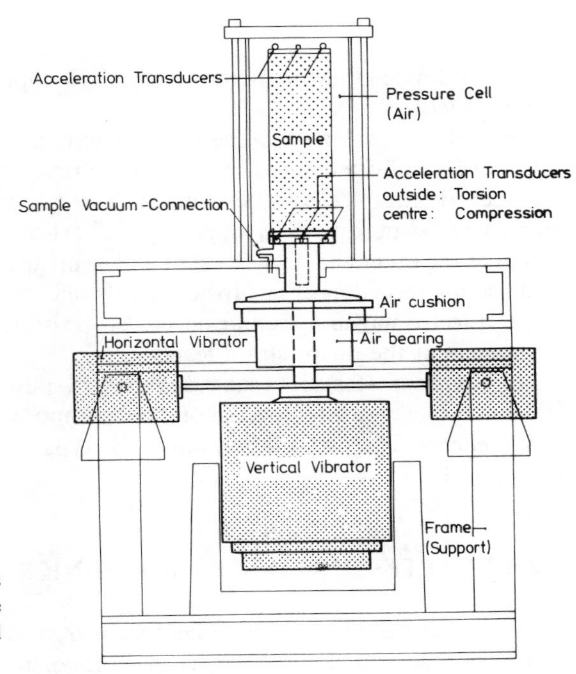

Figure 4.21. Resonant column apparatus with control at base of sample (fixed-free conditions) (ETH-Z device for gravel samples)

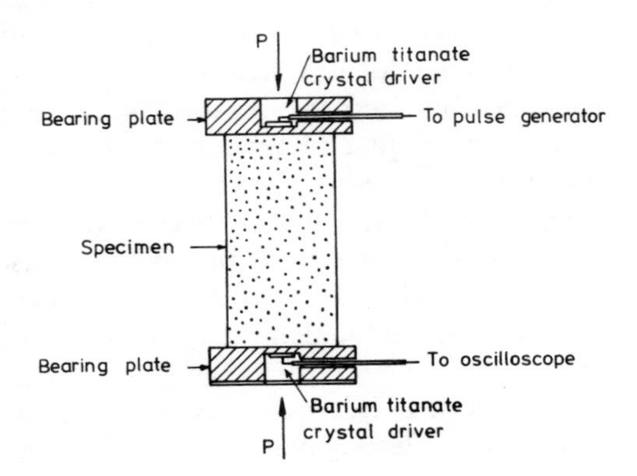

Figure 4.22. Stress pulse arrangement for low strain moduli determinations

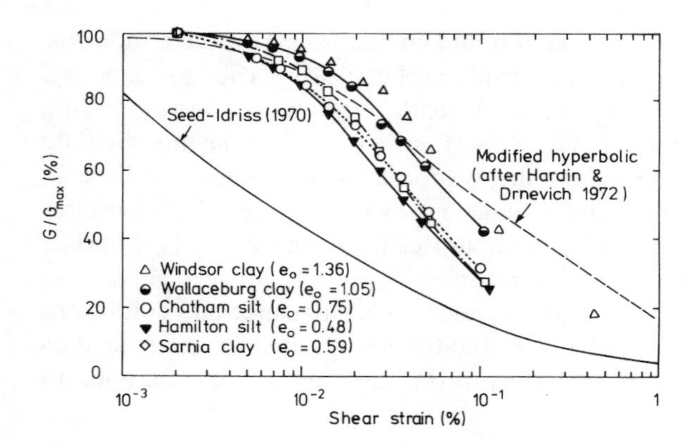

Figure 4.23. Degradation of shear modulus of clay in function of strain (after Kim & Novak 1981)

10^{-3}%. A device large enough to test gravel samples has been developed at the ETH–Zurich (Figure 4.21).

The pulse test is based on the speed of wave propagation in an elastic material. The pulses are usually generated by a piezoelectric crystal fixed to the end of a cylindrical specimen (Figure 4.22). Depending on how the crystal is cut compression or shear waves can be produced by applying a short-duration high voltage. A companion crystal at the other end acts as a signal pick-up and the travel time is obtained from an oscilloscope recording. For further details reference can be made to Obert & Duvall (1967). Electromagnetic excitation is also possible. The pulse test provides material parameters at the micro-strain level.

Up to shear strains of about $5 \cdot 10^{-4}$% purely elastic behaviour is obtained. Thereafter there is a degradation of the shear modulus, due to grain slippage and other plastic effects. The degradation curve for typical clay soils is of the form shown in Figure 4.23.

4.12 FIELD TESTING IN GEOMECHANICS

There are basically two types of field measurements in geomechanics, namely those whose purpose is to monitor the performance of full-size structures and those resulting from in-situ tests to determine such properties as shear strength, stiffness etc. In the case of the former an increased interest in observing the real behaviour of soil and rock structures is evident, which is probably connected with improved measuring systems enhanced by electronic sensors. Not only is it possible to compare theoretical and actual behaviour and conduct back-analyses to calibrate calculational models, the observational method also has the possibility of incorporation in the design process itself, e.g. in the design of underground openings with rock deformation measurements or for construction control purposes in consolidating soft foundation sediments, apart from the usual function of detecting impending danger. Such measurements include deformations and settlements, pore-water pressures etc. (ASTM: STP 584, 1975; Filliat 1981).

The scope of this section is restricted to the second type of measurement, namely field property testing devices and techniques. There is no up-to-date standard text, but reference may be made to Schultze & Muhs (1950), Leonards (1962), Filliat (1981) and to the conference proceedings *In-situ testing* (1983).

4.12.1 *In-situ tests in soil*

Today's market is replete with soil testing equipment. However, only tests and devices relevant to constitutive modelling are discussed here. Thus emphasis is on strength tests (block shear, vane shear, penetration, bearing capacity) and deformation tests (plate loading, borehole dilatometer). Soil dynamics investigations sometimes require the determination of elastic moduli at low strain values. Due to factors like sample recovery difficulties, in-situ soil structure effects and layering it is advantageous to conduct one of the various types of seismic refraction test (i.e. the cross-hole, down-hole or up-hole test).

Block shear tests

This test is usually carried out on materials like fissured clays whose sampling may be difficult or for which laboratory tests on small samples may be unrepresentative of more global behaviour in-situ. As in the laboratory direct shear tests the block of material must be loaded normal to the plane of shear. The test may be carried out in a test pit, shaft, drift or underground opening.

Vane shear tests

The vane, a device consisting of a rod with four thin blades at its head, is driven into the ground and a torque is applied to the rod, producing shear on the surfaces of the cylinder of soil enclosing the vane (Figure 4.24). The test is suitable in soft to medium-stiff clays and has been much used in sensitive clays, which are otherwise prone to sampling disturbances. It can be conducted in boreholes to depths exceeding 50 m, if the appropriate torque shaft equipment is available. As with the various types of penetrometer equipment the test is quick and economic to perform.

There are, however, certain problems, namely disturbance to ground during insertion of vane, effect of adhesion on the torque shaft (the usual practice being to insert the vane to a depth of five times its diameter to reach an undisturbed zone) and torque measurement. Nevertheless, there has been an accumulation of data evidencing the usefulness of the test.

Cone penetration tests

There are various penetration or sounding devices in use, e.g. the standard penetration test (SPT) device or the Dutch cone penetrometer. These tests serve essentially as index tests correlating soil strength with penetration resistance unlike the block shear and vane shear tests from which strength parameters are directly measured for the known stress conditions. Cone penetration tests may be of the dynamic or static types. The former is similar to the SPT-test in which the number of blows is recorded versus penetration. In the latter the cone is pushed into the ground at a constant velocity.

The Dutch cone has a point angle of 60° and is back-tapered. The projected end area is usually 10 cm^2. Devices have an inner rod with an outer jacket design (Figure 4.25). The drive-point penetrometer can be advanced by itself to find the bearing capacity or together with the jacket for the additional determination of shaft friction and adhesion.

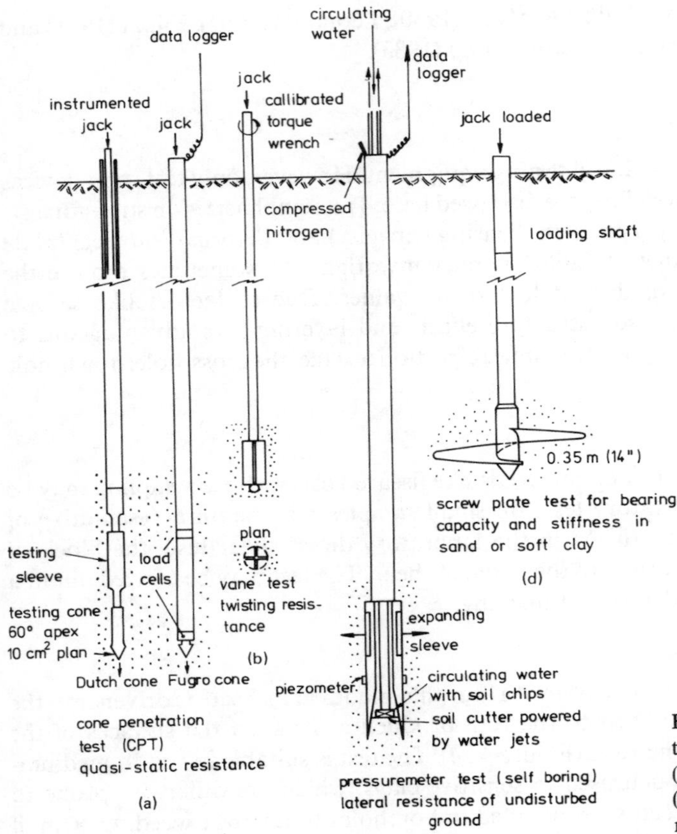

Figure 4.24. Some common types of in-situ soil probes; (a) cone penetrometer devices; (b) shear device; (c) pressuremeter device of self-boring type; (d) screw plate device

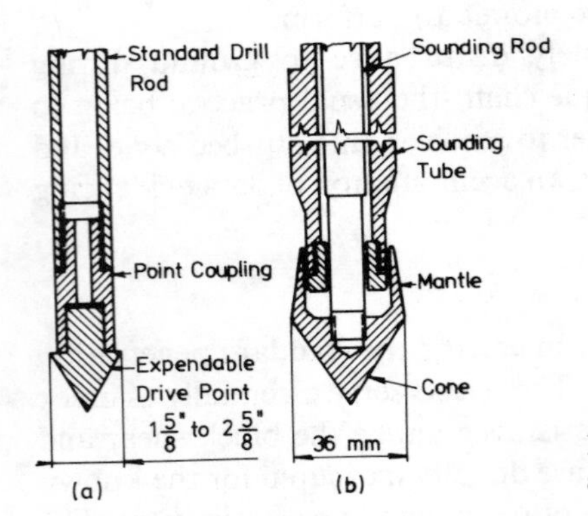

Figure 4.25. Details of different cone devices: (a) drive point; and (b) Dutch cone

Whereas the vane test is superior in soft clays it cannot be used in granular materials. In this respect penetrometers are more universal in application though the presence of gravel or boulders can limit test capability and invalidate the results. Some important uses are in the design of pile foundations, investigating the homogeneity of the subsoil and the control of compaction apart from strength and compressibility correlations.

Plate bearing tests

Foundation design has often been based on the results of rigid plate bearing tests conducted in a test pit at the grade level. Because of the limited amount of load that can be applied to a plate (usually by jacking against kentledge or anchors) its dimensions are usually small and thus the depth of influence is rather restricted. Thus factors such as soil layering and variation of properties with depth are inadequately investigated, so that extrapolation of the test results to actual-size foundations is uncertain. A further problem is that of the disturbance to the ground by excavating the test pit and accounting for the influence of the pit area, which amounts to an embedment condition. To overcome these problems the screw-plate load test (Schmertmann 1970) has been developed, which consists basically of a single flight earth auger simulating the plate. With this equipment tests can be carried out at successively greater depths. The data provided by the test are vertical compressibility and bearing capacity (see Figure 4.24).

One of the main shortcomings of plate bearing tests is that they do not give information on long-term settlement, especially due to the consolidation of underlying layers. Thus careful interpretation of the results is required.

Borehole tests

The original idea of plate bearing tests was to model foundation loading. The test may, in principle, also be carried out at the bottom of an exploration borehole, but the difficulties are correspondingly great, e.g. loosening and unevenness of soil surface and squeezing out of material around the plate, whose typical diameter is about 5 cm.

Devices to measure the deformation modulus in a horizontal direction are also available. There are basically two types: *jacking* and *dilatometer* devices. With the former a pair of plates or wedges are loaded against the cylindrical wall of the borehole, e.g. in the case of the Goodman jack two plates having a 90° arc section are forced apart by twelve hydraulic pistons. With such set-ups no reactive force is required as in the case of vertical loading. The disadvantage of jacking devices is the limited area of contact of the plates. This has been remedied by applying an all-round radial stress by means of a rubber jacket, albeit with a reduced stress level. In soils such devices are called *pressure-meters*. In the deflated state they can be lowered to any desired depth in the borehole.

The deformation modulus is obtained from the formula for an infinitely thick elastic cylinder assuming a value of Poisson's ratio. To account for bedding and other inelastic effects the mean change in diameter for a given pressure is measured usually in the third load cycle. Inflatable dilatometer or pressure-meter devices have been in use for many years. Today a widely marketed product in soil engineering is the Menard pressure-meter.

A fairly recent development is that of *self-boring pressure-meters*. Hughes et al. (1977) report on equipment which appears to have been used successfully in sand to determine the friction and dilatancy angles respectively, in addition to the shear modulus. The self-boring technique results in minimum disturbance to the ground in regard to pressure-meter testing. It is particularly useful in sands, in which the difficulties in avoiding sample disturbance for laboratory testing are well-known. In common with some recently developed penetration probes the pressure-meter is fitted with a pore pressure cell (see Figure 4.24).

4.12.2 *In-situ tests in rock*

Some of the techniques already described to obtain the in-situ properties of soil apply to rocks as well, so that a further discussion of the devices, i.e. plate bearing and borehole dilatometer test equipment, is unnecessary. Another type of jacking device which enjoys some popularity is the *flat jack* device, which consists of a pair of circular or square metal plates welded together at their periphery. The jacks are pressurized giving an extended area of contact with the rock surface, enabling load-deformation characteristics to be observed. In one test arrangement slots, e.g. horizontal slots in a mine pillar, are drilled into the rock and the flat jacks are cemented into place. The pressure is increased in stages with unload-reload cycles until failure is reached. To achieve more reliable results it is necessary to cut the slots with precision drilling equipment and dispense with the cementation. Even so there is still much uncertainty regarding the stress distribution at the jacking surface.

The large-scale direct shear test is often conducted in rock. A typical test set-up is shown in Figure 4.26.

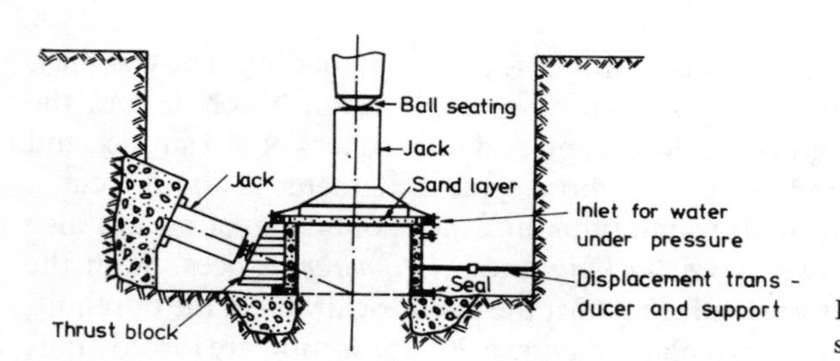

Figure 4.26. Set-up for large scale in-situ direct shear test on rock

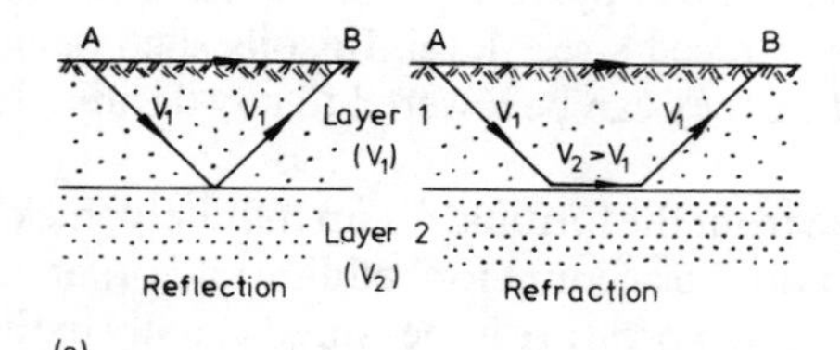

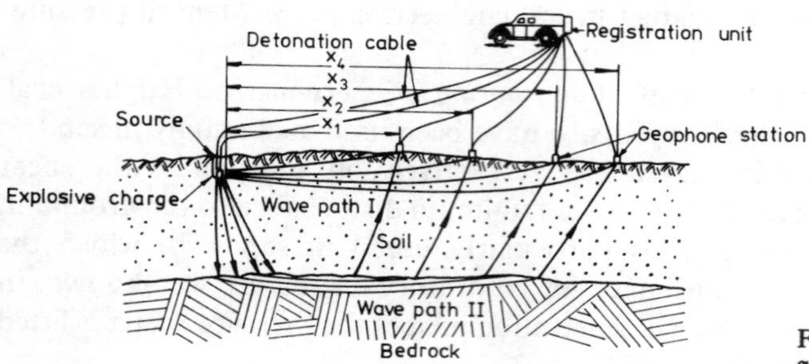

Figure 4.27. (a) Seismic surveys: reflection and refraction types; and (b) set-up for seismic traverse

4.12.3 *Seismic exploration and elastic modulus determination*

The use of seismic techniques, i.e. reflection and refraction surveys, is fairly extensive in geophysical and geotechnical exploration. The methods involve measuring the travel time of seismic waves from a source (impulsive or vibrating loading) to various receiver stations. The velocities of the body waves, i.e. the longitudinal compressional or P-waves and transverse shear or S-waves, depend on the respective elastic moduli, E and G, and the in-situ density.

The main purpose of seismic surveys is to determine the profile of the bedrock (Figure 4.27(a)). Since, in the case of reflected waves, their arrival time interferes with those of direct and refracted waves reflection surveys are only suitable for depths greater than about 600 m. With refraction surveys the wave is refracted, as shown in Figure 4.27(a), at the upper surface of the bedrock. A typical testing arrangement is shown in Figure 4.27(b) in which there are a number of geophone receivers. Normally, in geotechnical engineering, the signal is generated by an explosive charge or hammer blow.

Depending on the stiffness of the bedrock its 'rippability' characteristics can be estimated for rock excavation work. In recent years, however, another important use of seismic methods has been for earthquake analysis studies.

In order to obtain the elastic deformation moduli at low strain levels more accurate seismic techniques have been developed. Usually one is interested in the properties for a layered site condition. The source and the receivers are placed at various levels in bore-holes in the cross-hole (or inter-hole) method (Stokoe & Woods 1972), the spacing between the bore-holes depending upon the layer thickness (Figure 4.28). Both P and S waves can be generated. Reduced costs but with less reliable results can be achieved by

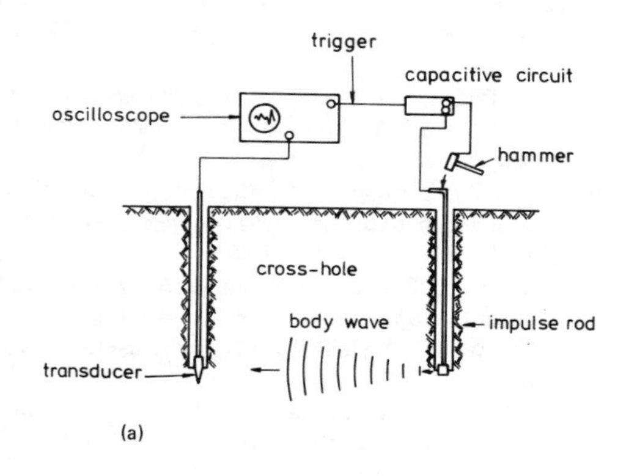

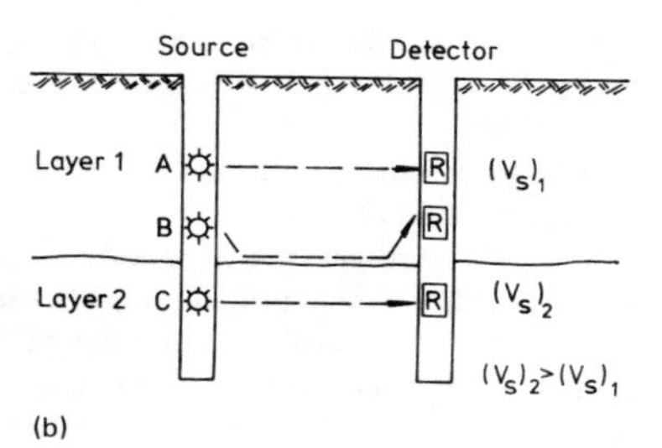

Figure 4.28. (a) A cross-hole test arrangement; and (b) effects of layering on test interpretation

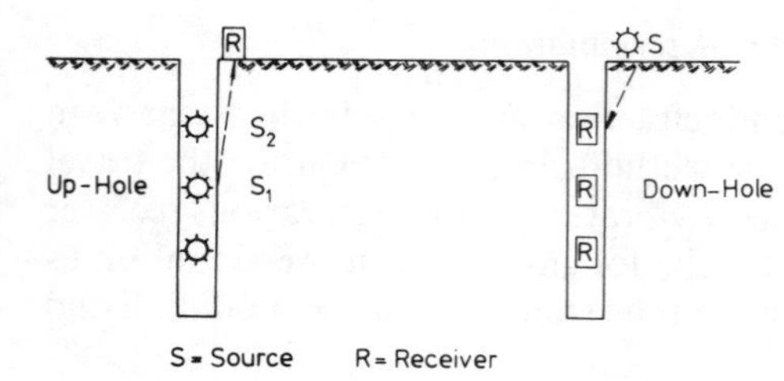

Figure 4.29. The up-hole and down-hole arrangements

drilling only one borehole and using the down-hole and up-hole techniques, see Figure 4.29 (Beeston & McEvilly 1977).

For accurate measurements in both cross-hole and down-hole tests careful test control is needed for such factors as bore-hole verticality, high energy content of shear waves (e.g. explosive charge is a poor source with too little S-wave input and not allowing directional reversing of input), and good coupling of geophone to wall of bore-hole.

REFERENCES

Al-Husseini, M.M. 1971. Investigation of plane strain shear testing. Report 5–71–2 (No. 1), US Army Engineers, cf. Horslev, p. 144, Waterways Experiment Station, Vicksburg, Miss.

Arthur, J.R.F., S. Beckenstein, J.T. Germaine & C.C. Ladd 1981. Stress path tests with controlled rotation of principal stress directions. In Yong/Townsend (eds.) *Laboratory shear strength of soil.* ASTM, STP 740; 516–540.

Arthur, J.R.F., K.S. Chua, T. Dunstan & C. del Rodriguez 1980. Principal stress rotation: a missing parameter. *J. Geotech. Engng. Div.* (ASCE) 106 (GT 4): 419–433.

ASTM (American Society of Testing Materials) 1975. *Performance monitoring for geotechnical construction.* STP 584.

Attewell, P.B. & I.W. Farmer 1976. *Principles of engineering geology.* London: Chapman & Hall.

Beeston, H.E. & T.V. McEvilly 1977. Shear wave velocities from downhole measurements. *Earthquake Engng. and Struct. Dynamics* 5: 181–190.

Bishop, A.W., I. Alpan, G.E. Blight & I.B. Donald 1960. Factors controlling the strength of partially saturated cohesive soils. In *Proceedings of ASCE Research Conference on Shear strength of Cohesive Soils* (Boulder, Colorado): 503–532.

Bishop, A.W., G.E. Green, V.K. Garga, A. Andersen & J.D. Brown 1971. A new ring shear apparatus and its application to the measurement of residual strength. *Geotechnique* 21 (4): 273–328.

Bishop, A.W. & D.J. Henkel 1962. *The measurement of soil properties in the triaxial test* (2nd ed.). London: Arnold.

Broms, B.B. & A.O. Casbarian 1965. Effects of the rotation of the principal stress axes and of the intermediate stress on shear strength. In *Proceedings of 6th International Conference on Soil Mechanics and Foundation Engineering* (Montreal), 1: 179–183.

Casagrande, A. & S.D. Wilson 1960. Moderator's report: testing equipment, techniques and errors. In *Proceedings of ASCE Research Conference on Shear Strength of Cohesive Soils* (Boulder, Colorado): 1123–1130.

Cornforth, D.H. 1964. Some experiments on the influence of strain conditions on the strength of sand. *Geotechnique* 14 (2): 143–167.

Deman, F. 1975. Achsensymmetrische Spannungs- und Verformungsfelder in trockenem Sand. Report No. 62, Institut für Bodenmechanik und Felsmechanik der Universität Fridericiana, Karlsruhe.

Drnevich, V.P., B.O. Hardin & D.J. Shippy 1978. Modulus and damping of soils by the resonant column method. In *Dynamic geotechnical testing.* ASTM, STP 654: 91–125.

Filliat, G. 1981. *La pratique des sols et fondations.* Paris: Moniteur.

Franklin, J.A. 1971. Triaxial strength of rock materials. *Rock Mech.* 3 (2): 86–98.

Franklin, J.A. & E. Hoek 1970. Developments in triaxial testing technique. *Rock Mech.* 2 (4): 223–228.

Goodman, R.E. 1980. *Introduction to rock mechanics.* New York: Wiley.

Hambly, E.C. 1972. Plane strain behaviour of remoulded normally consolidated kaolin. *Geotechnique* 22 (2): 301–317.

Hight, D.W., A. Gens & M.J. Symes 1983. The development of a new hollow cylinder apparatus for investigating the effects of principal stress rotation in soils. *Geotechnique* 33 (4): 355–383.

Hoek, E. 1968. Brittle failure of rock. In Stagg/Zienkiewicz (eds.), *Rock mechanics in engineering practice*: 99–124.

Hughes, J.M.O., P. Wroth & D. Windle 1977. Pressuremeter tests in sands. *Geotechnique* 27 (4): 455–477.

Hvorslev, M.J. 1937. On the strength properties of remoulded cohesive soils, Ph.D. diss. Tr. 1963 US Army Engineers, Waterways Experiment Station, Vicksburg, Miss.

In-situ testing. 1983. *Proceedings of International Symposium on Soil and Rock Investigations by In-Situ Testing* (Paris), 3 vols.

Ishibashi, I. & M.A. Sherif 1974. Soil liquefaction by torsional simple shear device. *J. Geotech. Engng. Div.* (ASCE) 100 (GT 8): 871–888.

Jaeger, J.C. & N.G.W. Cook 1976. *Fundamentals of rock mechanics* (2nd. ed.). London: Chapman & Hall.

Janbu, N., D. Tokheim & K. Senneset 1981. Consolidation tests with continuous loading. In *Proceedings of 10th International Conference on Soil Mechanics and Foundation Engineering* (Stockholm), 1: 645–654.

Kiekbusch, M. & B. Schuppener 1977. Membrane penetration and its effect on pore pressures. *J. Geotech. Engng. Div.* (ASCE) 103 (GT 11): 1267–1279.

Kim, T.C. & M. Novak 1981. Dynamic properties of some cohesive soils of Ontario. *Can. Geotech. J.* 18: 371–389.

Ko, H.Y. & R.F. Scott 1967. A new soil testing apparatus. *Geotechnique* 17 (1): 40–57.

Kolymbas, D. 1981. Bifurcation analysis for sand samples with a nonlinear constitutive equation. *Ingenieur-Archiv* 50: 131–140.

Kovari, K. & A. Tisa 1975. Multiple failure state and strain controlled triaxial tests. *Rock Mech.* 7 (1): 17–33.

Kuntsche, K. 1982. Materialverhalten von wassergesättigtem Ton bei ebenen und zylindrischen Verformungen. Report No. 91, Institut für Bodenmechanik und Felsmechanik der Universität Fridericiana, Karlsruhe.

Lade, P.V. 1981. Torsion shear apparatus for soil testing. In Yong/Townsend (eds.), *Laboratory shear strength of soil.* ASTM, STP 740; 145–163.

Lade, P.V. & J.M. Duncan 1973. Cubical tests on cohesionless soil. *J. Soil Mech. Fdns. Div.* (ASCE) 99 (SM 10): 793–812.

Lade, P.V. & S.B. Hernandez 1977. Membrane penetration effects in undrained tests. *J. Geotech. Engng. Div.* (ASCE) 103 (GT 2): 109–125.

Lee, K.L. 1978. End restraint effects on undrained static triaxial strength of sand. *J. Geotech. Engng. Div.* (ASCE) 104 (GT 6): 687–704.

Leonards, G.A. 1962. Foundation engineering. In W.L. Shannon, S.D. Wilson & R.H. Meese (eds.), *Field problems: field measurements*: 1025–1080. New York: McGraw-Hill.

Leroneil, S. & F. Tavenas 1981. Pitfalls of back-analysis. In *Proceedings of 10th International Conference on Soil Mechanics and Foundation Engineering* (Stockholm), 1: 185–190.

Lowe, J. III & L. Karafiath 1960. Effect of anisotropic consolidation on the undrained shear strength of compacted clays. In *Proceedings of ASCE Research Conference on Shear Strength of Cohesive Soils* (Boulder, Colorado): 837–858.

Marachi, N.D., J.M. Duncan, C.K. Chan & H.B. Seed 1981. Plane strain testing of sand. In Yong/Townsend (eds.), *Laboratory shear strength of soil.* ASTM, STP 740: 294–302.

Marr, W.A. & K. Hφeg 1979. *Stress-strain behaviour from stress path tests.* Publ. No. 125, Norwegian Geotechnical Inst., Oslo.

Martin, G.R., W.D.L. Finn & H.B. Seed 1978. Effects of system compliance in liquefaction tests. *J. Geotech. Engng. Div.* (ASCE) 104 (GT 4): 463–479.

Morgan, J.R. & P.J. Moore 1968. Application of soil dynamics to foundation design. In I.K. Lee (ed.), *Soil mechanics: selected topics*: 465–527. London: Butterworths.

Obert, L. & W.I. Duvall 1967. *Rock mechanics and the design of structures in rock.* New York: Wiley.

Pearce, J.A. 1971. A new true triaxial apparatus. In Parry (ed.), *Stress-strain behaviour of soils,* Roscoe Memorial Symposium (Cambridge, UK): 330–339.

Roberts, A. 1977. *Geotechnology.* New York: Pergamon Press.
Rowe, P.W. & L. Barden 1964. Importance of free ends in triaxial testing. *J. Soil Mech. Fdns. Div.* (ASCE) 90 (SM 1): 1–27.
Saada, A.S. & C.D. Ou 1973. Stress-strain relations and failure of anisotropic clays. *J. Soil Mech. Fdns. Div.* (ASCE) 99 (SM 12): 1091–1111.
Saada, A.S. & F.C. Townsend 1981. State of the art: laboratory strength testing of soils. In Yong/Townsend (eds.), *Laboratory shear strength of soil.* ASTM, STP 740: 7–77.
Sangrey, D.A., W.S. Pollard & J.A. Egan 1978. Errors associated with rate of undrained cyclic testing of clay soils. In *Dynamic geotechnical testing.* ASTM, STP 654: 280–294.
Schmertmann, J.H. 1970. Screw-plate load test. In *Special procedures for testing soil and rock for engineering purposes.* ASTM, STP 479: 81–85.
Schultze, E. & H. Muhs 1967. *Bodenuntersuchungen fur Ingenieurbauten* (2nd ed.). Berlin: Springer Verlag.
Smith, R.E. & H.E. Wahls 1969. Consolidation under constant rate of strain. *J. Soil Mech. Fdns. Div.* (ASCE) 95 (SM 2): 519–539.
Stokoe, K.M. II & R.H. Woods 1972. In situ shear wave velocity by crosshole method. *J. Soil Mech. Fdns. Div.* (ASCE) 98 (SM 5): 443–460.
Taylor, P.W. & I.M. Parton 1973. Dynamic torsion testing of soils. In *Proceedings of 8th International Conference on Soil Mechanics and Foundation Engineering* (Moscow), 1/2: 425–432.
Wawersik, W.R. & C. Fairhurst 1970. A study of brittle rock fracture in laboratory compression experiments. *Int. J. Rock Mech. Min. Sci.* 7: 561–575.
Wood, D.M. 1982. Laboratory investigations of the behaviour of soils under cyclic loading: a review. In Pande/Zienkiewicz (eds.), *Soil mechanics – transient and cyclic loads*: 513–581. Chichester: Wiley.
Yamada, Y. & K. Ishihara 1979. Anisotropic deformation characteristics of sand under three-dimensional stress conditions. *Soils and Foundations* 19 (2): 79–94.
Yoshimi, Y. & H. Oh-oka 1975. Influence of degree of shear stress reversal on the liquefaction potential of saturated sand. *Soils and Foundations* 15 (3): 27–40.

Part 2
Numerical modelling of selected engineering problems

The contribution of numerical analysis to the design of shallow tunnels

5

Z. EISENSTEIN

5.1 INTRODUCTION

Design of shallow tunnels includes dealing with three important problems:

1. Maintaining stability of face and wall of the tunnel before being supported by a lining.
2. Predicting displacements caused by excavation of the tunnel throughout the adjacent ground mass and on surface.
3. Predicting magnitude and distribution of earth pressures acting on the lining.

The first problem is usually dealt with by employing a tunnelling method suitable for a particular soil type and if necessary, combining it with appropriate ground control measures. Methods nowadays exist for driving tunnels through virtually every ground condition.

The other two problems are treated in current design practice most often separately. The prediction of displacements is limited to prediction of surface settlements. Essentially only one method is used. It is based on a semi-empirical approach, combining accumulated field observations with simple theoretical framework (Schmidt 1969, Peck 1969, Clough & Schmidt 1977). The method appears to provide satisfactory results in normally consolidated clays, but tends to be misleading in granular soils (O'Reilly & New 1982) and in overconsolidated clays (Eisenstein et al 1982).

The prediction of lining pressures for shallow tunnels has received considerable attention over the last half a century. Duddeck & Erdmann (1982) were able to review thirteen proposed methods. All of them consider the soil around the tunnel as linearly elastic and most of them do not provide for reduction of soil pressure with inward deflection of the lining. Among the exceptions are the method of Muir Wood (1975), who proposed an arbitrary reduction in the pressure and the more recently proposed continuum model (Duddeck 1980). Field measurements of lining loads indicate that the actual pressures developed are often smaller than those predicted or assumed by the above methods (El-Nahhas 1980; Branco 1981).

5.2 GROUND REACTION CURVE

The separation of the problem of displacement from the problem of lining pressure is convenient analytically, but is entirely artificial. In reality, soil displacements are

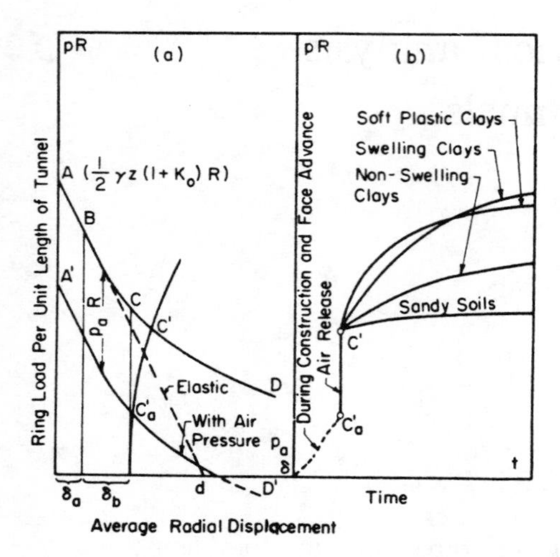

Figure 5.1. Typical ground reaction curves (after Peck 1969)

closely linked to soil pressures and vice versa. This has been recognized for some time and is best illustrated by the concept of ground reaction curve (Figure 5.1). These curves relate the release of radial in-situ stresses in the soil to radial displacement. They have been used fairly extensively in the theory of underground structures, but mainly as tools for conceptual explanations of soil-tunnel interaction processes. Rarely have these curves been quantified and used to indicate the individual sources of ground stress release and deformation in terms of their actual magnitude for soil tunnels. In order to illustrate this type of problem a ground reaction curve obtained from in-situ

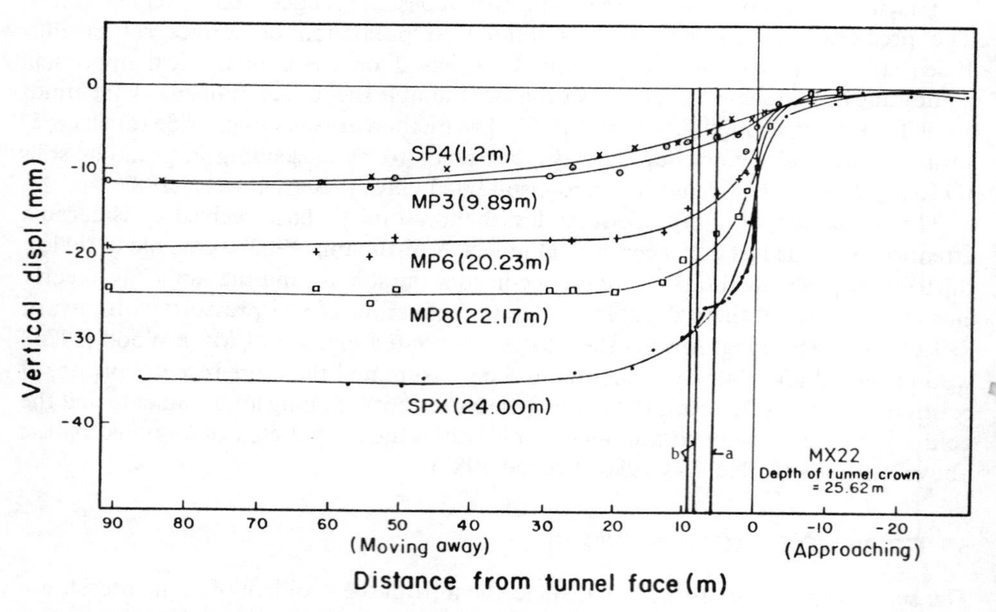

Figure 5.2. Measurement of vertical displacements versus tunnel advance (after Eisenstein et al. 1981b)

measurements will be introduced and discussed (Eisenstein et al. 1981b). The curve results from combining vertical displacements with the history of development of the lining stresses and loads, as measured simultaneously on an experimental tunnel project. The soil (stiff overconsolidated clay) was assumed to follow a linear elastic model. The tunnel was excavated by a tunnelling boring machine and lined with expanded precast segmented liner. The ground reaction curve refers to the crown of the tunnel. Figure 5.2 shows measurements of vertical displacements versus tunnel advance measured at different depths between surface and crown. Of interest is the lowest single point extensometer (SPX) located near the crown. The dashed curve, which was drawn by connecting the actual measurement points rather than plotting an average smooth curve, illustrates that there are two distinct stages of development of the soil displacement around the tunnel which are different from the portion occurring ahead of the tunnel face. First the soil moved radially to close the annular space resulting from soil overcut around the shield. As soon as this space was filled the soil displacement was stabilized by the shield itself. The soil was allowed to move again when the tailpiece of the shield cleared the section at line 'a'. This movement stabilized soon after the lining was expanded, at lines 'b'. Corresponding measurements were taken on the liner to monitor soil pressures.

Plotted in a ground reaction curve in Figure 5.3 (which is a straight line in this case because of the assumption of linear elasticity), several distinctive phases of lining activation are apparent. Initially, there is stress release due to excavation of the face, followed by shield clearance. During this phase the soil closed on the shield. After the shield clears the profile, the soil closes further on the yet unexpanded lining until its expansion. From this point, the segmented liner begins to counteract the soil pressures until an equilibrium is reached.

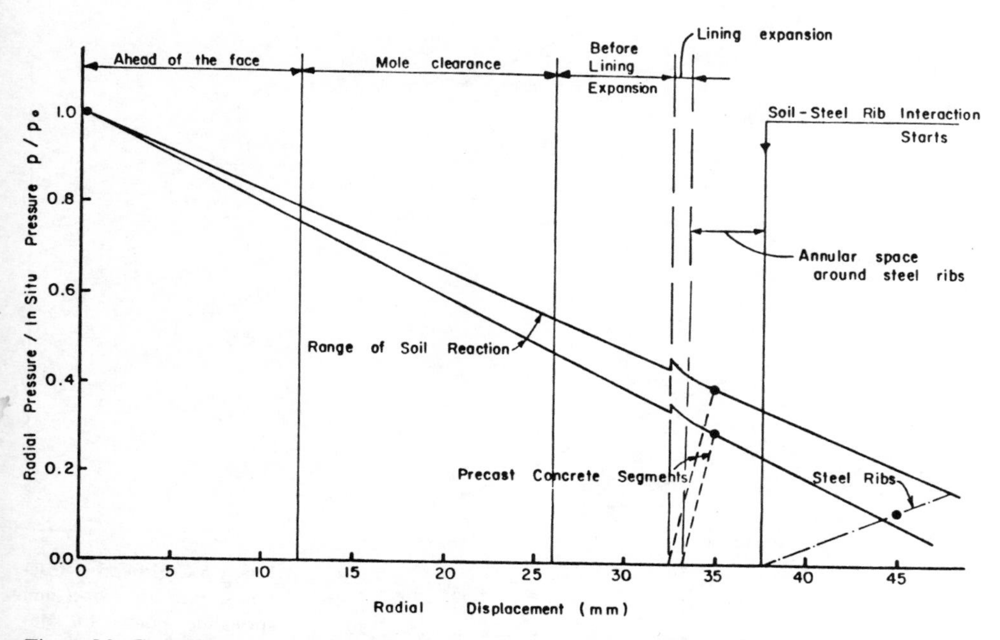

Figure 5.3. Ground reaction curve obtained from field measurements (after Eisenstein et al. 1981b)

The above example shows some of the key components participating in formation of a ground reaction curve. A different tunnelling method would result in different components, while a different soil might require a curved rather than a straight line. Also, for shallow tunnels with an initial stress field distinctly varying across the height of the tunnel, there is a different ground reaction curve for every point around the tunnel. Figure 5.4 illustrates this for the crown and springline points of two tunnels with different K_0 coefficients.

It is obvious that a quantitative prediction of ground reaction curves for at least two points (crown and springline) would form a basis, from which the predictions of both the lining pressures and the ground displacements could be made. The lining pressures are given directly by the soil-lining equilibrium points at the curves, and the difference of pressure between the crown and the springline is a measure of the bending moments and shear forces in the lining. In the case of the displacements, the final radial displacements at the tunnel wall, as represented again by the equilibrium points, can be regarded analytically as displacement boundary condition. Imposition of this boundary displacement on the soil continuum surrounding the tunnel then governs the final displacement field, including the surface settlement.

The question then is how to predict a ground reaction curve. In principle, a ground reaction curve is a stress-strain curve for loading along a special stress path and under

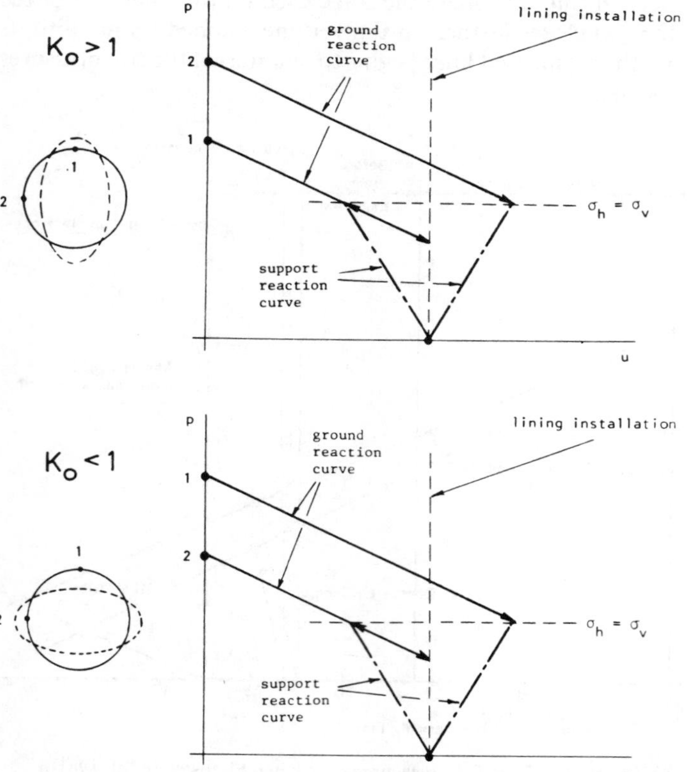

Figure 5.4. Ground reaction curves for crown and springline points for different K_0 values

special boundary conditions. Since it is an almost standard exercise in analytical soil mechanics to predict one stress-strain curve from another, one might expect that a prediction of a ground reaction curve from some conventional test curve should be possible.

An attempt in this direction has been done by Deere et al. (1969). They assumed a linearly elastic medium around the tunnel and a completely axisymmetric behaviour allowing the use of closed form formulas. The amount of the initial displacement (prior to activation of the lining) has to be estimated, including the face effect. To overcome these simplifications and to obtain a more realistic model of soil-tunnel interaction complex numerical methods have been employed. In the following some of these attempts will be reviewed. Particular attention will be given to the ability of various models to incorporate:

1. The three-dimensional effects which occur at and ahead of the tunnel face.
2. The various sources of loss of ground (displacement at the tunnel wall) as given by different tunnelling methods and technologies.
3. The complexity of stress-strain behaviour of soil around a shallow tunnel, including the effects of yielding.

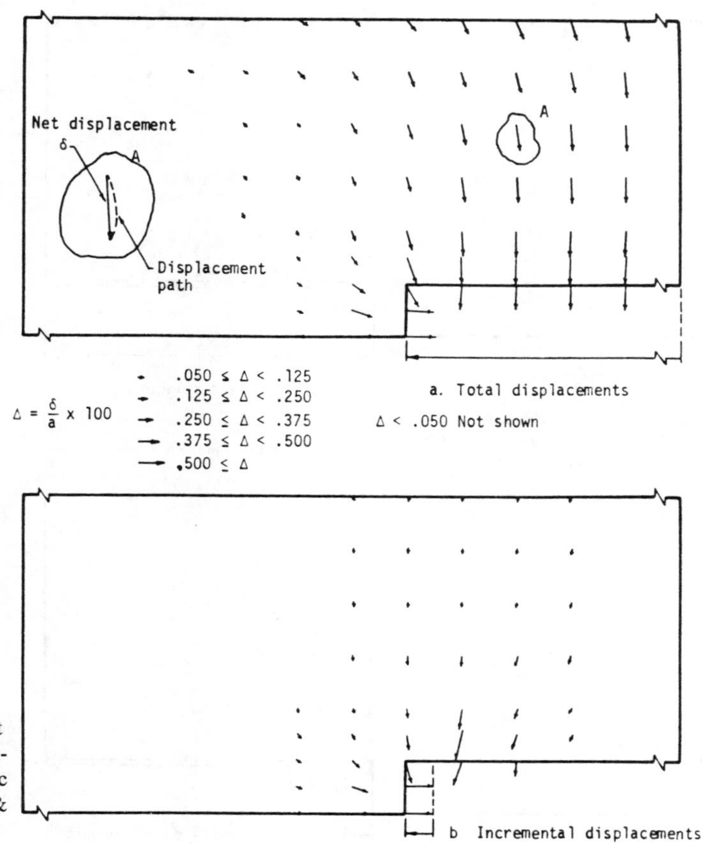

Figure 5.5(a). Development of displacements around unlined tunnel in linear elastic medium (after Ranken & Ghaboussi 1975)

5.3 THREE-DIMENSIONAL EFFECTS BY TWO-DIMENSIONAL ANALYSES

The response of soil above, below and beside an advancing tunnel is obviously a three-dimensional phenomenon. Nevertheless, many attempts have been made to employ simpler and less demanding two-dimensional models to study, if not the complete displacement field, at least some important features of it. These attempts can be divided into two categories: (a) plane strain analysis in a vertical section along the tunnel axis; (b) plane strain analysis in a vertical section perpendicular to the tunnel axis.

The category (a) is useful to investigate some features occurring ahead of an advancing tunnel as well as the role of the history of lining placement. The category (b) is applicable to studies of displacements and stresses at a certain distance behind the tunnel face, where the behaviour of a tunnel is assumed to settle into a condition of plane strain.

An example of the (a) category is given by Ranken & Ghaboussi (1975) who performed an extensive study of the soil-lining interaction. They used an axisymmetric finite element analysis for the special case, $K_0 = 1$, of the general three-dimensional problem considering unlined or lined tunnel. The soil behaviour was assumed either

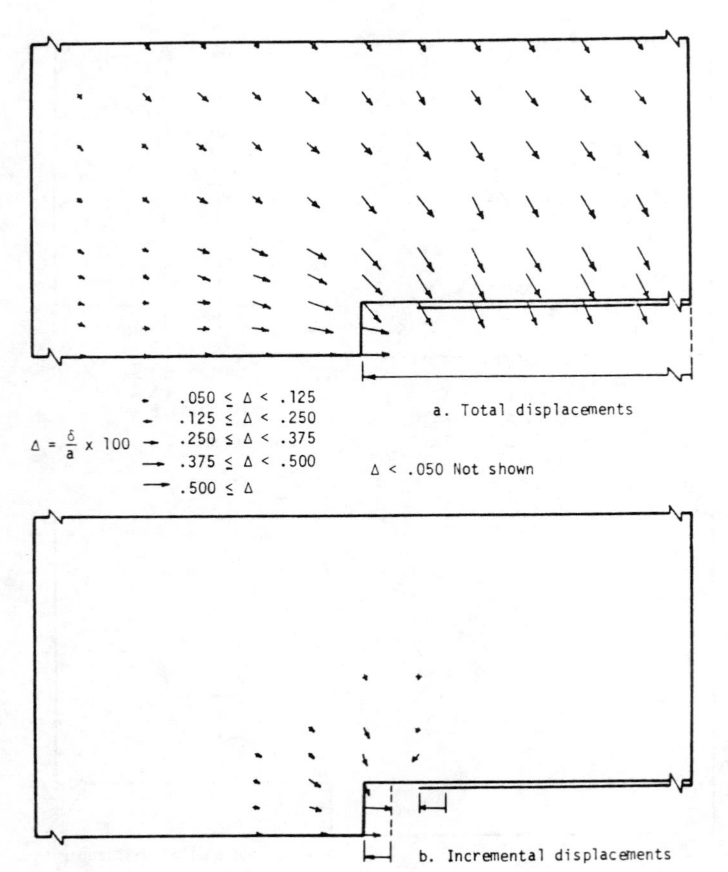

Figure 5.5(b). Development of displacements around lined tunnel in linear elastic medium (after Ranken & Ghaboussi 1975)

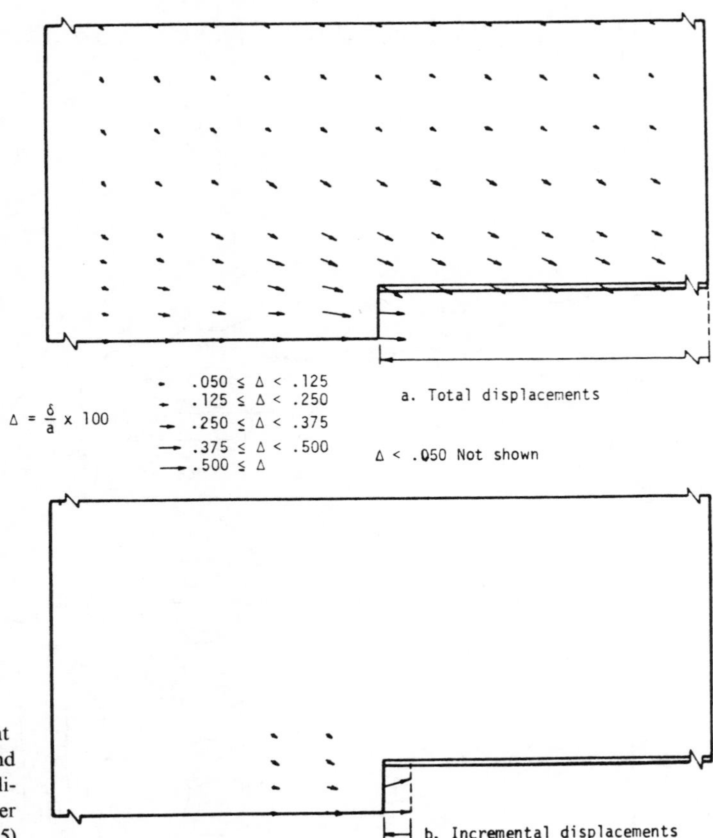

Figure 5.5(c). Development of displacements around partially lined tunnel in linear elastic medium (after Ranken & Ghaboussi 1975)

linearly elastic or linearly elasto-plastic. Despite these highly idealized assumptions some important factors emerged. One of them is the development of displacement ahead and above an advancing tunnel. Figures 5.5(a), 5.5(b) and 5.5(c) compare the distribution of displacements for an unlined, lined and partially lined tunnel. For the unlined tunnel, the displacement pattern settles soon into an entirely radial direction but the displacement path has a distinct longitudinal component. The rigidly lined tunnel prevents any radial displacement behind the face and increases the displacements ahead and through the face. The reality for shielded, lined tunnels is somewhere between these two extremes, as represented by the partially lined tunnel. There is a longitudinal component of displacement 'locked-in'. Its magnitude depends on how much loss of ground is allowed prior to lining erection and activation. If a large percentage of the total displacement can occur before the lining is activated, then a reasonable plane strain approximation perpendicular to the tunnel axis may be obtained and the final displacement field can be obtained from a two-dimensional analysis.

It is of interest to look at a displacement field of this type as measured in the field on an experimental tunnel (El-Nahhas 1980). Figure 5.6 provides contours of the vertical and horizontal displacements. The tunnel can be considered as partially lined. The

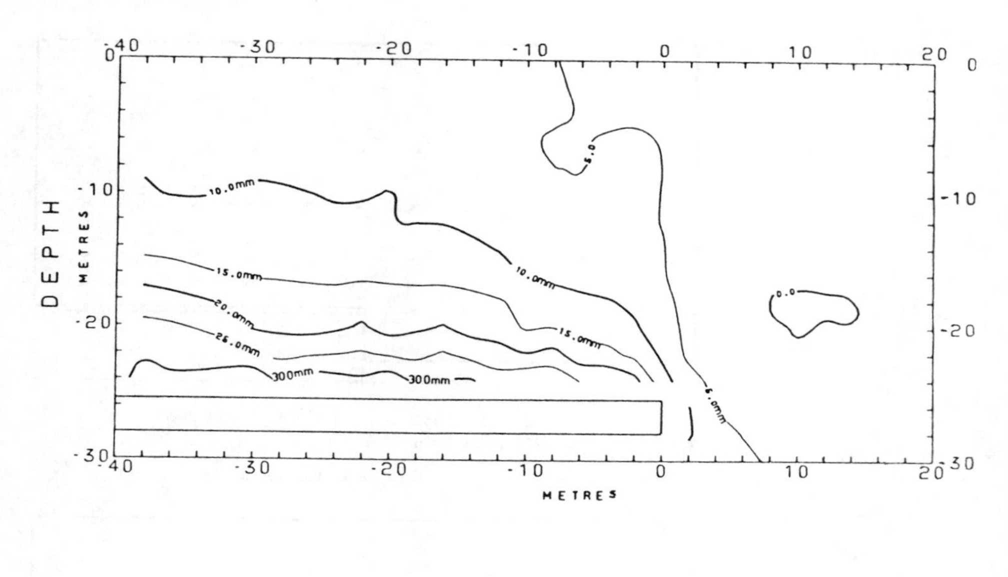

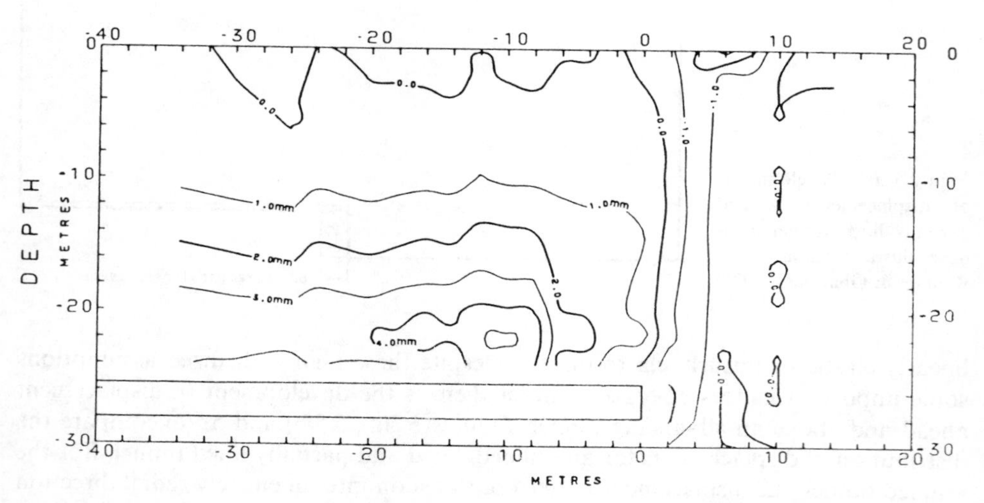

Figure 5.6. Contours of vertical and horizontal displacements around experimental tunnel (after El-Nahhas 1980)

longitudinal (horizontal) component of the final displacement is in the order of 10% of the vertical component. Similar comparison for longitudinal (horizontal) and vertical straining in Figure 5.7 supports this conclusion even more clearly. The horizontal strains disappear completely about four tunnel diameters behind the face.

The problem with the type of analysis performed by Ranken & Ghaboussi (1975) is that it simulates only an axisymmetric loading ($K_0 = 1$). Hanafy & Emery (1980, 1982) developed a method of modelling non-symmetric axial loading within the relative simplicity of a two-dimensional (axisymmetric) approach. The axisymmetric approach is adopted to consider the two most common loading conditions due to axisymmetric

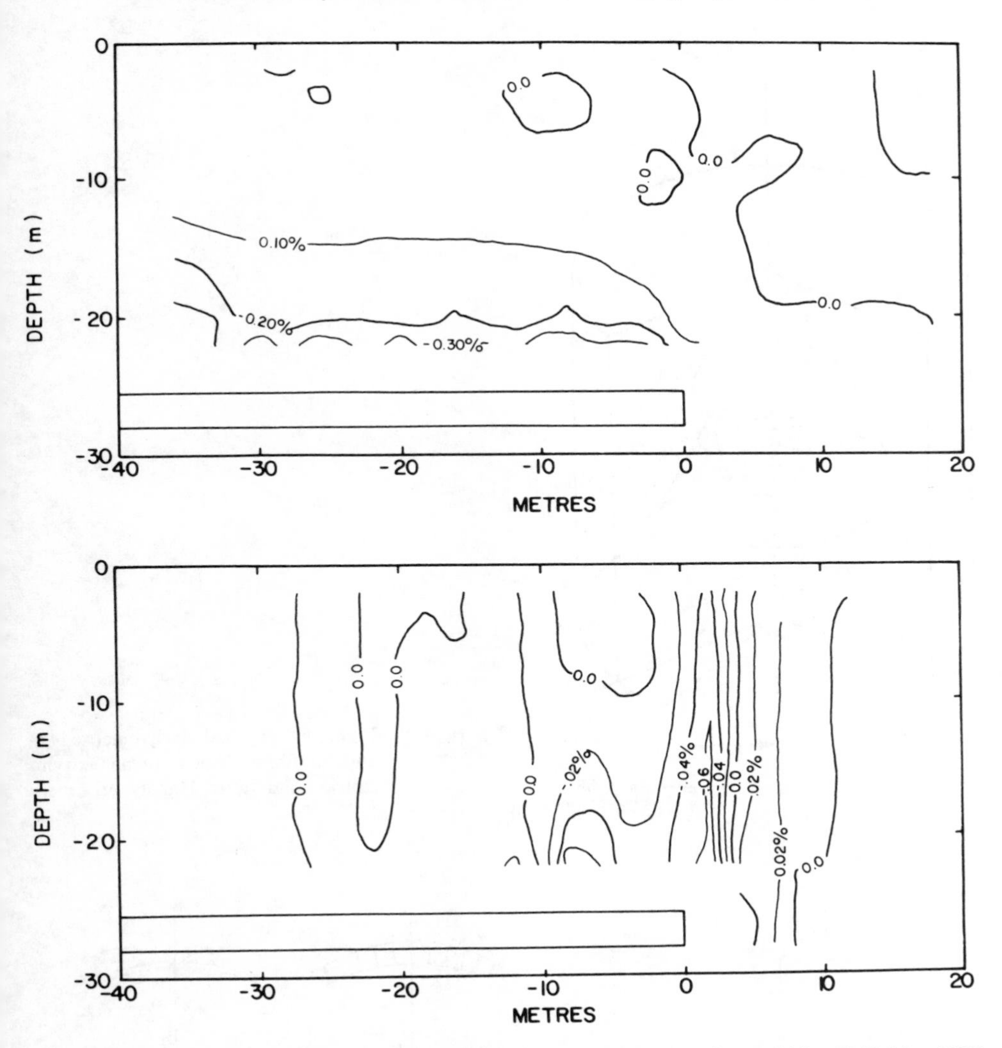

Figure 5.7. Contours of vertical and longitudinal strains around experimental tunnel (after El-Nahhas 1980)

and non-axisymmetric initial in-situ stresses. The axisymmetric condition is valid for situations with uniform radial loading, requiring no change of stress field across the tunnel and $K_0 = 1$. The finite element formulation is then similar to that for plane strain analysis. For cases with $K_0 \neq 1$ the radial loading is non-uniform. Hanafy & Emery (1980, 1982) solve this problem by separating the loading variables by expanding them into Fourier series with respect to the angular direction θ. The approach is thus identical to the analysis of solids of revolution subjected to non-axisymmetric loadings, as described by Zienkiewicz (1977). A typical, axisymmetric, triangular, finite element is shown in Figure 5.8. The most general case of non-axisymmetric loading can be resolved (expanded) into two components: symmetric and antisymmetric modes, as shown in Figure 5.9. Since most tunnels involve a plane of symmetry, the necessity of

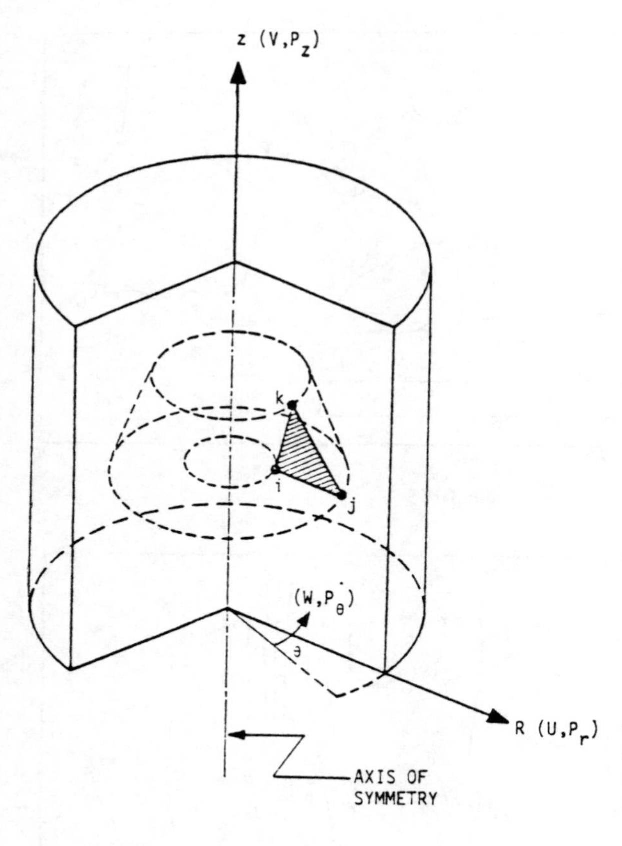

Figure 5.8. Typical axisymmetric, triangular finite element of an axisymmetric solid (after Hanafy & Emery 1982)

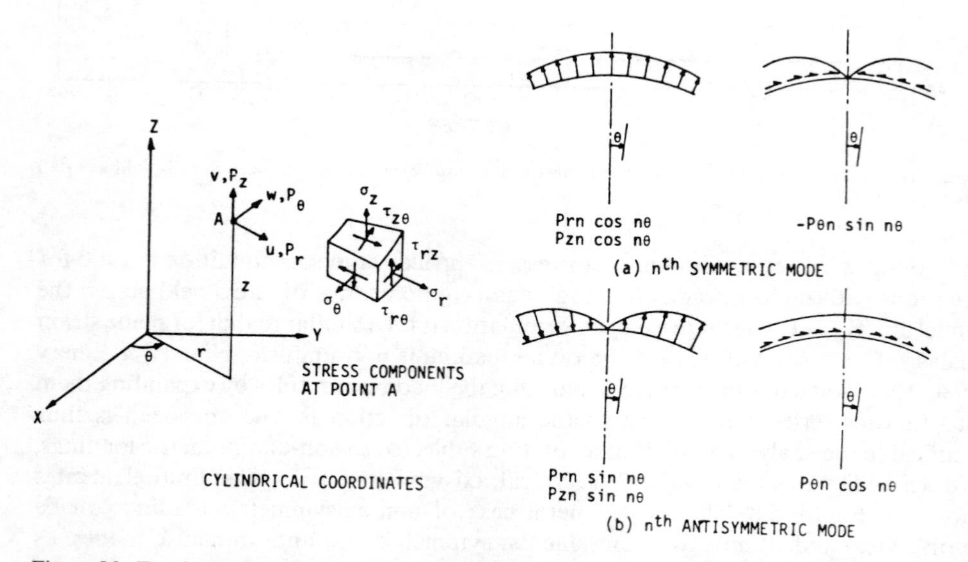

Figure 5.9. Fourier series representation of nonsymmetric loading (after Hanafy & Emery 1980)

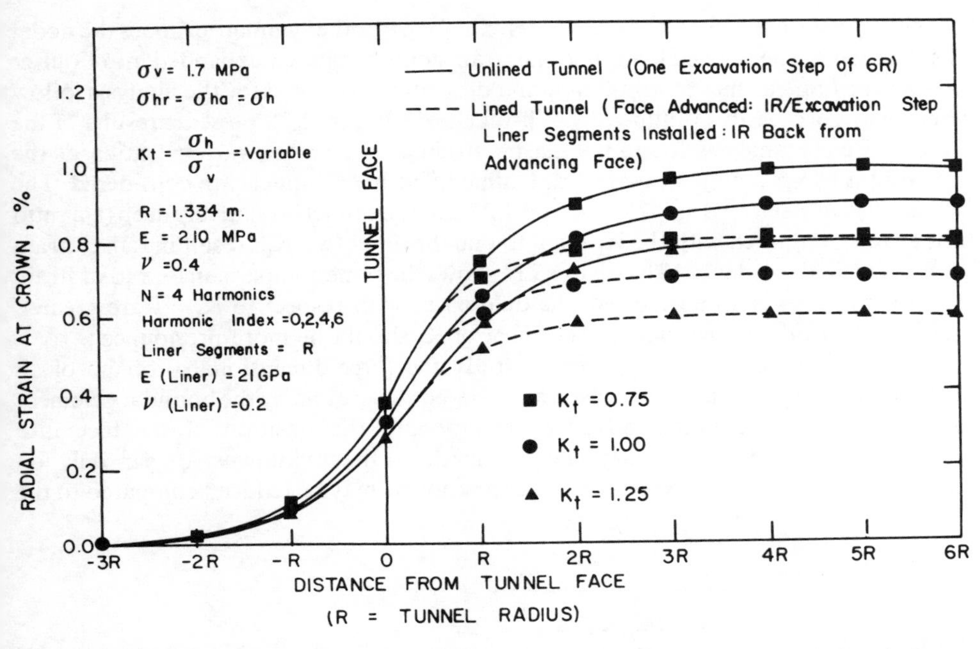

Figure 5.10. Radial elastic strain at crown for three different ratios of horizontal to vertical strain (K_t) at unlined and lined tunnel (after Hanafy & Emery 1982)

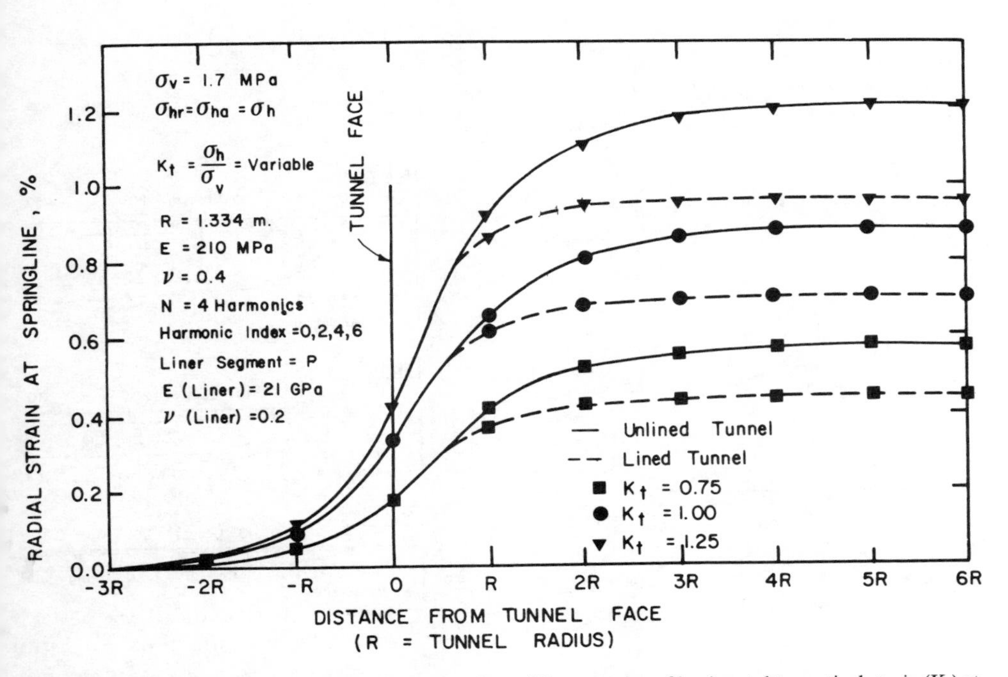

Figure 5.11. Radial elastic strain at springline for three different ratios of horizontal to vertical strain (K_t) at unlined and lined tunnel (after Hanafy & Emery 1982)

considering the antisymmetric modes is eliminated. For the symmetric modes the nodal displacements, the applied loads and the strain components are expanded into Fourier series. The finite element formulation and derivations adopted for the analysis follow the general procedures outlined by Wilson (1965). Figure 5.10 presents results of the Hanafy-Emery analysis for radial elastic strain at crown for various ratios of the horizontal to vertical in-situ stresses. Unlined and lined tunnels are considered. The values of K_t are selected as 0.75, 1.00 and 1.25 and the tunnel is relatively deep (z/a ratio about 70). Compared with the K_t equal to one the other two K_t's result in radial strain differences in the order of 12%. Figure 5.11 shows the same comparison for radial strain at springline. As can be expected, the differences with respect to $K_t = 1$ are greater, about 15%. For shallow tunnels, the effect of K_t should be more pronounced.

Ito & Hisatake (1982) proposed to study the three-dimensional problem of an advancing shallow tunnel in an elastic and non-elastic ground by the boundary element method, taking the tunnel advance velocity and the position of the face into consideration. In the boundary element method, the unknowns appear only on boundaries of a domain, so the number of unknowns may be reduced compared to the

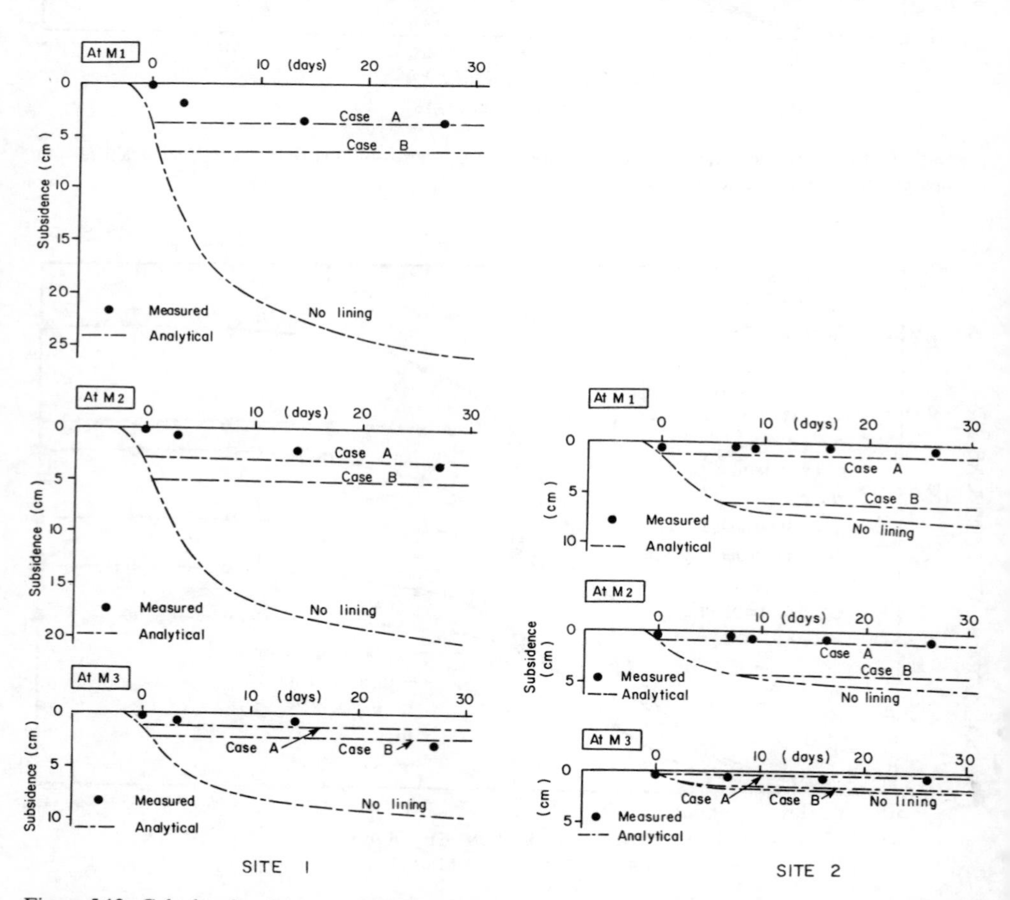

Figure 5.12. Calculated and measured time-dependent surface subsidence for two tunnels in Japan (after Ito & Hisatake 1982)

three-dimensional finite element method. This condition is well suited to tunnels, where the most significant unknown – the surface subsidence – appears on the boundary. Their method has been illustrated and verified on two sites where subsidence measurements were carried out. The tunnels were 6.9 m in diameter, with depths to centre 14.5 m and 15.5 m, driven in clay by shield with compressed air. The shield clearance was grouted but as the grouting was impossible to be grasped quantitatively the effects of the grouting were treated by two extreme cases:

– Case A: grouting is done instantaneously, no displacement of tunnel boundary allowed after passage of the face.

– Case B: grouting is neglected, the tunnel boundary is free to displace the full amount of the clearance.

Figure 5.12 shows calculated and measured time-dependent surface subsidences for points located at the tunnel axis (M_1) and 6.5 m (M_2) or 13.0 m (M_3) apart from the axis. The calculated results are shown for Cases A (immediate grouting of clearance), B (no grouting of clearance) and no lining (for comparative purposes). The grouting at these

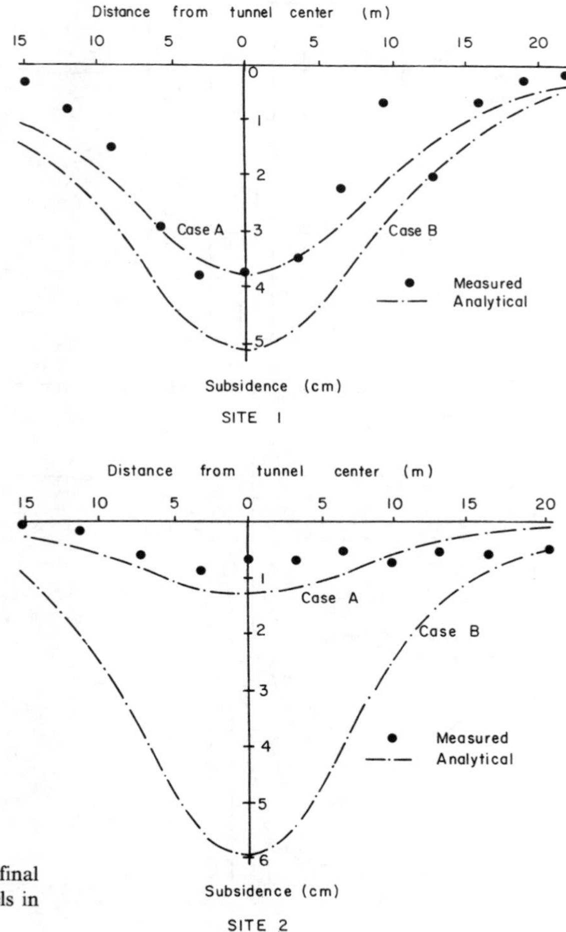

Figure 5.13. Calculated and measured final surface settlement troughs for two tunnels in Japan (after Ito & Hisatake 1982)

sites is known to be done very efficiently, and the measured values compare well with Case A. Similar conclusions can be drawn with respect to the surface settlement trough, as shown in Figure 5.13. The disadvantage of the boundary element method is that it does not deal with displacement inside the ground nor with corresponding changes in stresses. Thus the method cannot be used in connection with a ground reaction curve.

The problem with the category (b) approaches (analysis perpendicular to tunnel axis) is that it is difficult to model the construction history. An interesting method of overcoming this problem has been proposed by Ohnishi et al. (1982). They use the

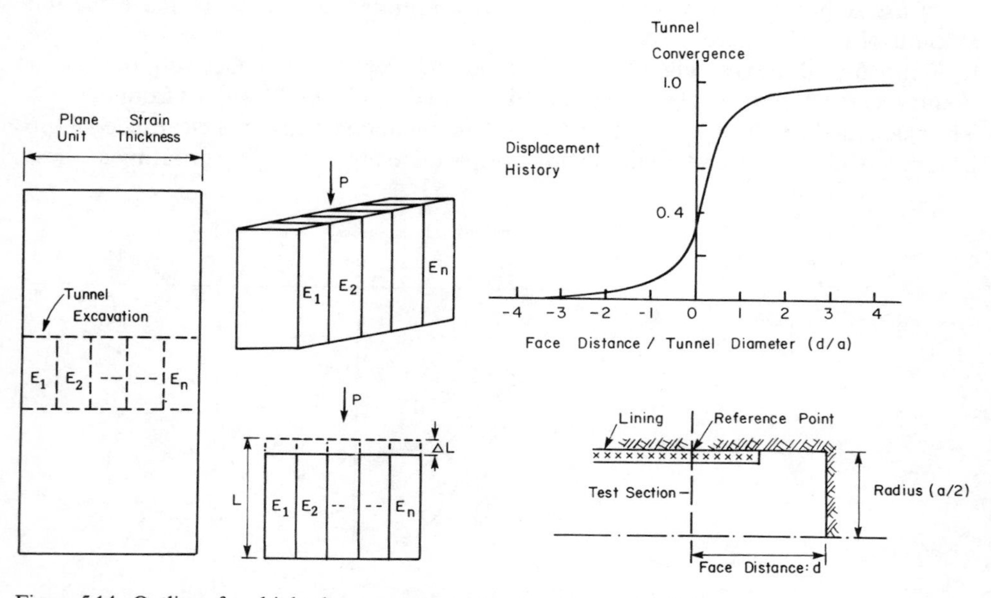

Figure 5.14. Outline of multiple element method (after Ohnishi et al. 1982)

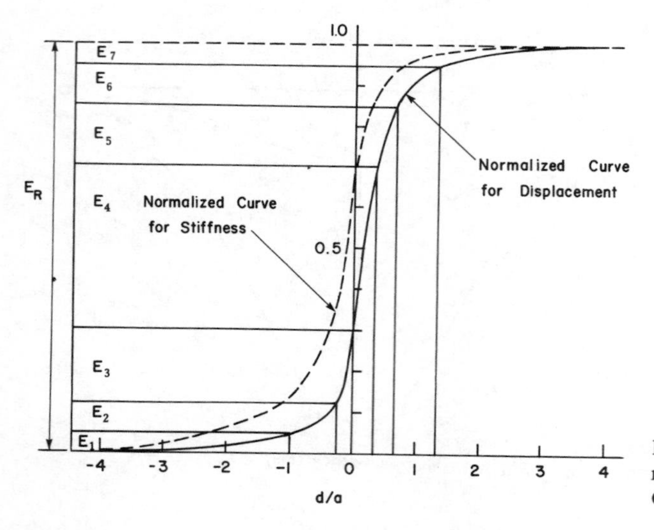

Figure 5.15. Normalized curve for multiple element method (after Ohnishi et al. 1982)

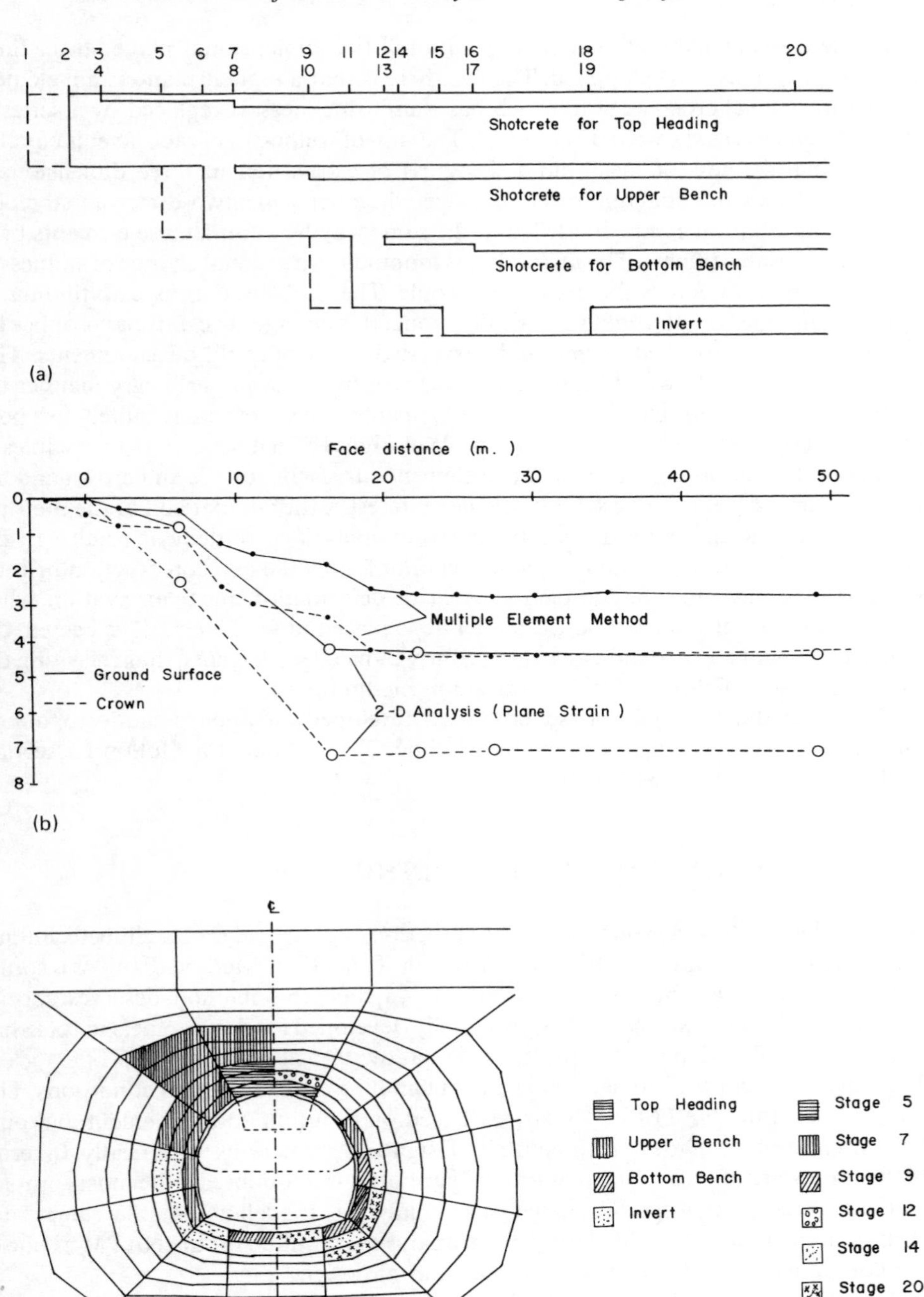

Figure 5.16. Example of multiple element method analysis for a shotcrete tunnel (after Ohnishi et al. 1982); (a) analytical procedure for construction sequence; (b) computed result; and (c) development of yielded zones at each construction stage

'multiple element method', where conventional two-dimensional plane strain finite element programs can be applied. The idea here is that a regular plane strain element within the tunnel cross-section, which has a unit thickness, is replaced by a series of layered elements as shown in Figure 5.14. The sum of stiffnesses of each layer is equal to the original stiffness of the ground. Progress of excavation in three dimensions is simulated by a sequential removal of the layered elements in a two-dimensional mode. Support installation can be modelled in the same way by adding these elements back with appropriate stiffness. The method thus amounts to a gradual change of stiffness at different locations across the excavated profile. The problem then is with finding an appropriate rule which would realistically model this change. The authors propose for this purpose the so-called normalized curve, deduced from field measurements. This normalized curve, shown in Figure 5.15, describes in somewhat arbitrary manner the stiffness fractions for individual layered elements. The method is suited for both shielded and shotcreted tunnels. An example in Figure 5.16 illustrates the principles of the method for a shotcrete tunnel, the settlement curves for surface and crown and the development of yielded zones. Of particular interest is the comparison of the multiple element method and conventional plane strain analysis in settlement results. In the conventional plane strain analysis, which cannot follow the real construction history, shotcrete is placed after the completion of elastic deformation due to excavation, while in the multiple element method the shotcrete is placed immediately. As expected, the multiple element method indicates significantly reduced settlements, thus reflecting the three-dimensional effects of the construction method used.

True three-dimensional analyses have been developed and applied mainly to tunnels built by the New Austrian Tunnelling Method. Their review thus follows a section dealing with this particular method.

5.4 NEW AUSTRIAN TUNNELLING METHOD

Analytical modelling of ground losses as a function of a particular tunnelling technique is especially important with the New Austrian Tunnelling Method (NATM), sometimes referred to as the Shotcrete Method. As such this method deserves special attention in this review. The method, originally developed for deep tunnels in rock, has been successfully adapted for soft ground shallow tunnels. The method offers great flexibility in selection of tunnel profile and in handling difficult ground conditions. The application of the NATM began largely based on intuition and judgement and only later attempts were made to rationalize its design and predictions analytically. In terms of the procedure of excavation and ground support it is undoubtedly the most complex method, presenting a special challenge to analytical modelling. At the same time, perhaps more than for shielded tunnels, an analysis is required for the NATM as one of the components for the observational method of design.

Initially a need was felt to predict the development of loading on the shotcrete shell lining of an NATM tunnel. The problem has been dealt with by Swoboda (1979). Since only lining loads are of concern, the method uses a known displacement function to determine the load function, approximated by means of a step function as illustrated in Figure 5.17. This means that the interaction forces [F] acting in the shell are assumed to develop proportionally to the measured deformations. Since the three-dimensional

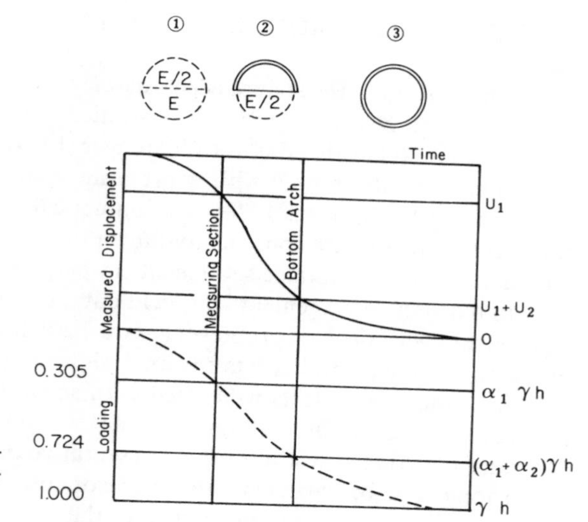

Figure 5.17. Relationship between measured displacement and load function (after Swoboda 1979)

effect of the face and of the sequential condition is already reflected in the displacement function, the analysis of the lining loads and the lining itself can be done using a two-dimensional model.

Displacement field caused by NATM has been studied by Wanninger (1979) for shallow tunnels in Frankfurt, where this method has been pioneered in soft ground urban environment. The availability of high quality field data makes this study particularly valuable. A two-dimensional model with nonlinear stress-strain description is used to simulate the excavation technique. Figure 5.18 illustrates this feature including the sequence of placing of shotcrete. The results of this analysis, when compared with field measurements, tend to overestimate the actual settlement by more than 100%. The question then arises, whether this overestimate is due to the neglect of the true three-dimensional effects in this analysis.

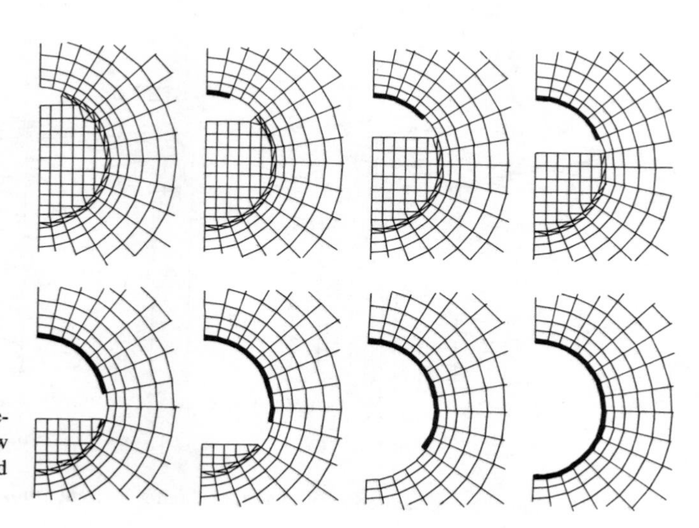

Figure 5.18. Simulation of sequence of excavation for New Austrian Tunnelling Method (after Wanninger 1979)

5.5 THREE-DIMENSIONAL ANALYSES

An attempt to study the spatial effects associated with the NATM has been presented by Katzenbach & Breth (1981). They studied the same NATM tunnels in Frankfurt with the same soil testing data as Wanninger (1979). They carried out, however, a fully three-dimensional analysis with a program system developed at the University in Darmstadt (Czapla et al. 1978). The numerical simulation of the complex NATM construction procedure consisted of fifteen steps, with some of them shown in Figure 5.19. The calculated surface settlements in perpendicular and longitudinal directions were in reasonable agreement with field data, not only in their magnitude but also in relationship with the face progress (Figure 5.20). It is obvious that apart from the true modelling of the spatial effects the analysis employed a realistic stress-strain model. Triaxial compression tests were used with separate relationships for moduli during primary loading and unloading-reloading. While the primary loading moduli from a conventional triaxial test are not representative of stress paths around a tunnel and thus should not be expected to give reasonable results, they probably were not so important in this analysis. Since most of the soil region around a tunnel is unloaded

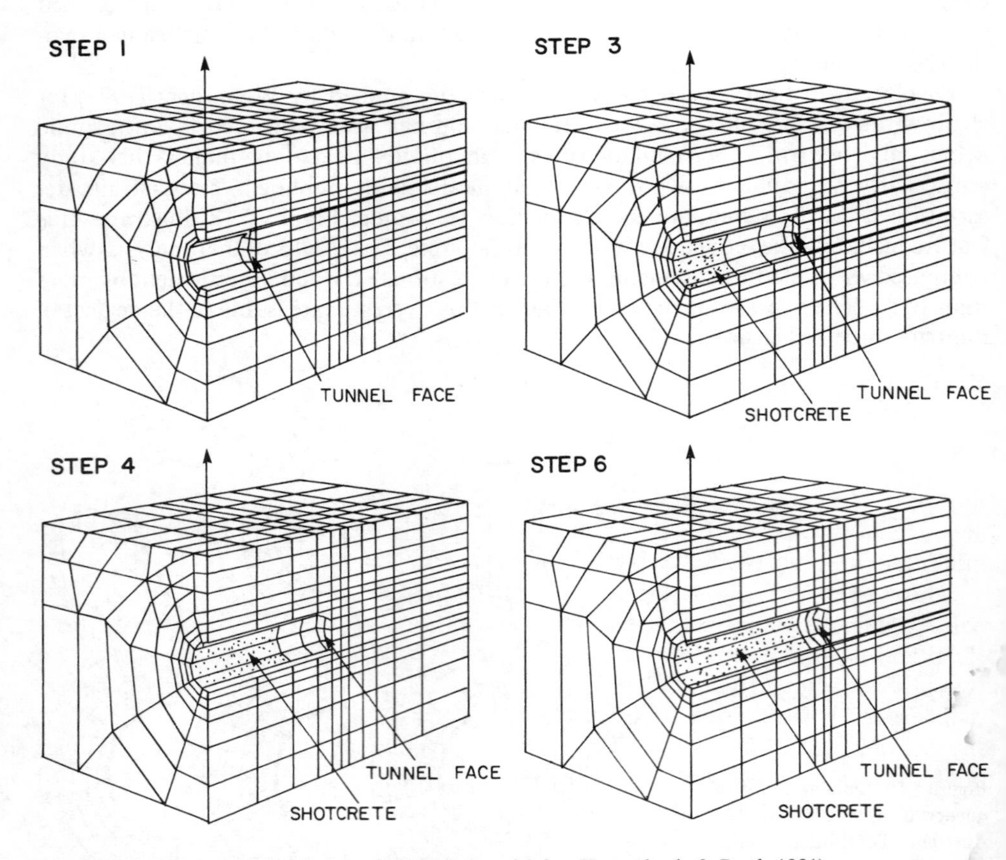

Figure 5.19. Numerical simulation of NATM tunnel (after Katzenbach & Breth 1981)

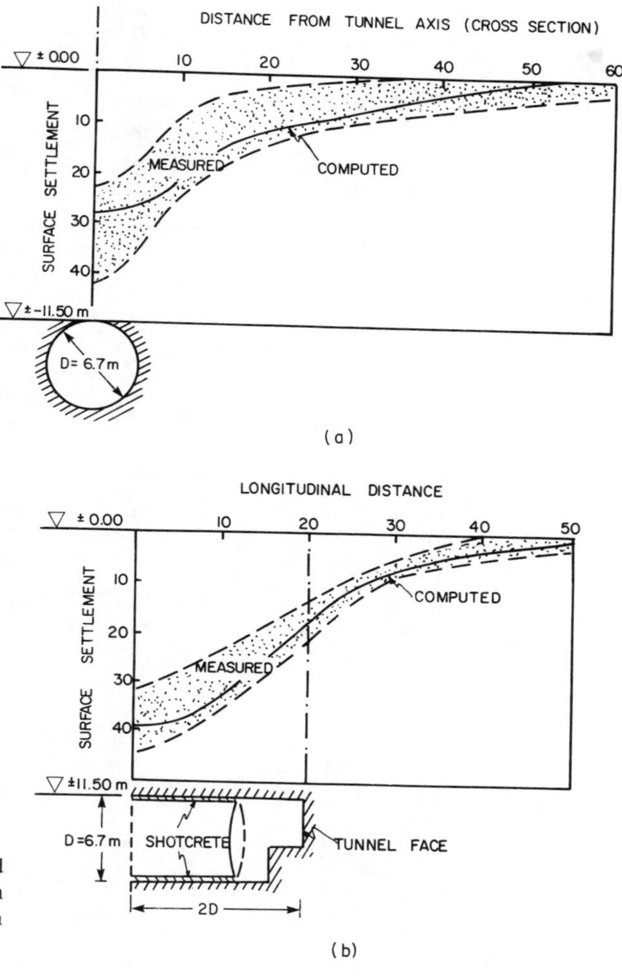

Figure 5.20. Measured and computed settlements for NATM tunnel in Frankfurt (after Katzenbach & Breth 1981)

during excavation, the unloading-reloading moduli played a more important role. These moduli represent the elastic portion of soil behaviour and as such are stress path independent – thus the triaxial test can produce reliable soil representation.

Another comparison favourable for this analysis can be made for strain fields calculated in the longitudinal section of the tunnel. Contours of computed vertical and longitudinal strains are shown in Figure 5.21. Figure 5.7 showed contours of the same two types of strains, derived from an extensive field instrumentation program at an experimental tunnel in Edmonton (El-Nahhas 1980). The Edmonton tunnel was driven by a shield in a stiff, overconsolidated clay (glacial till). Despite the difference in construction methods, the results are qualitatively very similar and lend further credibility to this analysis.

One of the first successful applications of a truly three-dimensional analysis to an NATM shallow tunnel was carried out by Wittke & Gell (1980). The analysis was

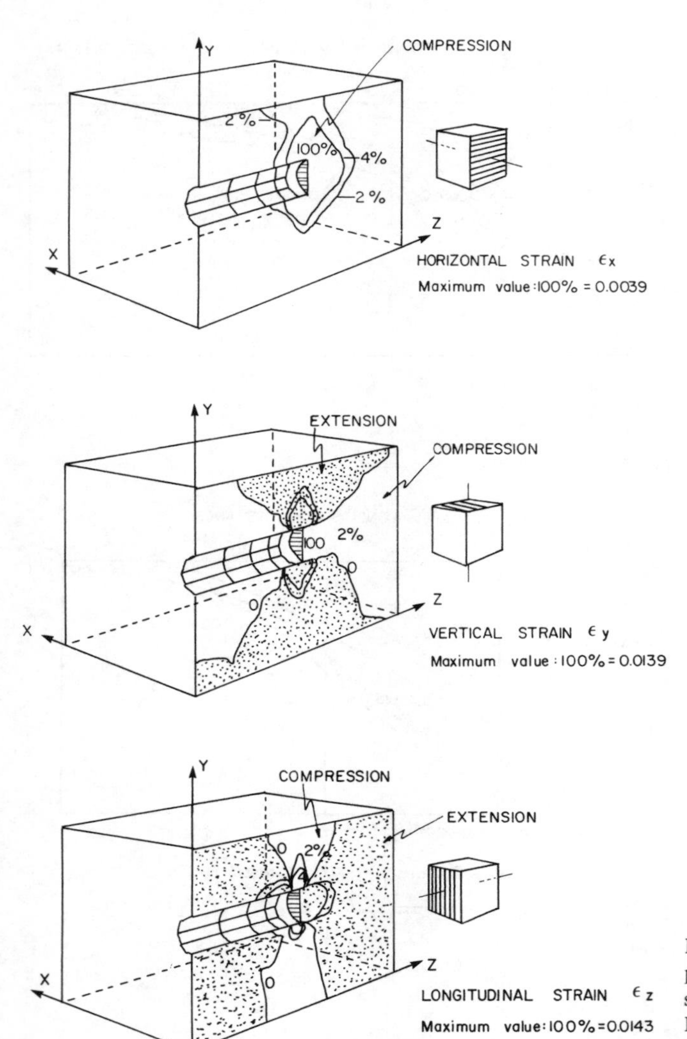

Figure 5.21. Contours of computed vertical and longitudinal strains for NATM tunnel in Frankfurt (after Katzenbach & Breth 1981)

performed as part of an observational method of design, along with field observations for a tunnel in extremely difficult environment. A linearly elastic/perfectly plastic stress-strain law was used. The three-dimensional finite element model is shown in Figure 5.22. The analysis studied the effect of different modes of excavation on ground control and on surface settlement. The three types of face excavation (bench and heading, inclined face and vertical face) are compared in terms of their crown and surface settlement and shear strength mobilization in Figure 5.23. The tunnel was eventually excavated with a face inclination of 65°. The soil conditions actually found on the site differed from those assumed in the analysis. A new analysis, taking into account these changes, but using the same soil properties, resulted in encouraging agreement between field data and computations.

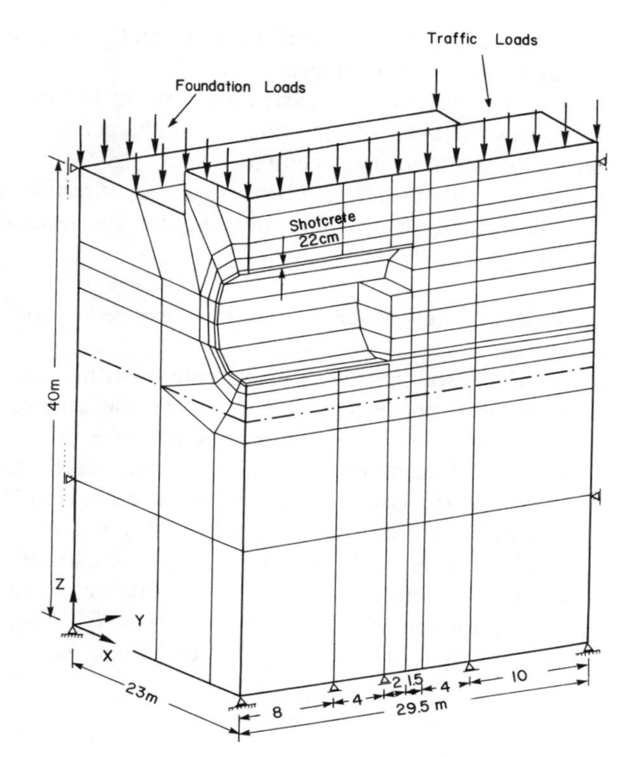

Figure 5.22. Three-dimensional finite element mesh for NATM tunnel in Bochum (after Wittke & Gell 1980)

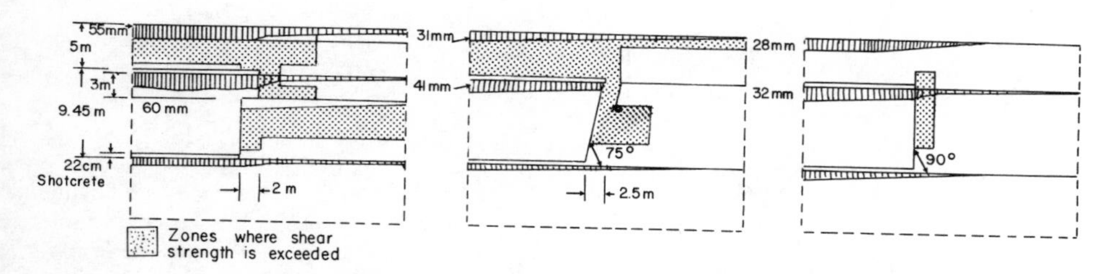

Figure 5.23. Comparison of three modes of NATM tunnel excavation in terms of settlement and shear strength mobilization (after Wittke & Gell 1980)

5.6 SOIL REPRESENTATION

The stress changes occurring in the soil around an excavated tunnel are mostly of an unloading type along a variety of stress paths. The success of relating these stress changes to realistic displacements depends on the ability of a numerical model to incorporate the complex relationships which exist between stresses and strains in soils, namely:

– increments in deformation are not linearly proportional to increments in load (stress-strain nonlinearity);

– magnitude of deformation depends on the stress path followed during the history of loading (stress path dependency);

– deformation or their parts do not occur immediately after load application, but only after certain time periods – either dependent (consolidation) or independent (creep) of effective stress change;

– same loads in different directions result in different deformations (anisotropy);

– shear loads cause not only angular displacements but also volume changes (dilatancy).

Analytical models which deal with the above features can be divided into three basic categories:

1. *Pseudo-elastic models* which remain within the framework of elasticity and incorporate one or several of the nonelastic features in a phenomenological way (e.g. nonlinearity modelled as piecewise incremental linearity, time dependency as series of repeated calculations at different time steps, etc.);
2. *Elasto-plastic models* which are based on some of the comprehensive constitutive relationships proposed for soils (e.g. Cam Clay model, Lade's model, etc.) and in which features such as nonlinearity (strain hardening or softening), stress path dependency and plastic phenomena are embraced in one compact model.
3. *Stress path dependent pseudo-elastic models* which are a compromise between the previous two. To account for the stress path dependency of soils (which is

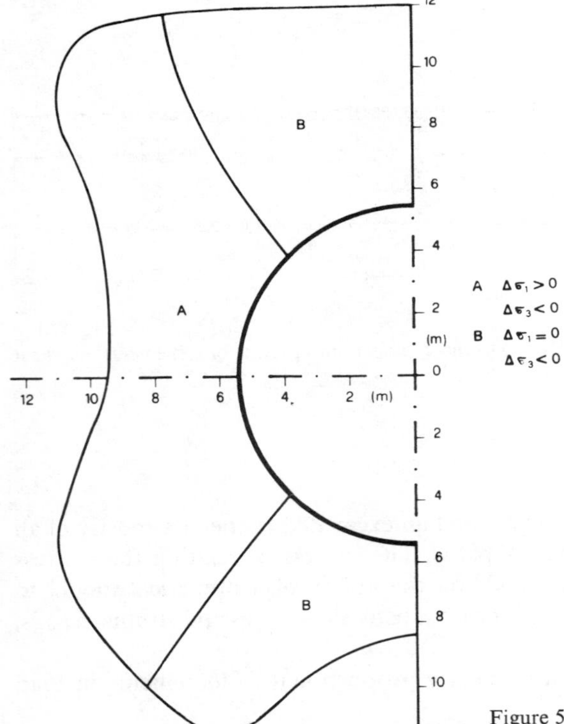

Figure 5.24. Main stress path zones around tunnel (after Cappellari & Ottaviani 1982)

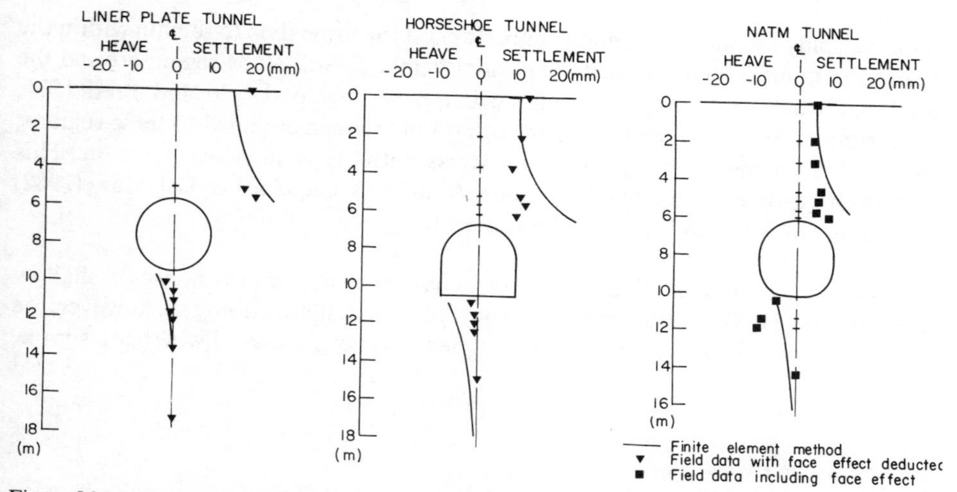

Figure 5.25. Calculated and observed vertical displacement profile at experimental tunnels in Sao Paulo (after Negro & Eisenstein 1981)

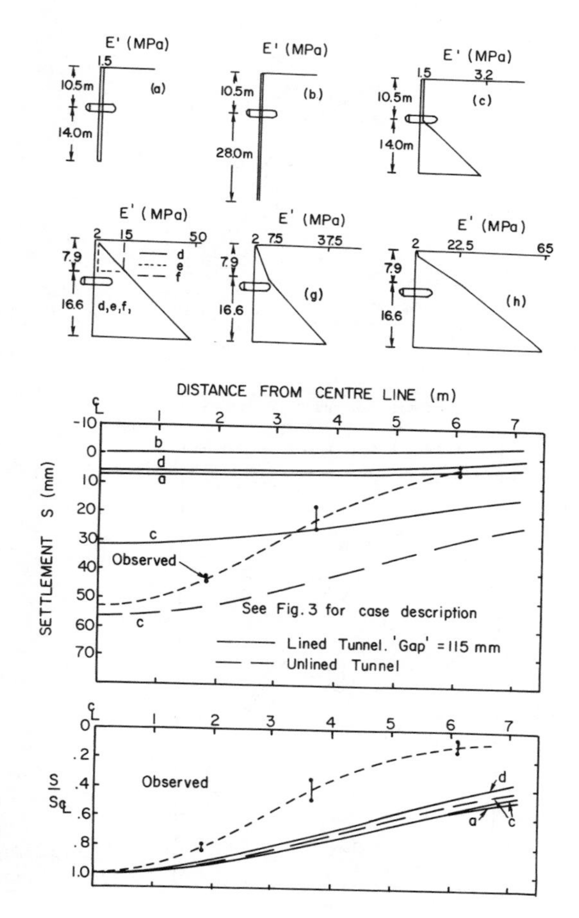

Figure 5.26. Comparison of settlement profiles for several variations of modulus profile with depth, Thunder Bay Tunnel (after Rowe et al. 1981)

automatically satisfied with models sub 2) and at the same time to remain within the relative computational simplicity of elasticity (models sub 1) the region around the analyzed structure is divided into zones with typical stress paths (Stroh & Breth 1976, Eisenstein & Medeiros 1983). Nonlinear moduli are then assigned to these regions, obtained from tests carried out along stress paths typical for each region. This approach has been suggested for shallow tunnels by Capellari & Ottaviani (1982) who have shown such a zoning (Figure 5.24).

Finite element programs of all three categories have been applied to analysis of shallow tunnels and some of these attempts will be now reviewed. Before doing so, however, the question must be asked whether all this complexity is necessary and whether a simple,

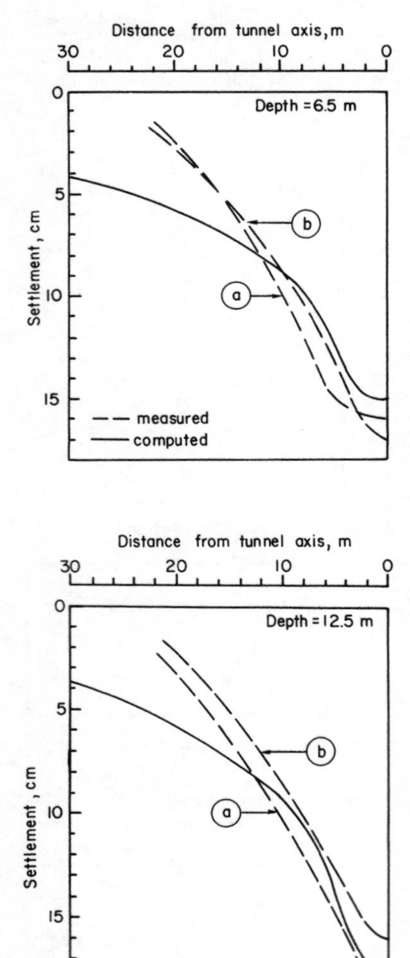

Figure 5.27. Comparison of measured and computed settlements at tunnel in Mexico City (after Romo & Resendiz 1982)

perhaps a linearly elastic model could provide an answer sufficient for engineering purposes. As opposed to other conventional geotechnical structures (foundations, embankments and slopes, retaining walls) shallow tunnels induce a stress field which almost always brings the soil to shear failure. This happens in the most significant zone – at the tunnel's wall. Ward & Pender (1981) claim that there are no elastic tunnels. The writer's own experience supporting this claim can be illustrated by an example of a back-analysis of three experimental tunnels (Negro & Eisenstein 1981). Figure 5.25 shows that by a judicial selection of elastic parameters one can match either the surface settlement or the crown displacement but not both and certainly not the whole displacement field. Rowe et al. (1981) back-analyzed a sewer tunnel in soft clay in Thunder Bay in Canada using a linearly elastic/perfectly plastic model with six different variations of the elastic modulus with depth. Figure 5.26 indicates that while they were able to reproduce the maximum settlement for one of the selected modulus profile, they could not follow the settlement trough.

Using a nonlinear stress-strain relationship Resendiz & Romo (1981) developed a method of analysis based on finite elements for which the results are organized and generalized with the aid of dimensional analysis and a theorem of similitude. Analytical expressions were developed for calculating ground settlements induced by stress release at the tunnel face, by inward movement due to wall closure and due to consolidation of soil around the tunnel. The finite element solutions are expressed in formulae for use without a computer. The method was verified at an experimental tunnel section in N.C. clay in Mexico City. The total settlements for the 6.28 m diameter tunnel at 21 m depth are compared in Figure 5.27. While the maximum settlement

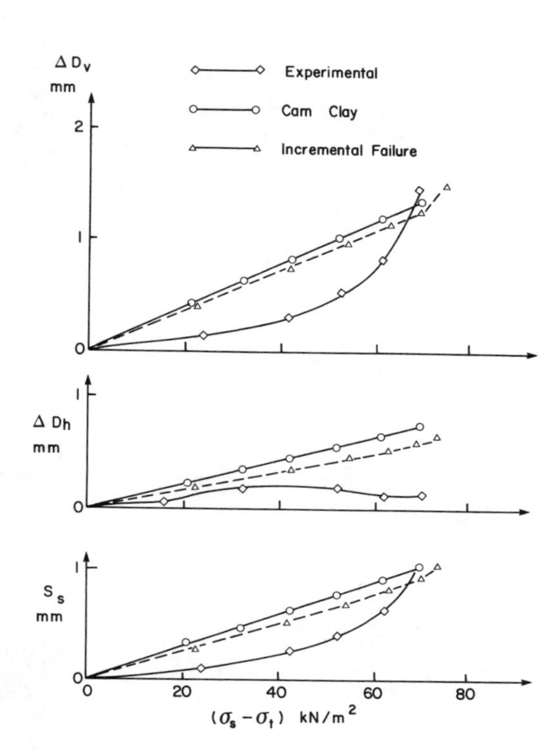

Figure 5.28. Ground reaction curves and surface settlement, Cambridge static model (after Orr et al. 1978)

values for the measured and calculated settlement agree, the shape of the trough is not completely matched.

Orr et al. (1978) presented a detailed comparison between the observed displacements around two model tunnels and finite element computations based on two nonlinear models of soil behaviour. One of the models was of the incremental pseudo-elastic type while the other one was an elasto-plastic model of the Cam clay type. The computed displacements and strains were compared with those observed on a laboratory model of a tunnel, loaded up to failure. Figure 5.28 presents these comparisons for two ground reaction curves (crown and springline) and for surface settlement. It is of interest to note that this has been so far the only example of an analytical prediction of a complete ground reaction curve, as defined earlier in this chapter. Figure 5.29 compares the surface settlement profiles at two different stages of unloading of the model tunnel. From the comparisons in Figures 5.28 and 5.29 it becomes obvious that the two soil representation models do not produce very different results, but their agreement with the physical observations is not encouraging.

Mair et al. (1981) continued comparisons between calculations and models of a tunnel, only this time the loading on the tunnel has been produced by increasing the soil gravity in a centrifuge. The analytical model used was again of an elasto-plastic type (Cam clay). The observed surface settlements for several centrifuge tests are plotted in Figure 5.30. Unfortunately the authors have not included the calculated data but they claim that their analysis poorly predicted the shape of the surface settlement profile, the predicted profile being considerably wider than observed.

Kawamoto & Okuzono (1977) used a two-dimensional finite element model to

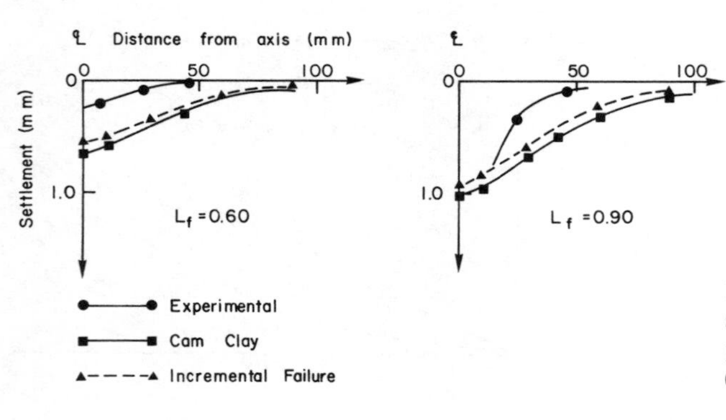

Figure 5.29. Comparison of measured and calculated settlement profiles, Cambridge static model (after Orr et al. 1978)

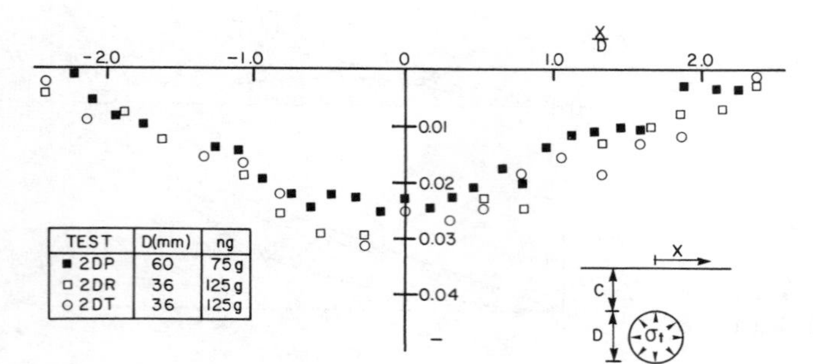

Figure 5.30. Observed settlement profiles above model tunnels (after Mair et al. 1981)

analyze surface settlements above a shallow tunnel at Nagoya subway. The tunnel was excavated in sand and gravel by a shield, with measurements of settlement at the surface. Several types of analysis were employed: linearly elastic, linearly elastic with added effect of local yielding and nonlinear analysis. Results of these calculations are compared with field data in Figure 5.31. Linearly elastic analysis failed to predict the magnitude and shape of the settlement trough, while the nonlinear analysis showed a reasonable agreement. If only the maximum settlement is of interest, the linearly elastic model with superimposed local yielding can also provide an acceptable answer.

Ghaboussi et al. (1978) studied the problem of subsidence above a shallow circular tunnel, comparing the results of a model test with a two-dimensional finite element analysis. The test was carried out in sand and the effect of tunnelling was represented by

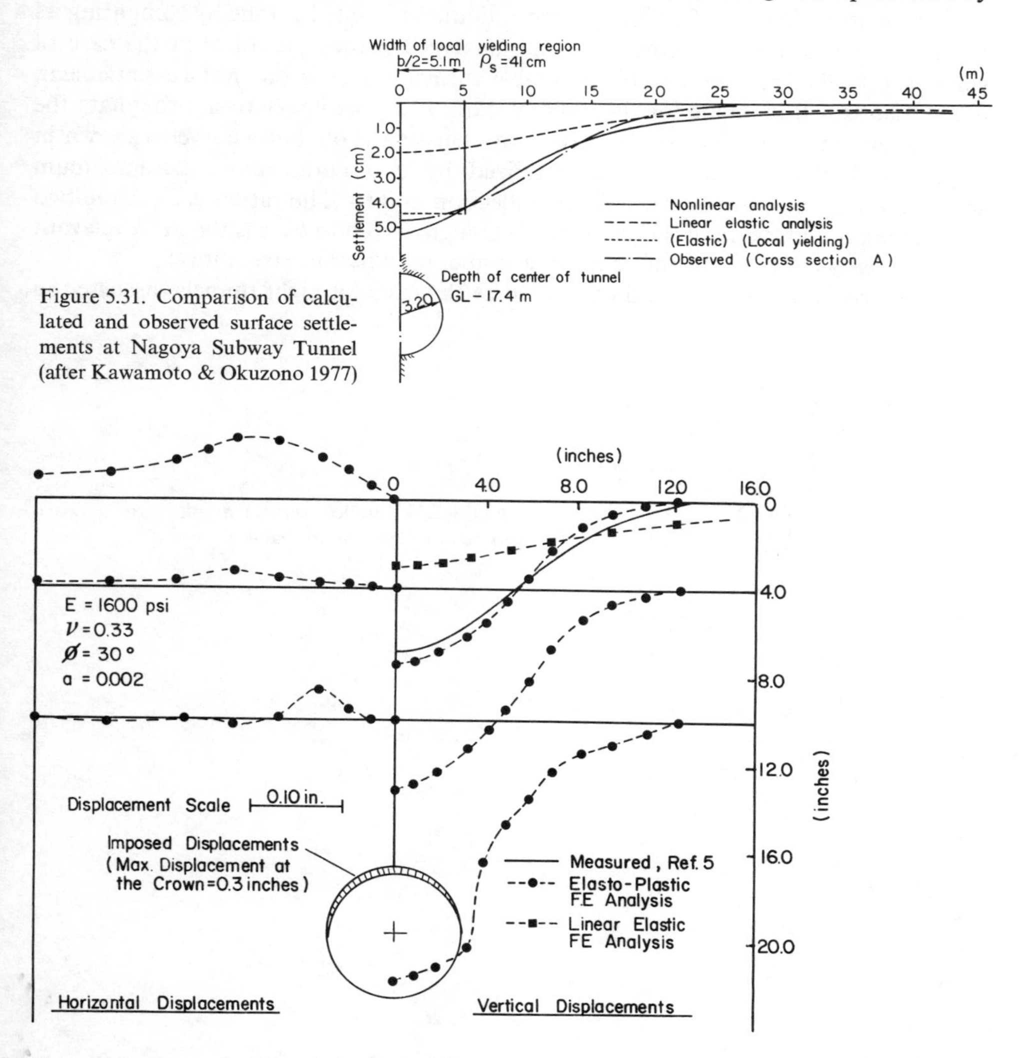

Figure 5.31. Comparison of calculated and observed surface settlements at Nagoya Subway Tunnel (after Kawamoto & Okuzono 1977)

Figure 5.32. Comparison of calculated and observed settlements at model tunnel (after Ghaboussi et al. 1978)

imposed displacements at the tunnel wall. This analytical technique simulates closely the conditions at a shield where soil is allowed to displace in an amount equal to the overcut. Two stress-strain models were used – a linearly elastic one (type 1) and an elasto-plastic model based on the concept of critical state (type 2). The results are presented in Figure 5.32. As with the previous example the linearly elastic model underpredicts both the maximum settlement and the slope of the settlement trough. The elasto-plastic model is much closer to the observed values.

5.7 CONCLUSIONS

The ultimate evaluation of an analytical technique can only be done by comparing its results with in-situ measurements on the analyzed prototype. Since in the case of shallow tunnels, the most readily available measurement is the surface settlement, comparisons will be made for this type of data. It is customary to approximate the surface settlement trough over tunnels by a Gaussian probability curve as shown in Figure 5.33. Such a curve is characterized by two parameters – the maximum settlement S_{max} and the distance to the inflection point i. The ratio between the two parameters is a measure of the slope of the trough, which in turn is the most relevant factor in assessment of the influence of a tunnel on adjacent structures.

Comparison of observed and calculated settlements for eight tunnels discussed in

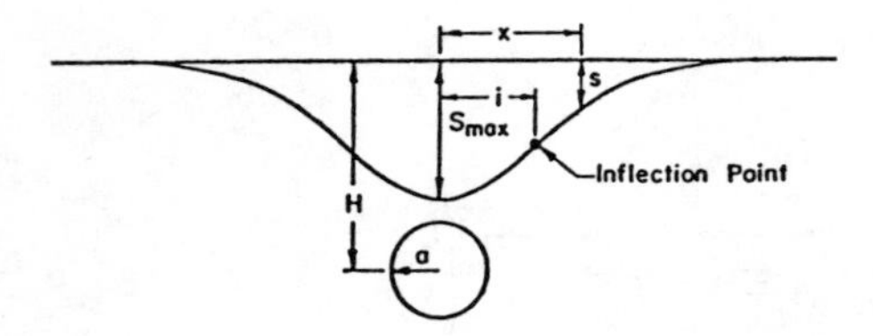

Figure 5.33. Surface settlement through approximated by Gaussian probability curve

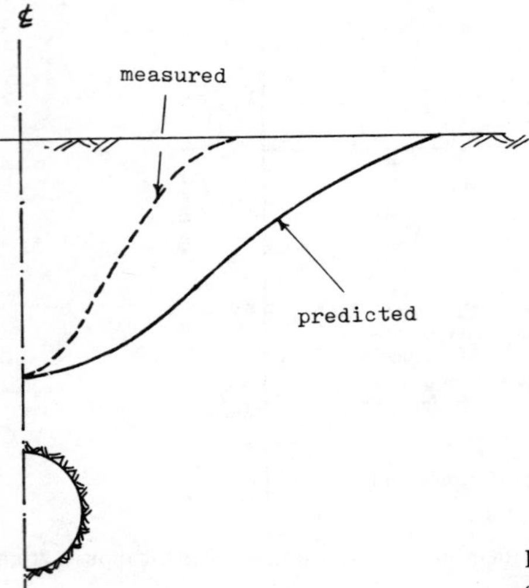

Figure 5.34. Tendency between measured and predicted surface settlements for tunnels in cohesive soils

Table 5.1. Comparison of observed and calculated settlement

Tunnel	Author(s) of analysis	Type of soil	Tunnelling method	Dimensions of analysis	Type of stress-strain model	Comparison of: Max. settlement	Comparison of: Slope of through
Nagoya Subway, Japan	Kawamoto & Okuzono (1977)	Sand and gravel	Shield	2D	Type 1 (plus yielding) (a) linear (b) nonlinear	$S_0 > S_c$ $S_0 \simeq S_c$	$I_0 < I_c$ $I_0 \simeq I_c$
University of Illinois Model, USA	Ghaboussi et al. (1978)	Sand	Induced displacement	2D	Type 1 (linear) Type 2	$S_0 > S_c$ $S_0 \simeq S_c$	$i_0 < i_c$ $i_0 \simeq I_c$
Cambridge Univ. static model, UK	Orr et al. (1978)	N.C. clay	External loading	2D	Type 1 (nonlinear) Type 2	$S_0 < S_c$ $S_0 < S_c$	$i_0 < i_c$ $i_0 < i_c$
Bochum Subway, West Germany	Wittke & Gell (1980)	Sand and gravel	NATM	3D	Type 1 (linear with yielding)	$S_0 \simeq S_c$	No data
Cambridge Univ. centrifuge model, UK	Mair et al. (1981)	N.C. clay	Gravitational loading	2D	Type 2	No data	$i_0 < i_c$
Frankfurt Subway, West Germany	Katzenbach & Breth (1981)	O.C. clay	NATM	3D	Type 3	$S_0 \simeq S_c$	$i_0 < i_c$
Thunder Bay Sewer, Canada	Rowe et al. (1981)	N.C. clay	Shield	2D	Type 1 (linear, modulus varies with depth)	$S_0 \simeq S_c$	$i_0 < i_c$
Mexico City storm Sewer, Mexico	Romo & Resendiz (1982)	N.C. clay	Shield	2D (with face effect)	Type 1 (nonlinear)	$S_0 \simeq S_c$	$i_0 < i_c$

this review is summarized in Table 5.1. It is of interest to note that reasonably good predictions have been achieved for tunnels in granular soils, regardless whether using a non-linear or an elasto-plastic soil model. The situation is different for cohesive soils. Regardless whether the analysis was two- or three-dimensional, or which type of stress-strain model was used and regardless of the method by which the tunnel was built the observed settlement trough is always narrower than the calculated one. This tendency is schematically illustrated in Figure 5.34. The significance of the deficiency is amplified by the fact that these predictions are on the unsafe side because they indicate a slope less critical with respect to assessment of differential settlement of nearby structures.

The fact that the observed trough is steeper indicates that the shear strains developing above a tunnel are higher than the stress-strain models for cohesive soils are able to predict. It is well known from field measurements (Eisenstein et al. 1981a) that shear stresses and strains develop in relatively narrow bands extending from the sides of a shallow tunnel upwards. It is thus obvious that further progress in analysis of shallow tunnels in cohesive soils would require a stress-strain model capable of modelling shear bands.

ACKNOWLEDGEMENT

The author is grateful to Mr. Heinrich Heinz for valuable assistance with preparation of this contribution.

REFERENCES

Branco, P. 1981. Behaviour of shallow tunnel in till. M.Sc. thesis, University of Alberta, Edmonton.

Cappellari, G. & M. Ottaviani 1982. Predicted surface settlements due to shield tunnelling with compressed air. In *Proceedings of 4th International Conference on Numerical Methods in Geomechanics* (Edmonton), 2: 531–544.

Clough, G.W. & B. Schmidt 1977. Design and performance of excavations and tunnels in soft clay. State-of-the-Art Report, International Symposium on Soft Clay (Bangkok). Also in E.W. Brand & R.P. Brenner (eds.) 1981. *Soft clay engineering*: 569–534. Amsterdam: Elsevier.

Czapla, H., R. Katzenbach, H. Ruckel & R. Wanninger 1978. *STATAN-15 Geotechnical version manual*. TH Darmstadt.

Deere, D.U., R.B. Peck, J.E. Monsees & B. Schmidt 1969. Design of tunnel liners and support systems. Report No. PB 183 799 for US Dept. of Transportation, NTIS.

Duddeck, H. 1980. Empfehlungen zur Berechnung von Tunneln im Lockergestein. *Bautechnik* 7:10: 349–356.

Duddeck, H. & J. Erdmann, 1982. Structural design models for tunnels. In *Proceedings Tunnelling* (Brighton): 83–91.

Eisenstein, Z., F. El-Nahhas & S. Thomson 1981a. Strain field around a tunnel in stiff soil. In *Proceedings of 10th International Conference on Soil Mechanics and Foundation Engineering* (Stockholm), 1: 283–288.

Eisenstein, Z., F. El-Nahhas & S. Thomson, 1981b. Pressure-displacement relations in two systems of tunnel lining. In *Soft-ground tunnelling, failures and displacements*: 85–94. Rotterdam: Balkema.

Eisenstein, Z. & L. Medeiros, 1983. A deep retaining structure in till and sand. Part II: Performance and analysis. *Canad. Geotech. J.* 20: 131–140.

Eisenstein, Z., S. Thomson & P. Branco 1982. Jasper Avenue Twin Tunnel–Instrumentation Section at 102th Street. Report to LRT Project Office, City of Edmonton.

El-Nahhas, F. 1980. The behaviour of tunnels in stiff soils. Ph.D. thesis, University of Alberta, Edmonton.

Ghaboussi, J., R.E. Ranken & M. Karshenas 1978. Analysis of subsidence over soft-ground tunnels. In

Proceedings of ASCE International Conference on Evaluation and Prediction of Subsidence (Pensacola Beach, Florida): 182–196.

Hanafy, E.A. & J.J. Emery 1980. Advancing face simulation of tunnel excavations and lining placement. In *Proceedings of 13th Canadian Rock Mechanics Symposium* (Toronto): 119–125.

Hanafy, E.A. & J.J. Emery 1982. Three-dimensional simulation of tunnel excavation in squeezing ground. In *Proceedings of 4th International Conference on Numerical Methods in Geomechanics* (Edmonton), 3: 1203–1210.

Ito, T. & M. Hisatake 1982. Three-dimensional surface subsidence caused by tunnel driving. In *Proceedings of 4th International Conference on Numerical Methods in Geomechanics* (Edmonton), 2: 551–559.

Katzenbach, R. & H. Breth 1981. Nonlinear 3-D analysis for NATM in Frankfurt Clay. In *Proceedings of 10th International Conference on Soil Mechanics and Foundation Engineering* (Stockholm), 1: 315–318.

Mair, R.J., M.J. Gunn & M.P. O'Reilly 1981. Ground movements around shallow tunnels in soft clay. In *Proceedings of 10th International Conference on Soil Mechanics and Foundation Engineering* (Stockholm), 1: 323–328. Also in *Tunnels and Tunnelling* 14 (June 1982): 45–48.

Muir Wood, A.M. 1975. The circular tunnel in elastic ground. *Geotechnique* 25 (1): 115–127.

Negro, A. & Z. Eisenstein 1981. Ground control techniques compared in three Brazilian water tunnels. *Tunnels and Tunnelling* 13 (10, 11, 12).

Ohnishi, Y., Y. Nishagaki, H. Kishimoto & Y. Tanaka 1982. Analysis of advancing tunnel by 2-dimensional FEM. In *Proceedings of 4th International Conference on Numerical Methods in Geomechanics* (Edmonton), 2: 571–578.

O'Reilly, M.P. & B.M. New 1982. Settlements above tunnels in the United Kingdom–their magnitude and predictions. In *Proceedings of Tunnelling* (Brighton): 173–181.

Orr, T.L.L., J.H. Hutchinson & C.P. Wroth 1978. Finite element calculations for the deformation around model tunnels. In *Computer methods in tunnel design*: 121–144. London: Institution of Civil Engineers.

Peck, R.B. 1969. Deep excavations and tunnelling in soft ground. In *Proceedings of 7th International Soil Mechanics and Foundation Engineering* (Mexico City), State-of-the-Art Volume: 225–290.

Ranken, R.E. & J. Ghaboussi 1975. Tunnel design considerations: analysis of stresses and deformations around advancing tunnels. Report for US Dept. of Transportation, UILU–ENG 75–2016.

Resendiz, D. & M.P. Romo 1981. Settlements upon soft-ground tunnelling – theoretical solution. In *Soft-ground tunnelling, failures and displacements*: 65–74. Rotterdam: Balkema.

Romo, M.P. & D. Resendiz 1982. Observed and computed settlements in case of soft-ground tunnelling. In *Proceedings of 4th International Conference on Numerical Methods in Geomechanics* (Edmonton), 2: 597–604.

Rowe, R.K., K.Y. Lo & G.J. Kack 1981. The prediction of subsidence above shallow tunnels in soft soil. In *Proceedings of Symposium on Implementation of Computer Procedures and Stress-Strain Laws in Geotechnical Engineering* (Chicago), 1: 266–280.

Schmidt, B. 1969. Settlements and ground movements associated with tunnelling in soil. Ph.D. thesis, University of Illinois, Urbana.

Stroh, D. & H. Breth 1976. Deformations of deep excavations. In *Proceedings of 2nd International Conference on Numerical Methods in Geomechanics* (Blacksburg), 2: 686–700.

Swoboda, G. 1979. Finite element analysis of the New Austrian Tunnelling Method (NATM). In *Proceedings of 3rd International Conference on Numerical Methods in Geomechanics* (Aachen), 2: 581–586.

Wanninger, R. 1979. New Austrian Tunnelling Method and finite elements. In *Proceedings of 3rd International Conference on Numerical Methods in Geomechanics* (Aachen), 2: 587–597.

Ward, W.H. & M.J. Pender 1981. Tunnelling in soft ground – General report. In *Proceedings of 10th-International Conference on Soil Mechanics and Foundation Engineering* (Stockholm), State-of-the-Art volume: 19–52.

Wilson, E.L. 1965. Structural analysis of axisymmetric solids. *AIAA Journal* 3: 2269–2274.

Wittke, W. & K. Gell 1980. Three-dimensional stability analyses for a shallow tunnel section of the subway of the City of Bochum, Construction Lot B3 (in German). *Geotechnik* 3: 111–119.

Zienkiewicz, O.C. 1977. *The finite element method* (3rd. ed.). New York: McGraw-Hill.

The use of linear elastic and piecewise linear models in finite element analyses

6

ROBERT PYKE

6.1 INTRODUCTION

The finite element method and other methods of numerical analysis have provided geotechnical engineers with powerful tools for analysis and design, but it is unlikely that such numerical analyses will ever provide totally reliable predictive tools because of the difficulties involved in modelling both soil properties and the geometry of the field problems. The difficulties involved in characterizing in situ soil properties and then modelling them in numerical analyses are widely recognized. The difficulties involved in modelling the geometry, including the boundary conditions, and the loading conditions for field problems are not so widely admitted, but they can nonetheless be significant. Definition of groundwater conditions (see, for example, Osaimi & Clough 1979) and construction-induced stresses (see, for example, Katona 1981 and Felix et al. 1982) are among the problems which are frequently encountered in modelling field loading conditions. The analyst may obtain a nice solution for the problem expressed by his or her model, but this problem may not coincide very well with the real problem.

Thus, numerical analyses should be regarded as tools for guiding the engineer's final judgment rather than as tools for making precise predictions. Various assumptions regarding both soil properties and problem geometry should be explored and the possible effects of factors that cannot be incorporated in the analyses should be considered in at least a qualitative fashion. Duncan (1979) and Mana & Clough (1981) provide excellent examples of the use of numerical analyses to extend simple analyses and field experience relative to flexible culverts and braced excavations, respectively.

In this context then, there would appear to be a continuing role for analyses which use linear elastic soil properties and simplified nonlinear soil properties, which are actually piecewise linear elastic, even though soils generally exhibit linear elastic behaviour only at very small strains, say at shear strains of $10^{-3}\%$ and less, and clearly exhibit plastic behaviour as failure is approached.

The usefulness of linear elastic analyses is, however, generally restricted to those cases where the engineer is interested only in gaining an idea of the stress distribution and is not interested in the computed displacements. This usually occurs when the stresses are required as part of the input to a second analysis. For instance, the conventional procedure for dynamic analysis of embankment dams (Seed, Chapter 8 in this volume) requires the determination of the initial static stresses prior to the dynamic analysis itself. The initial static stresses are commonly determined using simplified nonlinear analyses of the kind described subsequently, but Lee & Idriss (1975) have shown that the static stress distribution can normally be determined with sufficient accuracy by linear elastic, turn-on-gravity analyses.

Early experiences with the use of linear elastic finite element analyses in geotechnical engineering quickly led to the obvious conclusion that they are not so satisfactory for computing deformations. Then, bilinear or linear elastic perfectly plastic analyses were explored (see, for example, Dunlop & Duncan 1970, D'Appolonia & Lambe 1970 and the summary of early work and comments by Smith & Kay 1971). Finally, procedures for simplified nonlinear analyses which use hyperbolas to represent the stress-strain relationship were developed (see, for example, Duncan & Chang 1970 and Simon et al. 1974) and these have been used quite widely in practice. The reasons for the popularity of this approach have been summarized by Christian (1982) as follows:

> The hyperbolic models have been available for about ten years and are incorporated into a number of computer programs. The models and the programs are understood by many people, including those who do not work primarily with numerical methods or constitutive relations. The parameters are reasonably comprehensible in the context of soil mechanics. They can also be measured without extraordinary testing equipment in most cases.
>
> A more subtle and fundamental reason for the success of the hyperbolic model arises from its range of practical use: at stress levels less than about 75% of failure. This is the range in which a great many geotechnical problems occur.

The use of simplified nonlinear, or hyperbolic, models is described in more detail in the remainder of this chapter.

6.2 INCREMENTAL FINITE ELEMENT ANALYSES

The kind of simplified nonlinear properties that are discussed in this chapter are generally used in 'incremental' finite element analyses in which the load is applied in increments and tangent moduli, which are based on the stress levels for that increment, are used to reform the stiffness matrix. Hooke's law is assumed to apply within each load increment so that the analysis is piecewise linear elastic, rather than truly nonlinear. Commonly, an initial calculation is performed for each load increment using tangent moduli based on the stresses at the end of the previous load increment. One or more further iterations are then carried out using tangent moduli based on the average stresses within the load increment. While programs can be written so that they iterate until a specified convergence criterion is met, it is generally more desirable to use load increments small enough that acceptable convergence is obtained in two or three iterations.

Analyses of this kind are usually conducted using effective stress or drained soil properties and assuming fully drained loading conditions, or using total stress soil properties for undrained loading conditions. While it is possible to formulate programs in terms of effective stresses (Christian 1968, Byrne & Janzen 1981) and to construct schemes in which solution of the consolidation problem is loosely coupled with the finite element solution for stresses and deformations, the pore pressures that are used in such schemes should be computed using empirical or semi-empirical methods since linear elastic or simplified nonlinear soil models do not normally allow calculation of the volume change tendencies of the soil with sufficient accuracy to allow their use in fully coupled solutions.

Undrained analyses of fully saturated soils are usually carried out using somewhat arbitrary values of the Poisson's ratio approaching 0.5. While it is possible to conduct constant volume analyses using more sophisticated finite element formulations (Hughes & Malkus 1977) the bulk moduli of saturated soils are in fact finite, with the corresponding values of Poisson's ratio being close to, but less than, 0.5 (Bishop & Hight 1977) so that the extra expense of the constrained formulation may not be warranted.

6.3 THE DUNCAN & CHANG MODEL

As indicated in the introduction, the most widely used simplified nonlinear models have been based on the approach taken by Duncan & Chang (1970) who, following Konder (1963) and Konder & Zelaski (1963), adopted a plain hyperbola for the shape of the stress-strain relationship for primary loading. In Duncan & Chang (1970) and the following discussion, it is the axial stress-strain relationship obtained from a conventional triaxial test that is modelled with the hyperbolic shape, but similar formulations in terms of shear stress and shear strain are also possible (Simon et al. 1974, Drnevich 1975).

As the first step in determining the tangent Young's modulus for use in the finite element calculation, the axial stress-strain relationship is described as follows:

$$(\sigma_1 - \sigma_3) = \frac{\varepsilon}{\dfrac{1}{E_{max}} + \dfrac{\varepsilon}{(\sigma_1 - \sigma_3)_{max}}} \tag{6.1}$$

where E_{max} is the initial tangent modulus and $(\sigma_1 - \sigma_3)_{max}$ is the asymptotic value of the stress difference as shown in the upper part of Figure 6.1. If the hyperbolic equation is transformed as shown in the lower part of Figure 6.1, it represents a linear relationship between $\varepsilon/(\sigma_1 - \sigma_3)$ and ε. Thus, to determine the best-fit hyperbola for the stress-strain curve, values of $\varepsilon/(\sigma_1 - \sigma_3)$ are calculated from test data and are plotted against ε. The best-fit straight line on this transformed plot corresponds to the best-fit hyperbola on the stress-strain plot. When data from actual tests are plotted on the transformed plot, the points frequently are found to deviate from the ideal linear relationship. The data for stiff soils such as dense sands usually plot on a mild curve which is concave upward, whereas the data for soft soils such as loose sands usually plot on a mild curve which is concave downward. Experience with stress-strain curves for many soils indicates that a good match is usually achieved by selecting the straight line so that it passes through the points where 70% and 95% of the strength are mobilized. Examples of the goodness of fit of the plain hyperbolic shape are given by Daniel & Olson (1974) and Hardin (1983).

The modified hyperbola of Hardin & Drnevich (1972) generally provides a more exact fit to actual stress-strain relationships but the full expression for stress in terms of strain used by Hardin & Drnevich cannot be differentiated in order to obtain the

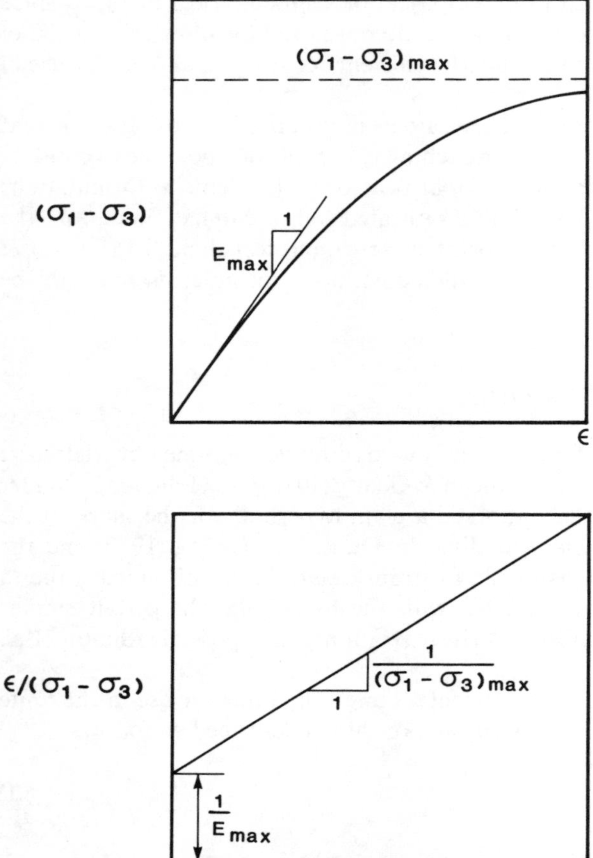

Figure 6.1. Hyperbolic representation of stress-strain curve

tangent modulus. However, if the secondary parameter b is dropped from the Hardin & Drnevich expression and only the parameter a is used, quite a good fit to most soil stress-strain relationships can be obtained over a wide range of stress levels. In this approach one makes use of a reference strain, ε_r, defined as:

$$\varepsilon_r = \frac{(\sigma_1 - \sigma_3)_{max}}{E_{max}} \tag{6.2}$$

Then the scale on the strain axis is distorted by substituting a quantity called the hyperbolic strain, ε_h, for the reference strain. The hyperbolic strain is given by:

$$\varepsilon_h = \frac{\varepsilon_r}{1 + a} \tag{6.3}$$

Using the notation:

$$\sigma_d = \sigma_1 - \sigma_3$$

and rearranging Equation 6.1, the expression for stress in terms of strain becomes:

$$\frac{\sigma_d}{\sigma_{max}} = \frac{\varepsilon/\varepsilon_h}{1+\varepsilon/\varepsilon_h} \qquad (6.4)$$

which makes the hyperbolic form more obvious.

Examples of the modified hyperbolic shapes described by this relationship are shown in Figure 6.2. Typically the parameter a is negative for sands and positive for clays. The Duncan & Chang procedure corresponds to setting the parameter a equal to zero.

The tangent Young's modulus, as required in incremental finite element calculations, can then be obtained by differentiating the expressions for stress given by either Equation 6.1 or 6.4 with respect to strain. The result of this differentiation is shown subsequently.

For all soils except fully saturated soils tested under unconsolidated-undrained conditions, an increase in confining pressure will result in a steeper stress-strain curve and a higher strength, and the values of E_{max} and $(\sigma_1 - \sigma_3)_{max}$ therefore increase with increasing confining pressure. This stress-dependency is taken into account by using empirical equations to represent the variations of E_{max} and $(\sigma_1 - \sigma_3)_{max}$ with confining pressure.

The variation of E_{max} with σ_3 is represented by an equation of the following form:

$$E_{max} = K_E p_a \left(\frac{\sigma_3}{p_a}\right)^n \qquad (6.5)$$

where p_a is the atmospheric pressure. The parameters K_E and n are both dimensionless numbers and the units of E_{max} are the same as the units of p_a. The constants K_E and n can readily be determined by plotting E_{max}/p_a against σ_3/p_a on logarithmic paper.

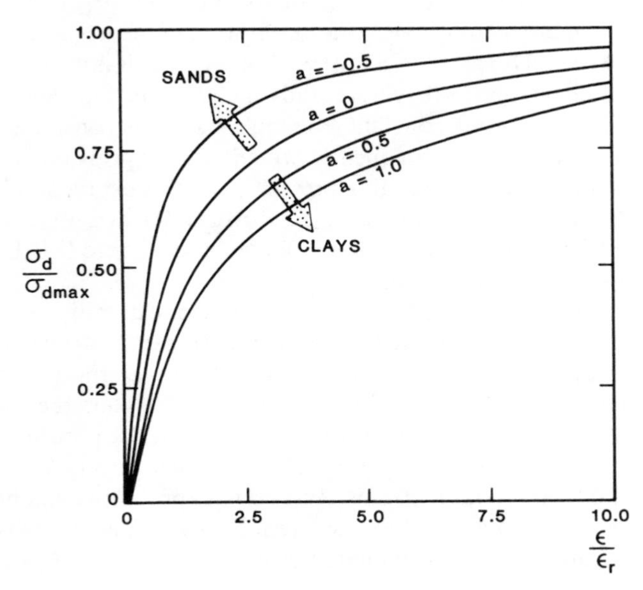

Figure 6.2. Hardin & Drnevich hyperbolas

The variation of $(\sigma_1 - \sigma_3)_{max}$ with σ_3 is accounted for in the Duncan & Chang procedure by relating $(\sigma_1 - \sigma_3)_{max}$ to the compressive strength or stress difference at failure, $(\sigma_1 - \sigma_3)_f$, and then using the Mohr-Coulomb strength equation to relate $(\sigma_1 - \sigma_3)_f$ to σ_3. The values of $(\sigma_1 - \sigma_3)_{max}$ and $(\sigma_1 - \sigma_3)_f$ are related by:

$$(\sigma_1 - \sigma_3)_f = R_f(\sigma_1 - \sigma_3)_{max} \tag{6.6}$$

in which R_f is the failure ratio. Because $(\sigma_1 - \sigma_3)_f$ is always found to be smaller than $(\sigma_1 - \sigma_3)_{max}$, the value of R_f is always smaller than unity, and varies from 0.5 to 0.9 for most soils. In the Hardin & Drnevich procedure $(\sigma_1 - \sigma_3)_f$ and $(\sigma_1 - \sigma_3)_{max}$ are synonymous and the need for R_f disappears.

The variation of $(\sigma_1 - \sigma_3)_f$ with σ_3 can be expressed as follows:

$$(\sigma_1 - \sigma_3)_f = \frac{2c\cos\phi + 2\sigma_3 \sin\phi}{1 - \sin\phi} \tag{6.7}$$

in which c and ϕ are the cohesion intercept and the friction angle. In general, the friction angle, ϕ, is not constant but decreases with increasing stress level as follows:

$$\phi = \phi_1 - \Delta\phi \log\frac{\sigma_3}{p_a} \tag{6.8}$$

in which ϕ_1 = the value of the friction angle at a confining stress of 1 atmosphere and $\Delta\phi$ = the reduction in friction angle for a ten-fold increase in confining stress.

For unloading and reloading, Duncan & Chang (1970) suggested using a constant Young's modulus, E_{UR}, which can be expressed in terms of σ_3 using the same form as is used in Equation 6.5:

$$E_{UR} = K_{EUR} p_a \left(\frac{\sigma_3}{p_a}\right)^n \tag{6.9}$$

Values of K_{EUR} are typically two to three times greater than values of K_E.

In order to complete the description of the basic model, a second elastic constant is needed. Duncan & Chang (1970) noted that it is difficult to characterize the behaviour of sands such as they were studying by a constant Poisson's ratio, but for the purposes of their study a constant Poisson's ratio was considered to be adequate. Kulhawy & Duncan (1972) subsequently presented an expression for Poisson's ratio in terms of the stress level and confining pressure, and this expression has been used in University of California programs such as ISBILD (Ozawa & Duncan 1973). In the program FEADAM (Duncan et al. 1980b), the successor to ISBILD, the bulk modulus is used as the second elastic constant.

In the program FEADAM and in Duncan et al. (1980a), it is assumed that the bulk modulus appropriate for use in analysis can be determined by increasing the deviator stress in a conventional triaxial test. However, this procedure tends to mix the volume changes caused by the increase in mean stress and the volume changes caused by shear stress-volumetric strain coupling. Further, the procedures used in FEADAM are not capable of modelling the dilation which occurs for many soils when sheared to moderate to large strains. An alternate procedure can be used for drained analyses in which the bulk modulus is determined as a function of isotropic changes in stress and shear-dilatancy is simulated using the procedure described in Section 6.5.3.

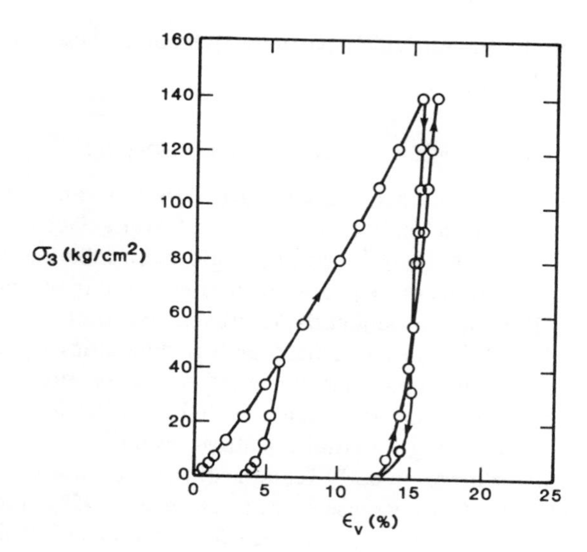

Figure 6.3. Stress-strain curves for isotropic compression (from Lade 1977)

The isotropic stress versus volumetric strain relationship for soils can readily be determined from the isotropic consolidation stage of a normal triaxial test. Typical curves of isotropic stress versus volumetric strain for primary loading and unloading-reloading are shown in Figure 6.3 (from Lade 1977). It can be seen that the curves for unloading and reloading are almost identical and that this behaviour can be taken to be nonlinear elastic. If the tangent bulk modulus is measured for a number of values of isotropic stress it is found that expressions of the same form used by Lade (1977) and Duncan et al. (1980a) can be used to obtain both the bulk modulus for primary loading, K, and the bulk modulus on unloading/reloading, K_{UR}, in terms of the isotropic stress, σ_3:

$$K = K_K p_a \left(\frac{\sigma_3}{p_a}\right)^{nk}$$

$$\text{and } K_{UR} = K_{KUR} p_a \left(\frac{\sigma_3}{p_a}\right)^{nkur} \tag{6.10}$$

As before, the constants K_K, nk, K_{KUR} and nkur can be obtained by plotting K/p_a and K_{UR}/p_a against σ_3/p_a on logarithmic paper. Because the typical values of K_K and nk given by Duncan et al. (1980a) (that is K_b and m in their terminology) are determined at relatively low strain levels for which dilatancy is small, these values may generally be used in the absence of values obtained from isotropic loadings.

6.4 LIMITATIONS OF THE SIMPLE HYPERBOLIC MODEL

While the simple hyperbolic model has the advantages noted in the introduction, it does, of course, have limitations in representing real soil behaviour. The main limitations of the basic model are discussed in the following subsections and extensions of the model which partially overcome these limitations are discussed in the next section.

6.4.1 *Behaviour on unloading and reloading*

Following Duncan & Chang (1970) most analyses using hyperbolic shapes for the stress-strain relationship on initial loading have used a constant value for the modulus on unloading and reloading. This is not a bad assumption although an even simpler and more accurate procedure for determining the modulus on unloading and reloading is discussed subsequently. However, an aspect of behaviour in unloading and reloading which has caused some practical difficulties is the choice of the criterion used to determine when an element of soil goes into unloading. In earlier University of California programs such as ISBILD, the criterion used to determine that unloading was occurring was that the stress level $(\sigma_1 - \sigma_3)/(\sigma_1 - \sigma_3)_{max}$ decreased by a value of 0.05 or more. In early versions of the program FEADAM, however, the criterion was changed to a decrease in the stress level to 95% or less of the previous maximum stress level. While these criteria are similar, and in fact are equal if the previous maximum stress level is unity, for values of the previous maximum stress level less than unity the second criterion will always be met sooner than the first criterion, and it was observed in some FEADAM analyses that the stiffer modulus on unloading was being selected for some elements, even when additional loads were being applied and it was expected that these elements would be subjected to continuous loading. The reason for these apparently strange results is simply that if the minor principal stress, σ_3, increases more than the major principal stress, σ_1, on any load increment, the stress level will decrease, even though σ_1 may have increased. To some extent this problem may have occurred with ISBILD (see, for example Bossoney & Dungar 1980) but adoption of the tighter criterion in FEADAM made the problems more noticeable.

While use of the earlier, looser, criterion will minimize the occurrence of this problem, a more positive fix follows from the concept that, since the hyperbolic stress-strain relationship is normally obtained from conventional triaxial tests in which σ_3 is held constant, the changes in σ_3 from one load increment to another should be viewed as causing a change from one 'material' to another. Thus the properties used in the analysis are viewed as only being 'correct' in the direction of the major principal stress, with discontinuous behaviour as the principal stress directions rotate in field problems and σ_3 jumps from one value to another. Unloading and reloading is then keyed to the value of σ_1, rather than the stress level, and the modulus in unloading is chosen if σ_1 decreases by any amount. Use of this concept will still not yield entirely accurate results for loading conditions such as the proportional or constant stress level loading discussed by Duncan & Chang (1972) and Lade & Duncan (1976) – following such loadings is simply beyond the capability of a simplified model – but it does lead to consistent results and should more generally help users understand the behaviour of the model. Even in those cases where tests other than conventional triaxial tests are used to obtain the stress-strain relationship for primary loading, it is believed that the concept of viewing changes in the minor principal stress from one value to another as constituting a change from one 'material' to another is helpful and that unloading and reloading should still be keyed to the value of the major principal stress.

6.4.2 *Anisotropy*

The combination of the hyperbolic model and the incremental finite element method with which it is normally used, does not allow for the complete modelling of anisotropy since properties are assumed to be isotropic, linear elastic on each load increment. However, if use is made of the concept that the model is fitted to the properties in the direction of the major principal stress, anisotropy can be accounted for as discussed in Section 6.5.2. However, no matter how exactly the properties are fitted in the direction of the major principal stress, the assumption of isotropic behaviour on each load increment will lead to some degree of error in the properties in all other directions.

6.4.3 *Tension failure*

It is well known that soils take little or no tension and to some extent this will automatically be taken care of by resetting σ_3 to a small positive value whenever it goes negative on the first iteration of a new load increment. A small value of σ_3 will result in small values for both Young's modulus and bulk modulus and stress-redistribution tends to eliminate any tensile stresses. Note, however, that this implies that the soil is equally soft in all directions whereas, in fact, while cracking may occur in the direction to which the minor principal stress is normal, in cohesive soils the soil may still carry a significant load in the direction parallel to cracking. This kind of behaviour can probably best be modelled by including properly formulated interface elements between all the solid elements in regions of expected tensile stresses. The interface elements will then tend to gap in the direction of the minor principal stress but will remain closed in the direction of the major principal stress.

6.4.4 *Residual strengths*

Both the hyperbolic formulation of the stress-strain relationship and the incremental finite element method prevent the direct modelling of stress-strain relationships which exhibit a drop in the strength once the deformation corresponding to the peak strength is exceeded. By using other procedures (see, for example Hoeg 1972, Prevost & Hoeg 1975 and Desai 1977) it is possible to follow post-peak decreases in strength, however, there are strong indications that residual strengths do not in fact develop in soil masses but occur only along surfaces along which large monotonic deformations or reversing deformations take place. Lade (1972) for example, showed that while brittle behaviour is observed in conventional 'long' triaxial tests on dense sands, plastic behaviour is observed when the premature failure along one plane is prevented by the use of 'short' specimens and lubricated platens. Thus, it might be that the need for following post-peak decreases in strength is restricted to interface elements which should be used to accommodate slip along surfaces on which failure occurs.

6.4.5 *Shear-dilatancy coupling*

Use of standard incremental finite element analyses does not allow use of values of Poisson's ratio equal to 0.5 or more and the dilatancy that is observed when dense sands or overconsolidated clays are subjected to shearing cannot be modelled. This limitation will often have only a minor effect on computed stresses and displacements but if the soil is constrained from movement, say by the presence of a rigid wall, the calculated stresses can be significantly in error. A simple scheme which can be used to model dilation is described in Section 6.5.3.

6.4.6 *Effect of intermediate principal stress*

The simple hyperbolic model uses the Mohr-Coulomb failure criterion and neglects the effect of the intermediate principal stress (see, for example Lade & Duncan 1973). This remains so even if dilatancy is included as suggested in Section 6.5.3. The inclusion of dilatancy will lead to higher values of σ_2 being inferred from the elastic stress-strain relationships in plane strain, which is the condition normally assumed in analysis, but the higher values of σ_2 will not be reflected in the stress-strain relationships and strengths if they are obtained from conventional triaxial tests. In general, improved results should be obtained by using plane strain tests in the laboratory to obtain the stress-strain relationships for use in plane strain analyses, but otherwise it is not possible to include the effect of the intermediate principal stress in the hyperbolic soil model.

6.4.7 *Behaviour at large strains*

Perhaps the most significant limitation of simplified nonlinear soil models, which assume that material behaviour is isotropic, linear elastic within each load step, is that computed deformations will become less accurate as failure is approached. This discrepancy results from the fact that plastic behaviour dominates in soils loaded to high stress levels and the strain increments should then be applied in the direction of the stresses, rather than the direction of the stress increments as is the case when Hooke's law is used. Nice examples of the differences obtained using the simple hyperbolic model and various plasticity theories are given by Ozawa & Duncan (1976), Almeida & Ramalho-Ortigao (1982) and Dickin & King (1982).

6.5 EXTENSION OF SIMPLIFIED NONLINEAR MODEL

6.5.1 *Nonlinear unloading and reloading*

As noted in Section 6.4.1, most analyses using hyperbolic shapes for the stress-strain relationship on initial loading have used a constant value for the modulus on unloading and reloading. However, a simple procedure is available for describing nonlinear unloading and reloading which matches actual soil behaviour more closely. Use of this procedure also allows the effects of anisotropic consolidation to be taken into account without using special formulations such as those suggested by Byrne et al. (1982) or Hansen & Clough (1982). As discussed in Section 6.4.1, it is recommended that unloading and reloading be keyed to decreases or increases in σ_1, rather than the stress level. Using the suggestion regarding rules for stress-strain behaviour under cyclic loadings advanced by Cundall (1975) and Pyke (1979), the general stress-strain relationship equivalent to Equation 6.4 becomes:

$$\sigma_d = \sigma_{dc} + \sigma_{dmax}\left(\frac{\varepsilon - \varepsilon_c}{\varepsilon_h}\right)\frac{1}{1 + \dfrac{|\varepsilon - \varepsilon_c|}{c\varepsilon_h}} \tag{6.11}$$

where:

ε_h = the hyperbolic strain, as before;
ε_c = strain at last reversal;
$\sigma_{dc} = \sigma_{1c} - \sigma_{3c}$ where σ_{1c} is the major principal stress at the last reversal and σ_{3c} is the minor principal stress at the last reversal;

$$c = \left| \pm 1 - \frac{\sigma_{dc}}{\sigma_{dmax}} \right|.$$

where the first term is negative for loading or reloading and positive for unloading.

For small unloading/reloading loops superimposed on monotonic loading a stress-strain relationship of the form shown in Figure 6.4 is obtained. The behaviour shown agrees rather well with published data such as that shown by Sangrey et al. (1969) and Taylor (1971). If many unloading/reloading cycles are carried out it is necessary to introduce additional parameters in order to follow stress-strain history effects, but this is not normally required for one or two unloading/reloading cycles.

By differentiation of Equation 6.11 with respect to ε, an expression for the tangent Young's modulus can be obtained as follows:

$$E_t = \frac{E_{max}}{1 + a}\left[\pm 1 - \frac{(\sigma_d - \sigma_{dc})}{c\sigma_{dmax}} \right] \tag{6.12}$$

in which the sign on the first term is negative for loading or reloading and positive for unloading. In the Duncan & Chang procedure the parameter a is equal to zero and σ_{dmax} is a function of the parameter R_f. Since the hyperbolic stress-strain curve is asymptotic to the fictitious value $(\sigma_1 - \sigma_3)_{max}$, the tangent modulus is usually set equal to a reduced value such as $E_{max}/100$ if the value $(\sigma_1 - \sigma_3)_f$ is equalled or exceeded. In the Hardin & Drnevich procedure $(\sigma_1 - \sigma_3)_{max}$ and $(\sigma_1 - \sigma_3)_f$ are equal and no cut-off is necessary.

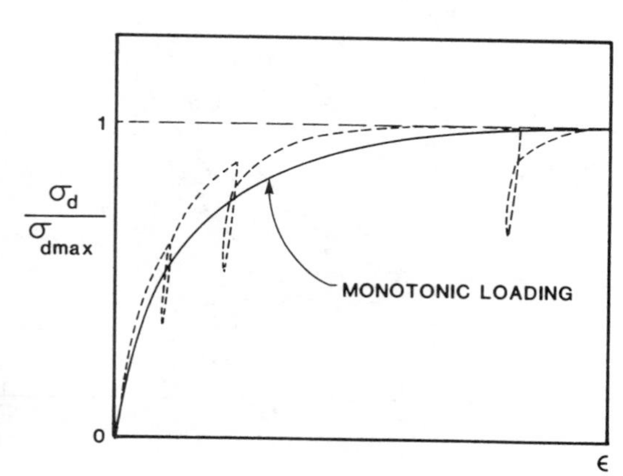

Figure 6.4. Behaviour in unloading and reloading in accordance with Cundall-Pyke hypothesis

6.5.2 *Inclusion of anisotropy*

It is not possible to fully represent material anisotropy in the simplified model because the use of only two elastic constants requires that isotropic, linear elastic properties be used for each load increment. However, since the properties used on each load increment are already keyed to the direction of the major principal stress, it is relatively easy to take anisotropy into account to some extent by varying the stiffness and/or the strength as a function of the direction of the major principal stress. The inclusion of some degree of anisotropy in simplified nonlinear analyses appears to have been restricted in the past to undrained analyses of clays (e.g. Simon et al. 1974, Hansen & Clough 1982 and Byrne et al. 1982) but generally similar procedures can also be applied in analysis of drained loadings of clays and sands.

The literature on anisotropy of soils is somewhat confusing because of differences in terminology and differences in the expressions that have been used to describe anisotropy, but two kinds of anisotropy have generally been described: inherent anisotropy and induced anisotropy. Inherent anisotropy is anisotropy of stiffness and/or strength that results from the fabric of the soil prior to the loading of interest (e.g. Oda 1972a, Arthur & Menzies 1972). The purest form of induced anisotropy results from changes in the fabric of the soil during the loading of interest (e.g. Oda 1972b, Arthur et al. 1977 and Arthur et al. 1980). Additionally, for undrained loadings where the behaviour is characterized in terms of the conditions at consolidation, there can be a phenomenon referred to as stress-system induced anisotropy (e.g., Ladd et al. 1977). In practical applications, however, it is more convenient to ignore these distinctions and to simply model their combined effects.

A convenient form for modelling anisotropy is based on that used by Duncan & Dunlop (1969) and Hansen & Clough (1982). Any or all of the initial Young's modulus, E_{max}, the cohesion, c (equivalent to the undrained strength, S_u, in undrained analyses),

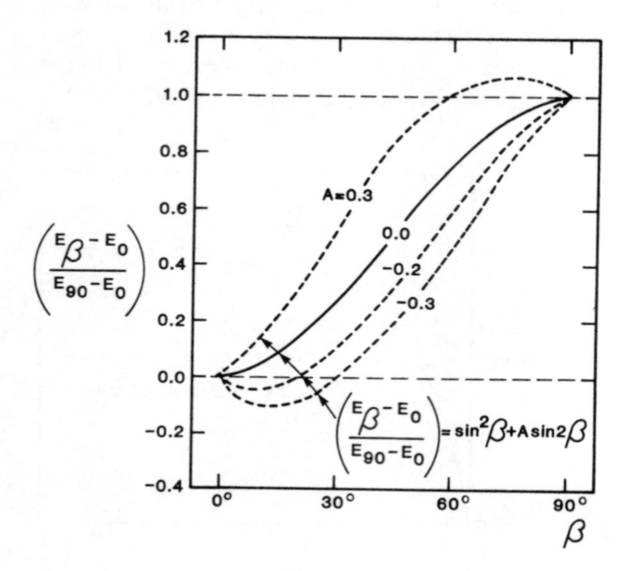

Figure 6.5. Variation of modulus with direction (from Hansen & Clough 1982)

or the friction angle, ϕ, can be varied with direction in accordance with the formula:

$$X_\beta = X_0[1 + (k - 1)(\sin^2 \beta + A \sin 2\beta)] \tag{6.13}$$

where:

X_β = the value of the property at an inclination β from a reference direction which is β_0 degrees clockwise from vertical;
x_0 = the value of the property in the reference direction;
k = the ratio of the values of the property in the directions β_{90} and β_0; and
A = an empirical constant.

If A is set equal to zero, this expression reduces to the simpler form used by some other workers.

For undrained clays, taking the reference direction to be vertical, Hansen & Clough suggest that $k_{su} = S_{u90}/S_{uo}$ is usually less than unity and $k_E = E_{max90}/E_{maxo}$ is usually greater than unity. Typical variations of the undrained Young's modulus with direction, expressed as a normalized modulus difference, are shown in Figure 6.5. Additional data on both stiffness and strength anisotropy can be found in the other references cited previously. In the case of sands, it is frequently assumed that the initial stiffness shows rather more anisotropy than the ultimate strength, but Oda (1981) shows that there may nonetheless be some strength anistropy for cohesionless soils.

6.5.3 *Simulation of dilatancy*

Volumetric strains are induced by changes in shear stress as well as by changes in the isotropic stress as can be seen from the simple shear test data in Figure 6.6. The incremental shear-induced volume change, $\Delta\varepsilon_v$, can be expressed in terms of a tangent

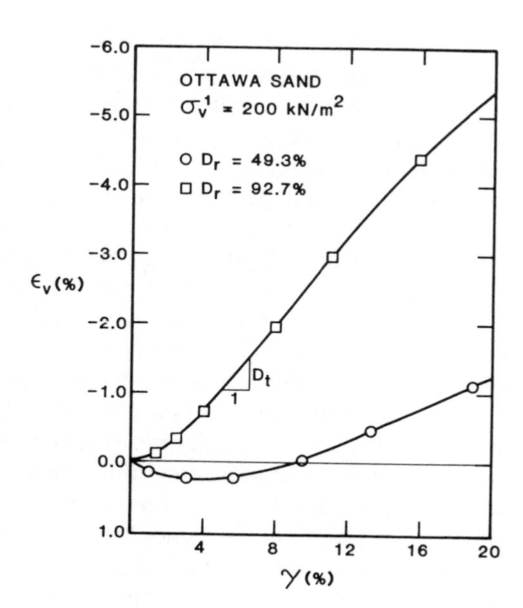

Figure 6.6. Volumetric strains in simple shear tests (from Vaid et al. 1981)

dilation parameter, D_t, as follows:

$$\Delta\varepsilon_v = D_t \Delta\gamma \tag{6.14}$$

in which $\Delta\gamma$ = the increment of maximum shear strain.

The dilation term, D_t, depends upon the stress level and the density of the sand and it can be conveniently expressed in terms of the dilation angle, v, as suggested by Hansen (1958), as follows:

$$D_t = -\sin v \frac{\Delta\varepsilon_v}{\Delta\gamma} \tag{6.15}$$

The dilation angle in turn can be computed from a modified form of Rowe's stress dilatancy theory suggested by Hughes et al. (1977), as follows:

$$\tan\left(45 + \frac{v}{2}\right) = \tan\left(45 + \frac{\phi}{2}/\tan\left(45 + \frac{\phi_{cv}}{2}\right)\right) \tag{6.16}$$

in which:

ϕ = the friction angle;
ϕ_{ev} = the constant volume friction angle.

The above equation is in good agreement with laboratory test data when ϕ is greater than ϕ_{ev}. When ϕ is less than ϕ_{ev} the equation predicts volume contractions that are too large. For example when ϕ equals zero, the equation predicts that the dilation angle, v, equals $-\phi_{ev}$, whereas in fact no shear-induced volume change occurs for this condition, and v should equal zero. Byrne & Eldridge (1982) indicate that a good estimate of dilation can be obtained by assuming that the dilation angle increases linearly from zero when ϕ equals zero to the value given by Equation 6.16 when ϕ equals $2/3\ \phi_{ev}$.

To use this procedure to simulate dilatancy, the shear-induced volume change is computed as described above using the shear strains obtained for the previous load increment on the first iteration for each new load increment. Then the nodal point forces required to cause this volume change for each element are computed as a function of the current bulk modulus of that element and the finite element problem is solved applying these nodal point forces in addition to the nodal point forces that define the load increment. On the second and any subsequent iterations, the shear-induced volume change is computed using the shear strains obtained on the previous iteration. If the material is not subjected to deformation constraints, the additional nodal point forces will cause an expansion of the volume but will have little or no effect on the computed stress increments. If the material is confined, however, the additional nodal point forces will result in an increase in the hydrostatic component of the stress increments. Byrne & Eldridge (1982) show a nice example of the use of this scheme to simulate a conventional triaxial test on dense sand.

If the maximum shear strain decreases, the sign of the computed volume change should be changed until such time as the maximum shear strain starts increasing again. Under triaxial test loading conditions, this will cause cycling of the volumetric strain with no net decrease in the volume as a result of shear stress-volumetric strain coupling which is not entirely correct since in reality a small net decrease in volume should occur for most soils; however, a small decrease in the volume will result from the manner in which the Young's modulus changes with cycling of the deviator stress.

6.6 BOUNDARY CONDITIONS

While this chapter devotes relatively more space to the limitations of simplified nonlinear models and some techniques for overcoming these limitations, it turns out that in practice the accuracy of numerical analyses of geotechnical problems is limited just as much by our ability to correctly model boundary conditions as it is by our ability to correctly model soil behaviour.

Perhaps the most significant limitation to date is that, for reasons of economy, analyses have usually been restricted to two-dimensional plane strain sections, even through in the field problems are necessarily three-dimensional. However, with containuing improvements in the performance-price ratio of computer hardware, it is likely that three-dimensional analyses will become more common in the future. In general, the kind of simplified nonlinear properties discussed in this chapter can be readily extended to three dimensions.

In both two- and three-dimensional analyses, reasonably correct modelling of field problems can be greatly facilitated by the use of two special kinds of elements, namely, interface and infinite elements.

6.6.1 *Interface elements*

Various types of joint or interface elements are described by Desai (1981) and Heuze & Barbour (1982). Katona et al. (1976) describe the two basic approaches to formulating interface elements as follows:

> There are two fundamental approaches to treat interfaces in the context of a finite element formulation. The first is the method of stiffnesses, and the second is the method of constraints.
>
> The method of stiffnesses is basically the simple concept of using 'bar' elements (or directionally stiff elements) across the interface in both the normal and tangential directions. For example, if it is desired to model frictionless slippage across an interface the normal stiffness would be specified arbitrarily large to force near compatibility of normal displacements, while the tangent stiffness would be specified extremely small (or zero) to allow independent movement in the tangential direction. Although this method has been used successfully, it has certain inconsistencies that are different to overcome. For example, nodes on either side of a zero-width interface will penetrate each other under compressive loading, because the normal stiffness is finite. Moreover, this relative movement (penetration) is required in order to recover the normal force in the interface element. However, if the normal stiffness is selected too large with respect to computer word length, the significant digits of the relative displacement become truncated, and the calculation of interface forces is in error. On the other hand, it the normal stiffness is selected too low, significant penetration will occur, and the kinematics will be in error. These same inconsistencies apply in the tangential direction when frictional resistance is modelled. That is, the stiffness approach requires some relative tangential slip to occur at each load step whether or not the frictional resistance has been exceeded.
>
> The alternative approach, the method of constraints (Hughes et al. 1976) eliminates the above inconsistencies and provides a more general capability for modelling interface interactions.

Interface elements based on the method of constraints can be used to 'connect' adjacent nodes in a finite element mesh and will develop complete bond unless tensile separation occurs or unless the interface shear strength is exceeded and slip occurs. Normally these elements will be placed between elements representing soil and elements representing structures but they also provide a powerful means of modelling cracking in zones within the soil mass where tensile stresses tend to occur.

6.6.2 *Infinite elements*

'Infinite' elements have not as yet been used widely in the analysis of static problems but they are likely to become more common. In particular, infinite elements can be used to help keep three-dimensional numerical models down to tractable sizes. Booker & Small (1981) and Beer (1983) describe such elements.

6.7 APPLICATIONS

Simplified nonlinear properties of the kind described in this chapter have been used in analyses of many different kinds of geotechnical analyses. A number of these are summarized in the following subsections. Some references to bilinear and other simple elasto-plastic schemes are also included.

6.7.1 *Embankments*

The initial work conducted at the University of California, Berkeley, is described by Kulhawy & Duncan (1972) and Nobari & Duncan (1972). Eisenstein et al. (1972) and Lefebvre et al. (1973) describe early three-dimensional analyses. Smith & Hobbs (1974) summarize earlier work and provide a comparison of the use of numerical and physical models. Kulhawy (1972) provides a summary of the state-of-the-art at that time. Bossoney & Dunger (1980) and Adikari & Parkin (1982) provide more recent examples of applications to practical problems.

6.7.2 *Embankments on soft ground*

Analyses of the construction stability of embankments on soft ground are usually conducted with the soil properties fitted to the results of undrained tests, rather than drained tests as is the case with the majority of the applications discussed in this section. Simon et al. (1974), Foott & Ladd (1977) and Almeida & Ramalho-Ortigao (1982) provide excellent examples of the analysis of soft ground problems.

6.7.3 *Excavation*

Dunlop & Duncan (1970) and Chang & Duncan (1970) describe early work at the University of California, Berkeley. Christian & Wong (1973) and Chandrasekaran & King (1974) point to problems encountered in some analyses. Kulhawy (1977) again provides a summary of the state-of-the-art at that time. Popescu (1982) describes an interesting recent analysis and highlights the weakness of conventional limit-equilibrium slope stability analyses for progressive failure problems.

6.7.4 *Braced excavations*

A series of papers by G.W. Clough and his colleagues provides a good example of the gradual evolution of the numerical analysis techniques applied to this rather complex problem. These papers include Tsui & Clough (1974), Murphy et al. (1975), Clough & Denby (1977), Clough & Tsui (1977), Osaimi & Clough (1979), Mana & Clough (1981) and Clough & Hansen (1981).

6.7.5 *Retaining walls*

Accurate analysis of earth pressures on retaining walls using elastic or simplified nonlinear analysis is difficult whenever the soil is brought close to failure (see, for example Ozawa & Duncan 1976). Nonetheless, instructive examples of the analysis of retaining walls are described by Morgenstern & Eisenstein (1970), Clough & Duncan (1971), Kulhawy (1974) and Roth & Crandall (1981).

6.7.6 *Culverts*

The modelling of structural elements which exhibit nonlinear behaviour is not discussed in this chapter but flexible metal culverts, such as are often used under highway fills, provide a good example of a soil-structure interaction problem for which nonlinearity in both the soil and the structural behaviour may be important. Examples of the analysis of flexible culverts are given by Duncan (1979), Chang et al. (1980), Katona (1981) and Rude (1982).

6.7.7 *Footings*

Finite element analyses of the settlement of building or tank foundations are described by D'Appolonia & Lambe (1970), Domaschuk & Valliappan (1975), Roth et al. (1981) and Byrne et al. (1982).

6.7.8 *Piles*

Piles represent a class of soil-structure interaction problems in which it is adequate to model the structure with linear elastic properties but great care must be given to modelling the boundary conditions since field behaviour is very dependent on the load transfer mechanism along the shaft of the pile and numerical results may be sensitive to the distance to the outer boundaries of the numerical model. Both dilation and the tendency to develop tensile stresses at the base of the pile may also be important. Examples of the analyses of pile behaviour are given by Ellison et al. (1971), Desai (1974) and Withiam & Kulhawy (1979).

6.8 SUMMARY

Linear elastic or simplified nonlinear analyses which are actually piecewise linear elastic cannot hope to fully model all the aspects of soil behaviour. However, for load levels short of failure they can often be used to help the engineer better understand the problem at hand and to guide the engineer's final judgment. To the extent that characteristics of soil behaviour such as anisotropy, low tensile strength, residual strength and dilation can be incorporated in analyses, the need for guessing the effects of these characteristics is reduced and the engineer's attention can be directed more to other parts of the problem. Uncertainty regarding the actual field loading conditions and proper modelling of the boundary conditions are noted as two pieces of the overall problem to which sufficient attention has not always been paid.

REFERENCES

Adikari, G.S.N. & A.K. Parkin 1982. Deformation behaviour of Talbingo Dam. *Int. J. for Numer. and Analyt. meth. in geomech.* 6 (3): 353–382.

Almeida, M.S.S. & J.A. Ramalho-Ortigao 1982. Performance and finite element analyses of a trial embankment on soft clay. In R. Dungar, G.N. Pande & J.A. Studer (eds.), *Proceedings of International Symposium on Numerical Models in Geomechanics* (Zurich): 548–558. Rotterdam: Balkema.

Arthur, J.R.F., K.S. Chua & T. Dunston 1977. Induced anisotropy in a sand. *Geotechnique* 27 (1): 13–30.

Arthur, J.R.F., K.S. Chua, T. Dunston & J.I. Rodriguez 1980. Principal stress rotation: a missing parameter. In *Proceedings of ASCE* 106 (GT 4): 419–434.

Arthur, J.R.F. & B.K. Menzies 1972. Inherent anisotropy in a sand. *Geotechnique* 22 (1): 115–128.

Beer, G. 1983. Infinite domain elements in finite element analysis of underground excavations. *Int. J. for Numer. and Analyt. meth. in Geomech.* 7 (1): 1–7.

Bishop, A.W. & D.W. Hight 1977. The value of Poisson's ratio in saturated soils and rocks stressed under drained conditions. *Geotechnique* 27 (3): 369–384.

Booker, J.R. & J.C. Small 1981. Finite element analysis of problems with infinitely distant boundaries. *Int. J. for Numer. and Analyt. meth. in Geomech.* 5 (4): 345–368.

Bossoney, C. & R. Dungar 1980. Past experience and future demands related to dynamic analysis of soil and rockfill structures and foundations. In *Proceedings of International Symposium on Soils under Cyclic and Transient and Loading* (Swansea), 2: 809–820.

Byrne, P.M. & T.L. Eldridge 1982. A three parameter dilatant elastic stress-strain model for sand. In R. Dungar, G.N. Pande & J.A. Studer (eds.), *Proceedings of International Symposium on Numerical Models in Geomechanics* (Zurich): 73–80. Rotterdam: Balkema.

Byrne, P.M. & W. Janzen 1981. SOILSTRESS: A computer program for nonlinear analysis of stresses and deformations in soil. Soil Mechanics, Report No. 52, University of British Columbia, Vancouver.

Byrne, P.M., Y.P. Vaid & L. Samarasekera 1982. Undrained deformation analysis using path dependent material properties. In R. Dungar, G.N. Pande & J.A. Studer (eds.), *Proceedings of International Symposium on Numerical Models in Geomechanics* (Zurich): 294–302. Rotterdam: Balkema.

Chandrasekaran, V. & G.J.W. King 1974. Simulation of excavation using finite elements. In *Proceedings of ASCE* 100 (GT 9): 1086–1089.

Chang, C.Y. & J.M. Duncan 1970. Analysis of soil movement around a deep excavation. In *Proceedings of ASCE* 96 (SM 5): 1655–1681.

Chang, C.S., J.M. Espinoza & E.T. Selig 1980. Computer analysis of Newtown Creek Culvert. In *Proceedings of ASCE* 106 (GT 5): 531–558.

Christian, J.T. 1968. Undrained stress distribution by numerical methods. In *Proceedings of ASCE* 94 (SM 6): 1333–1345.

Christian, J.T. 1982. The application of generalized stress-strain relations. In R.N. Yong & E.T. Selig (eds.), *Application of plasticity and generalized stress-strain in geotechnical engineering*: 182–204. New York: ASCE.

Christian, J.T. & I.H. Wong 1973. Errors in simulating excavation in elastic media by finite elements. *Soils and Foundations* 13 (1): 1–10.

Clough, G.W. & G.M. Denby 1977. Stabilizing berm design for temporary walls in clay. In *Proceedings of ASCE* 103 (GT 2): 75–90.

Clough, G.W. & J.M. Duncan 1971. Finite element analyses of retaining wall behaviour. In *Proceedings of ASCE* 97 (SM 12): 1657–1673.

Clough, G.W. & L.A. Hansen 1981. Clay anisotropy and braced wall behavior. In *Proceedings of ASCE* 107 (GT 7): 893–913.

Clough, G.W. & Y. Tsui 1977. Static analysis of earth retaining structures. In C.S. Desai & J.T. Christian (eds.), *Numerical methods in geotechnical engineering*: 506–527. New York: McGraw-Hill.
Cundall, P.A. 1975. QUAKE–A computer program to model shear-wave propagation in a hysteretic, layered site. Technical Note, Dames & Moore, ATC London.
Daniel, D.E. & R.E. Olson 1974. Stress-strain properties of compacted clays. In *Proceedings of ASCE* 100 (GT 10): 1123–1136.
D'Appolonia, E. & T.W. Lambe 1970. Method for predicting initial settlement. In *Proceedings of ASCE* 96 (SM 2): 523–544.
Desia, C.S. 1974. Numerical design-analysis for piles in sands. In *Proceedings of ASCE* 100 (GT 6): 613–635.
Desai, C.S. 1977. Deep foundations. In C.S. Desai & J.T. Christian (eds.), *Numerical methods in geotechnical engineering*: 235–271. New York: McGraw-Hill.
Desai, C.S. 1981. Behavior of interfaces between structural and geologic media. In *Proceedings of International Conference on Recent Advances in Geotechnical Earthquake Engineering* (St. Louis), 2: 619–638.
Dickin, E.A. & G.J.W. King 1982. The behavior of hyperbolic stress-strain models in triaxial and plane strain compression. In R. Dungar, G.N. Pande & J.A. Studer (eds.), *Proceedings of International Symposium on Numerical Models in Geomechanics* (Zurich): 303–311. Rotterdam: Balkema.
Domaschuk, L. & P. Valliappan 1975. Nonlinear settlement analysis by finite element. In *Proceedings of ASCE* 101 (GT 7): 601–614.
Drnevich, V.P. 1975. Constrained and shear moduli for finite elements. In *Proceedings of ASCE* 101 (GT 5): 459–474.
Duncan, J.M. 1979. Behavior and design for long-span metal curves. In *Proceedings of ASCE* 105 (GT 3): 399–418.
Duncan, J.M., P. Byrne, K.S. Wong & P. Mabry 1980a. Strength, stress-strain and bulk modulus parameters for finite element analyses of stresses and movements in soil masses. Report No. UCB/GT/80–01, University of California, Berkeley.
Duncan, J.M. & C.Y. Chang 1970. Nonlinear analysis of stress and strain in soils. In *Proceedings of ASCE* 96 (SM 5): 1629–1653.
Duncan, J.M. & C.Y. Chang 1972. Closure to nonlinear analysis of stress and strain in soils. In *Proceedings of ASCE* 98 (SM 5): 495–497.
Duncan, J.M. & G.W. Clough 1971. Finite element analyses of Port Allen Lock. In *Proceedings of ASCE* 97 (SM 8): 1053–1067.
Duncan, J.M. & P. Dunlop 1969. The behaviour of soils in simple shear tests. In *Proceedings of 7th International Conference on Soil Mechanics and Foundation Engineering* (Mexico City), 1: 101–109.
Duncan, J.M., K.S. Wong & Y. Ozawa 1980b. FEADAM: A computer program for finite element analysis of dams. Report No. UCB/GT/80–02, University of California, Berkeley.
Dunlop, P. & J.M. Duncan 1970. Development of failure around excavated slopes. In *Proceedings of ASCE* 96 (SM 2): 471–493.
Dysli, M. & A. Fontana 1982. Deformations around excavations in clayey soils. In R. Dungar, G.N. Pande & J.A. Studer (eds.), *Proceedings of International Symposium on Numerical Models in Geomechanics* (Zurich): 634–642. Rotterdam: Balkema.
Eisenstein, Z., A.V.G. Krishnayya & N.R. Morgenstern 1972. An analysis of the cracking at Duncan Dam. In *Proceedings of ASCE Specialty Conference on Performance of Earth and Earth-Supported Structures* (Purdue University, Lafayette, Indiana), 1 (1): 765–778.
Ellison, R.D., E. D'Appolonia & G.R. Thiers 1971. Load-reformation mechanism for bored piles. In *Proceedings of ASCE* 37 (SM 4): 661–678.
Felix, B., R. Frank & M. Kutniak 1982. F.E.M. calculations of a diaphragm wall, influence of the initial pressures and of the contact laws. In R. Dungar, G.N. Pande & J.A. Studer (eds.), *Proceedings of International Symposium on Numerical Models in Geomechanics* 643–652. Rotterdam: Balkema.
Foott, R. & C.C. Ladd 1977. Behavior of Atchafalaya levees during construction. *Geotechnique* 27 (2): 137–160.
Hansen, B. 1958. Line ruptures regarded as narrow rupture zones: basic equation based on kinematic considerations. In *Proceedings of Brussels Conference on Earth Pressure Problems* (Brussels), 1: 39–48.
Hansen, L.A. & G.W. Clough 1982. Characterization of the undrained anisotropy of clays. In R.N. Yong & E.T. Selig (eds.), *Application of plasticity and generalized stress-strain in geotechnical engineering*: 253–276. New York: ASCE.

Hardin, B.O. 1983. Plane strain constitutive equations for soils. *J. Geotech. Engng*. 109 (3): 388–407.
Hardin, B.O. & V.P. Drnevich 1972. Shear modulus and damping in soils: design equations and curves. In *Proceedings of ASCE* 98 (SM 7): 667–692.
Heuze, F.E. & T.G. Barbour 1982. New models for rock joints and interfaces. In *Proceedings of ASCE* 108 (GT 5): 757–776.
Hoeg, K. 1972. Finite element analysis of strain-softening clay. In *Proceedings of ASCE* 98 (SM 1): 43–58.
Hughes, T.J.R. & D.S. Malkus 1977. On the equivalence of mixed finite element methods with reduced/selective integration displacement methods. In *Proceedings of Symposium on Applications of Computer Methods in Engineering* (University of Southern California, Los Angeles), 1: 23–32.
Hughes, J.M.D., C.P. Wroth & D. Windle 1977. Pressuremeter tests in sand. *Geotechnique* 27 (f): 455–478.
Katona, M.G. 1981. A simple contact-friction interface element with applications to buried culverts. In C.S. Desai & S.K. Saxena (eds.), *Proceedings of Symposium on Implementation of Computer Procedures and Stress-Strain Laws in Geotechnical Engineering* (Chicago), 1: 45–63. Durham: Acorn Press.
Katona, M.G., J.M. Smith, R.S. Odello & J.R. Allgood 1976. CANDE–A modern approach for the structural design and analysis of buried culverts. Report No. FHWA-RD-77-5, Federal Highway Administration, Washington, D.C.
Kondner, R.L. 1963. Hyperbolic stress-strain response: cohesive soils. In *Proceedings of ASCE* 89 (SM 1): 115–143.
Kondner, R.L. & J.S. Zelasko 1963. A hyperbolic stress-strain formulation for sands. In *Proceedings of 2nd Pan-American Conference on Soil Mechanics and Foundation Engineering* (Brazil), 1: 289–324.
Kulhawy, F.H. 1974. Analysis of a high gravity retaining wall. In *Proceedings of ASCE Specialty Conference on Numerical Analysis and Design in Geotechnical Engineering* (Austin), 2: 159–172.
Kulhawy, F.H. 1977. Embankments and excavations. In C.S. Desai & J.T. Christian (eds.), *Numerical methods in geotechnical engineering*: 528–555. New York: McGraw-Hill.
Kulhawy, F.H. & J.M. Duncan 1972. Stresses and movements in Oroville Dam. In *Proceedings of ASCE* 98 (SM 7): 653–665.
Ladd, C.C. et al. 1977. Stress-deformation and strength characteristics. In *Proceedings of 9th International Conference on Soil Mechanics and Foundation Engineering* (Tokyo).
Lade, P.V. 1972. The stress-strain and strength characteristics of cohesionless soils. Ph.D. thesis, University of California, Berkeley.
Lade, P.V. & J.M. Duncan 1973. Cubical triaxial tests on cohesionless soil. In *Proceedings of ASCE* 99 (SM 10): 793–812.
Lade, P.V. & J.M. Duncan 1976. Stress-path dependent behavior of cohesionless soil. In *Proceedings of ASCE* 102 (GT 1): 51–68.
Lee, K.L. & I.M. Idriss 1975. Static stresses by linear and nonlinear methods. In *Proceedings of ASCE* 101 (GT 9): 871–888.
Lefebvre, G., J.M. Duncan & E.L. Wilson 1973. Three-dimensional finite element analyses of dams. In *Proceedings of ASCE* 99 (SM 7): 495–507.
Mana, A.I. & G.W. Clough 1981. Prediction of movements for braced cuts in clay. In *Proceedings of ASCE* 107 (GT 6): 759–777.
Morgenstern, N.R. & Z. Eisenstein 1970. Methods of estimating lateral loads and deformations. In *Proceedings of ASCE Specialty Conference on Lateral Stresses in the Ground* (Cornell University, Ithaca, New York): 51–102.
Murphy, D.J., G.W. Clough & R.S. Woolworth 1975. Temporary excavation in varved clay. In *Proceedings of ASCE* 101 (GT 3): 279–295.
Nobari, E.S. & J.M. Duncan 1972. Movements in dams due to reservoir filling. In *Proceedings of ASCE Specialty Conference on Performance of Earth and Earth-Supported Structures* (Purdue University, Lafayette, Indiana), 1 (1): 797–814.
Oda, M. 1972a. Initial fabrics and their relations to mechanical properties of granular material. *Soils and Foundations* 12 (1): 17–36.
Oda, M. 1972b. The mechanism of fabric changes during compressional deformation of sand. *Soils and Foundations* 12 (2): 1–18.
Oda, M. 1981. Anisotropic strength of cohesionless sands. In *Proceedings of ASCE* 107 (GT 9): 1219–1231.
Osaimi, A.E. & G.W. Clough 1979. Pore press dissipation during excavation. In *Proceedings of ASCE* 105 (GT 4): 481–498.
Ozawa, Y. & J.M. Duncan 1973. ISBILD: A computer program for analysis of static stresses and movements in embankments. Report No. TE-73-4, University of California, Berkeley.

Ozawa, Y. & J.M. Duncan 1976. Elasto-plastic finite element analyses of sand deformations. In *Proceedings of 2nd International Conference on Numerical Methods in Geomechanics* (Blacksburg), 1: 293–262.

Popescu, M.E. 1982. Stability analysis of deep excavations in expansive clays. In R. Dungar, G.N. Pande & J.A. Studer (eds.), *Proceedings of International Symposium on Numerical Models in Geomechanics* (Zurich): 660–667. Rotterdam: Balkema.

Prevost, J.H. & K. Hoeg 1975. Soil mechanics and plasticity analysis of strain softening. *Geotechnique* 25 (2): 279–297.

Pyke, R.M. 1979. Nonlinear models for irregular cyclic loadings. In *Proceedings of ASCE* 105 (GT 6): 715–726.

Roth, W.H. & L. Crandall 1981. Nonlinear elastic finite element analysis of lateral earth pressures against basement wall. *Int. J. for Numer. and Analyt. Meth. in Geomech.* 5 (4): 327–344.

Roth, W.H., T.D. Swantko & L.D. Handfelt 1981. Finite element analysis of displacement patterns in a sand fill on soft ground. In C.S. Desai & S.K. Saxena (eds.), *Proceedings of Symposium on Implementation of Computer Procedures and Stress-Strain Laws in Geotechnical Engineering* (Chicago), 2: 403–418. Durham: Acorn Press.

Rude, L.C. 1982. Computer modeling of a cross canyon culvert. In R. Dungar, G.N. Pande & J.A. Studer (eds.), *Proceedings of International Symposium on Numerical Models in Geomechanics* (Zurich): 668–676. Rotterdam: Balkema.

Sangrey, D.A., D.J. Henkel & M.I. Esrig 1969. The effective stress response of a saturated clay soil to repeated loading. *Can. Geotech. J.* 6 (3): 241–252.

Seed, H.B. The use of numerical modelling in the practical design of earth dams for seismic loading. This volume.

Simon, R.M., J.T. Christian & C.C. Ladd 1974. Analysis of undrained behavior of loads on clay. In *Proceedings of ASCE Specialty Conference on Numerical Analysis and Design in Geotechnical Engineering* (Austin), 1: 51–84.

Smith, I.M. & R. Hobbs 1974. Finite element analysis of centrifuged and built-up slopes. *Geotechnique* 24 (4): 531–559.

Smith, I.M. & S. Kay 1971. Stress analysis of contractive or dilative soil. In *Proceedings of ASCE* 97 (SM 7): 981–996.

Taylor, P.W. 1971. The properties of soils under dynamic stress conditions. Report 79, School of Engineering, University of Auckland.

Tsui, Y. & G.W. Clough 1974. Plane strain approximations in finite element analyses of temporary walls. In *Proceedings of ASCE Specialty Conference on Numerical Analysis and Design in Geotechnical Engineering* (Austin), 2: 173–198.

Vaid, T.P., P.M. Byrne & J.M.O. Hughes 1981. Dilation angle and liquefaction potential. In *Proceedings of International Conference on Recent Advances in Geotechnical Earthquake Engineering* (St, Louis) 1: 161–166.

Withiam, J.L. & F.H. Kulhway 1979. Analytical model for drilled shaft foundations. In *Proceedings of 3rd International Conference on Numerical Methods in Geomechanics* (Aachen) 3: 1115–1122.

Analysis of construction pore pressures in embankment dams

7

J. GHABOUSSI & K.J. KIM

7.1 INTRODUCTION

The excess pore pressures generated during construction can have important implications is some geotechnical problems. An example is the pore pressures generated during the construction of earth dams. The subsequent dissipation of such pore pressures may lead to unsatisfactory stress condition due to arching in the core of zoned dams. Another example is the excess pore pressures generated during tunnelling. The stresses in the ground, in the vicinity of the tunnel heading, are modified as a result of tunnel construction. In some types of grounds, the change in the state of stress results in generation of excess pore pressures. Later in this chapter, only some results of analysis of pore pressures in earth dams will be presented. However, the methodology presented in this chapter is general and can also be applied to other problems, such as study of excess pore pressures during tunnel construction.

The formulation is presented in a general form such that it can be used in analysis of fully saturated and partially saturated soils. Following, a brief discussion of the field equations and the finite element formulation, a constitutive model for the compressibility of pore air-water mixture in partially saturated soils is presented. The remainder of the paper is devoted to analysis of pore pressures during construction of earth dam and in the last section the results of analysis are compared with field measurements.

7.2 SATURATED AND PARTIALLY SATURATED SOILS

Saturated soils can be modelled as coupled fluid saturated porous deformable media. The beginning of treatment of saturated soils as coupled two-phase media can probably be traced to Terzaghi's consolidation theory. In recent years methodology for finite element analysis of saturated soils, modelled as coupled two-phase media, have been proposed and well established. A method of analysis which included the compressibility of the pore fluid was proposed by Ghaboussi & Wilson (1973) and extended by Ghaboussi & Karshenas (1978) to consider the nonlinear material behaviour. These two works form the starting point of this chapter. Although in fully saturated soils the compressibility of pore fluid can be neglected, it must be taken into account in partially saturated soils. The pore space in partially saturated soils is assumed to be saturated by an equivalent fluid with compressibility equal to that of pore air-water mixture.

A nonlinear finite element method of analysis of coupled two-phase media is presented in this chapter for unified treatment of saturated and partially saturated soils. Nonlinear material behaviour as well as the nonlinear compressibility of pore air-water mixture are taken into account.

The fundamental equation in modelling the saturated and partially saturated soils as coupled two-phase media is Terzaghi's effective stress equation. This equation indicates that total stresses σ_{ij} are the sum of effective stresses σ'_{ij} and pore pressure π. Terzaghi's effective stress equation is also valid in incremental form, needed in nonlinear analysis:

$$d\sigma_{ij} = d\sigma'_{ij} + \delta_{ij} d\pi \tag{7.1}$$

(δ_{ij} = Kronecker's delta)

The effective stress equation is also valid for partially saturated soils, provided the pore air is in occluded state, in the form of air bubbles. This occurs at water contents equal to and greater than the standard optimum water content.

The behaviour of solid skeleton (in drained condition) is represented by some elasto-plastic material model (see for example: Schofield & Wroth 1968, Ghaboussi & Karshenas 1978 and Karshenas & Ghaboussi 1979). The incremental stress-strain relation is given in the following equation:

$$d\sigma'_{ij} = D^{ep}_{ijkl} d\varepsilon_{kl} \tag{7.2}$$

in which ε_{ij} is the strain tensor:

$$\varepsilon_{ij} = \tfrac{1}{2}(u_{i,j} + u_{j,i}) \tag{7.3}$$

Displacements are denoted by u_i and comma denotes differentiation with respect to the subsequent indices.

The continuity of pore fluid (pore water or pore air-water mixture) is expressed through the storage equation. The following is the incremental form of the storage equations:

$$d\pi = (1 - n)\alpha d\varepsilon_v + n\alpha d\zeta \tag{7.4}$$

in which:

$$\frac{1}{\alpha} = (1 - n)C_g + nC_w \tag{7.5}$$

and n = porosity; C_g = compressibility of solid grain; C_w = compressibility of pore fluid; ε_v = volumetric strain; and ζ = volumetric strain of the pore fluid. For fully saturated soils, it is often assumed that the pore fluid is incompressible, which is equivalent to ignoring the compressibility of the solid grains and the pore fluid ($C_g = C_w = 0$). However, for partially saturated soils C_w must be taken into account but C_g, which is smaller than C_w by several orders of magnitude, can be neglected. The determination of C_w for partially saturated soils will be discussed in a later section.

The flow of pore fluid is assumed to be governed by the generalized D'Arcy flow law:

$$n(\dot{U}_i - \dot{u}_i) = k_{ij}(\pi_{,j} + \rho_f b_j) \tag{7.6}$$

in which k_{ij} = permeability tensor; U_i = displacement of pore fluid; ρ_f = mass density of pore fluid; b_j = body forces; and the superposed dot denotes time rate.

7.3 FINITE ELEMENT EQUATIONS

Finite element formulation of coupled two-phase media for linearly elastic materials and compressible fluid is given by Ghaboussi & Wilson (1973). Extensions to nonlinear materials and nonlinear fluid compressibility are presented by Ghaboussi & Karshenas (1977) and Kim (1982), which will be briefly described in this section.

The incremental form of the matrix equation for the nonlinear two-phase problem can be written as follows:

$$\left[\begin{array}{c|c} K_t & C \\ \hline C^T & -E-1*H \end{array}\right]\left\{\begin{array}{c} \Delta u_n \\ \hline \pi_n \end{array}\right\}=\left\{\begin{array}{c} P_n-R_{n-1} \\ \hline -1*Q_n-S_{n-1} \end{array}\right\} \tag{7.7}$$

where subscript n denotes the value of the function at time t_n, superposed T denotes transpose and the symbol * designates the convolution integral. The vectors R and S are internal resisting force vectors which are computed as follows:

$$R_n=\sum\int_v B^T\sigma'_n dv \tag{7.8}$$

$$S_n=C^T u_n \tag{7.9}$$

The definitions of the other terms of Equation 7.7 are as follows:

Δu = increment of displacement vector;
π = vector of pore pressures;
P = vector of nodal forces;
Q = vector of nodal flow rates;
K_t = tangent stiffness matrix;
C = coupling matrix between solid skeleton and pore fluid;
E = matrix of compressibility of pore fluid;
H = dissipation resistance matrix.

Equation 7.7 is an integral equation. To obtain the solution of this equation some form of discretization in time must be used. When the simple assumption of linear variation of displacements and pore pressures within a time step Δt is used, the following time-marching scheme results:

$$\left[\begin{array}{c|c} K_t & C \\ \hline C^T & -E-\frac{1}{2}\Delta tH \end{array}\right]\left\{\begin{array}{c} \Delta u_n \\ \hline \pi_n \end{array}\right\}=\left\{\begin{array}{c} P_n-R_{n-1} \\ \hline \bar{Q}_n \end{array}\right\} \tag{7.10}$$

The total displacements are obtained by accumulation of incremental displacements, $u_n=u_{n-1}+\Delta u_n$. The equivalent flow vector $\bar{Q}_n$ is given by the following equation:

$$\bar{Q}_n=\tfrac{1}{2}\Delta t(Q_n+Q_{n-1})+(\tfrac{1}{2}\Delta tH-E)\pi_{n-1} \tag{7.11}$$

Due to dependence of tangent stiffness matrix on the displacements, iterations at each time step may be required. The methodology available for nonlinear analysis of solids is also applicable to Equation 7.10.

7.4 PARTIALLY SATURATED SOILS

In analysis of partially saturated soils, it is assumed that the pore space is filled with an equivalent fluid whose compressibility is the same as that of the air-water mixture. Therefore, the main object of this section is to present a constitutive model for compressibility of the air-water mixture and to discuss methods of determining the material parameters from laboratory tests on samples of partially saturated soils.

The physics of pore air-water mixture is complex. The compressibility depends mainly on the degree of saturation, S. The treatment in this section is restricted to ranges of degree of saturation corresponding to standard optimum or above optimum water contents. At these water contents the pore air is likely to exist in occluded state, in the form of air bubbles.

In earlier treatments of compressibility of air-water mixture by Hilf (1948), Bishop (1957) and Skempton & Bishop (1954), the basic principles of Boyle's law and Henry's solubility are used to arrive at an expression for compressibility, C_w:

$$C_w = (1 - S_0 + H_c S_0)\frac{\pi_{a0}}{\pi^2} \tag{7.12}$$

in which C_w = compressibility of air-water mixture, S_0 = initial degree of saturation, H_c = coefficient of solubility, π_{a0} = initial pore air pressure and π = pore water pressure. For the sake of simplicity, the surface tension between the air and water phases has been neglected in Equation 7.12. The surface tension, which is equal to one-half of the difference between pore air pressure and pore water pressure multiplied by the radius of the air bubble, has been explicitly introduced in an expression for compressibility of air-water mixture proposed by Schuurman (1966). Apart from complexity of this expression, the initial radius of the air bubbles has to be determined from the initial negative pore pressure under the assumption that the initial pore pressure is atmospheric. Thus, the accuracy of the whole expression depends on the degree of accuracy in the measurement of the initial negative pore pressure.

A simpler model has been proposed by Kim (1982) which is more amenable to determination of parameters from laboratory experiments. The difference between the air and water pressures, $T = \pi_a - \pi$, is assumed to remain constant over the pressure ranges encountered in practice. The pressure difference T slightly increases with increasing pore water pressure since the radius of the air bubbles decreases as the pressure in the surrounding pore water increases. Since the compressibility is approximately inversely proportional to the square of the pore air pressure, the slight variations in the pressure difference T will cause a very slight change in compressibility C_w. In short, the assumption of constant pressure difference is a reasonable assumption as long as the value of T can be determined from laboratory tests. With the assumption of constant T, Kim (1982) has derived the following expressions for the compressibility of air-water mixture:

$$C_w = (1 - S_0 + H_c S_0)\frac{\pi_{a0}}{(\pi + T)^2} \tag{7.13}$$

The value of pressure difference T can be determined from trial simulation of undrained isotropic compression tests with pore pressure and volume change measurements.

The volume change in undrained isotropic test is solely due to contraction (or expansion) of the pore air space. Therefore, the increments of effective pressure dp' and pore pressure $d\pi$ can be related to the increment of volume change $d\varepsilon_v$ independently of each other:

$$dp' = (Bp')d\varepsilon_v \tag{7.14}$$

$$d\pi = \left(\frac{1}{nC_w}\right)d\varepsilon_v \tag{7.15}$$

in which n = initial porosity and $B = B_c$ in virgin compression or $B = B_s$ in swelling and recompression. B_c and B_s are related to virgin compressibility C_c and swelling index C_s, respectively:

$$B_c = 2.3(1 + e_0)\frac{1}{C_c} \tag{7.16}$$

$$B_s = 2.3(1 + e_0)\frac{1}{C_s} \tag{7.17}$$

Test results by Gilbert (1959), Garlanger (1970) and Campbell (1973) show that, when the water content is on or above the optimum value, the pore air remains in occluded state and thus Terzaghi's effective stress principle is applicable:

$$dp = dp' + d\pi \tag{7.18}$$

The relation between applied pressure increment and the volumetric strain increment is obtained by substituting Equations 7.14 and 7.15 into Equation 7.18:

$$dp = \left(Bp' + \frac{1}{nC_s}\right)d\varepsilon_v \tag{7.19}$$

The following equations can be used to compute the increments of pore pressure and volumetric strain in terms of increments of applied pressure in undrained isotropic compression tests on partially saturated soils:

$$d\pi = \lambda dp \tag{7.20}$$

$$d\varepsilon_v = \lambda nC_w dp \tag{7.21}$$

$$\frac{1}{\lambda} = 1 + \frac{2.3np'}{1-n}\frac{C_w}{C_c}$$

The value of the pressure difference T is determined from undrained isotropic compression tests on samples of partially saturated soil. The results of such tests are presented in the form of plots of pore pressure π versus applied pressure p and volumetric strain ε_v versus applied pressure p. For a fully saturated sample the pore pressure will be equal to the applied pressure and volumetric strain will be zero. In partially saturated samples, the pore pressures are lower than the applied pressure and volumetric strains increase with applied pressure, approaching a constant value as the volume of pore air decreases. Eventually, at higher applied pressures, the air bubbles collapse and the sample behaves similar to a fully saturated one.

The results of an undrained isotropic compression test are shown in Figure 7.1 by open circles. This is one of a number of tests performed by Garlanger (1970) on

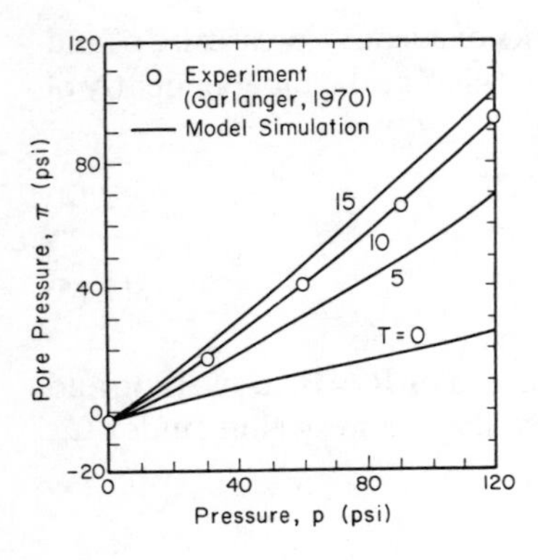

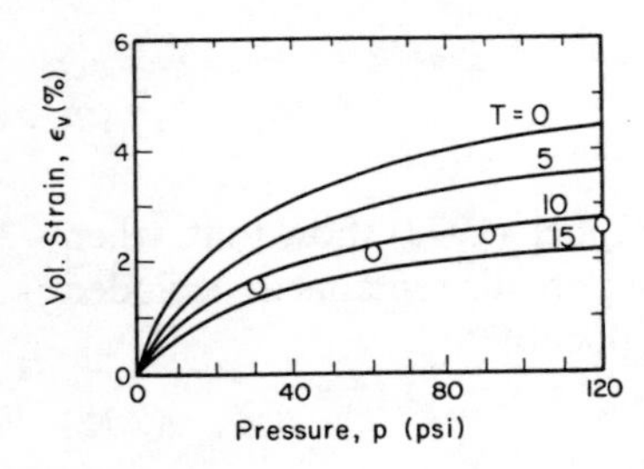

Figure 7.1. Simulation of undrained isotropic compression test on Champaign Till with water content, $w = 13\%$

Champaign Till, which is a well graded sandy silty clay. The water content was $W = 13\%$ (standard optimum $W = 12\%$) with a degree of saturation $S = 85\%$ and a porosity of $n = 29\%$. The soil was compacted by kneading under 90 psi confining pressure. Thus, the sample is recompressed up to 90 psi pressure. The weighted average of the coefficient of recompression bulk modulus was determined to be $B_s = 82$. The sample exhibits an initial negative pore pressure of 3 psi under atmospheric confining pressure. Therefore, it can be expected that the pressure difference T would be greater than 3 psi. The simulated results with four values of $T = 0, 5, 10, 15$ psi are shown in Figure 7.1. The curve for $T = 0$ corresponds to the case of neglecting the surface tension, which in this case is not a good assumption. It can be seen that the most likely value of pressure difference is $T = 10$ psi.

Shown in Figure 7.2 are the results of one of the tests performed by Campbell (1973) on Peorian Loess which is classified as inorganic silt. The sample water content is $W = 21\%$ (standard optimum $W = 18\%$) with a degree of saturation $S = 84\%$ and a porosity of $n = 41\%$. The soil was statically compacted under 40 psi confining pressure and coefficients of bulk modulus were estimated to be $B_c = 41$ and $B_s = 105$. The sample does not exhibit any initial negative pore pressure. As can be seen in Figure 7.2 the simulation result with $T = 0$, corresponding to neglecting of surface tension is reasonably close to the measurements. This seems to indicate that, in absence of initial negative pore pressure, the surface tension can be neglected and the value of pressure difference $T = 0$ is a reasonable approximation. Additional experimental evidence is given by Kim (1982) to support this conclusion.

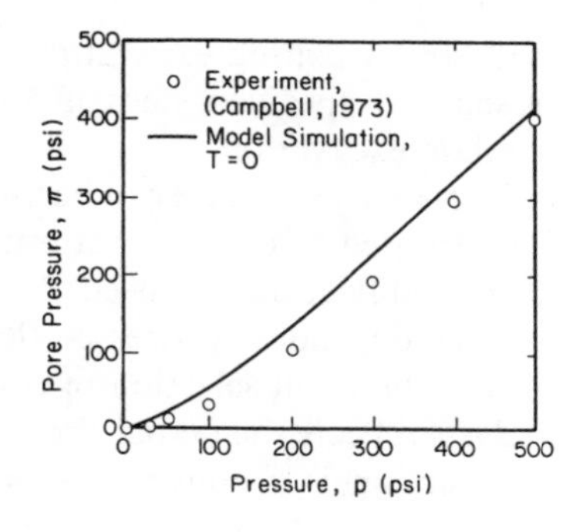

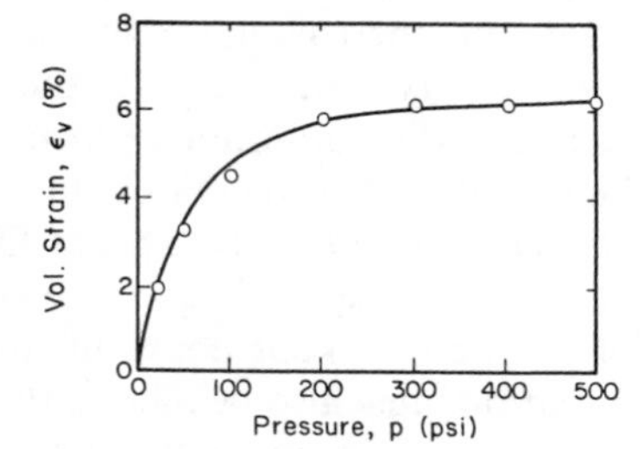

Figure 7.2. Simulation of undrained isotropic compression test on Peorian Loess with water content, w = 21%

7.5 PORE PRESSURES DURING CONSTRUCTION

The first finite element simulation of earth dam construction was reported in literature by Clough & Woodward (1967). Later, similar but independently developed methodology was proposed for finite element simulation of excavation. However, numerical problems have been reported in finite element simulation of excavation (Christian & Wong 1972). A simple method of analysis for finite element simulation of construction and excavation has been developed by the first author. This method treats the excavation and construction as two aspects of the more general process of construction and avoids the numerical problems encountered in the earlier method of simulation of excavation. This general method has been extensively applied by the first author to problems of simulation of underground excavation and construction (Ghaboussi & Ranken 1977, Ghaboussi & Gioda 1977, Ghaboussi et al. 1978, and Ghaboussi et al. 1981). This method differs from earlier methods (for example, see Christian & Wong 1972) in which excavation is simulated by applying, to the newly created boundary surface in post-excavation configuration, the negative of tractions which existed on such boundary in pre-excavation configuration.

In the more general approach proposed by the first author, the construction and excavation process is treated as a form of nonlinearity, the change in geometry being the source of nonlinearity. Similar to other nonlinear problems, an incremental approach such as equation 7.10 is used. For the regions to be excavated or constructed elements are assigned initially. Construction is simulated by activating appropriate elements at specified steps in the process of incremental analysis. A reverse process is used for simulating excavation. Elements in the region to be excavated are initially present. At specified steps in the incremental analysis elements are deactivated to simulate excavation.

The same procedure can be used in simulation of excavation and construction when the saturated or partially saturated soils are modelled as two-phase media. However, some modifications are needed. Since the generation and dissipation of pore pressures are time-dependent processes, the state equations are integrated in real time and care must be taken in selection of construction increments and time intervals so that the analysis reflects the actual progress of construction as closely as possible. The actual construction is a continuous process corresponding to small increments. In finite element simulations the whole construction process is simulated in a few discrete stages, each corresponding to a finite increment. After the simulation of construction of each finite increment, the geometry of the system must be kept constant for a period of time approximately equal to the actual construction time for such an increment. As an example, the continuous process of pore pressure generation and dissipation as the height of a dam increases during the actual construction, is replaced in the finite element simulation by abrupt finite increases in the height of the dam, resulting in abrupt increases in the pore pressures. During the following interval of time the height of the dam is kept constant and pore pressures are allowed to dissipate. Special attention must also be paid to the boundary conditions. For example, in simulation of construction of an earth dam, the surface of a newly placed lift is a free drainage boundary surface (zero pore pressure). However, after adding the next lift, the nodes on such a surface become interior nodes, with their unknown pore pressures as independent variables.

The simplest case which has been studied is the one-dimensional problem of pore pressures in a saturated infinite layer, while the height of the layer h increases at a constant rate m ($h = mt$). The governing differential equation for this problem, expressed in terms of excess pore pressure π_e is as follows:

$$C_v \frac{\partial^2 \pi_e}{\partial x^2} = \frac{\partial \pi_e}{\partial t} - \gamma' \frac{dh}{dt} \qquad (7.22)$$

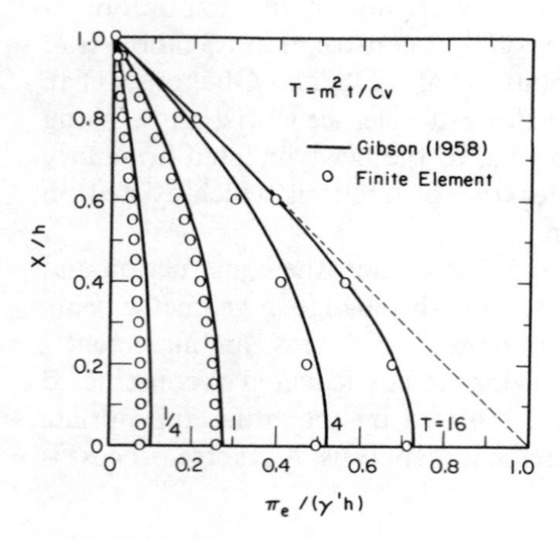

Figure 7.3. Excess pore pressure in fully saturated layer with increasing height, normalized by the current height, h

subject to boundary conditions:

$$\pi_e = 0 \qquad \text{at } x = h$$

$$\frac{\partial \pi_e}{\partial x} = 0 \qquad \text{at } x = 0$$

$$\gamma' = \gamma - \gamma_w$$

The solution to this problem is given by Gibson (1958). The results of finite element simulation of this one-dimensional problem are compared with Gibson's solution in Figure 7.3 for several values of time factor T. The agreement between finite element simulations and Gibson's solution is good. The results shown in Figure 7.3 are normalized with respect to the current height, h. It is obvious that there are some differences between finite element simulation results and Gibson's solution in the early time response.

7.6 CONSTRUCTION PORE PRESSURES IN EARTH DAMS

The problem of pore pressures during the construction of earth dams has attracted the attention of many researchers. Various analytical methods for evaluation of such pore pressures have been proposed and comparisons have been made with field measurements. The stress redistribution resulting from dissipation of construction pore pressures may result in arching and zones of low stress in the core. Such low stress zones in the core may be vulnerable to hydraulic fracturing.

Probably the first analytical method for evaluation of construction pore pressures in earth dams was developed by Hilf (1948). Most of the early methods, including Hilf's formulation, are based on the assumption of one-dimensional vertical flow. Moreover, surface tension between pore air and pore water, as well as dissipation were neglected in Hilf's solution. Bishop (1957) and Li (1959) presented an improved solution by allowing certain amounts of pore pressure dissipation. Li (1959) divided a construction season into a number of incremental steps. At the beginning of each step pore pressures are generated with the assumption of no drainage and use of Hilf's formula. At the end of each step, the pore pressures are allowed to dissipate by a pre-determined fraction of the current pore pressures, thus necessitating the subjective use of engineering judgment.

Gibson (1958) proposed a more rigorous method by considering a moving boundary and allowing dissipation at such boundary. Terzaghi's consolidation theory was modified by including the moving boundary (see Equation 7.22). Solutions for arbitrary rate of construction can be obtained from Gibson's equation by finite difference method. Sheppard (1957) noted that this one-dimensional solution gives good agreement with field measurements, except towards the end of the construction when lateral flow effects are more pronounced.

The lateral flow effects which have been neglected in the earlier methods become specially important in case of dams with clayey cores. In such dams the high construction pore pressures are concentrated in the core and the geometry of cores dictate significant lateral flows. A number of two-dimensional solutions have been proposed. Koppula (1970) and Osaimi & Hoeg (1977) proposed finite difference based

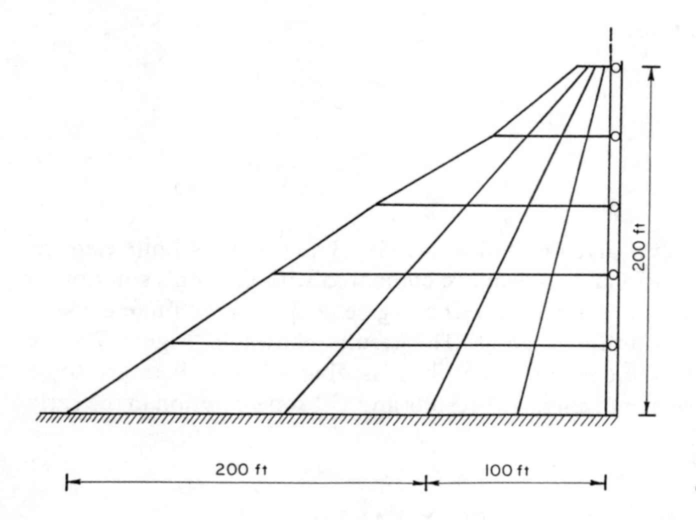

Figure 7.4. Finite element mesh for one-half of the dam with central core

methods, while Eisenstein et al. (1976), Eisenstein & Steven (1977), Cavounidis (1975) and Cavounidis & Hoeg (1977) have proposed finite element based methods. Various simplifying assumptions are implicit in these methods.

A method of analysis was recently proposed by Kim (1982). This method considers the coupling between solid and fluid phases and uses the previously described incremental construction method. The compressibility of pore air-water mixture, including the surface tension effects, as well as elasto-plastic material behaviour for solid skeleton are taken into account. This incremental solution uses the fully coupled Equation 7.7

Shown in Figure 7.4 is the finite element mesh for one-half of a hypothetical earth dam with clayey central core. The construction of this dam is simulated in five lifts, each layer of elements comprising a construction lift. The rate of construction is assumed to be $m = 0.5$ ft per day. After adding a layer of elements, representing a 40 ft lift, the height of the dam was kept constant for 80 days which is the actual construction time for a 40 ft lift at the assumed rate of construction. The material properties used in the analysis are summarized below:

Initial condition:
  Porosity, $n_0 = 0.30$
  Degree of saturation, $S_0 = 0.95$
  Pressure difference, $T = 8$ psi
Elastic properties:
  Modulus of elasticity, $E = E_0 p'$
  for core, $E_0 = 150 \quad \nu = 0.4$
  for shell, $E_0 = 375 \quad \nu = 0.3$
Coefficient of permeability:
  for core, $k = 0.2$ ft/year
  for shell, $k = 6.0$ ft/year

The pore pressure distributions at 80 days after the addition of each lift are shown in Figure 7.5. Also shown in this figure is the pore pressure distribution at about one year

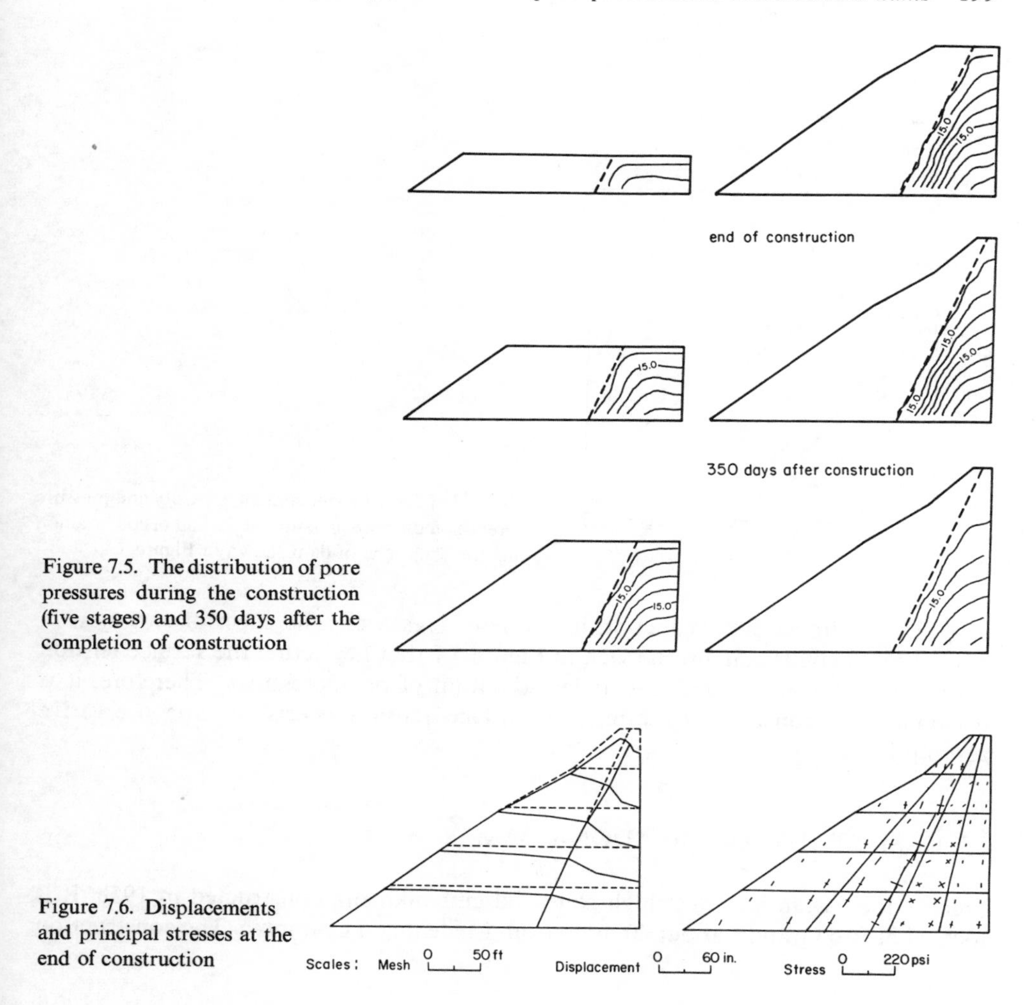

Figure 7.5. The distribution of pore pressures during the construction (five stages) and 350 days after the completion of construction

Figure 7.6. Displacements and principal stresses at the end of construction

after the completion of the construction. The effective principal stresses and the deformed shape are shown in Figure 7.6. It is clear from these figures that as the pore pressures in the core dissipate, more vertical stresses are transferred from the core to the shells. Stresses in the core are generally low, specially around the mid-height of the dam, where the highest settlements occur. The stress transfer from core to shells also results in high shear stresses in the areas of the core adjacent to the shells.

The construction pore pressures in earth dams are strongly dependent on soil water content. This point is well illustrated in Figure 7.7, which shows the pore pressures at the end of construction along the centerline of the dam shown in Figure 7.4. For all the cases shown in Figure 7.7 the construction is simulated in five stages and all the material properties are the same except porosity, degree of saturation and pressure difference, T. The computed pore pressures are normalized with respect to those of fully saturated case ($S_0 = 100\%$), labelled 'incompressible fluid', which develops the highest pore pressures. It can be seen that a reduction of degree of saturation S_0 from 95% to

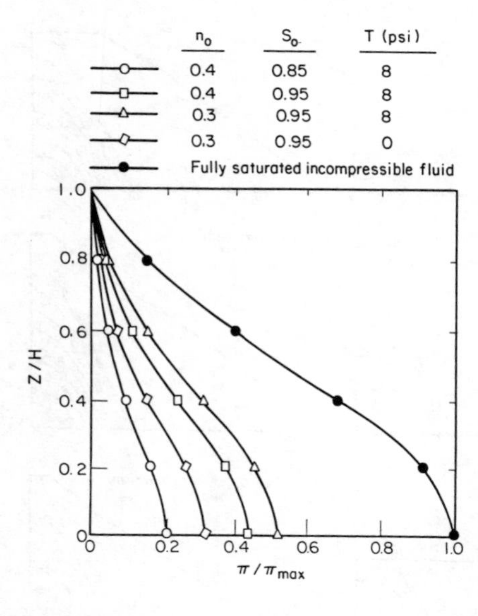

Figure 7.7. Effect of water content, porosity and pressure difference on the pore pressures at the end of construction along the centerline of dam shown in Figure 7.4

85% causes almost 50% reduction in the pore pressures. The effect of variation in porosity is small. It can also be seen in Figure 7.7 that neglecting the surface tension effects (T = 0) causes significant underestimation of pore pressures. Therefore, it is reasonable to consider neglecting the surface tension effects an unconservative assumption.

7.7 CASE STUDY: QUEBRADONA DAM

Quebradona Dam is a 98 ft high rolled fill embankment, constructed in 1958. It is located at Rio Grande, about 40 miles from Medellin, in Colombia. The construction

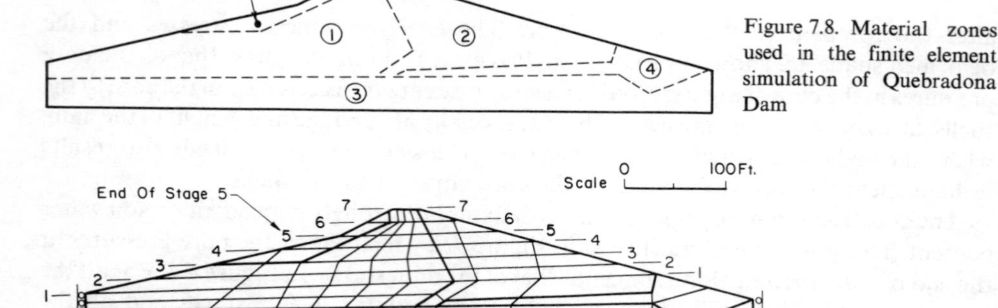

Figure 7.8. Material zones used in the finite element simulation of Quebradona Dam

Figure 7.9. Finite element mesh and seven construction stages for analysis of Quebradona Dam

Table 7.1. Material properties used in finite element solution of Quebradona Dam construction*

Zone	B_0	ν	k (ft/year)	γ (pcf)	n_0 (%)	S_0 (%)	T (psi)
1. Core	115	0.35	2	131	43	92	8
2. Shell	340	0.30	6	124	39	78	8
3. Foundation	1000	0.30	10	124	39	78	8
4. Drainage	340	0.30	20	124	39	78	8

*Assumed that all materials are isotropic and their elastic moduli are dependent on the effective mean pressure; bulk modulus $B = B_0P'$

history of Quebradona Dam as well as extensive field measurements of pore pressures and displacements during the construction have been reported by Li (1959, 1967). The observed field data are thought to be reliable since the climatic conditions were favourable. The actual construction of the embankment took place during a short period of time in the dry season, from December 10, 1957 to March 15, 1958, with an average construction rate of one ft per day.

As shown in Figure 7.8, the cross section of the dam is composed of four separate material zones. The earth fill consisted of a compacted sandy-silt core (zone 1) and relatively more pervious compacted decomposed rock as shell material (zone 2). The earth fill was placed on the residual hard decomposed rock (zone 3) overlying the impervious sound rock base. A sand and gravel drainage blanket over the decomposed rock foundation leads to a downstream rock toe (zone 4).

The finite element mesh used in the analysis is shown in Figure 7.9. Also shown in this figure are the seven construction stages. The material properties for the four zones are given in Table 7.1. These material properties are based on the laboratory test data reported by Li (1967). However, the coefficients of permeability in Table 7.1 are somewhat lower than laboratory test measurements, which were reported by Li (1959) to be 10 ft/year for the core material and 30 ft/year for the shell material. The lower values of the coefficient of permeability were judged to be more appropriate on the basis of a parametric study by Kim (1982). The laboratory test measurements of the coefficient of permeability are thought to differ from field values due to a number of factors which are difficult to reproduce in the laboratory, such as stratification, boundary conditions, degree of saturation, etc.

The actual rate of construction is compared in Figure 7.10 with the stepwise

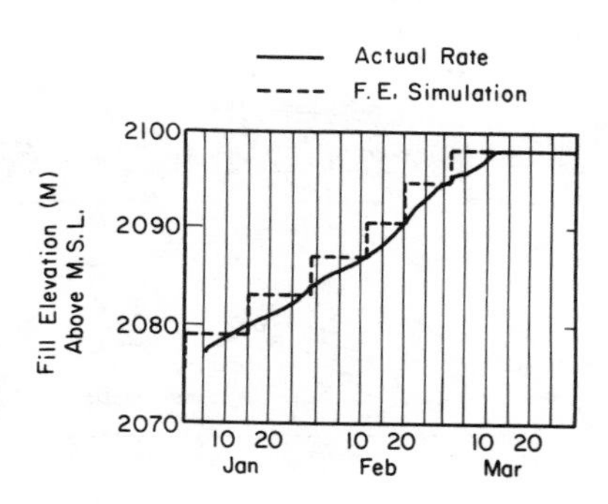

Figure 7.10. Comparison of actual and finite element construction rates

construction rate used in the finite element analysis. It is obvious from this figure that the finite element simulation results can be expected to be comparable to field values at the end of each of the seven construction stages, immediately prior to addition of the next lift. The computed pore pressure distributions at two intermediate construction stages and at the completion of construction are shown in Figure 7.11. The measured pore pressure distributions at roughly the comparable times, reported by Li (1959), are shown in Figure 7.12. The computed pore pressures are slightly higher than the field values. The details of the computed pore pressure distributions are somewhat different, but on the whole they are in reasonable agreement with the measured pore pressure distribution. It is also obvious from Figures 7.11 and 7.12 that at the earlier construction stages the pore pressure distributions indicate nearly one-dimensional

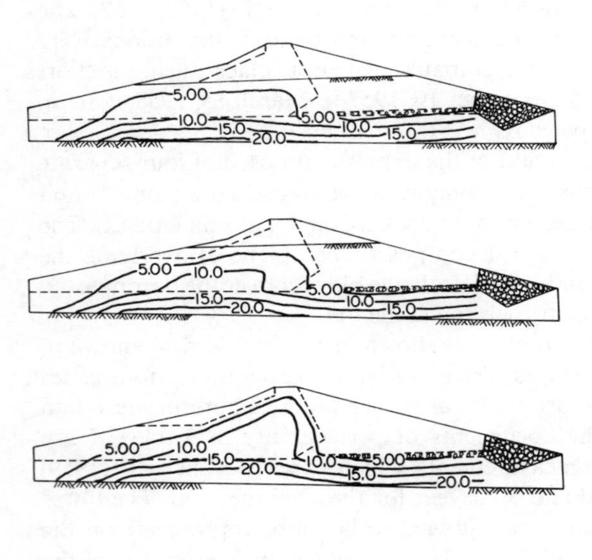

Figure 7.11. Computed pore pressures at two intermediate stages and at completion of construction for Quebradona Dam (pressure units in psi)

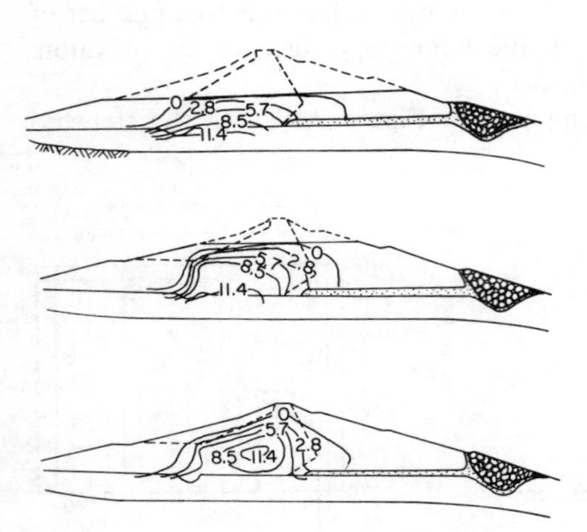

Figure 7.12. Measured pore pressures at two intermediate stages and at completion of construction for Quebradona Dam (pore pressures in psi)

vertical flow. However, towards the end of construction the lateral flow is more pronounced.

The computed and measured pore pressure time histories at a point in the core are shown in Figure 7.13. The agreement is reasonable, although the computed pore pressures are somewhat higher. Also shown in Figure 7.13 is the time history of the computed effective stress. The effect of the dissipation of the construction pore pressure is obvious. The relative displacement between two cross-arms are compared with the

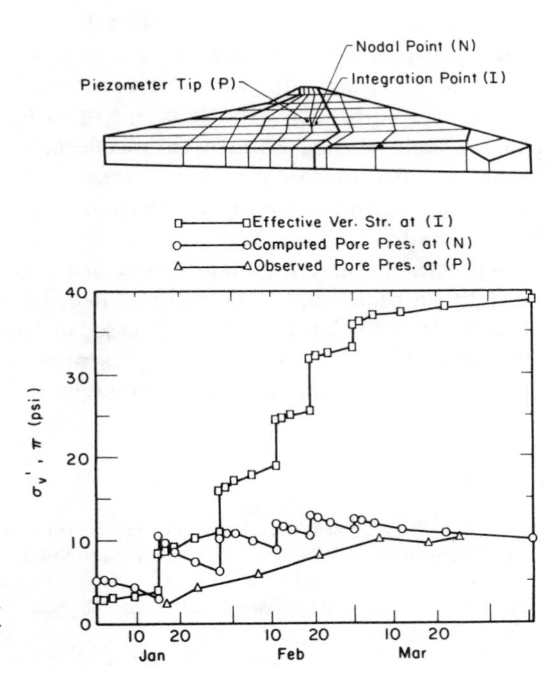

Figure 7.13. Comparison of measured and computed pore pressure time histories for Quebradona Dam

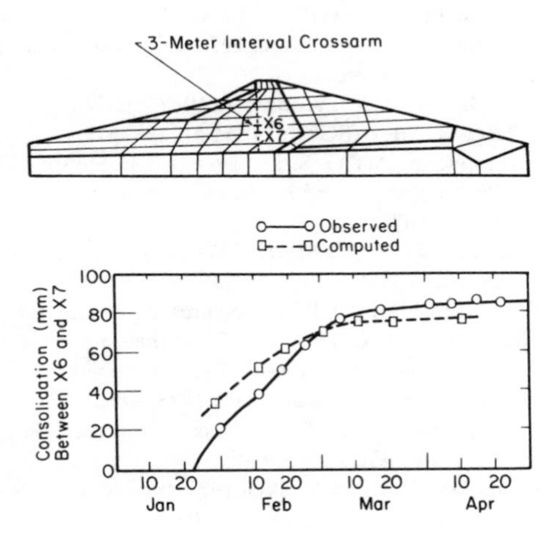

Figure 7.14. Comparison of measured and computed consolidation between cross-arms X6 and X7 for Quebradona Dam

field measurements in Figure 7.14. Also shown in this figure are the locations of these cross-arms. The computed values of the relative displacement are in reasonably close agreement with the field measurements.

7.8 CONCLUDING REMARKS

A general method of finite element analysis of saturated and partially saturated soils has been presented. Two sources of non-liner material behaviour are identified. The nonlinear material behaviour under fully drained conditions is represented by an elasto-plastic material model formulated in terms of effective stresses. For partially saturated soils a constitutive model is presented for representing the compressibility of air-water mixture. The important role of surface tension between pore air and pore water is demonstrated. Surface tension effects are included in the proposed formulation in the form of a constant pressure difference between the pore air and the pore water. A method of determination of this pressure difference from laboratory experiments has been presented.

A general method of simulation of construction and excavation with saturated and partially saturated soils has been presented. The specific problem of construction pore pressures in earth dams has been discussed in detail and some analysis results have been presented. Finally, a case study is presented in which the computed and measured construction pore pressures are compared.

REFERENCES

Bishop, A.W. 1957. Some factors controlling the pore pressure set-up during the construction of earth dams. In *Proceedings of 4th International Conference on Soil Mechanics and Foundation Engineering* (London), 2: 294–300.

Campbell, J.D. 1973. Pore pressures and volume changes in unsaturated soils. Ph.D. thesis, University of Illinois, Urbana.

Cavounidis, S. 1975. Effective stress-strain analysis of earth dams during construction. Ph.D. thesis, Stanford University, Stanford, California.

Cavounidis, S. & K. Hoeg 1977. Consolidation during construction of earth dams. *J. Geotech. Engng. Div.* (ASCE) 103 (GT 10): 1055–1067.

Christian, J.T. & I.H. Wong 1972. Errors in simulating excavations in elastic media by finite elements. *Soils and Foundations* 13 (1).

Clough, R.W. & R.J. Woodward 1967. Analysis of embankment stresses and deformations. *J. Soil Mech. Fdn. Engng. Div.* (ASCE) 93 (SM 4): 529–549.

Eisenstein, Z., A.V.G. Krishnayya & T.C. Steven 1976. Analysis of consolidation in cores of earth dams. In *Proceedings of 2nd International Conference on Numerical Methods in Geomechanics* (Blacksburg), 1: 1089–1107.

Eisenstein, Z. & T.C. Steven 1977. Analysis of consolidation behavior of mica dam. *J. Geotech. Engng. Div.* (ASCE) 103 (GT 8): 879–895.

Garlanger, J.E. 1970. Pore pressures in partially saturated soils. Ph.D. thesis, University of Illinois, Urbana.

Ghaboussi, J. & G. Gioda 1977. On the time-dependent effects in advancing tunnels. *Int. J. for Numer. and Analyt. meth. in Geomech.* 1 (3).

Ghaboussi, J. & M. Karshenas 1978. On the finite element analysis of certain material nonlinearities in geomechanics. In *Proceedings of International Conference on Finite Elements in Nonlinear Solids and Structures* (Geilo, Norway).

Ghaboussi, J. & R.E. Ranken 1977. Interaction between two parallel tunnels. *Int. J. for Numer. and Analyt. meth. in Geomech.* 1 (1): 75–103.

Ghaboussi, J., R.E. Ranken & M. Karshenas 1978. Analysis of subsidence over soft-ground tunnels. In *Proceedings of ASCE International Conference on Evaluation and Prediction of Subsidence* (Pensacola Beach, Florida): 182–196.

Ghaboussi, J., R.E. Ranken & A.J. Hendron 1981. Time-dependent behavior of solution caverns in salt. *J. Geotech. Engng. Div.* (ASCE) 107 (GT 10): 1379–1401.

Ghaboussi, J. & R.E. Ranken 1981. Finite element simulation of underground construction. In *Proceedings of Symposium on Implementation of Computer Procedures and Stress-Strain Laws in Geotechnical Engineering* (Chicago)

Ghaboussi, J. & E.L. Wilson 1973. Flow of compressible fluid in porous elastic media. *Int. J. for Numer. meth. in Engng.* 5 (3).

Gibson, R.E. 1958. The progress of consolidation in a clay layer increasing in thickness with time. *Geotechnique* 8 (4): 171–182.

Gilbert, O.H. 1959. The influence of negative pore water pressures on the strength of compacted clays. M.Sc. thesis, Massachusetts Inst. of Technology, Cambridge, Massachusetts.

Hilf, J.W. 1948. Estimating construction pore pressures in rolled earth dams. In *Proceedings of 2nd International Conference on Soil Mechanics and Foundation Engineering*, 3: 234–240.

Karshenas, M. & J. Ghaboussi 1979. Modeling and finite element analysis of soil behavior. Report No. UILU-ENG-79-2020, Dept. of Civil Engineering, University of Illinois, Urbana.

Kim, K.J. 1982. Finite element analysis of nonlinear consolidation. Ph.D. thesis, University of Illinois, Urbana.

Koppula, S.D. 1970. The consolidation of soil in two dimensions and with moving boundary. Ph.D. thesis, University of Alberta, Edmonton.

Li, C.Y. 1959. Construction pore pressures in an earth dam. *J. Soil mech. Fdns. Div.* (ASCE) 85 (SM 5): 43–59.

Li, C.Y. 1967. Construction pore pressures in three earth dams. *J. Soil mech. Fdn. Engng. Div.* (ASCE) 93 (SM 2): 1–26.

Osaimi, A.E. 1973. Consolidation of earth dams during construction. Ph.D. thesis, Stanford University, Stanford, California.

Osaimi, A.E. & K. Hoeg 1977. Construction pore pressures in earth dams. Ph.D. thesis, Stanford University, Stanford, California.

Schofield, A.N. & C.P. Wroth 1968. *Critical state soil mechanics*. McGraw-Hill, New York.

Schuurman, E. 1966. The compressibility of an air/water mixture and a theoretical relation between the air and water pressures. *Geotechnique* 16: 269–281.

Sheppard, G.A.R. & L.B. Aylen 1957. The usk scheme for the water supply of Swansea. In *Proceedings of Institution of Civil Engineers* 7: 246–265.

Skempton, A.W. & A.W. Bishop 1954. Building materials, their elasticity and inelasticity. *Soils*: Chapter 10. Amsterdam: North Holland Publ. Co.

The use of numerical modelling in the practical design of earth dams for seismic loading 8

H.B. SEED

8.1 INTRODUCTION

Since the near-failure of the Lower San Fernando Dam in California during an earthquake just north of Los Angeles in 1971, an event which necessitated the immediate evacuation of over 80,000 people whose lives were endangered, the design of earth dams to resist earthquake effects has assumed a position of much greater significance among design engineers. Prior to this event it was generally believed that earth dams were inherently resistant to earthquake shaking but the major slide in the Lower San Fernando Dam, which involved the upstream shell, the crest of the dam, and 30 ft of the downstream slope, with a resulting loss of 30 ft of freeboard, provided dramatic evidence that this is not necessarily so. As a result regulatory agencies became more stringent in their requirements for demonstration of adequate seismic stability, and design engineers responded by developing new and more convincing design approaches than had previously been used. Thus the past twelve years have seen a major change in interest and attitude towards this aspect of design.

Another reason for the increased interest of design engineers in the seismic design problem is the increasing construction of major dams close to densely populated areas, where a slope failure or even major cracking, leading to a release of water from the reservoir, could have devastating consequences. In such cases it may be necessary to assess the risk associated with the very remote chance of a seismically induced failure and the catastrophic consequences if it should occur. Such considerations invariably lead to more detailed design studies by the engineer.

Finally the problem is not of limited interest. Many parts of the world are subjected to the potentially hazardous effects of earthquakes and stringent earthquake-resistant design criteria have already been adopted for the design of dams in many different countries. Thus it is not uncommon to find designs being required to withstand ground shaking levels of 0.5 g or more, and the response of earth structures to such strong shaking requires the most critical evaluation on the part of the designer. However ground motions much smaller than this can present major hazards for many existing structures.

Because of the severe risks involved, seismic safety evaluation merits careful attention in design and re-evaluation studies, and this now seems to be widely recognized throughout the profession.

8.2 COMPLEXITY OF THE SEISMIC DESIGN PROBLEM

A rigorous analysis of the response of an earth dam to earthquake excitation is a formidably complex problem because it necessarily involves:

1. A structure with a highly complicated three-dimensional geometrical configuration. The cross-section of a major dam often involves many different zones of different materials (see Figure 8.2 for example) and in a relatively narrow valley this section will vary across the length of the dam. Cases where the embankment can be treated as a homogeneous section are rare.
2. Construction materials which have widely different characteristics in different zones, whose properties are strongly non-linear and often strain-softening, and which vary not only from zone to zone but also in any one zone with the magnitude of the effective confining pressure, which itself may vary throughout the period of earthquake shaking as pore water pressures are generated in the different soils.
3. A highly complex excitation, consisting of random motions in three dimensions which may vary spatially across the base of the embankment.
4. Potentially complicated problems of interaction between the soil and the rock of the valley walls and floor, between the soil and adjacent concrete structures against which it may be placed and between the dam and the water in the reservoir.

Because of the complexity of the problem it is essential to simplify the field situation for analysis purposes in order to gain some insight into the nature of the response and the stability and deformations of the dam. This involves considerable judgment, since over-simplification can lead to erroneous conclusions while attempts at over-elaborate analyses may be equally misleading; good design requires the use of as simple an analytical approach as possible without losing the essential elements of the problem. It is around this question that some of the greatest challenges in seismic analyses are generated.

The earliest attempts at simplification of the seismic response problem involved the representation of the effects of earthquake shaking on the stability of the slopes of a dam by an equivalent static horizontal force determined as the product of a seismic coefficient, k or n_g, and the weight of the potential slide mass (the pseudo-static analysis procedure). It is not clear who originated this procedure for earth dams but it seems to have been first described in writing by Terzaghi (1950), together with a clear statement of his concern regarding its reliability. Subsequent developments have justified his views in this regard. Earthquake-induced failures have occurred in several dams which were indicated by the pseudo-static approach to have adequate seismic stability, largely because the method provides no means for taking into account the effects of pore-pressures generated by earthquake shaking and the resulting reduction in strength of the soils on the stability of the embankment. In recent years this approach has been used mainly as a preliminary guide in the selection of embankment sections and not as a reliable basis for safety evaluations of earth dams involving saturated cohesionless soils in the embankments or foundations. A detailed discussion of the usefulness of the method is presented elsewhere (Seed 1979).

8.3 USE OF NUMERICAL METHODS

With the increasing criticism of the pseudo-static analysis procedure it was necessary for designers to adopt a more meaningful approach, and in view of the complexity of the structures involved it was quickly recognized that the most logical choice for practical evaluations was a numerical approach based on the finite element analysis procedure. This method makes possible an acceptably reliable evaluation of embankment response for most practical purposes. In order to obtain meaningful evaluations of probable performance, however, it is necessary to model adequately the essential elements of the problem including the geometrical configuration, the properties of the soils comprising the embankment and its foundation, and the earthquake excitation, as well as ensuring that the computational procedure is capable of providing accurate results. Some of the considerations involved in modelling these factors are discussed below.

8.3.1 *Geometrical configuration*

Because of the characteristic cross-sectional shape of earth dams and the limited storage capacity of earlier computers, the first response analyses were all developed for two-dimensional models of plane strain conditions. Thus a typical analytical cross-section for a dam was similar to that shown in Figure 8.1, and the results of a response analysis for a given base motion, in this case an earthquake motion representative of that resulting from a Magnitude 7 event occurring at a distance of about 20 miles, is shown in Figure 8.2. The difference between the motions at the base of the dam and those developed at the crest is readily apparent. Such analyses are generally performed to compute the stresses induced in the embankment by the earthquake shaking, and the computed stresses are used in turn to evaluate the induced pore-water pressures and the stability of the slopes or the resulting permanent deformations (Seed 1979). Such analyses are appropriate for long dams where the crest length/height ratio is greater than about 6.

More recent studies have shown that plane strain analyses may lead to substantial errors if they are used to compute the stresses induced by earthquake shaking in dams constructed in steep-walled canyons where the crest length/height ratio may be less than 3 or 4 (Mejia & Seed 1983). For such dams it is necessary to use a three-dimensional analytical model similar to that shown in Figure 8.3. A comparison of the stresses computed using two-dimensional and three-dimensional models for dams with identical maximum sections but with crest length/height ratios of 7 and 2 is shown in Figure 8.4. For the greater crest length, the error in computed stresses determined by a plane strain analysis was only about 10% in this case, but for the crest length/height ratio of 2 the corresponding error in computed stresses ranged from 30 to several hundred percent near the base of the embankment. Clearly the selection of an appropriate model is important in such analyses.

Equally important, however, is the finding that in many cases, the stresses developed by a given earthquake motion in a vertical column of elements in a dam cross-section can often be computed with an adequate degree of accuracy using a simple one-dimensional or shear beam analysis (Vrymoed & Galzascia 1978). A comparison of the

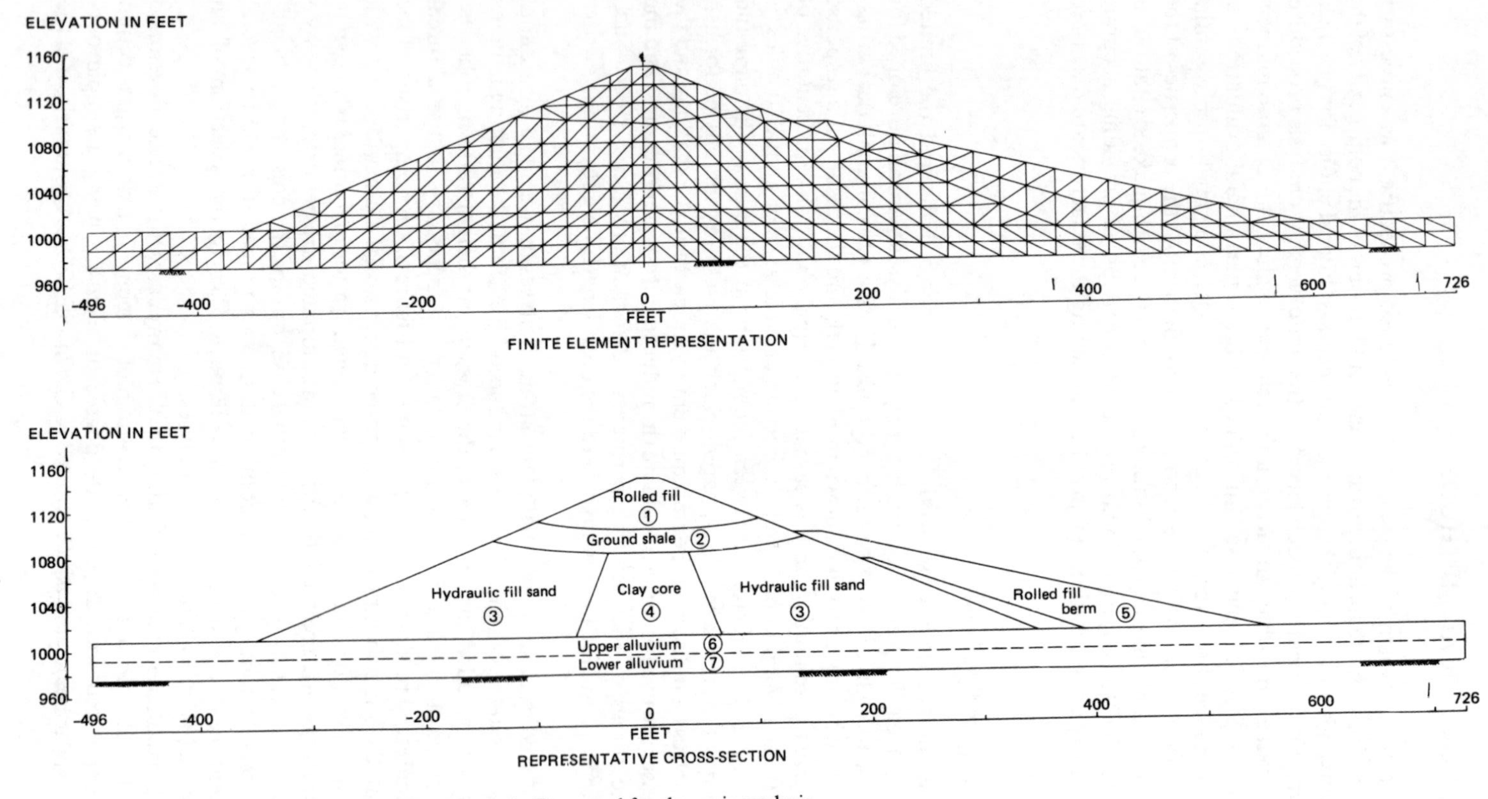

Figure 8.1. Cross-section through Lower San Fernando Dam used for dynamic analysis

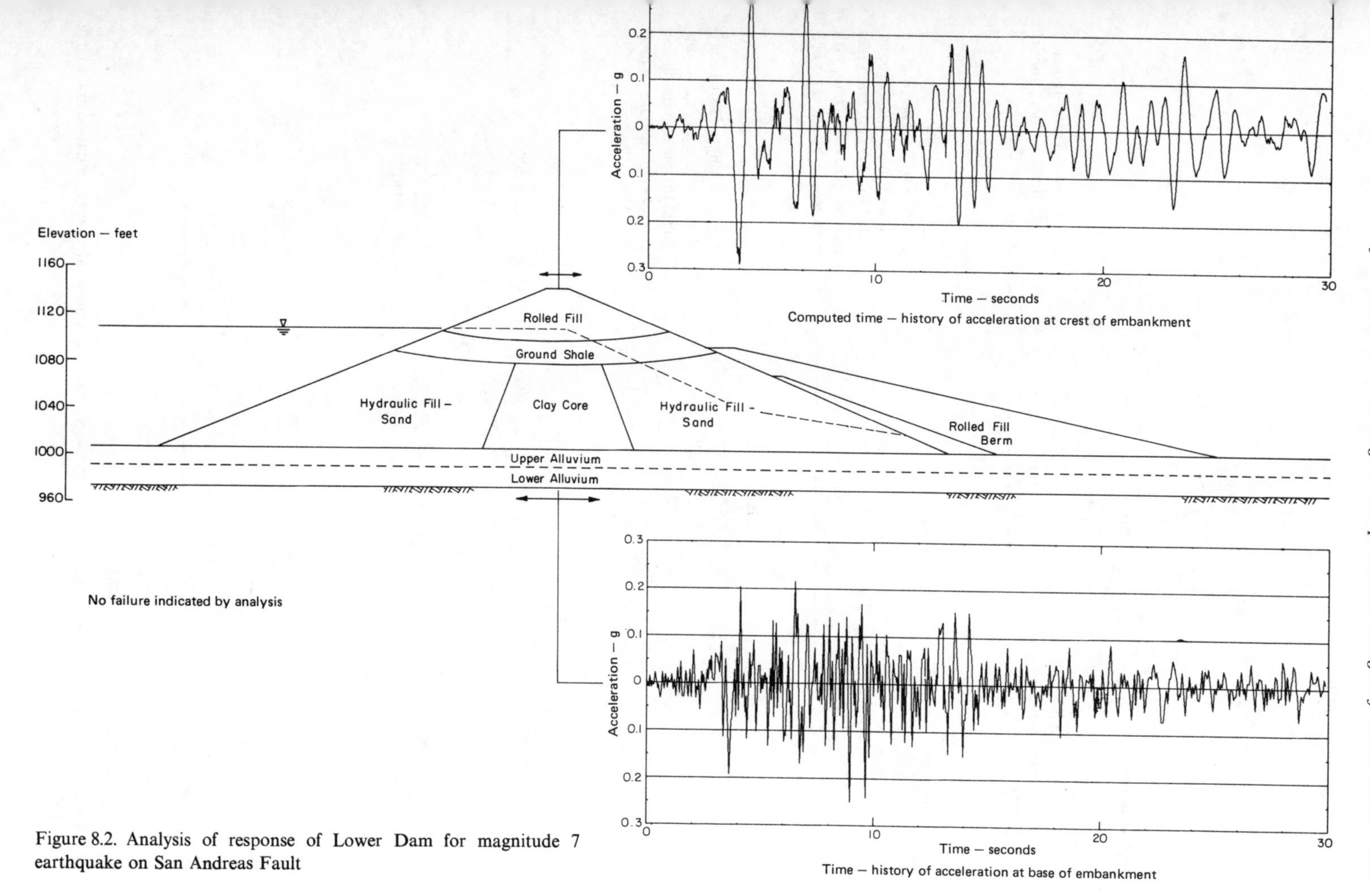

Figure 8.2. Analysis of response of Lower Dam for magnitude 7 earthquake on San Andreas Fault

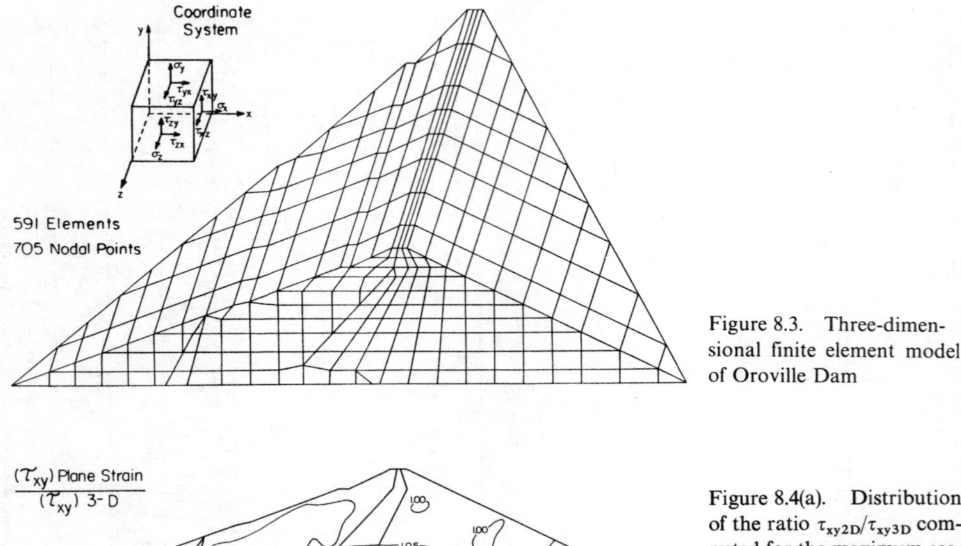

Figure 8.3. Three-dimensional finite element model of Oroville Dam

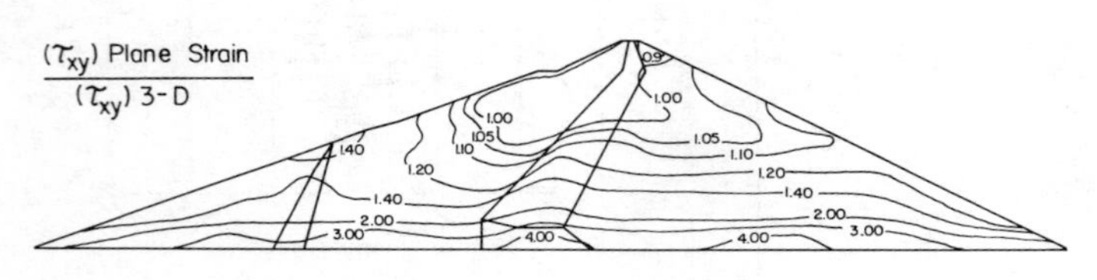

Figure 8.4(a). Distribution of the ratio τ_{xy2D}/τ_{xy3D} computed for the maximum section of Oroville Dam with crest length/height ratio of 7

Figure 8.4(b). Distribution of the ratio τ_{xy2D}/τ_{xy3D} computed for the maximum section of Oroville Dam if crest length/height ratio = 2

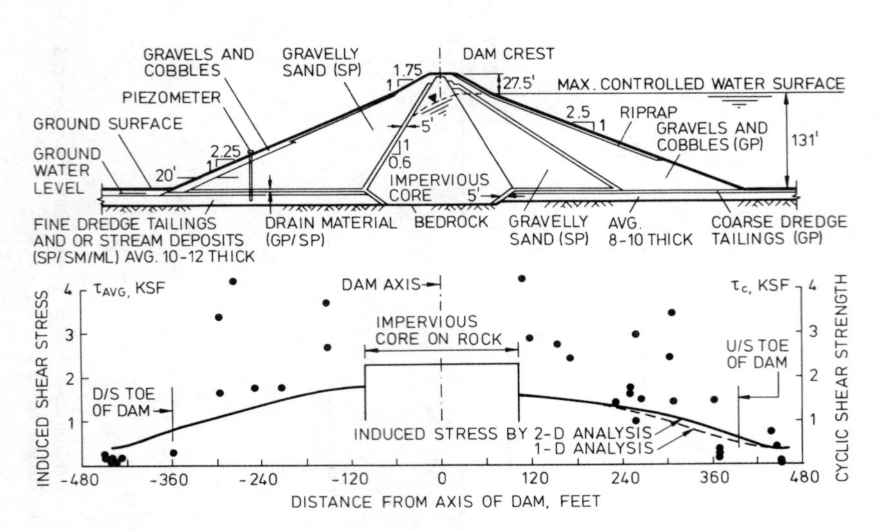

Figure 8.5. Comparison of earthquake induced stresses with soil resistance values for foundation material of major dam

stresses induced in the foundation soils of a major dam using one- and two-dimensional analysis is shown in Figure 8.5. There is very little difference between the two sets of results in this case. Similar good agreement has been found in a number of other analyses but sometimes the difference may be several tens of percent. In view of the small cost differential between performing one- and two-dimensional studies, the uncertainty of the possible error does not seem to warrant making one-dimensional response studies despite their greater simplicity. It is only where a two-dimensional analysis capability does not exist and may be difficult to develop that recourse to one-dimensional response studies seems to be justified.

Nevertheless it is an important part of the required analytical capability of the engineer to know where 1-D, 2-D and 3-D analyses can appropriately be used or must be used to obtain the required level of accuracy in the computed response.

8.3.2 *Earthquake excitation*

No matter what geometrical form of analytical model is used, it can only provide meaningful indications of the behaviour of a prototype structure if it is used in conjunction with representative values of material properties for the various soils in the section and with a realistic representation of the earthquake motions to which the dam may be subjected. The selection of an appropriate time-history of earthquake excitation at the base of the dam or in an adjacent rock outcrop is thus an important part of the analysis procedure.

Although earthquakes produce multi-directional motions, it is usually sufficiently accurate to use only one horizontal component of motion as excitation for the analytical model in a design study. In some cases involving long dams, a phase difference between motions developed at different points along the base of the dam has been introduced to represent the effects of spatial variations in rock motions (Udaka 1975) while in some studies, different motions have been applied at different points along the base for the same purpose (Altinisik & Severn 1980). Nevertheless reasonable results can usually be obtained using a single motion at all points along the base of the dam, so long as this motion is itself a reasonable representation of an actual earthquake motion. Unreasonable responses will inevitably be obtained by the use of unreasonable base motions such as for example, using an acceleration time-history with a peak acceleration greater than 1 g to represent a moderate nearby earthquake or using an uncharacteristically low acceleration level, say less than 0.2 g, for the same purpose. A good understanding of the general characteristics of strong earthquake shaking is thus an essential component of any analysis to be used for practical purposes.

8.3.3 *Soil properties for use in the analysis*

Since soils have nonlinear stress-strain characteristics under the levels of earthquake motions likely to threaten the safety of earth dams (say greater than about 0.2 g), it has long been recognized that this aspect of soil behaviour must be incorporated into the analysis in some way. There are basically two ways to do this:

1. To use equivalent-linear material properties in which the modulus and damping ratio of any soil element are selected to be compatible with the strain induced in the soil element. Tests show that the effective modulus of a soil sample decreases with

Table 8.1. Methods of modelling soil properties for use in seismic analyses of earth dams

1. Equivalent-linear stress-strain
   - (a) with no explicit pore pressure generation or dissipation (total stress analysis)
   - (b) with some allowance for pore pressure generation and dissipation (judgment, strength degradation, etc.)
   - (c) with computed pore pressure generation and dissipation (effective stress analysis)
2. Nonlinear stress-strain
   - (a) with no explicit pore pressure generation or dissipation (total stress analysis)
   - (b) with some allowance for pore pressure generation and dissipation (judgment, strength degradation, etc.)
   - (c) with computed pore pressure generation and dissipation (effective stress analysis) with
     - (1) pore pressure generation an integral part of the soil model, or
     - (2) pore pressure generation determined by a separate analysis procedure

increasing strain while the effective damping increases with increasing strain. Thus these properties can be represented as strain-dependent characteristics and an iterative procedure used to select the appropriate values for any response study.

2. To use a nonlinear stress-strain law directly in the analytical procedure. This greatly increases the complexity of both the analysis and the testing procedures. At the present time such methods have not been developed and validated sufficiently to justify their use in practical design studies.

In addition to considering the nonlinearity problem, it is also important to take into account the variation of material properties with effective confining pressure and to make allowance, in some way, for the effects of possible pore pressure generation and dissipation during earthquake shaking which will in turn change the effective confining pressure as the shaking progresses. There are basically three ways to consider the effects of pore pressure generation and dissipation:

1. To make no explicit allowance for pore pressure generation and dissipation during the period of earthquake shaking, but to make an evaluation of the magnitude of the pore pressures developed when the shaking is completed (analogous to a total stress analysis).
2. To make some allowance for the effects of pore pressure generation and dissipation on soil properties on the basis of judgment or by incorporating some level of strength and stiffness degradation in the analysis.
3. To compute the magnitude of the changes in pore pressure due to generation and dissipation effects with time and include their effects on the other significant material characteristics as the analysis proceeds (analogous to an effective stress analysis).

Since all of these methods can be used in conjunction with either an equivalent-linear or a nonlinear representation of stress-strain relationships, there are potentially a wide variety of methods available for modelling soil properties as listed in Table 8.1.

8.4 EVALUATION OF IN-SITU SOIL CHARACTERISTICS

Whichever procedure is used to represent the soil characteristics in an analytical model, it is important to recognize that the accuracy with which the analytical results can predict the performance of an actual structure is necessarily limited by the accuracy with which the in-situ properties of the soils are determined and represented in the modelling procedure. Failure or inability to determine the in-situ soil properties correctly is often a potentially greater source of error in predictions of field performance than are the details of the property characterization method incorporated in the analysis. It is essential to understand this fact since even the most sophisticated model of soil properties which might be used in any analysis will necessarily provide a poor prediction of performance if in fact it is used to represent an incorrect evaluation of the actual characteristics. For example, use of an effective stress approach for soil properties with a full consideration of pore pressure generation and dissipation effects may change the computed response of an embankment dam by 10 to 20%, but errors in determining the correct properties of the soils in the embankment may easily change the predicted performance by 50% or more. This aspect of property characterization puts a premium on the understanding of factors affecting the measured properties of soils. Important aspects of this problem which necessarily limit the accuracy of the results obtained include:

1. The ability of the engineer to select representative samples of soil on which to perform the necessary testing program. This may be extremely difficult in a non-uniform soil deposit (see for example, the variability of the properties of the foundation sand for the dam shown in Figure 8.5).
2. The ability of the engineer to obtain truly undisturbed samples of natural deposits or constructed works, or to prepare samples of soil having the same properties as those of an embankment which has yet to be constructed. Test data for very dense sands (see Figure 8.6) show, for example, that the cyclic loading resistance of good quality undisturbed samples may be only about 35% of that of the deposit from which they

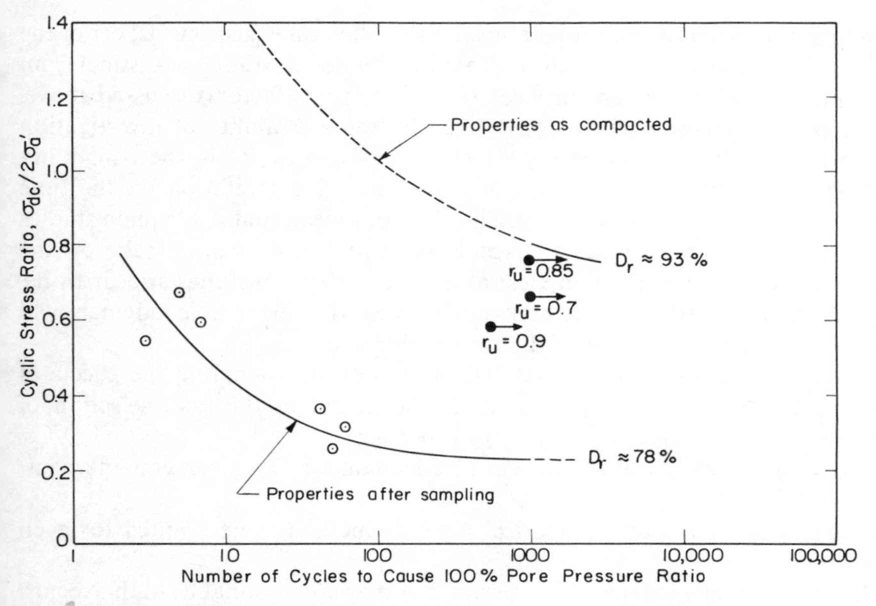

Figure 8.6. Comparison of cyclic loading resistance of undisturbed samples with resistance of original deposit

were extracted (Seed et al. 1982), while different methods of taking 'undisturbed' samples may lead to results differing by 100% (Marcuson & Franklin 1979). Good judgment or simplified in-situ testing may be required to overcome these deficiencies in sampling techniques.
3. The ability to obtain and test adequately large samples of soils containing gravel and cobble size particles. There are no undisturbed sampling procedures for these types of soil and the largest cyclic loading testing devices are only capable of testing samples up to about 38 cms diameter with particles up to about 6 cms maximum size. Many construction materials contain larger particles and cannot therefore be tested in available equipment.
4. The ability to perform meaningful tests, which reproduce the correct initial stress conditions, stress path and other loading conditions in the laboratory.
5. The ability to interpret laboratory test data in a meaningful way with appropriate allowance for the variability of the results.

In the light of these difficulties, it may well be desirable to sacrifice analytical sophistication based on an idealized concept of soil properties determined from tests on unrepresentative soil samples in favour of a more simplified approach involving a larger and more representative testing program. At the very least it is important to recognize that in engineering practice there is little merit in using an evaluation procedure which involves a significant imbalance in the accuracies of analytical modelling and soil property determinations.

8.5 SELECTION OF ANALYTICAL PROCEDURE

The preceding discussion of soil property evaluation illustrates just one aspect of the problems involved in selecting an analysis procedure for design studies as distinct from an analysis procedure for research purposes. In addition to the factors discussed above, design studies of any type may be constrained by the availability of investigation equipment in the field, by the availability of facilities at the work site, by the availability of skilled personnel, by the availability of funds and by a limitation on the time available in which to perform the studies. This does not mean that inadequate studies can be accepted but it does mean that each individual study cannot take on the character of a research investigation, desirable as this may sometimes appear to be.

Because of these and other factors necessarily involved, considerable judgment will inevitably be required on the part of the responsible engineer:
1. In deciding how to characterize material properties, in evaluating the effects of different factors on these properties in the field, and in the final selection of appropriate values for use in the analysis procedure.
2. In deciding which analytical model will be adequate for any given embankment: 3-D, 2-D or 1-D.
3. In deciding which particular numerical analysis method is best suited for each particular job.
4. In evaluating the analytical results to ensure that they are reasonable. In this regard it is important to recognize that the more factors that are built into a computer

program, the more difficult it is likely to be to judge the reasonableness of the results it produces.

Thus the most elaborate analysis procedures may not offer the maximum benefits for use in design studies.

8.6 VALIDATION OF ANALYSIS PROCEDURES

Probably the most important part of developing the ability to provide meaningful results for engineering purposes is to establish the credibility of any given numerical analysis procedure by using it to predict the performance of structures whose field performance is known, or if no field case studies are available, verifying the applicability of a given analytical procedure by using performance data from tests on small-scale structures. In the final analysis, however, the ability to predict field performance is the true test of the usefulness of any analysis procedure and every opportunity should be taken to make such checks. It is more difficult to do this in the earthquake field than in other areas of geotechnical engineering because of the more limited field case studies available for examination, but it is an essential step in developing a correct understanding of numerical analysis capabilities in the design field.

8.7 ESSENTIAL ELEMENTS OF SEISMIC SAFETY EVALUATION PROCEDURE

In recognition of the many factors involved, the writer has for many years used an analysis approach which seems to embody most of the essential elements of a seismic safety evaluation procedure for embankment dams. The details of the general procedure have undergone many improvements since it was first proposed (Seed 1966) – primarily through the development and application of improved finite-element procedures with the aid of I.M. Idriss, J.M. Duncan, F.I. Makdisi, N. Serff, J.R. Booker, M.S. Rahman and W.D.L. Finn and also through the development of an improved understanding and capability with regard to testing procedures, accomplished with the help of K.L. Lee, P. DeAlba, R.M. Pyke and N. Banerjee.

In spite of improvements, however, the basic principles of the procedure have remained unchanged and involve a series of steps which might be summarized simply as follows:

1. Determine the cross-section of the dam to be used for analysis.
2. Determine, with the cooperation of geologists and seismologists, the maximum time history of base excitation to which the dam and its foundation might be subjected.
3. Determine, as accurately as possible, the stresses existing in the embankment before the earthquake; this is probably done most effectively at the present time using finite element analysis procedures.
4. Determine the dynamic properties of the soils comprising the dam, such as shear modulus, damping characteristics and bulk modulus or Poisson's ratio, which determine its response to dynamic excitation. Since the material characteristics are nonlinear, it is also necessary to determine how the properties vary with strain.

5. Compute, using an appropriate dynamic finite element analysis procedure, the stresses induced in the embankment by the selected base excitation.
6. Subject represenative samples of the embankment materials to the combined effects of the initial static stresses and the superimposed dynamic stresses and determine their effects in terms of the generation of pore water pressures and the development of strains. Perform a sufficient number of these tests to permit similar evaluations to be made, by interpolation, for all elements comprising the embankment.
7. From the knowledge of the pore pressures generated by the earthquake, the soil deformation characteristics and the strength characteristics, evaluate the factor of safety against failure of the embankment either during or following the earthquake.
8. If the embankment is found to be safe against failure, use the strains induced by the combined effects of static and dynamic loads to assess the overall deformations of the embankment.
9. Be sure to incorporate the requisite amount of judgment in each of steps 1 to 8 as well as in the final assessment of probable performance, being guided by a thorough knowledge of typical soil characteristics, the essential details of finite element analysis procedures, and a detailed knowledge of the past performance of embankments in other earthquakes (Seed 1979).

This procedure may seem rather long and cumbersome but it also seems to incorporate the essential steps in evaluating such a complex problem as the response of earth dams to earthquake effects.

It lends itself naturally, however, to somewhat simplified versions of the method, which have often been used for reason of time and economy (e.g., Finn 1967, Klohn et al. 1978, Lee & Walters 1972, Lee 1978, Leps et al. 1978a and b, Vrymoed & Galzascia 1978; etc.). The ultimate simplification is, of course, the total elimination of all analysis procedures and a simple evaluation, based on a knowledge of the materials comprising the dam and the judgment resulting from conducting many previous analyses and observing the performance of existing dams (Seed 1981). However, it should be noted that each of the steps is an essential element of the procedure and if one of them is performed incorrectly, the results of the analysis may be grossly misleading. In such cases, where the job cannot be done properly, it may be better not to do it at all rather than to be misled by the erroneous results which may ensue. It is for this reason that judgment is necessary at each step in the development.

In the most modern verions of the method, the assessment of pore water pressures during and following the earthquake shaking may involve studies of simultaneous pore pressure generation and dissipation using appropriate computer programs (Booker et al. 1976, Finn et al. 1978) and the evaluation of the final configuration of the structure using a strain-harmonizing technique, again involving finite element procedures (Lee 1974, Serff et al. 1976).

The particular procedure used in any given case should depend on the complexity of the case being considered, the margin of safety provided for the level of earthquake shaking likely to develop, and the judgment and experience of the engineer responsible for the study.

An interesting example of the use of this method to analyze a slope failure is provided by the analysis of the Lower San Fernando Dam in the San Fernando earthquake of 1971 (Seed 1979). An example of a seismic stability analysis where failure

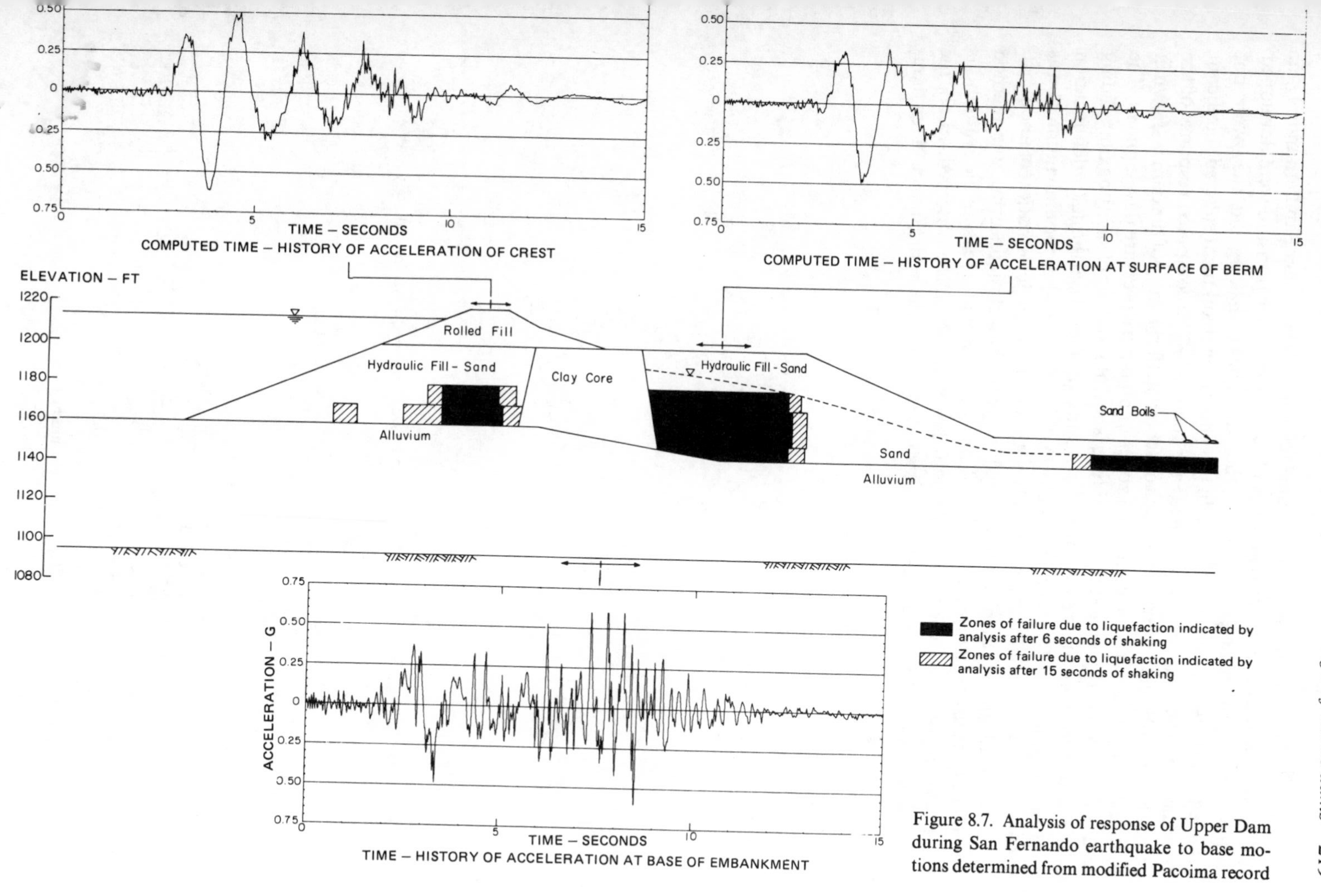

Figure 8.7. Analysis of response of Upper Dam during San Fernando earthquake to base motions determined from modified Pacoima record

did not occur is illustrated by the computed response of the Upper San Fernando Dam in the same earthquake. Extensive zones of high pore-water pressure were developed within the embankment as a result of the earthquake shaking but they were not sufficiently extensive to cause a failure. In this case the embankment suffered significant deformations, the crest moving downstream about 5 ft. The computed response of the embankment subject to the rock motions observed in the vicinity of the dam is shown in Figure 8.7, together with the computed zones of high pore pressure development. The results of a post-earthquake stability analysis, showing a factor of safety against sliding of 1.8 are shown in Figure 8.8, and the results of a post-earthquake deformation analysis in Figure 8.9. The computed deformation of the crest of the embankment was 3.8 ft, a value which compares favourably with the observed deformation of about 5 ft.

As with all analytical procedures used in geotechnical engineering, the method should only be used if it is found to work – that is, if it provides reasonable evaluations of behaviour for cases where the behaviour has been or can be observed. In all, the general procedure described above has been used to study the performance of ten dams

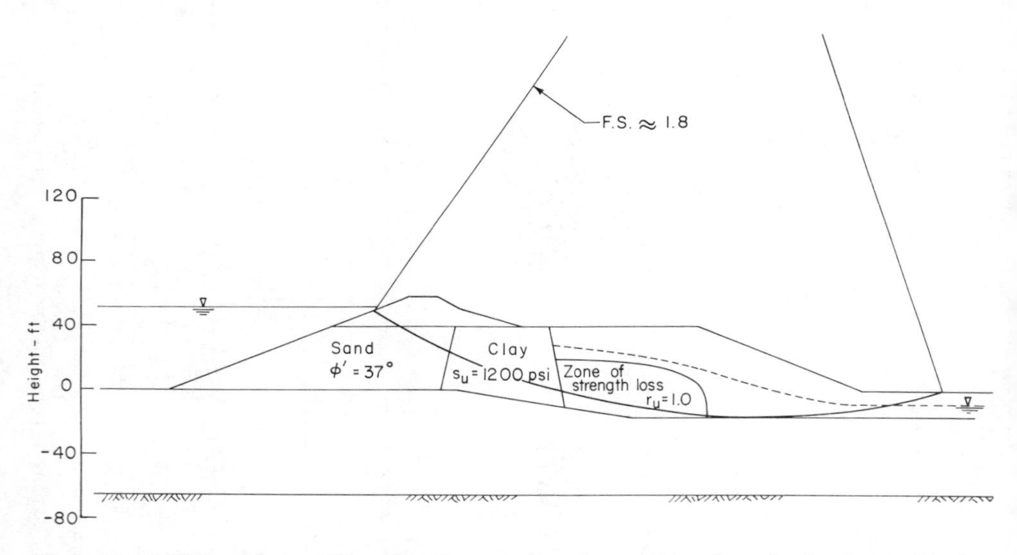

Figure 8.8. Stability analysis of Upper San Fernando Dam for conditions immediately after earthquake

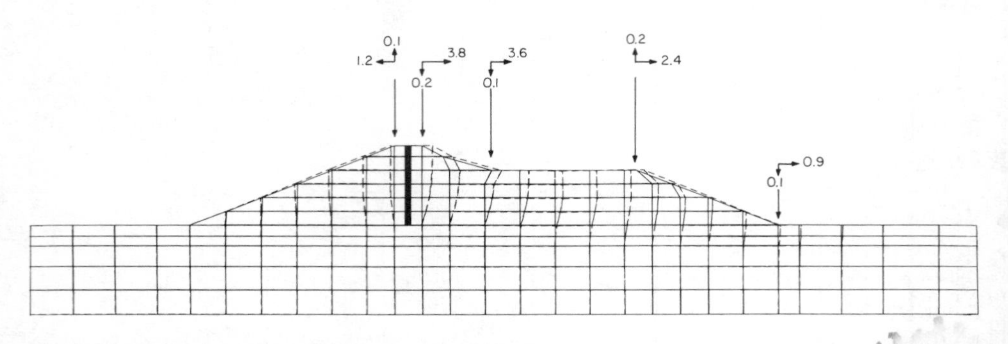

Figure 8.9. Deformation analysis of Upper San Fernando Dam

whose performance during earthquakes is known. Two of these had major slides, one underwent large deformations and six had no discernible damage. The behaviour predicted by the analysis was similar to that observed in the field in each case, and while it is true that each of these cases was in fact studied after the event involved, it seems that the procedure has the capability of giving considerable insight into the possible behaviour of embankments subjected to earthquake effects. For this reason, presumably, it has been adopted in studies of many dams throughout the world as a guide to final assessment of their probable performance during earthquakes (e.g., Seed et al. 1969, 1973 and 1975, Gordon et al. 1974, Kramer et al. 1975, Marcuson et al. 1977, Makdisi et al. 1978, Sadigh et al. 1978, State of California Department of Water Resources 1979; etc.)

8.8 CONCLUSION

The main purposes of this chapter have been to show that:

1. Numerical methods play an essential role in developing our understanding of the behaviour of embankment dams during earthquakes.
2. With the aid of these methods and with repeated opportunities to check the ability of analytical procedures to predict known field performance, it is possible to develop procedures which can make meaningful assessments of the probable future performance of new structures and to guide in the design of these structures.
3. In selecting an analytical method for use in engineering practice it is necessary to have a full understanding of all aspects of the design procedure including selection of the earthquake shaking intensity, field investigations, laboratory testing method and available analytical techniques, in order that a suitably practical study can be made.

One of America's most famous scientists, Joel H. Hildebrand, once said:

> To be successful in unlocking the doors concerning Nature's secrets, a person must have ingenuity. If he does not have the key to the lock, he must not hesitate to try to pick it, to climb in a window or even to kick in a panel.

This might equally well be applied to geotechnical engineering, and particularly in dealing with large projects, which tend to embody within them more of Nature's secrets than we encounter in, say, a building foundation problem where the soil conditions can be studied in much greater detail so that less is left to geological interpretation, or in dealing with Nature's deeper secrets such as earthquakes and the engineering provisions which must be taken to understand their effects on soils and mitigate the hazards they present.

In such cases it is essential that an engineer have ingenuity and insight – and since we can rarely have the key to the lock we must use all the tricks at our disposal to gain the insight required to achieve a satisfactory solution. This may well involve special field investigations, unique types of laboratory tests, novel types of analyses, the use of empirical rules, the exercise of good judgment, and the use of sophisticated numerical methods of analysis – all aimed at gaining a greater understanding of Nature's secrets that she might otherwise have been willing to reveal. In major engineering undertakings, all methods which can improve the knowledge of the engineer and give him the insights required to inspire what might well be considered ingenuity are essential to the consummation of a satisfactory project – and if some are less sophisticated than others, they are none the less valuable for this. The important thing is that all are required – and while numerical methods of analysis alone are no more likely to provide the key to the lock than any other technique, they are just as likely to fill the role of climbing in a window or kicking in a panel as any of the other engineering tools at our disposal. They have already shown that they can play a vital role in this respect by providing results and insights which could never otherwise have been obtained – and they will undoubtedly continue to fill such a role in the future. Viewed in this light they must be regarded as one of the essential tools of modern engineering practice in the practical design of critical structures for complex loading conditions such as those produced by earthquakes.

REFERENCES

Altinisik, D. & R.T. Severn 1980. Natural frequencies and response characteristics of gravity dams. In *Proceedings of Conference on Design of Dams to Resist Earthquake* (Institution of Civil Engineers, London).

Booker, J.R., M.S. Rahman & H.B. Seed 1976. GADFLEA: A computer program for the analysis of pore pressure generation and dissipation during cyclic or earthquake loading. Report No. EERC 76–24, Earthquake Engineering Research Centre, University of California, Berkeley.

Finn, W.D.L. 1967. Behaviour of dams during earthquakes. In *Proceedings of 9th International Congress on Large Dams* (Istanbul), 4: 355–367.

Finn, W.D.L., K.W. Lee, C.H. Maartman & R. Lo 1978. Cyclic pore pressures under anisotropic conditions. In *Proceedings of ASCE Specialty Conference on Earthquake Engineering and Soil Dynamics* (Pasadena)

Gordon, B.B., D.J. Dayton & K. Sadigh 1974. Seismic stability of Upper San Leandro Dam, *J. Geotech. Engng. Div.* (ASCE) 100 (GT 5): 523–545.

Klohn, E.J., C.H. Maartman, R.C.Y. Lo & W.D.L. Finn 1978. Simplified seismic analysis for tailings dams. In *Proceedings of ASCE Specialty Conference on Earthquake Engineering and Soil Dynamics* (Pasadena): 540–556.

Kramer, R.W., R.B. MacDonald, D.A. Tiedmann & A. Viksne 1975. Dynamic analysis of Tsengwen Dam, Taiwan, Republic of China. U.S. Dept. of the Interior, Bureau of Reclamation.

Lee, K.L. 1974. Seismic permanent deformations in earth dams. Report No. UCLAENG-7497, School of Engrg. & Appl. Sci., University of California, Los Angeles.

Lee, K.L. 1978. Seismics stability considerations for tailings dams adjacent to San Andreas Fault. In *Proceedings of 1st Central American Conference on Earthquake Engineering* (San Salvador):

Lee, K.L. & H.G. Walters 1972. Earthquake induced cracking of Dry Canyon Dam. ASCE Annual & Natl. Environmental Engrg. Meeting, Houston, TX, Preprint 1794.

Leps, T.M., A.G. Strassburger & R.L. Meehan 1978a. Seismic stability of hydraulic fill dams. Part 1, Water Power and Dam Construction, Oct. 27–36.

Leps, T.M., A.G. Strassburger & R.L. Meehan 1978b. Seismic stability of hydraulic fill dams. Part 2, Water Power and Dam Construction, Oct. 27–36.

Makdisi, F.I., H.B. Seed & I.M. Idriss 1978. Analysis of Chabot Dam during the 1906 earthquake. In *Proceedings of ASCE Specialty Conference on Earthquake Engineering and Soil Dynamics* (Pasadena): 569–587.

Marcuson, W.F. III & G. Franklin 1979. Undisturbed sampling of cohesionless soils. State-of-the-Art Report (Singapore).

Marcuson, W.F. III, E.L. Krinitzky & E.R. Kovanic 1977. Earthquake analysis of Fort Peck Dam, Montana. In *Proceedings of 9th International Conference Soil mechanics and Foundation Engineering* (Tokyo):

Mejia, Lelio H. & H. Bolton Seed 1983. Comparison of 2-D and 3-D dynamic analyses of earth dams. *J. Geotech. Engng. Div.* (ASCE)

Sadigh, K., I.M. Idriss & R.R. Youngs 1978. Drainage effects on seismic stability of rockfill dams. In *Proceedings of ASCE Specialty Conference on Earthquake Engineering and Soil Dynamics* (Pasadena): 802–818.

Seed, H. Bolton 1966. A method for earthquake-resistant design of earth dams. *J. Soil Mech. Fdns. Div.* (ASCE) 92 (SM 1).

Seed, H.B. 1979. Considerations in the earthquake-resistant design of earth and rockfill dams. *Geotechnique* XXIX (3).

Seed, H. Bolton 1981. Earthquake resistant design of earth dams. In *Proceedings of International Conference on Recent Advances in Geotechnical Earthquake Engineering and Soil Dynamics* (University of Missouri, Rolla).

Seed, H.B., I.M. Idriss, K.L. Lee & F.I. Makdisi 1975. Dynamic analysis of the slide in the Lower San Fernando Dam during the earthquake of February 9, 1971. *J. Geotech. Engng. Div.* (ASCE) 101 (GT 9): 889–911.

Seed, H.B., K.L. Lee & I.M. Idriss 1969. Analysis of Sheffield Dam failure. *J. Geotech. Engng. Div.* (ASCE) 95 (SM 6): 1453–1490.

Seed, H.B., K.L. Lee, I.M. Idriss & F. Makdisi 1973. Analysis of the slides in the San Fernando Dams during the earthquake of February 9, 1971. Report No. EERC73-2, Earthquake Engineering Research Centre, University of California, Berkeley.

Seed, H.B., S. Singh, C.K. Chan & T.F. Vilela 1982. Considerations in undisturbed sampling of sands. *J. Geotech. Engng. Div.* (ASCE) 108 (GT 2): 265–283.

Serff, N., H.B. Seed, F.I. Makdisi & C.K. Chan 1976. Earthquake induced deformations of earth dams. Report No. EERC 76-4, Earthquake Engineering Research Centre, University of California, Berkeley.

State of Calif. Dept. of Water Resources 1979. The August 1, 1975 Oroville Earthquake Investigations. *Dept. of Water Resources Bulletin*: 203–278.

Terzaghi, K. 1950. Mechanisms of landslides. In *The Geological Survey of America*, Engineering Geology (Berkeley) volume.

Udaka, T. 1975. Analysis of response of large embankments to travelling base motions. Ph.D. thesis, University of California, Berkeley.

Vrymoed, J.L. & E.R. Galzascia 1978. Simplified determination of dynamic stresses in earth dams. In *Proceedings of ASCE Speciality Conference on Earthquake Engineering and Soil Dynamics* (Pasadena): 991–1006.

Modelling of sedimentation and large strain consolidation 9

R.N. YONG

9.1 INTRODUCTION

The phenomenon of settling of suspended fines is one which is common to (a) natural processes in soil sedimentation such as the initial stages for formation of sedimentary soils, and (b) management of tailings discharge from mineral resource industries. In both general types of situations, one of the major items of interest is the problem of prediction of the rate of settling of the suspended fines and the consolidation of the sediment layer. In the case of the tailings problem, the excess tailings water volume more than fills the extracted volume previously occupied by the host rock – thus creating serious issues in regard to land and environmental management.

9.1.1 *The physical problem*

The fines (solids) concentration in a soil suspension problem, or as initially discharged from the extraction/process plant (i.e. initial condition), is considered to be very dilute –

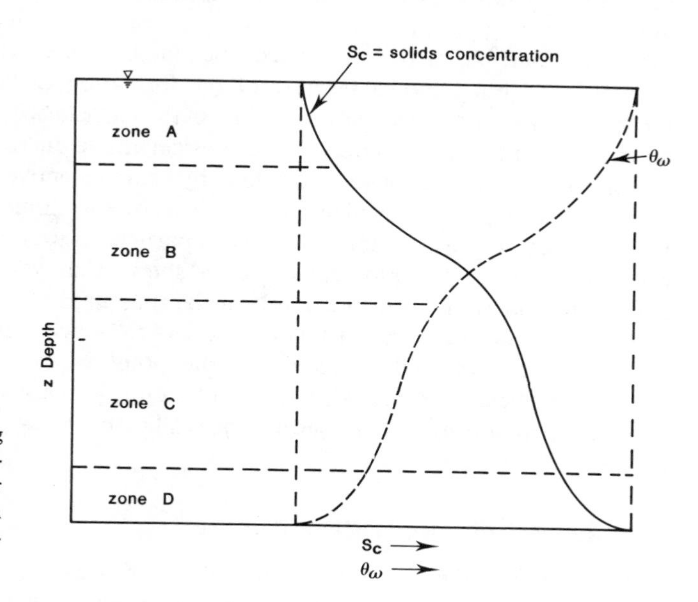

Figure 9.1. Schematic showing distribution of solids concentration and volumetric water content with depth. Zones A through D describe settling performance characteristics

generally of the order of 1 to 3% by weight. If the suspended solids are non-surface-active, and if they are silt-sized or larger, it is likely that they will settle in accord with the general predictions advanced by the simple Stockesian model. However, if the particles are clay-sized, and if they are surface-active in nature, simple Stokesian predictions will not accurately portray the settling behaviour of the solids.

As the solids settle, their concentration will increase to the point where proximal hindrances become significant. Thus, even if initial settling of the suspended solids can be predicted by the simple Stokesian model, subsequent settling of the solids will render the model invalid in application. The problem of the inability of the solids to settle in accord with gravitational mechanisms has been documented previously – e.g. Yong et al. (1979), Been & Sills (1981) and Yong et al. (1983) – and will not be repeated here. Figure 9.1, which shows the essence of the problem, portrays the solids concentrations and volumetric water content profiles a settling pond. These profiles are common to the settling performance of initially dilute suspensions from tailings discharge or from natural processes after a period of several years.

9.2 SETTLING/SEDIMENTATION PROCESS

The simplest procedure for description of the total settling/sedimentation and consolidation process of the fines suspension is to trace the 'life' or status of a typical representative fines particle, beginning with its initial state. Following introduction of the representative particle as one of a multitude of particles in the suspension at the 1 to 3% concentration, the particle will settle in a fashion more or less controlled by either gravitational forces or by interactive forces dictated by surface-active relationships – as the case may be. The concentration of particles at this time is not sufficient to account for proximal hindrances, and is identified in Figure 9.1 as the material contained in zone A. When initial settling of the particle has proceeded to the stage where neighbouring particles begin to interfere because of their highly active surfaces, a hindered settling performance characteristic becomes evident – shown as Zone B in Figure 9.1. Settling of the suspended fines in this zone cannot be modelled by the standard Stokesian model because of the formation of flocs and also because of hindrances. Beyond this stage, the settling of the particle becomes tediously slow and is apparently constrained not only by the physical interferences of neighbouring particles, but also by interactions controlled by surface-active relationships (zone C). Interparticle forces developed at this stage behave in a manner akin to dispersive forces. Even though no physical contact between particles is estabilished in this zone – void ratios of 4 to 6 – the physical evidence shows that some small value of excess hydrostatic pressure (pore pressure) can be measured (Yong et al. 1983). A form of compression settling of the particle can be said to be occurring at this stage. When the particle undergoes further settling to the point where physical contact between adjacent particles is achieved, the consolidation process similar to the mechanism considered in soil mechanics settlement studies becomes the dominant mechanism (Zone D).

9.2.1 *Reconciliation with analytical models*

Table 9.1 shows the elements of the settling process in relation to representative analytic models. The comments made in this table are by no means meant to be

Tables 9.1. Settling and theory applications

	Stokesian	Hindered theories	Consolidation theories
Zone A: No measurable pore pressures (no excess hydro-static pressures)	Limited to non surface-active particle suspensions	Empirical models require extensive laboratory testing for derivation of correction coefficients	Inapplicable because of absence of relationships defining effective stress and material properties
Zone B: No measurable pore pressures	Not directly applicable	Same restrictions as for Zone A	Same restrictions as for Zone A
Zone C: Detectable pore pressures	Not applicable	Not applicable	Large strain consolidation phenomenon. Requires accountability for non-linear relationships and self-weight
Zone D: Pore pressures and effective stresses	Not applicable	Not applicable	Classical theories applicable if self-weight is accounted for

comprehensive; only the highlights are listed. In viewing the table, it is obvious that at the present time, total analysis of the complete problem of fines settling can only be achieved by breaking up the problem into four characteristic zones of fines settling for separate analytical treatment using available classical models – or their variants thereof. Whilst this procedure might be expedient, it is by no means satisfactory. The first significant problem that comes to mind, with this procedure, is the definition of initial conditions for each particular mode of analysis. This might not be especially critical in analysis. It is however decidedly so if predictions are to be made. Where does one transit from zone B to zone C? What are the material output values forthcoming from zone B – for use as input values for modelling of performance in zone C?

Of the several choices in seeking prediction of settling rates of the suspensed solids, the most common centres around the classical sedimentation and consolidation models. Theories relying on Stokes' principles and particle concentration appear to have been widely used in application to the fines settling in zones A and B identified in Figure 9.1. By and large, these theories/models rely on particle concentration for derivation of sedimentation relationships and accountability for hindrance is made through the appropriate changes in 'viscosity'. As such, gravity and free-fall constitute the major items – with no acknowledgement for the surface-active relationships of the fines – i.e. clay minerals.

Throughout the 'sedimentation zones' A and B, the principal parameters considered are:

– density of particles: d_p,
– density of liquid through which the particles are 'falling': d_l,
– nominal radius of the particles: r,
– absolute viscosity of the liquid: η.

The general relationship obtained provides the velocity V in terms of the above parameters as:

$$V = \frac{2(d_p - d_l)gr^2}{9_\eta} \tag{9.1}$$

where g = acceleration due to gravity.

In applying the predictive model to an actual tar sands sludge pond, Yong & Sethi (1978) have shown that the standard theories cannot adequately model the higher percentage of solids remaining in suspension over the same period of time. Because of the inability of the standard models to properly account for the surface and physico-chemical interparticle actions, the general predictions grossly overpredict the time-rate of sedimentation of the solids in zones A and B shown in Figure 9.1.

In regard to the prediction of time-rate of settlement of the solids in the sludge zone below zone A shown in Figure 9.1, common practice models the settling phenomenon as a 'consolidation' problem – replete with development of effective stresses and pore pressures. The classical concept of volume change associated directly with the amount of water extended is used – as is the time-rate effects. Recognition of the high volume change phenomenon is demonstrated through accountability for (a) permeability variation with void ratio, (b) nonlinearity in the effective stress-void ratio relationship, and (c) self-weight. As developed by Gibson et al. (1981), solution of the problem as a large strain or finite strain consolidation problem is more realistically achieved by using the Lagrangian coordinate system.

The difficulties that immediately present themselves are:

– Reconciliation of the effective stress principle with a soil-suspension system where initial void ratios are in the order of 9 or greater. Note that the solids concentration by weight at this void ratio is approximately 0.3, and the volume ratio as indicated in the void ratio shows the volume of voids to be nine times the volume of solids!

– Determining an appropriate and realistic void ratio-permeability relationship beginning with a void ratio of about 9.

– Determining a 'correct' effective stress-void ratio relationship.

Compatibility requirements between analytical and physical models demand that one can define a proper and realistic initial condition. If the initial void ratio of 9 (i.e. fines:water ratio of 0.3) is not appropriate, it would be interesting to determine when and at what void ratio effective stresses and pore pressures can be developed – for determination or specification of the appropriate initial conditions.

In addition, it is not entirely clear how or what permeability measurements mean if standard techniques are used, since Darcian requirements impose a no-volume change condition on permeability measurements. Thus one's ability to define the necessary sets of input values and relationships are hampered not only by the lack of appropriate tools for determination of the values, but also by the rather difficult problem of establishing initial conditions.

Obviously, until a unified (continuous) theory can be successfully developed which trespasses the various boundaries, we will have to work within the limitations of available theories to seek a broader range of applicability. The two theories presented herein are in part overlapping in their treatment of the problem in zone C. However, each on its own can handle problem analysis over two zones which define the settling performance of the suspended fines. The two theories are:

1. A relative flux model which encompasses particle settling performance in zones A through C, and
2. large or finite strain consolidation which covers settling performances in zones C and D.

9.3 RELATIVE FLUX MODEL

The detailed elements of the relative flux model have recently been presented as a simple diffusion model by Yong & Elmonayeri (1983). In essence, the particle settling problem is viewed as particle movement relative to the movement of elements of the 'suspending' fluid. The flux of the fluid $\vec{q}_w$ relative to a stationary coordinate system is written as:

$$\vec{q}_w = \vec{q}_{ws} + \theta\vec{q}_s \tag{9.2}$$

where:

$\vec{q}_s$ = particle flux;
θ = volumetric fluid content = volume of fluid in element under consideration divided by total volume of the same element;
$\vec{q}_{ws}$ = velocity of fluid relative to the moving particle.

Note that if θ is known, the solids concentration can be easily calculated. With the continuity condition:

$$\operatorname{div}\vec{q}_w = -\frac{\partial\theta}{\partial t} \tag{9.3}$$

We obtain:

$$-\frac{\partial\theta}{\partial t} = \operatorname{div}\vec{q}_{ws} + \theta\operatorname{div}\vec{q}_s + \vec{q}_s\cdot\operatorname{grad}\theta \tag{9.4}$$

Considering the conservation of particles within a volume fixed in the same stationary coordinate system, the equation of continuity in regard to the particles is given as:

$$\frac{\partial\rho}{\partial t} + \operatorname{div}(\rho\vec{q}_s) = 0 \tag{9.5}$$

where ρ = bulk density.

If v is the volumetric strain, it can be shown that:

$$\operatorname{div}\vec{q}_s = \frac{\partial v}{\partial t} + \vec{q}_s\cdot\operatorname{grad} v \tag{9.6}$$

The combination of Equations 9.4 and 9.6 will lead to:

$$-\frac{\partial\theta}{\partial t} = \operatorname{div}\vec{q}_{ws} + \theta\left(\frac{\partial v}{\partial t} + \vec{q}_s\cdot\operatorname{grad} v\right) + \vec{q}_s\cdot\operatorname{grad}\theta \tag{9.7}$$

Applying Fick's first law of diffusion, and respecting the relative motion between particles and fluid:

$$\vec{q}_{ws} = -K \text{ grad } \psi \tag{9.8}$$

where K = proportionality constant, and ψ = fluid potential.

Using Equation 9.8 and specializing the situation to a one-dimensional problem, we obtain:

$$-\frac{\partial\theta}{\partial t} = -\frac{\partial}{\partial z}\left[D(\theta)\frac{\partial\theta}{\partial z}\right] + \theta\frac{\partial\vec{q}_s}{\partial z} + \vec{q}_s\frac{\partial\theta}{\partial z} \tag{9.9}$$

where z = vertical spatial coordinate, and $D(\theta) = K\,\partial\psi/\partial\theta$ = diffusivity coefficient.

Equation 9.9 can now be used to predict the time rate of change of volumetric water content directly – and obviously the rate of solids accumulation per unit volume. The diffusivity coefficient $D(\theta)$ can be calculated with experimental information on settling behaviour. This constitutes a mandatory requirement since realistic material property information is considered essential to the successful prediction of the diffusion of the suspended fines – as a large strain settling/consolidation problem.

To obtain $D(\theta)$ from experimental measurements, settling column tests need to be conducted – similar to those described by Been & Sills (1981) or Yong & Elmonayeri (1983). The material contained in the columns consists of fines suspensions at specified initial solids concentrations. Sampling of the suspension at various locations after stated time intervals, together with measurements of settling rate of the solids/supernatant interface will provide the information needed to calculate the rate of water or solids diffusion. The procedure for calculation of $D(\theta)$ which has been described in detail by Yong & Warkentin (1975) takes the experimentally measured θ-h (i.e. volumetric fluid content (θ) varying with depth (h) of settling column) from laboratory experiments and divides the θ-h profile into n equal layers for analysis. Obviously, the thinner the layers are, the more accurate will be the answers calculated for $D(\theta)$. In a later section, the prediction of solids distribution with depth, using material input information obtained from the laboratory procedure described, will be compared with actual values obtained from a large operative stagnant containment pond to show how well the relative flux model presented herein can predict the low solids suspension settling performance.

9.4 LARGE STRAIN CONSOLIDATION MODELLING

The classical format for solution using the standard Terzaghi consolidation formulation restricts the successful application of the standard model to problems concerned with infinitesimal strain. As such, problems concerned with large volume change performance in settling behaviour such as sedimentation and settling of low concentration solids cannot be treated in the standard manner generally associated with the classical Terzaghi formulation. This problem has been recognized in many recent studies, e.g. Gibson et al. (1981), Been & Sills (1981), Olson & Ladd (1979), Koppula & Morgenstern (1982) and Yong et al. (1983).

If the phenomenon of consolidation is recognized as being applicable to the problem of settling of the solids in zones C and D, it is essential that definable relationships for material properties or characteristics expressed in relation to effective stresses be obtained. It is acknowledged that in zone C, the void ratio of about 6 provides no real basis for construction of a physical model that accepts actual physical contacts between the suspended solids. However, experience and laboratory experiments have confirmed that some small value of pore pressure (excess hydrostatic pressure) can indeed be measured. On that basis, the 'effective' stress obtained should be interpreted *solely* as the equivalent effective stress, the usual form for determination of this equivalent effective stress applies: i.e. equivalent effective stress is equal to the difference between total stress and *measured* pore pressure.

Two choices are available for computational solution of the large strain consolidation problem; these are (a) finite strain consolidation analyses, and (b) piecewise iterative (linear) consolidation analyses. In the former, the lucid development given by Gibson et al. (1981) is generally cited. The interested reader is referred to this reference for a detailed treatment of this development. The technique can be more precise than the infinitesimal strain procedure since the actual strains are not small. In addition, the use of a moving (convective) coordinate system is very convenient for 'book-keeping'. By and large, the computational effort required should be less and the solution should be sensitive to changes in constitutive properties a_v and C_v. It is this latter point which poses a problem if a_v and C_v are indeed highly variable with time – a fact which is indeed real in the case of dilute solids suspensions. In the derivation for finite strain consolidation, performed with respect to the Lagrangian coordinate system, with consideration given to one-dimensional loading through self-weight, the vertical equilibrium is written in the following form:

$$\frac{\partial \sigma}{\partial m} \pm (e\rho_w + \rho_s) = 0 \tag{9.10}$$

where:

σ = total vertical stress;
m = reduced coordinate;
ρ_w = unit weight of fluid;
ρ_s = unit weight of solids;
e = void ratio.

Taking into account that the equilibrium condition of the pore fluid requires:

$$\frac{\partial P}{\partial m} - \frac{\partial u}{\partial m} \pm \rho_w \frac{\partial \xi}{\partial m} = 0 \tag{9.11}$$

where:

P = pore pressure,
u = excess pore pressure, and
ξ = convective coordinate,

the governing equation is obtained (Gibson et al. 1981):

$$\pm\left(\frac{\rho_s}{\rho_w} - 1\right)\frac{d}{de}\left[\frac{k(e)}{1+e}\right]\frac{\partial e}{\partial m} + \frac{\partial}{\partial m}\left[\frac{k(e)}{\rho_w(1+e)}\frac{d\sigma'}{de}\frac{\partial e}{\partial m}\right] + \frac{\partial e}{\partial t} = 0 \tag{9.12}$$

where:

k = k(e) = void ratio;
σ' = vertical effective stress.

It should be noted that the term:

$$g(e) = -\frac{k(e)}{\rho_w}\frac{1}{(1+e)}\frac{d\sigma'}{de} \tag{9.13}$$

is highly nonlinear. Even if this term is taken to be a constant for simplicity in problem solution, the simplified governing relationship given now as:

$$\frac{\partial^2 e}{\partial m^2} \pm (\rho_s - \rho_w)\frac{d}{de}\left(\frac{de}{d\sigma'}\right)\frac{\partial e}{\partial m} = \frac{1}{g}\frac{\partial e}{\partial t} \tag{9.14}$$

still contains a nonlinear term. For dilute suspensions, it is not always easy to define a proper (and simple) relationship without sacrificing a measure of accuracy in prediction of settling performance.

The severe restrictions or unrealistic simplifications that need to be imposed because of the nonlinearity of the governing equations of the finite strain models can be avoided with the use of piecewise linear consolidation models – with the proper accounting for surcharge loading. In this technique, which is based on assumptions of infinitesimal strains, the solution is solved by time increments Δt, where Δt must be small to ensure stable solutions. The material properties are assumed to be constant for each layer and time increment, and are updated at the end of each increment. This piecewise linear approach has been used by Olson & Ladd (1979) and more recently by Yong et al. (1983) for application to prediction of two actual field pond studies. With the piecewise linear iterative approach, nonlinear material properties and non-homogeneous materials can be accommodated in the analysis. Considering the vertical profile in zones C and D in Figure 9.1 to be composed of elemental layers – much in the same manner as the n layers used in the computations used for determination of the diffusion coefficient in the relative flux model – the change in volume of the elemental volume defined is written as:

$$\frac{\Delta V}{\Delta t} = [k_1(u(P) - u(P-1)) - k_2(u(P+1) - u(P))]/(\Delta z \gamma_w) \tag{9.15}$$

where:

ΔV = change of volume
Δt = time interval
u = excess pore pressure
a_v = coefficient of compressibility
γ_w = density of water (fluid)
Δz = elemental thickness

The above relationship assumes that the amount of consolidation occurring is equal to the difference in the amount of fluid flow between the two planes defining the elemental layer, and that the flow of fluid is governed by Darcy's relationship.

Writing the change in volume as follows:

$$\frac{\Delta V}{\Delta t} = \frac{a_v \Delta z}{1+e}\frac{u(P)}{\Delta t} \tag{9.16}$$

the combination of Equations 9.15 and 9.16 will give the finite difference form of the well-known Terzaghi consolidation formulation, given as Equation 9.17:

$$u(P) = \frac{(1+e)\Delta t}{\gamma_w a_v \Delta z^2}[k_1(u(P) - u(P-1)) - k_2(u(P+1) - u(P))] \tag{9.17}$$

This is the form used by Yong et al. (1983) in their analysis, with the following procedure for implementation:

1. Divide the depth of the settling profile (sedimentation layer) into n layers and compute the elevation of each layer.
2. Determine the initial void ratio for each layer from a knowledge of either initial filling of pond, or from experimental information – or from field sampling program. This constitutes a vital part of the initial value problem. Simple experimentation with laboratory settling columns of fines suspension material representative of the sediment under study will give an indicative value for initial e.
3. Compute submerged unit weight at the bottom of the layer under consideration from a knowledge of initial e and position in the vertical profile, and determine the effective stress using experimental information relating effective stress to void ratio. The settling column tests required to establish this relationship is most crucial since pore pressures must be detected at the lowest possible solids concentration. As stated previously, experiments conducted have shown that pore pressures were measured for the pond material examined (in the next section) – where void ratios were about 5.
4. Establish pore pressure profile from information in step 2 and laboratory data from step 3.
5. Establish permeability-compressibility relationship from the same experimental studies discussed in step 3.
6. Calculate time interval Δt.
7. Use Equation 9.17 to compute Δu after time Δt.
8. Compute new effective stress and void ratio after time interval Δt.
9. Compute new elevation of layers and final thickness of sediment layer.
10. Return to step 1 if time is less than maximum time.

9.5 PREDICTION OF SETTLING/CONSOLIDATION IN SEDIMENT

The two predictive models discussed in the previous two sections are used herein to predict the settling behaviour of an initially low fines concentration suspension. In the field test situation chosen and reported previously by Roberts et al. (1980), the pond size measures 600,000 square meters at the top, with a depth of about 41 meters. The total contained volume of suspension material at the time of field sampling was 8.5 million cubic meters. The type of fines and total compositional characteristics of the suspension material – derived from processing of tar sands – have been detailed previously by Yong & Sethi (1978). The particle-size distribution at the time of sampling – two and one half years after complete filling of the pond – is given in Table 9.2.

Figure 9.2 shows the filling (loading) sequence of the pond described in Table 9.2. As noted, the filling period occupied one year. The concentration of suspended solids in the filling sequence was about 20%.

Figure 9.3 shows the solids concentration profile developed two and one half years

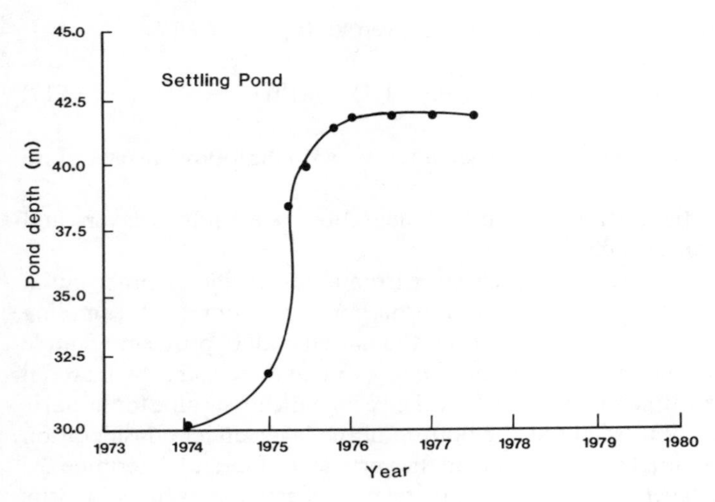

Figure 9.2. Filling sequence for tailings pond

Table 9.2. Particle-size distribution with depth

Depth (m)	% Sand greater than 44 microns	Silt 2-44 microns	Clay less than 2 microns
6.1	1.9	50.1	48.0
13.7	13.5	51.6	34.9
18.3	44.6	31.2	24.1
21.3	47.5	34.5	18.0

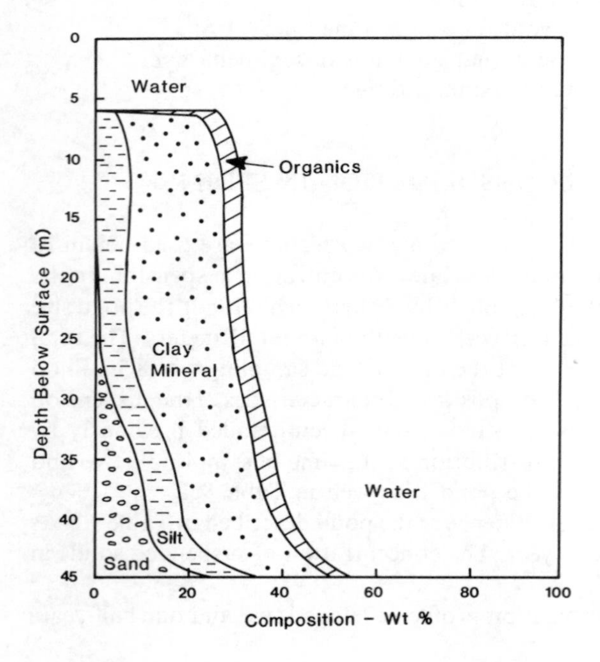

Figure 9.3. Measured composition of profile in tailings pond

after the filling sequence. The particle size distributions of the samples taken from the pond were determined experimentally – to serve as input for comparison with predictions from the two models discussed in this study.

9.5.1 *Comparison with relative flux and piecewise linear models*

Comparison with field information can be performed in two ways:

1. Comparing solids concentration profile using the relative flux model – showing thereby the applicability of analysis of the problem in terms of particle and fluid fluxes. Computations of the diffusion coefficient variation with depth from experimental measurements, implicitly account for the compositional characteristics of the suspended solids.

   The results obtained from solution of Equation 9.9 can be interpreted in terms of

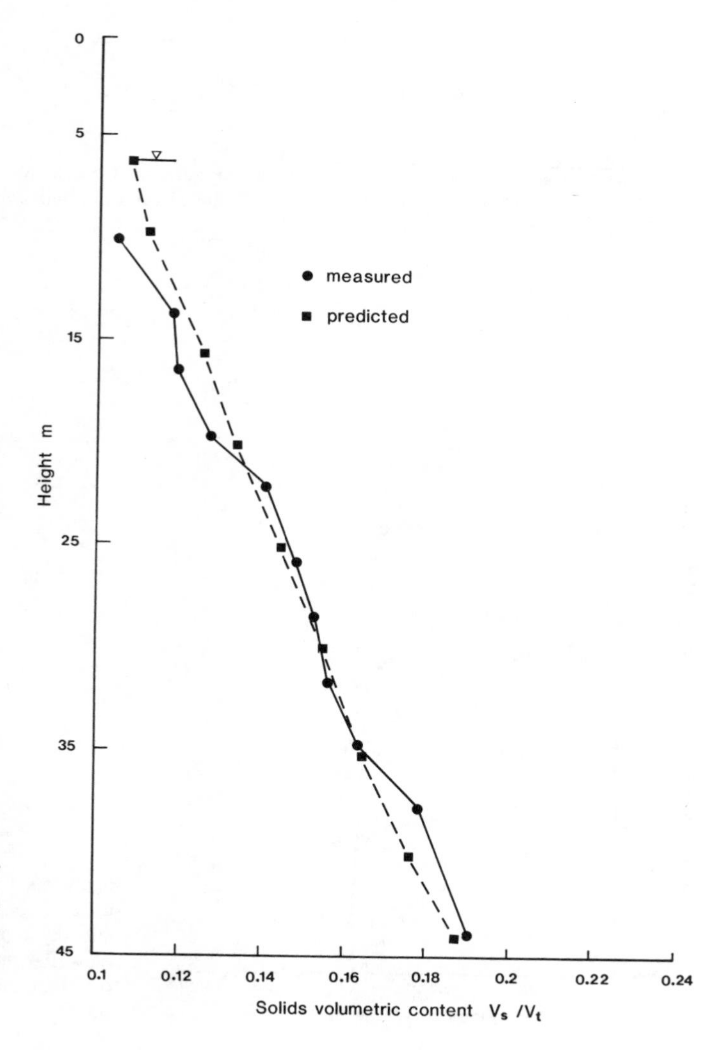

Figure 9.4. Comparison of measured and predicted solids volumetric content in tailings pond. Note that V_s is volume of solids, and V_t is total volume

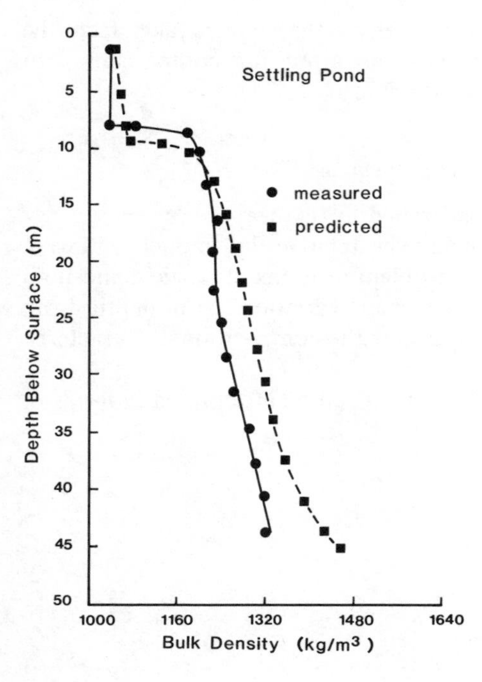

Figure 9.5. Comparison between measured and predicted values in bulk density in the tailings pond

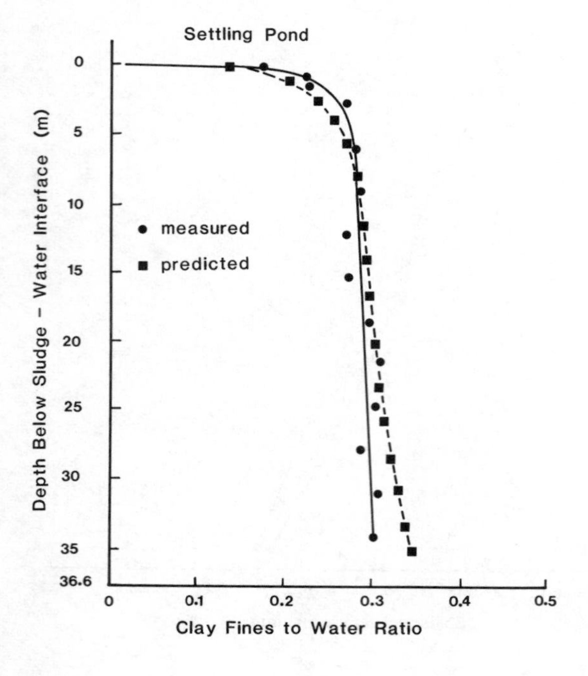

Figure 9.6. Comparison between measured and predicted values for fines: water ratio (by weight) as a function of depth below the settling interface of the pond. Note that the fines are indicated as clay fines

solids concentration for comparison with the measured profile given in Figure 9.3– taking the total composition as solids material. The comparison between predicted and measured values for the field pond using the relative flux model is shown in Figure 9.4 (Yong & Elmonayeri 1983).

2. Comparison of bulk density and fines:water ratio profiles using the piecewise linear consolidation model. Figures 9.5 and 9.6 show the comparison between predicted and measured values. Note that the clay fines:water ratio is used to more properly reflect the fines influence in the establishment of the dispersion stability of the sediment layer – since the sand fraction separates from the total solids and settles under gravitational forces. The sand fraction is not considered in the consolidation model and must therefore be separated from the comparative procedure instituted between predicted and measured values.

9.6 CONCLUDING REMARKS

The problem of large strain consolidation or settling of suspended fines where the volume change is extremely large cannot be handled using standard sedimentation theories if surface-active solids are involved. The reasons for this lie in the fact that settling behaviour is not governed by gravitational forces, but by forces controlled by surface activity of the solids comprising the suspension.

In the procedures developed herein, it is seen that at least two methods for prediction of the settling of the solids can be used. In the first, the relative flux model, the model developed allows one to trace the motion of representative particles all the way from the very dilute solids concentration to the stage where particle contact occurs – zones A through C. When particles come into intimate contact with each other, particle flux can no longer be considered the valid mechanism for analysis – within the framework established in this study. This is not to say that the model cannot be extended to account for particle contact. However, at this particular time, it is considered expedient to utilize the consolidation model for situations where actual consolidation processes are fully developed – as in zone D.

In the comparison showing the predicted solids concentration profile obtained with the relative flux model and the measured values, we should note that the total sand, silt and clay components have all been lumped into the general 'solids' nomenclature for convenience. This procedure can be refined to consider the various fractions separately. However, the experimental procedures required would also demand that the suspensions studied must consider those fractions separately – for determination of the respective diffusion coefficients. The results shown in Figure 9.4 indicate that using the general 'solids' classification, the predicted values compare favourably with the measured values – except near the bottom where the sand fraction has indeed settled out, because of gravitational settling. This latter situation is not covered by the relative flux (diffusion) model.

The comparisons made between measured and predicted values using the piecewise linear consolidation (iterative) procedure, show good comparisons in the case of the clay fines:water ratio profile. The technique used permits one to overcome the restrictions imposed on the finite strain consolidation models because of the nonlinearity of the governing equations.

Finally, it should again be noted that the analytical modelling of settling of suspended fines in sedimentation processes for prediction of time-rate of settling of the suspended solids is complicated by several physical issues related to:

– non-gravitational interactions, i.e. surface-active phenomena;

– complications in laboratory measurement (or field measurement) of properties in suspensions;

– defining initial conditions and effective stress-void ratio relationships.

ACKNOWLEDGEMENTS

This study constitutes part of the general study on *Stability and settling of suspended fines* supported by a Grant-in-Aid of research from the Natural Sciences and Engineering Research Council (of Canada), NSERC, under Grant No. A.882. The input provided by the various researchers in the Geotechnical Research Centre is acknowledged.

REFERENCES

Been, K. & G.C. Sills 1981. Self-weight consolidation of soft soils: an experimental and theoretical study. *Geotechnique* 31 (4): 519–535.

Gibson, R.E., R.L. Schiffmann & K.W. Cargill 1981. The theory of one-dimensional consolidation of saturated clays. II. Finite nonlinear consolidation of thick homogeneous layers. *J. Can. Geotech.* 18: 280–293.

Koppula, S.D. & N.R. Morgenstern 1982. On the consolidation of sedimenting clays. *J. Can. Geotech.* 19: 260–268.

Olson, R.E. & C.C. Ladd 1979. One-dimensional consolidation problems. *J. Geotech. Engng. Div.* (ASCE) 105 (1): 11–30.

Roberts, J.O.L., R.N. Yong & H.L. Erskine 1980. Survey of some tar sand sludge ponds. In *Proceedings of Applied Oilsands Geoscience Conference* (Edmonton): – .

Yong, R.N. & D.S. Elmonayeri 1983. On the stability and settling of suspended solids in settling ponds. Part II. Diffusion analysis of initial settling of suspended solids. *J. Can. Geotech.*

Yong, R.N. & A.J. Sethi 1978. Mineral particle interaction control of tar sand sludge stability. *J. Can. Petr. Technol.* 17 (4): 1–8.

Yong, R.N., A.J. Sethi, H.P. Ludwig & M.A. Jorgensen 1979. Interparticle action and rheology of dispersive clays. *J. Geotech. Engng. Div.* (ASCE) 105 (10): 1193–1209.

Yong, R.N., S.K.H. Siu & D.E. Sheeran 1983. On the stability and settling of suspended solids in settling ponds. Part I. Piecewise linear consolidation analysis of sediment layer. *J. Can. Geotech.*

Yong, R.N. & B.P. Warkentin 1975. *Soil properties and behaviour*. Amsterdam: Elsevier.

Part 3
Specific numerical models and parameter evaluation

Evaluation of soil properties for use in earthquake response analysis

10

K. ISHIHARA

10.1 INTRODUCTION

Dynamic analyses to evaluate the response of soil deposits and earth structures to seismic load applications have been finding increased application in geotechnical engineering practices. Various idealized models and analytical techniques are being used to represent a soil deposit and to evaluate its response, but whatever procedure is followed, it is necessary to determine the appropriate stress-strain and energy absorbing properties of the material in the deposit. In recognition of such needs, many attempts have been made in recent years to evaluate the dynamic properties of soil materials both in the laboratory and in the field. As a result, a comprehensive amount of information has been produced and publicized on this topic with progress having now advanced to the point where more reliable laboratory test data on undisturbed soil samples and more direct in-situ test data can be obtained in the design practice on an almost routine basis. Further investigations are being undertaken to identify dynamic properties of coarse-grained materials such as gravels and crushed rocks and also of soft rocks. With effective use of such information it becomes possible to increase the accuracy of dynamic response analysis. Seismic response analysis, taking into account the effects of the nonlinear nature of soil deformation, can be made either by the equivalent linear method or the step-by-step integration method. The equivalent linear method developed by Seed & Idriss (1969) assumes that a solution to the problem of soil deposits involving nonlinear deformation can be obtained approximately by a linear analysis provided the stiffness and damping used in the analysis are compatible with the effective shear strain amplitudes at all points of the system being analyzed. For this type of analysis, soil properties such as the shear modulus and damping ratio determined from the tests need to be expressed as functions of the shear strain amplitude either in graphic form or in the form of an analytical function. Since the use of an analytical function is not mandatory, it is not necessarily required when using the equivalent linear method to have an established material model that fits the soil properties determined from the tests.

It should be mentioned here that the equivalent linear method is capable of giving a reasonably accurate solution only when the shear strain involved in the analysis is less than about 1%. In cases where larger shear strains are induced in the soil deposits, the equivalent linear method fails to yield an acceptably good solution, and the step-by-step integration procedure is instead preferable.

The response analysis to such high-intensity seismic loading often produces shear strains in the soil deposits that are greater than a few percent and are almost on the verge of failure. For analysis of such a problem, the step-by-step integration procedure is badly in need.

Another aspect becoming increasingly important in recent times is the development of the effective stress approach for making response analyses of saturated sand deposits in which progressive build-up of pore water pressures is considered to lead to a significant deterioration of soil stiffness in the course of seismic loading. In such problems, the soils tend to induce a large amount of shear strain when the effective confining stress becomes very small approaching the onset of liquefaction. Therefore, the use of the step-by-step integration procedure is almost mandatory to successfully conduct the seismic response analyses.

When using the step-by-step method, it is necessary to feed the soil properties into the computer code at each time step of integration. This complicated procedure can be practically handled only when the stress-strain relation is expressed in an analytical form and the information on instantaneous soil properties can be continuously derived from the analytical relation. For this purpose, it is necessary to have an established material model for describing the soil properties under dynamic loading conditions.

In the following pages of this chapter, recently reported information on the dynamic properties of soil materials will be briefly reviewed with emphasis on the shear modulus and damping values obtained from tests on undisturbed samples. In view of the importance of establishing a soil model to be incorporated into the step-by-step response analysis procedure, an attempt will be made to fit available test data to those models known as the Ramberg-Osgood model and the hyperbolic model. Procedures for determining the parameters appearing in these models will be suggested, with particular emphasis on the representation of material behaviour at large strain levels near failure.

10.2 GENERAL CHARACTERISTICS OF SOIL BEHAVIOUR

It has been known that the stress-deformation characteristics of soils vary to a large extent depending upon the magnitude of shear strains to which soils are subjected. Overall changes in soil behaviour with changes in shear strain are illustrated in Figure 10.1, in which approximate ranges of the shear strain producing elastic, elasto-plastic and failure states of stress are indicated. In the infinitesimal strain range below the order of 10^{-5}, the deformations exhibited by most soils are purely elastic and recoverable. The phenomena associated with such small strains would be vibration or wave propagation through soil grounds. Over the intermediate range of strain between 10^{-4} and 10^{-2}, the behaviour of soils is elasto-plastic and produces irrecoverable permanent deformation. Development of cracks or differential settlements in soil structures appear to be associated with the elasto-plastic nature of soils that is exhibited within such a range of strain. When large strains exceeding a few percent are imposed on soils, the strains tend to become considerably large without a further increase in shear stress and failure takes place in the soils. Slope slides or compaction and liquefaction of cohesionless soils are associated with the failure-inducing large strains.

Another feature to be noted in soil behaviour is the dilatancy, i.e., the tendency of

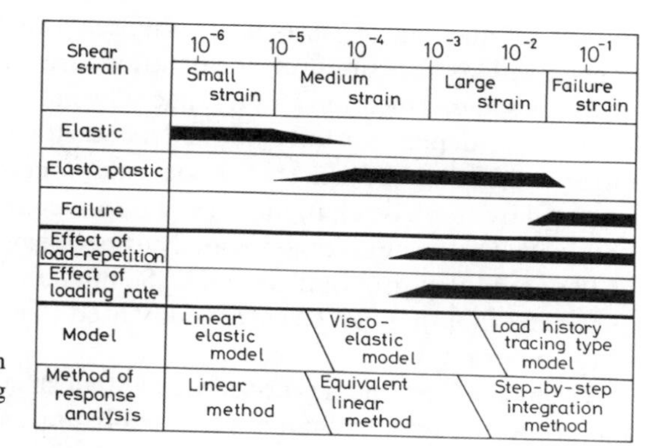

Figure 10.1. Changes in soil properties with shear strain and corresponding modelling principle and method of response analysis

soils to dilate or to contract during drained shear or pore water pressure changes during undrained shear. The dilatancy in the repetition of load does not come out in the infinitesimal and intermediate strain ranges. Its effect begins to appear when the magnitude of shear strain increases above a level of about 10^{-4} to 10^{-3} (Dobry et al. 1980) as indicated in Figure 10.1. It should be borne in mind that the progressive changes in soil properties during load repetition such as degradation in stiffness of saturated soils or hardening of dry or partially saturated soils can occur as a consequence of the dilatancy effect being manifested during shear.

Still another important aspect of soil deformation characteristics observed under dynamic loading conditions is the influence of the speed with which loads are applied to soils. Laboratory tests have shown that the resistance to deformation of soils under monotonic loading conditions generally increases as the speed of loading is increased, and also that the strength of soils increases with an increase in time to failure. It is to be noted that the effect of the loading speed does not appear when the shear strain is very small. It has been shown that the threshold shear strain between where the rate effect does or does not come out is in the order of about 10^{-3}. A detailed account of the effect of loading speed on the behaviour of soils is given elsewhere (Ishihara 1981).

Modelling of soil behaviour under cyclic or random loading conditions must be made so that the model can duplicate the different deformation characteristics varying over the wide range of shear strain as described above. When a soil behaviour is expected to stay within the range of the small strain, the use of an elastic model is justified and the wave propagation method based on the linear elastic theory can be employed to obtain an exact solution for any problem concerned. In this simplest case, the shear modulus is a key parameter to properly model the soil behaviour. When a given problem is associated with the medium range of strain, approximately below the level of 10^{-3}, the soil behaviour becomes elasto-plastic and the shear modulus tends to decrease as the shear strain increases. At the same time, energy dissipation occurs during cycles of load application. The energy dissipation in soils is mostly rate-independent and of a hysteretic nature, and damping ratio can be used to represent the energy absorbing properties of soils. Since the strain level concerned is still small enough not to cause any progressive change in soil properties, the shear modulus and damping ratio do not change with the progression of cycles in load application. Such steady-stage soil characteristics can be represented to a reasonable degree of accuracy by use of the linear viscoelastic theory. The shear modulus and damping ratio determined as functions of shear strain are the key parameters to represent soil properties in this medium strain range. The most useful analytical tool accommodating these strain-dependent but cycle-independent soil properties would be the equivalent linear method based on the viscoelastic concept. Generally, the linear analysis is repeated by stepwise changing in soil parameters until a strain-compatible solution is obtained. The seismic response analysis performed for horizontally layered soil by use of the computer program SHAKE (Schnabel et al. 1972) is a typical example of an analytical tool that can be successfully used to clarify the soil response in the medium range of strain.

When a soil problem is concerned with a shear strain level larger than about 10^{-2}, soil properties tend to change appreciably not only with shear strain but also with the progression of cycles. The manner in which the shear modulus and damping ratio change with cycles is considered to depend upon the manner of change in the effective confining stress during irregular time histories of shear stress application. When the law of changing effective stress is established, it is then necessary to have a hysteresis law in which stress-strain relations can be specified at each step of the loading, unloading and reloading phases. The concept most commonly used at present for this purpose is what is referred to as the Masing law. For analysis of a soil response accommodating such a stress-strain law covering large strain levels near failure, it is necessary to employ a numerical procedure involving the step-by-step integration technique as indicated in Figure 10.1.

Since the primary objective of this chapter is to present ideas for a proper representation of soil properties at larger shear strains conforming to the step-by-step method of seismic response analyses, it will be assumed throughout this chapter that usage of the Masing rule for the proposed material model is a mandatory premise for successful application of the model. Furthermore, the soil properties discussed in the following pages are assumed to be those associated with cyclic loading involving no change in the effective confining stress, and, therefore, to be independent of the cycles of load application.

10.3 STRUCTURE OF MODEL FOR CYCLIC LOADING

Modelling of soil behaviour in cyclic loading is usually made by first specifying the stress-strain relation in the virgin or initial loading. Let this relation be given by:

$$\tau = f(\gamma) \tag{10.1}$$

where τ and γ are appropriate shear stress and shear strain, respectively. The curve described by Equation 10.1 is called the skeleton curve or backbone curve as illustrated in Figure 10.2. If loading reversal occurs at point A where $\gamma = \gamma_a$ and $\tau = \tau_a$, then the equation of the stress-strain curve during subsequent unloading from the reversal point is assumed to be given by:

$$\frac{\tau - \tau_a}{2} = f\left(\frac{\gamma - \gamma_a}{2}\right) \tag{10.2}$$

If the curve defined by Equation 10.2 reaches point B on the skeleton curve on the

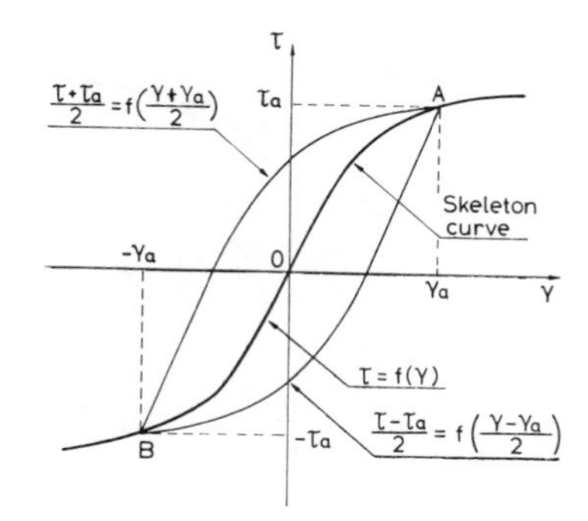

Figure 10.2. Construction of unloading and reloading curves based on the Masing rule

opposite side, the stress-strain curve is assumed to follow the skeleton curve farther on the negative side. If reloading is executed at point B, the stress-strain curve for the reloading is given by an equation similar to Equation 10.2 in which the signs of τ_a and γ_a are changed. If the skeleton curve is intersected again at point A during the reloading, further loading is assumed to follow the skeleton curve on the positive side. The rule for constructing the unloading and reloading branches of the stress-strain curves as above is referred to as the Masing rule. The rule can be applied to form stress-strain curves for any phase of loading during complicated load histories. It is to be noted that the stress-strain curves for unloading and reloading branches given by Equation 10.2 have the same shape as that of the skeleton curve, except that the scale is enlarged by a factor of 2 and the starting point of each curve is translated at the point of stress reversal. It is also noted that the unloading curve starting from point A in Figure 10.2 intersects the skeleton curve at point, B, which is symmetric with respect to the origin, O.

If soil properties are assumed not to change with the progression of cycles, then the stress-strain curve stays unchanged for constant-amplitude unload-reload cycles. For such steady-state cyclic loading conditions, the shear characteristics of soils can be represented by the secant modulus, G, which is defined by:

$$G=\frac{\tau_a}{\gamma_a}=\frac{f(\gamma_a)}{\gamma_a} \tag{10.3}$$

where τ_a and γ_a denote the amplitude of shear stress and shear strain, respectively. The damping characteristics of soils is represented by the damping ratio, D, which is defined by:

$$D=\frac{1}{4\pi}\frac{\Delta W}{W} \tag{10.4}$$

where ΔW is the damping energy, i.e., the area within the hysteresis loop shown in Figure 10.3, and W is the equivalent strain energy defined as:

$$W=\tfrac{1}{2}f(\gamma_a)\gamma_a \tag{10.5}$$

Since the hysteresis loop is obtained by enlarging the skeleton curve by a factor of 2 in γ and τ directions the half-moon section ABE has the same shape as the half-moon portion AOC in Figure 10.3 and, therefore, the area ABE is four times the area of AOC.

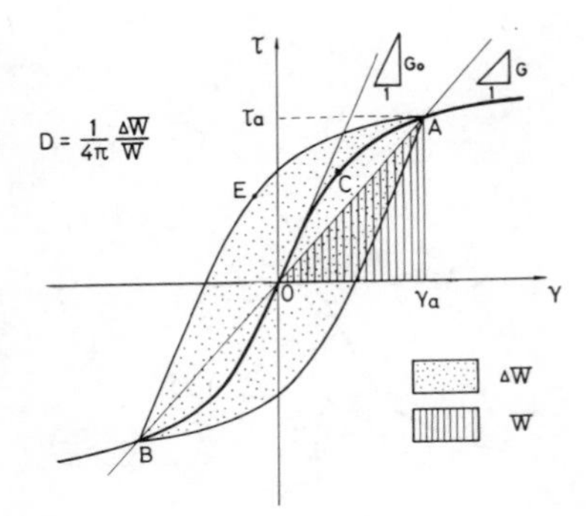

Figure 10.3. Definition of secant shear modulus, G, and damping ratio, D

Hence, the damping energy, ΔW, at the strain amplitude, γ_a, can be computed by:

$$\Delta W = 8\left[\int_0^{\gamma_a} f(\gamma)d\gamma - W\right] \tag{10.6}$$

Substituting Equations 10.5 and 10.6 in Equation 10.4, one obtains:

$$D = \frac{2}{\pi}\left[\frac{2\int_0^{\gamma_a} f(\gamma)d\gamma}{f(\gamma_a)\gamma_a} - 1\right] \tag{10.7}$$

In some cases, the constitutive model is constructed so as to express the shear strain in terms of the shear stress. Let the skeleton curve of such a model be expressed by:

$$\gamma = g(\tau) \tag{10.8}$$

Then, the secant shear modulus and damping ratio are given, respectively, by the formulae:

$$G = \frac{\tau_a}{\gamma_a} = \frac{\tau_a}{g(\tau_a)} \tag{10.9}$$

$$D = \frac{\pi}{2}\left[1 - \frac{2\int_0^{\tau_a} g(\tau)d\tau}{g(\tau_a)\cdot\tau_a}\right] \tag{10.10}$$

It is to be noted that both the shear modulus, G, and damping ratio, D, are expressed as functions of the shear strain amplitude, γ_a, or the shear stress amplitude, τ_a, and that these functional forms are determined if the relation of Equation 10.1 or Equation 10.8 for the skeleton curve is specified.

10.4 MODELS FOR DESCRIBING STRESS-STRAIN RELATIONS OF SOILS

The most commonly used models for describing nonlinear stress-strain relations of soils are the hyperbolic model proposed by Hardin & Drnevich (1972b) and the Ramberg-Osgood model (R–O model). Modified versions of these models and other models are also proposed, but in the following sections the basic nature of the above two models will be discussed in some detail.

10.4.1 *Hyperbolic model*

The stress-strain relation for the skeleton curve is assumed to be represented by the hyperbolic equation formulated by Kondner & Zelasko (1963), which is shown as:

$$\tau = \frac{G_0 \cdot \gamma}{1 + \dfrac{G_0}{\tau_f}\gamma} \qquad (10.11)$$

where G_0 is the initial tangent shear modulus at $\gamma = 0$ and τ_f is the shear strength of soils. The initial shear modulus, G_0, can be taken as being equal to the elastic shear modulus at a very small strain range. Hardin & Drnevich (1972b) defined the reference strain, γ_r, as:

$$\gamma_r = \frac{\tau_f}{G_0} \qquad (10.12)$$

The meaning of the reference strain is illustrated in Figure 10.4. Introducing Equation 10.12 into Equation 10.11, the expression of the shear modulus (secant modulus) for the hyperbolic model is obtained from Equation 10.3 as:

$$\frac{G}{G_0} = \frac{1}{1 + \gamma_a/\gamma_r} \qquad (10.13)$$

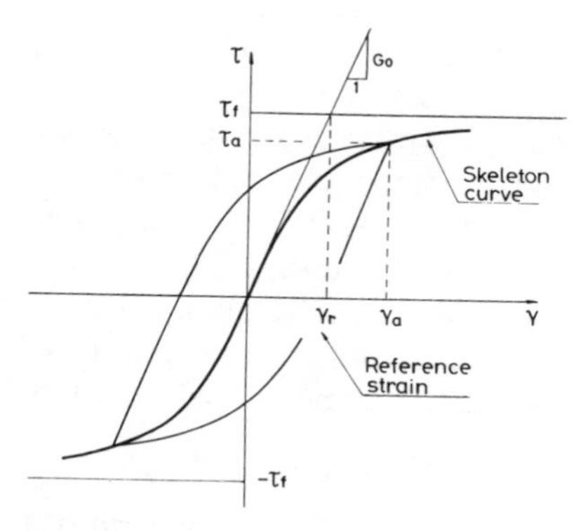

Figure 10.4. Definition of reference strain

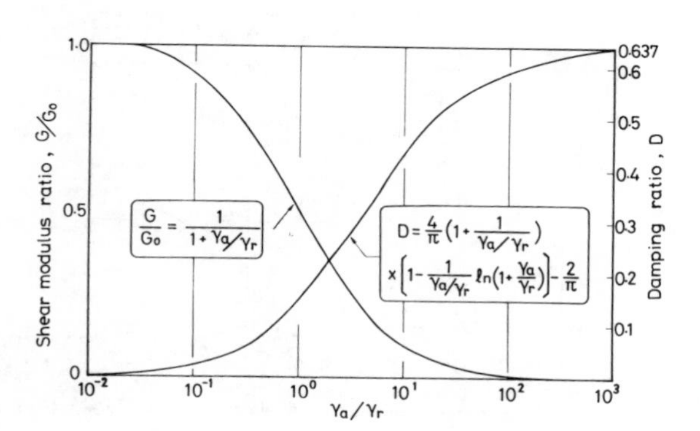

Figure 10.5. Shear modulus and damping ratio versus shear strain based on the hyperbolic model

The value of shear modulus ratio, G/G_0, computed from Equation 10.13 is plotted versus the shear strain ratio, γ_a/γ_r, in Figure 10.5. It is noted that the shear modulus is reduced to half of the initial shear modulus when the shear strain becomes equal to the reference strain. The expression for damping ratio can be derived by applying the Masing rule to the skeleton curve given by Equation 10.11. By introducing Equation 10.11 into Equation 10.7, the damping ratio for the hyperbolic model is calculated as:

$$D = \frac{4}{\pi}\left(1 + \frac{1}{\gamma_a/\gamma_r}\right)\left[1 - \frac{1}{\gamma_a/\gamma_r}\ln\left(1 + \frac{\gamma_a}{\gamma_r}\right)\right] - \frac{2}{\pi} \tag{10.14}$$

The value of the damping ratio computed from Equation 10.14 is plotted versus the shear strain ratio in Figure 10.5. It is noted that the damping ratio in the hyperbolic model converges to $2/\pi = 0.673$ when the shear strain becomes infinitely large. Since both shear modulus, G, and damping ratio, D, are expressed as functions of the strain amplitude, γ_a, it is possible to eliminate it between Equations 10.13 and 10.14. Then, a relationship between the shear modulus and damping ratio can be obtained as:

$$D = \frac{4}{\pi}\frac{1}{1 - G/G_0}\left[1 - \frac{G/G_0}{1 - G/G_0}\ln\left(\frac{1}{G/G_0}\right)\right] - \frac{2}{\pi} \tag{10.15}$$

The relationship of Equation 10.15 is numerically calculated and plotted in Figure 10.6.

In the hyperbolic model as formulated above, the parameters specifying the structure of the model are the initial shear modulus, G_0, and the shear strength, τ_f. In some cases, it is difficult to specify both the strain-dependent shear modulus and damping ratio by the use of only two parameters. Particularly inconvenient is the fact that, once the reference strain, γ_r, is determined from the strain-dependent characteristics of the shear modulus, the value of the strain-dependent damping ratio is automatically given and no adjustment of parameters is possible to achieve a good fit to the experimentally obtained damping data of soils.

In an attempt to overcome this drawback, Hardin & Drnevich (1972b) proposed the use of the empirical relationship, given below, instead of using Equation 10.15:

$$D = D_0(1 - G/G_0) \tag{10.16}$$

in which D_0 is the damping ratio at a large strain where the shear modulus, G, becomes

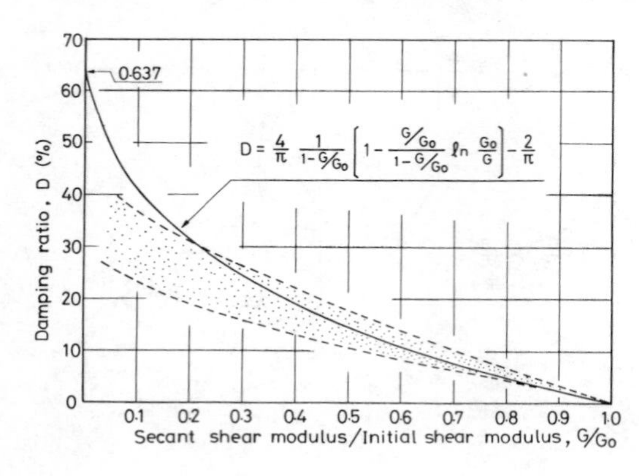

Figure 10.6. Relationship between damping ratio and shear modulus based on the hyperbolic model and Masing rule

very small as compared to the initial value, G_0, i.e., $G/G_0 = 0$. The use of Equation 10.16 is advantageous in that it introduces an additional parameter, D_0, to achieve a good fit of the model to experimental data. It is to be noted, however, that the relation of Equation 10.16 is not compatible with the Masing rule.

Therefore, when application of the Masing rule is necessary to conduct dynamic response analysis based on the hyperbolic model, it becomes impossible to establish a hysteresis loop satisfying the relationship of Equation 10.16, and there is no choice but to return to the use of Equation 10.14.

10.4.2 *Ramberg-Osgood model*

The stress-strain relation for the skeleton curve is assumed to be represented by:

$$\frac{\gamma}{\gamma_y} = \frac{\tau}{\tau_y}\left[1 + \alpha\left|\frac{\tau}{\tau_y}\right|^{r-1}\right] \tag{10.17}$$

where γ_y and τ_y are appropriate shear strain and shear stress, respectively, and α and r are constant which permit adjustment of the shape and position of the curve. When using Equation 10.17, it is desirable to choose as γ_y and τ_y a shear strain and shear stress that can be defined clearly with some physical meaning. Several possibilities have been suggested as to what quantity should be assigned to γ_y and τ_y.

Modified R–O model model by Richart

Richart (1975) proposed defining τ_y and γ_y as:

$$\tau_y = C_1\tau_f,\ \gamma_y = \frac{C_1\tau_f}{G_0} = C_1\gamma_r \tag{10.18}$$

where C_1 is a constant less than unity. In the above definition, the shear stress, τ_y, is taken to be a fraction of soil strength and the shear strain, γ_y, is chosen to be proportional to the reference strain by the same fraction. Introducing the definition of Equation 10.18 into Equation 10.17, the stress-strain relation for the modified R–O model is obtained as:

$$\tau = \frac{G_0\gamma}{1 + \alpha\left|\dfrac{\tau}{C_1\tau_f}\right|^{r-1}} \tag{10.19}$$

Modified R–O model by Hara

Hara (1980) proposed the use of the shear strength for τ_y and the reference strain for γ_y in the original form of the R–O model. This assumption is equivalent to putting $C_1 = 1.0$ in Equation 10.18. Then, the equation of the skeleton curve is written as:

$$\tau = \frac{G_0\gamma}{1 + \alpha\left|\dfrac{\tau}{\tau_f}\right|^{r-1}} \tag{10.20}$$

It is to be noted that the stress-strain relation in the form of Equation 10.20 is essentially the same as that of Equation 10.19, because the inclusion of parameter C_1

does not produce any structural change in the stress-strain relation represented by the R–O model. If parameter α in Equation 10.20 is chosen as being equal to α/C_1^{r-1} in Equation 10.19, both equations become identical. The stress-strain relation in the form of Equation 10.20 will be addressed in more detail in the following pages.

The expression for the strain-dependent modulus can be obtained immediately from Equation 10.20 as:

$$\frac{G}{G_0} = \frac{1}{1 + \alpha\left|\frac{G}{G_0}\frac{\gamma_a}{\gamma_r}\right|^{r-1}} \qquad (10.21)$$

The application of the Masing rule to the skeleton curve as represented by Equation 10.2 gives the functional expressions for the hysteresis loop during the unload-reload cycles. By introducing these expressions of the hysteresis loop in Equation 10.7, it is possible to derive a formula for the damping ratio as:

$$D = \frac{2}{\pi}\frac{r-1}{r+1}\,\alpha\frac{\left|\frac{G}{G_0}\frac{\gamma_a}{\gamma_r}\right|^{r-1}}{1 + \alpha\left|\frac{G}{G_0}\frac{\gamma_a}{\gamma_r}\right|^{r-1}} \qquad (10.22)$$

By eliminating the shear strain amplitude, γ_a, between Equations 10.21 and 10.22, it is possible to derive a relationship between the shear modulus and damping ratio as:

$$D = \frac{2}{\pi}\cdot\frac{r-1}{r+1}\cdot(1 - G/G_0) \qquad (10.23)$$

It should be noted that Equation 10.23 for the R–O model has the same form as Equation 10.16. It is to be remembered that, while Equation 10.16 was empirically assumed by Hardin & Drnevich (1972b), the relation of Equation 10.23 is derived analytically by constructing the hysteresis loop in accordance with the Masing rule. Therefore, when the R–O model is used to represent soil properties in the dynamic response analysis, application of the Masing rule to form the branches of the unload-reload cycles automatically leads to the use of the damping ratio defined by Equation 10.22.

Unlike the two-parameter hyperbolic model, R–O model has four parameters to represent the strain-dependent soil properties and, therefore, has more freedom to achieve a better fit of the model to experimental data.

10.5 APPLICABILITY AND LIMITATIONS OF THE MODELS

The soil models as described in the preceding section have advantages and disadvantages in properly representing soil properties varying over the broad spectrum of strain amplitude. The important features and conditions required of these models are summarized as follows.

1. A model should be able to express the secant shear modulus as a function of shear strain or shear stress. The function should be constructed so that it gives the tangent modulus at small strains, G_0, when the shear strain or shear stress becomes

infinitesimally small. The R–O model and the hyperbolic model as introduced above both satisfy these requirements.
2. When the shear strain becomes large, the shear stress expressed by the model should reach a certain limiting value which is equal to the shear strength of soils. This condition is satisfied by the hyperbolic model, but not by the R–O model. As easily verified from Equation 10.17 or 10.20 the shear stress can increase infinitely as the shear strain is increased.
3. When utilizing a model in the analysis of soil response in combination with the Masing rule, the structure of the model must be such that it can yield a reasonable value of the damping ratio especially for the large strain range. This requirement can be met in the R–O model, if the parameters in the model are properly chosen. However, the hyperbolic model is not designed to satisfy this condition. As can be easily seen in Figure 10.5, the hyperbolic model tends to give intolerably large damping ratio when the shear strains are large. The damping ratio in the hyperbolic model becomes equal to as much as $2/\pi = 63.7\%$ in the limit.

10.6 PROCEDURES FOR DETERMINING PARAMETERS IN THE MODELS

It is most preferable to be able to determine the parameters in the material model on the basis of some physically meaningful soil constants that can be determined easily by simple tests. In this context, the initial shear modulus, G_0, at very low levels of strain, and the shear strength, τ_f, appearing in the hyperbolic and R–O models are the most appropriate soil constants to be incorporated in the parameters in these models. The shear strain of soils at failure would also be a useful index property conforming to the determination of the parameters of the model.

10.6.1 *Hyperbolic model*

As explained in the foregoing section, only two parameters need to be determined in the hyperbolic model. The value of the shear strength at failure of soils is generally determined by laboratory tests using a triaxial or simple shear test apparatus. It is preferable to determine the strength parameters under cyclic loading conditions. In some cases, the Mohr-Coulomb type failure criterion with appropriate strength parameters may be used to estimate the shear strength of soils, as suggested by Hardin & Drnevich (1972b).

The value of the initial shear modulus at very low levels of strain is determined either by resonant column tests in the laboratory or by means of some in-situ wave propagation tests. Several empirical formulae may as well be used to evaluate the value of the initial shear modulus.

In the current practice of soil testing, the dynamic properties of soils are determined in most cases by means of resonant column tests. When performing this type of test, the amplitude of vibratory shear strain is increased stepwise, putting the sample into resonance in each step. The shear modulus is determined by monitoring the velocity of shear wave propagation at each resonant state and the damping value is obtained by observing attenuation characteristics of the sample in the free vibration test conducted

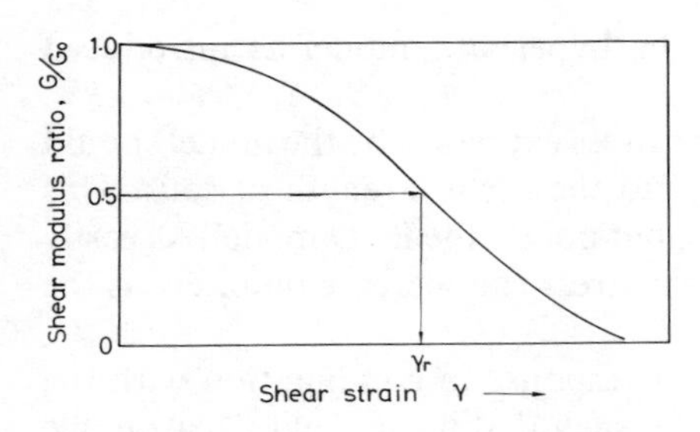

Figure 10.7. Determination of reference strain from shear modulus versus strain plot for the hyperbolic model

after each resonant test. Therefore, the values of the shear modulus and damping ratio can be routinely obtained as functions of the strain. By fitting the strain-dependent modulus formula of Equation 10.13 to the experimentally obtained curve, the reference strain can be determined. The reference strain, γ_r, can be easily obtained by locating, on the shear modulus versus strain plot, the strain point at which the shear modulus is reduced to one half of the initial modulus, as explained in Figure 10.7. It should be borne in mind that the above procedure for determining the reference strain is warranted only when the soil properties are to be represented by the hyperbolic function of Equation 10.13.

10.6.2 *Ramberg-Osgood model*

In the R–O model expressed by Equation 10.20, the initial shear modulus, G_0, and the shear strength, τ_f, can be determined by the same procedure as in the case of the hyperbolic model. With known values of these two soil constants, the reference strain, γ_r, is immediately determined from its definition of Equation 10.12. The remaining two parameters, α and r, may be determined in the following manner.

Parameter α

One of the methods for determining parameter α in the R–O model would be to utilize the shear strain at failure, γ_f, as suggested by Hara (1980). By putting $\tau = \tau_f$ when $\gamma = \gamma_f$ in Equation 10.20, a relation is obtained as:

$$\alpha = \frac{\gamma_f}{\gamma_r} - 1 \tag{10.24}$$

Parameter α can be alternatively expressed in terms of the shear modulus at failure, G_f, and the initial shear modulus, G_0, by putting $G = G_f$ and $\gamma_a = \gamma_f$ in Equation 10.21:

$$\alpha = \frac{G_0}{G_f} - 1 \tag{10.25}$$

where:

$$G_f = \tau_f / \gamma_f \tag{10.26}$$

Advantage of determining parameter α in the above fashion is that the soil properties can be represented with good accuracy, particularly in the large strain range, and that the potential drawback of the R–O model leading to an unrealistically large stress at large strains can be avoided.

Parameter r

It would be preferable to determine parameter, r, from a knowledge of the damping characteristics of soils. Since the damping ratio, D, in the R–O model is related with the shear modulus ratio, G/G_0, through Equation 10.23, the value of r may be determined if the values of D and G/G_0 at a certain strain level are known. When the damping ratio at failure, D_0 is known, the value of r is determined as:

$$r = \frac{1 + \dfrac{\pi D_0}{2} \dfrac{1}{1 - G_f/G_0}}{1 - \dfrac{\pi D_0}{2} \dfrac{1}{1 - G_f/G_0}} \tag{10.27}$$

10.7 INITIAL SHEAR MODULI AT SMALL STRAINS

A wealth of experimental data has been accumulated to evaluate the shear moduli of soils at very small levels of strain both in the laboratory and in situ. In laboratory tests, the most widely used procedure is the resonant column test, although some static triaxial tests are used with the precise strain measurement technique. In the field, the wave propagation methods such as the down-hole and cross-hole methods are widely used. In what follows, the main features of findings obtained from recent test results will be reviewed with emphasis being placed on in-situ soil investigations.

10.7.1 *Sandy soil*

According to the results of a comprehensive study by Iwasaki & Tatsuoka (1977), the shear moduli at low strains for specimens of clean sands reconstituted in the laboratory, are described most generally by an empirical formula:

$$G_0 = A(\gamma_a) \cdot B \cdot \frac{(2.17 - e)^2}{1 + e} (\sigma_0')^{m(\gamma_a)} \tag{10.28}$$

where e is the void ratio and B is a parameter showing the influence of grain characteristics of the sand. The initial shear modulus, G_0, and the effective confining stress, σ_0' in Equation 10.28 are expressed in kN/m^2.

For clean sands containing no fines, it was shown that as the shear strain amplitude increases from 10^{-6} to 10^{-4}, the value of A decreases correspondingly from 16,000 to 14,300. Therefore, a value of $A = 16{,}600$ may be taken as representative of that for clean sands, considering the shear strain amplitude of 10^{-6} as being small enough to produce the maximum modulus. It was also shown that the value of the exponent, m, may be taken to be approximately 0.4 for the small shear strain of 10^{-6}. For clean sands, the value of factor B is chosen to be unity, but for sands containing fines passing no. 200 mesh, factor B decreases below unity, as shown by the test data in Figure 10.8. It may be stated, therefore, that the shear modulus of fines-containing sands is generally smaller than that of clean sands.

In order to evaluate the shear modulus of in-situ sands, it would be necessary to conduct laboratory tests on undisturbed samples obtained from in-situ deposits. Such

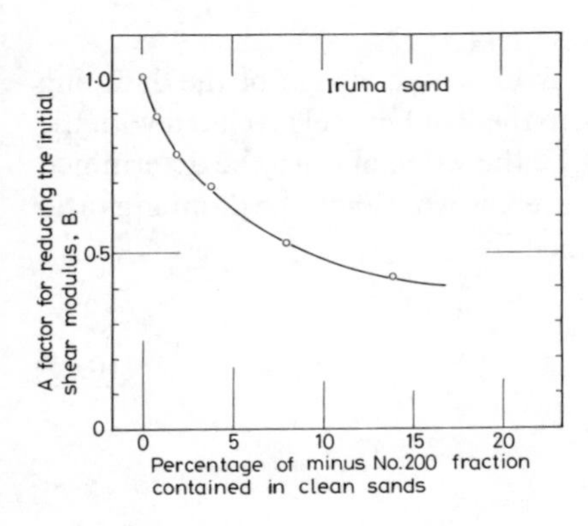

Figure 10.8. Factor for reducing initial shear modulus due to the effect of fines contained in clean sands (Iwasaki & Tatsuoka 1977)

an attempt was made by Higuchi et al. (1981), who performed a series of resonant column tests on undisturbed samples of sandy soils obtained from the seabed. Results of the tests are presented in Figure 10.9, in which the shear modulus, G_0, divided by a function of the void ratio is plotted versus the effective confining pressure, σ_0' The test data shown in the figure are those which were obtained with the small shear strain amplitude of $\gamma_a = 10^{-6}$ by means of a resonant column test. Also plotted in Figure 10.9 for reference is the empirical formula:

$$G = 16{,}600\frac{(2.17 - e)^2}{1 + e}(\sigma')^{0.4} \tag{10.29}$$

which was obtained for reconstituted specimens of clean sands by Iwasaki & Tatsuoka (1977). Another series of tests on undisturbed sands was also conducted by Kokusho & Esashi (1981), who used a cyclic triaxial shear test apparatus in which a pair of highly sensitive displacement sensors capable of detecting an axial strain as small as 10^{-6} were installed above the loading cap inside the triaxial cell. The results of these tests on undisturbed diluvial sands are also presented in Figure 10.9. This figure shows that the test data on the undisturbed specimens tend to give about 20% smaller moduli on the average than the test results on the reconstituted specimens of clean sands having similar grain-size characteristics. As indicated in Figure 10.9, the value of A corresponding to these reduced moduli lies between 7,900 and 14,300. The reason for such

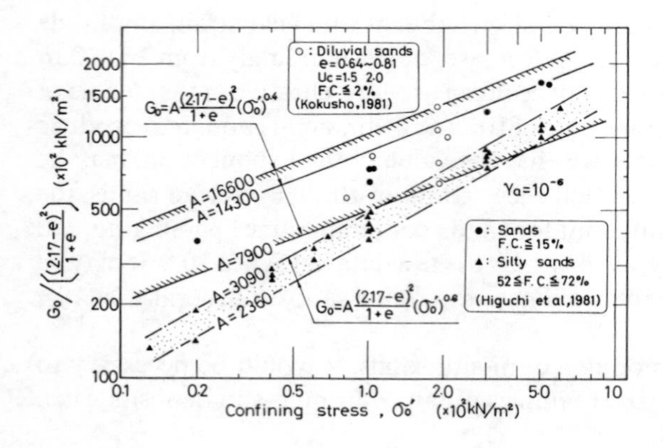

Figure 10.9. Shear modulus at small strains for undisturbed samples of in-situ sandy soils

reductions in shear modulus for undisturbed specimens is not precisely known, but cementation or aging effects may conceivably be accountable.

Higuchi et al. (1981) also conducted resonant column tests on undisturbed sands containing fines of more than 50% in weight. The results of these tests are also shown in Figure 10.9, where it may be seen that the initial shear modulus is generally smaller than that of the clean sand, giving a value of A between 2,360 and 3,090. It is to be noted that for undisturbed silty sands, the shear modulus at the strain level of $\gamma_a = 10^{-6}$ is proportional to the 0.6 power of the confining stress rather than to the 0.4 power for the clean sands.

10.7.2 *Cohesive soils*

Factors influencing the initial shear moduli of cohesive soils such as the void ratio, effective confining stress and overconsolidation have been investigated by Hardin & Black (1968, 1969), Humphries & Wahls (1968), Marcuson & Wahls (1972), Zen et al. (1978) and Kokusho et al. (1982).

An empirical formula for the initial shear modulus suggested by Hardin & Black (1968) is:

$$G_0 = 3{,}270\frac{(2.97 - e)^2}{1 + e}(\sigma_0')^{0.5} \tag{10.30}$$

This formula appears to be applicable, however, only for clays of low plasticity or relatively stiff clays having void ratios of approximately $e = 0.6 \sim 1.5$. For clays with a high plasticity index and hence, with high compressibility, the empirical formula by Marcuson & Wahls (1972), as follows, may be used.

$$G_0 = 445\frac{(4.4 - e)^2}{1 + e}(\sigma_0')^{0.5} \tag{10.31}$$

This formula was originally derived for more general cases in which the effects of overconsolidation and duration of consolidation time are included. Because Equation 10.31 was obtained on the basis of test results on bentonite clays, it is applicable for soft clays with a void ratio range of about $e = 1.5 \sim 2.5$.

Clay soils from natural deposits such as alluvial and reclaimed deposits are sometimes more compressible than those for which Equation 10.31 is applicable. For such soils, Kokusho et al. (1982) proposed the use of the formula:

$$G_0 = 90\frac{(7.32 - e)^2}{1 + e}(\sigma_0')^{0.6} \tag{10.32}$$

Equation 10.32 was derived on the basis of test results on undisturbed samples of soft alluvial clay from Teganuma, Japan. The plasticity index of this clay was in the range of $I_p = 40$ to 100, and the void ratio values ranged between $e = 1.5$ and 4.0. The values of the initial shear modulus computed by Equations 10.30, 10.31 and 10.32 are plotted in Figure 10.10, where it may be seen that the above three formulae give approximately the same shear modulus in the void ratio range around $e = 1.5$, but the formula by Hardin & Black gives unacceptably small shear modulus for soils with void ratio values exceeding $e = 2.5$.

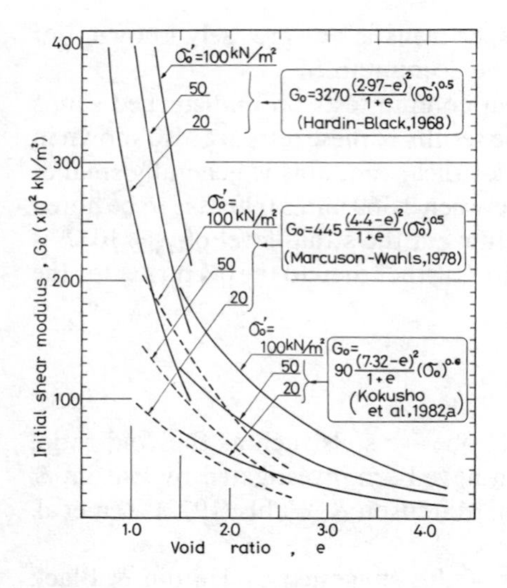

Figure 10.10. Shear modulus at small strains versus void ratio for cohesive soils

Another important variable influencing the initial shear modulus would be the length of time in which confining stress is applied to clays over a long period of time. The effects of the time of sustained stress application have been studied by Afifi & Woods (1971), Marcuson & Wahls (1972) and Afifi & Richart (1973) in the laboratory for prepared cohesive soils such as kaolinite and bentonite. In general, these authors showed that shear modulus increases in proportion to the logarithmic cycle of time when the confining stress is continuously applied subsequent to completion of consolidation. The rate of modulus change was found to be small for cohesionless soils, but as great as 17% per log cycle of time for some fine-grained soils. Similar time effects were also noted by Anderson & Woods (1976) in undisturbed samples of cohesive soils secured from some naturally occurring deposits.

10.7.3 *Coarse-grained materials*

Dynamic properties of coarse granular soils have not been thoroughly investigated because of the difficulty in constructing large-scale dynamic test equipments accommodating the large grain size of the test materials. However, in view of recent needs, some attempts have been made to measure the shear moduli and damping properties of coarse granular soils.

Prange (1981) reported the results of tests on a ballast material using a large resonant column test apparatus. Cylindrical samples 1 m in diameter and 2 m in height, consisting of railroad ballast material with a maximum particle size of 70 mm, were subjected to a torsional mode of vibration under isotropic confining stresses varying from 60 to 100 kN/m². The grain-size distribution curve of the ballast material used is shown in Figure 10.11. The results of the tests showed that the shear modulus at very small strains on the order of 10^{-6} is approximately given by:

$$G_0 = 7{,}230\frac{(2.97-e)^2}{1+e}(\sigma_0')^{0.38} \tag{10.33}$$

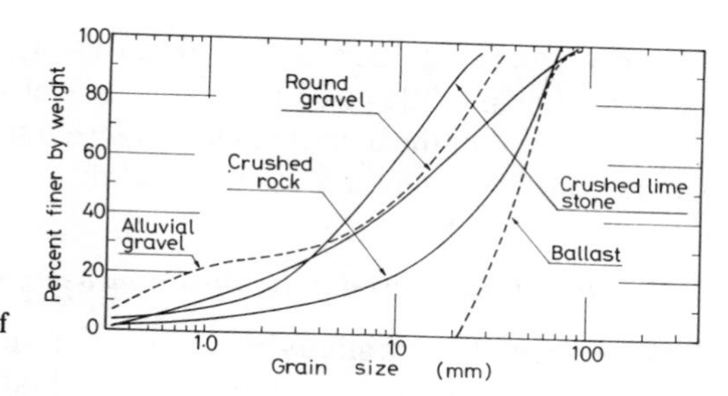

Figure 10.11. Grain-size distribution curves of coarse-grained soils used in the tests

where G_0 and σ_0' are expressed in terms of kN/m^2. The shear modulus calculated from Equation 10.33 for a confining stress of $\sigma_0' = 100\ kN/m^2$ is plotted versus the void ratio in Figure 10.12 for purposes of comparison with other test data on similar materials.

A comprehensive study was conducted by Kokusho & Esashi (1981) to investigate the dynamic properties of coarse-grained materials. A large cyclic triaxial test apparatus accommodating a sample size of 30 cm in diameter and 60 cm in height was utilized with special provision of a high precision displacement sensor monitoring axial strains. Crushed rock and round gravel were used for the tests. The grain-size distribution curves of these materials are shown in Figure 10.11. The test results showed that at strains on the order of 10^{-6}, the shear modulus for the crushed rock is approximately given by:

$$G_0 = 13{,}000\frac{(2.17 - e)^2}{1 + e}(\sigma_0')^{0.55} \qquad (10.34)$$

and for the round gravel, the shear modulus is approximated by:

$$G_0 = 8{,}400\frac{(2.17 - e)^2}{1 + e}(\sigma_0')^{0.60} \qquad (10.35)$$

The values of the shear modulus as computed from the above two formulae are plotted versus the void ratio in Figure 10.12 for the case of confining pressure of $100\ kN/m^2$. It may be seen in Figure 10.12 that the curve for the round gravel lies approximately on the extended portion of the curve for the Japanese standard sand (Toyoura sand). It is also noted that the shear modulus, G_0, for the crushed rock is about 40% larger than that of the round gravel. The comparison as demonstrated in Figure 10.12 appears to indicate that the initial shear modulus of coarse-grained materials is governed largely

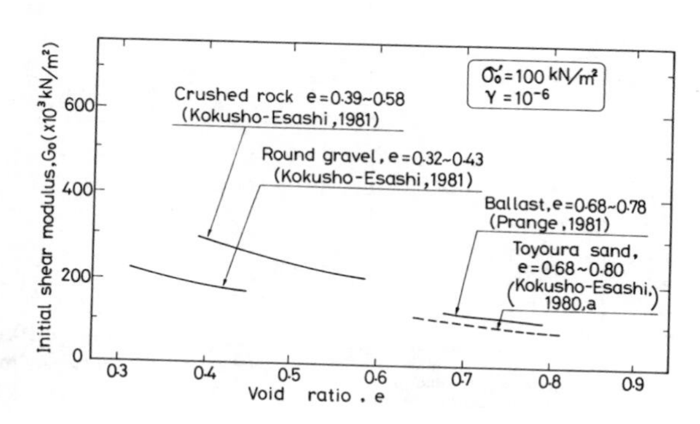

Figure 10.12. Initial shear modulus versus void ratio for coarse-grained materials

by the angularity of individual grains. In fact, the most angular crushed rock gives the largest shear modulus and the round gravel yields the smallest shear modulus. The shear modulus of the ballast could have been somewhere in between if the test had been carried out for a denser state of packing.

10.7.4 *Initial shear moduli by in-situ measurements*

In-situ methods commonly in use for determining the initial shear moduli are the seismic refraction method, down-hole method and cross-hole method. The refraction method has an advantage that it can be operated in the field without drilling holes, but its use is limited to the case where widespread land space is available such as in mountainous areas and sparsely populated areas. In densely populated areas, the down-hole method or cross-hole method is frequently used to determine the in-situ shear moduli of soil deposits.

An interesting comparison was made by Anderson & Woods (1975) between the values of initial shear moduli determined by the in-situ cross-hole method and the values determined by the resonant column test on undisturbed samples of soils secured from the same boreholes. The results of these tests are presented in Figure 10.13, in which the field values of shear wave velocity for six soils are plotted versus the corresponding values obtained from the laboratory resonant column test. In this figure, those data points with white marks pertain to the laboratory tests employing a short sustained time of 1000 minutes in applying a confining pressure after completion of consolidation. It may be seen in Figure 10.13 that except for Detroit clay all data points fall to the left of the 10% band indicating an underestimate of the values of shear wave velocity by the laboratory test. Plotted by black marks in Figure 10.13 are the data points corresponding to the shear wave velocity which were estimated by extrapolation as being the value after twenty years of sustained confining stress application. It may be seen in Figure 10.13 that the in-situ shear wave velocity coincides to a reasonable degree of accuracy with the value estimated to exist twenty years afterwards. Anderson & Woods (1975) thus concluded that especially for cohesive soils in-situ values of shear wave velocity tend to be larger than laboratory values of shear wave velocity by an amount grained by the secondary time effect.

Comparison of shear modulus between in-situ and laboratory values was also made by Yokota et al. (1981) for soils of alluvial and diluvial origins. In-situ shear moduli were determined by means of the down-hole technique and the moduli in the

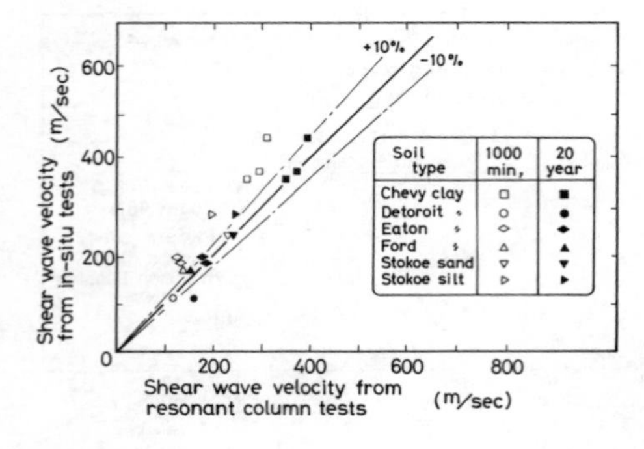

Figure 10.13. Comparison of shear wave velocity between in-situ and laboratory values (Anderson & Woods 1975)

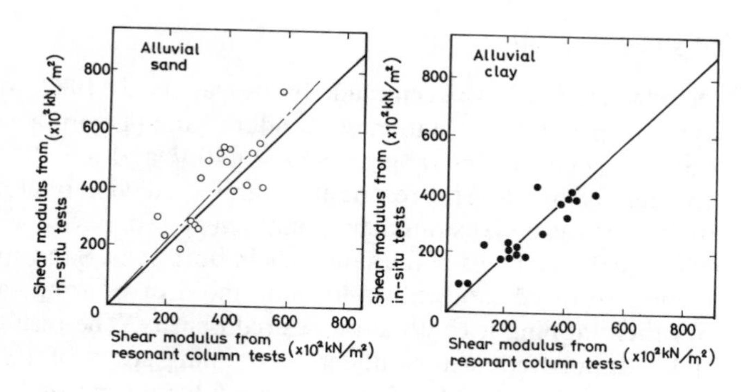

Figure 10.14. Comparisons of shear moduli between in-situ and laboratory values (Yokota et al. 1981)

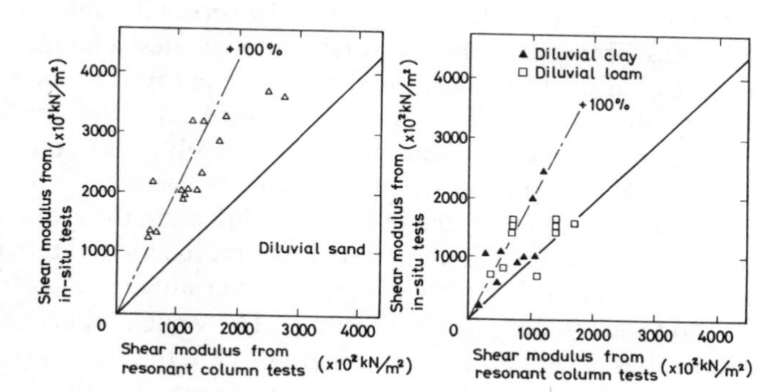

Figure 10.15. Comparisons of shear moduli between in-situ and laboratory values (Yokota et al. 1981)

laboratory were measured by the resonant column tests on undisturbed specimens recovered from the same borehole. Figure 10.14 shows a comparison of the shear modulus between in-situ and laboratory values for sand and clay soils from alluvial origin. It may be seen that, for soils deposited in a relatively recent era and hence with small shear moduli less than approximately 60,000 kN/m^2, the resonant column test tends to yield approximately the same values of shear modulus as those obtained in the field by the use of the down-hole method. Figure 10.15 shows a similar comparison for clay and sand soils of diluvial origin. Unlike the alluvial soils, the shear moduli from the laboratory test always give smaller values than those obtained by using the down-hole method. For soil deposits of diluvial origin having shear moduli larger than approximately 50,000 kN/m^2, the effects such as cementation and ageing are generally considered to act towards increasing the stiffness of soils. It appears likely that part of the stiffness due to these strengthening effects is lost by the disturbance incurred during sampling, transportation and handling of samples in the laboratory.

10.8 STRAIN-DEPENDENT SHEAR MODULUS AND DAMPING

Comprehensive efforts have been made recently to identify and study the strain-dependent nature of dynamic soil behaviour. The main results of these studies will be summarized in the following and some effort made to fit some of the test results to the material models described in the foregoing section.

10.8.1 *Sandy soils*

Detailed studies have been made by Iwasaki et al. (1978) to identify strain-dependent dynamic properties of Japanese standard sand (Toyoura sand: $D_{50}=0.16$ mm, Uc= 1.5). For small strains ranging between 10^{-6} and 3×10^{-4}, the sand behaviour was investigated by using the resonant column test device, but for larger strains up to about 10^{-2}, a statically operated torsional shear test device was employed with hollow cylindrical specimens of the same size in both tests. Sand specimens were prepared by placing air-dried sand into a forming mold or pouring saturated sand in the mold and then tapping it to obtain a desired density. The results of these tests conducted under drained conditions under a confining stress of 100 kN/m² are presented in Figures 10.16 and 10.17. The shear modulus normalized to the initial modulus at a strain of 10^{-6} is plotted in Figure 10.16 versus the shear strain amplitude. The shear modulus obtained from the torsional shear test changed to some extent with the progression of cycles, because the strain employed in the test was relatively large. In Figure 10.16 the modulus values obtained after the application of ten cycles were plotted in continuation from the data group obtained from the resonant column test. It is seen that the value of the shear modulus decreases with increasing strain down to about one-twentieth of the initial modulus when the strain increases to a value of 3% approaching failure. It may also be observed that the manner of decreasing shear modulus with strain is almost uniquely determined irrespective of the void ratio and the method used to prepare the specimen. The value of the damping ratio obtained in the same series of tests is shown in Figure 10.17 in the form of plots against the shear strain (Tatsuoka & Iwasaki 1978). This figure shows that the damping ratio increases with increasing shear strain to a value of about $D_0=0.38$ when the shear strain approaches

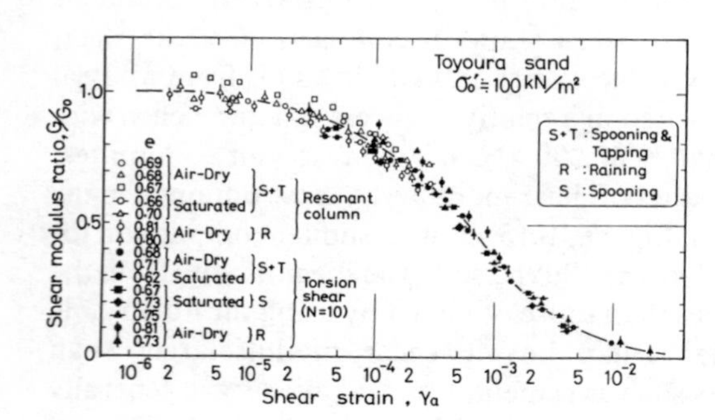

Figure 10.16. Shear modulus versus strain plot for Japanese standard sand (Iwasaki et al. 1978)

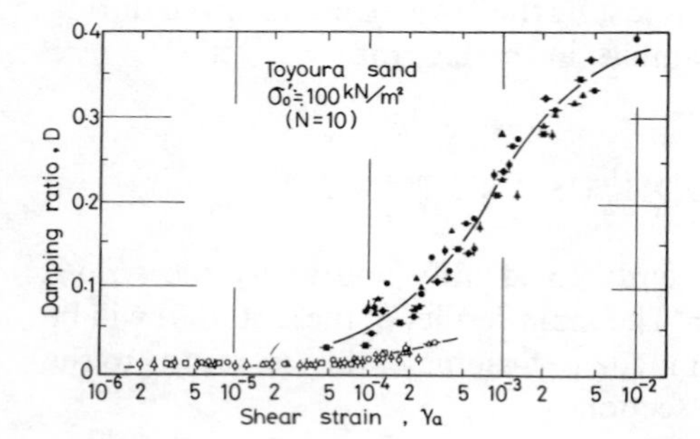

Figure 10.17. Damping ratio versus strain plot for Japanese standard sand (Tatsuoka & Iwasaki 1978)

a value near failure. Unlike the shear modulus, there is a break in the damping versus strain plot between the two classes of data obtained by the resonant column test and the torsional shear test. However, reasons for causing such discontinuity are not known. It is to be noted that the relationship between the damping and shear strain in Figure 10.17 is established irrespective of the density of the sample and the method used for preparing the laboratory specimen.

Effects of the number of cycles on the dynamic behaviour of the Toyoura sand were also investigated in the above cited test scheme. The test results showed that the shear modulus values obtained at the second and tenth cycles differ at most by 10% when the shear strain is larger than about 10^{-4}, and approximately the same percentage of difference was noted also in the damping characteristics. It was also noted that the effect of the number of cycles partially disappears when the stress is repeated more than ten cycles. Consequently, for all practical purposes, the changes in shear modulus and damping due to the progression of cycles may be disregarded, except for the case of undrained shear where pore water pressure build-up in the saturated sand is significant.

Iwasaki et al. (1978) and Tatsuoka & Iwasaki (1978) also investigated the influence of the confining stress on the strain-dependent dynamic properties of the Toyoura sand. Figure 10.18 shows a summary of the test results indicating the influence of the confining pressure on the shear modulus. It may be seen that, with decreasing confining stress, the modulus reduction curve is shifted towards the left of the figure. This fact implies that the reference strain decreases as a result of decreases in the confining stress. Let the strength of the sand be expressed by the Mohr–Coulomb failure criterion:

$$\tau_f = \sigma_0' \tan \phi' \tag{10.36}$$

where ϕ' is the angle of internal friction. Introducing this criterion into Equation 10.12 together with the formula for initial shear modulus given by Equation 10.29 for clean

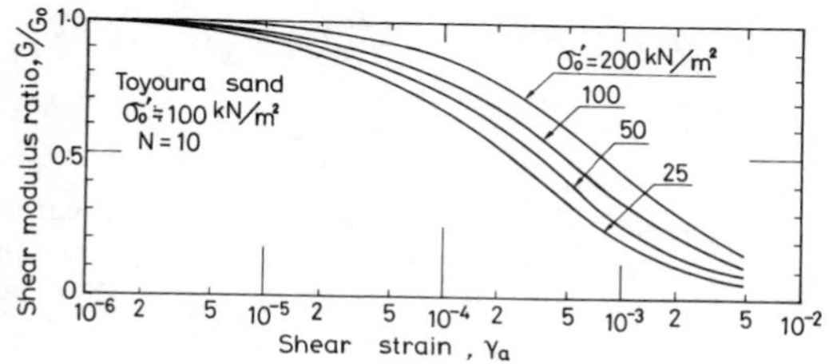

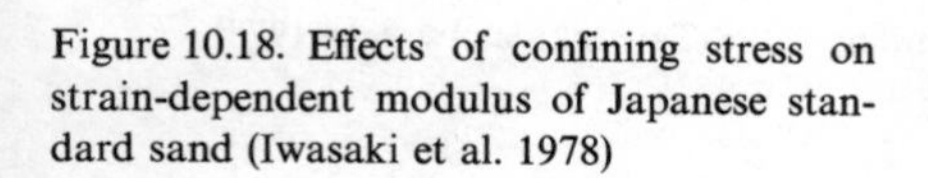
Figure 10.18. Effects of confining stress on strain-dependent modulus of Japanese standard sand (Iwasaki et al. 1978)

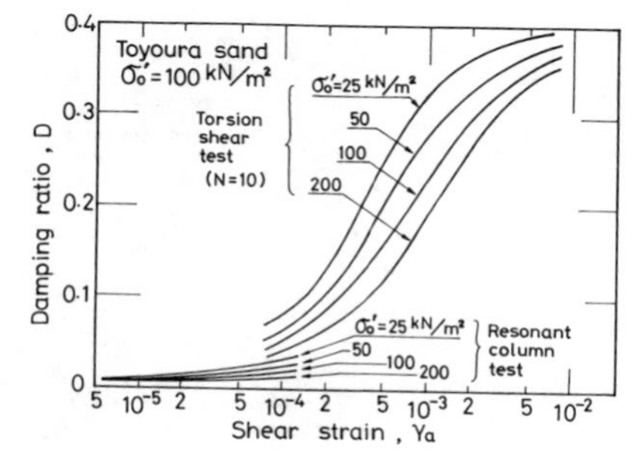

Figure 10.19. Effects of confining stress on strain-dependent damping ratio of Japanese standard sand (Tatsuoka & Iwasaki 1978)

sands, one obtains:

$$\gamma_r = \frac{\tan \phi'}{16{,}600} \frac{1 + e}{(2.17 - e)^2} (\sigma_0')^{0.6} \tag{10.37}$$

This expression clearly shows that the reference strain decreases with decreasing confining stress in proportion to $(\sigma_0')^{0.6}$. The influence of the effective confining stress on the strain-dependent shear modulus is particularly important when dealing with problems in which pore water pressure build-up occurs in the sand in the course of cyclic loading.

The damping of sand is also influenced to some extent by the confining pressure. Figure 10.19 shows a summary of the test results on the Toyoura sand. It is observed in this figure that the damping ratio becomes larger at all strain levels as the effective confining stress decreases.

The test results shown in Figures 10.16 and 10.17 seem to indicate that there exists a unique relationship between the shear modulus ratio, G/G_0, and the damping ratio, D, with the value of shear strain taken as a parameter. Reading off the value of these two quantities for each value of shear strain amplitude used in the test, Tatsuoka & Iwasaki (1978) constructed the curve shown in Figure 10.20, in which the damping ratio is plotted versus the modulus ratio with the strain value inscribed at each data point as a parameter. Similar plots were also made of test data by other investigators and are shown in Figure 10.21. It may be seen that, although there are some scatters among the individual sets of data, the damping ratio is roughly inversely proportional to the shear modulus ratio. The range of the data plotted in Figure 10.21 is also shown in Figure

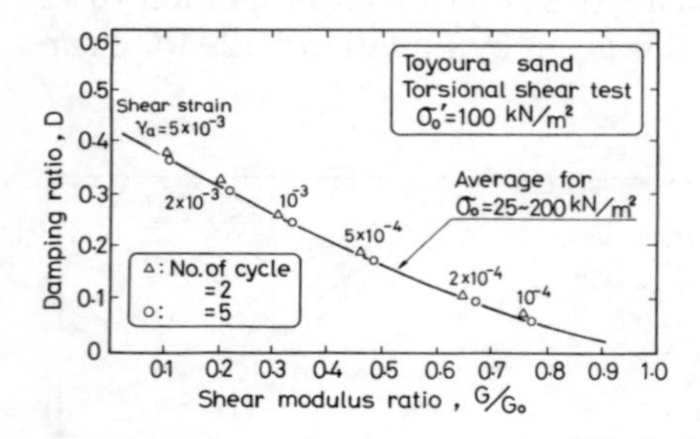

Figure 10.20. Relationship between shear modulus ratio and damping ratio (Tatsuoka & Iwasaki 1978)

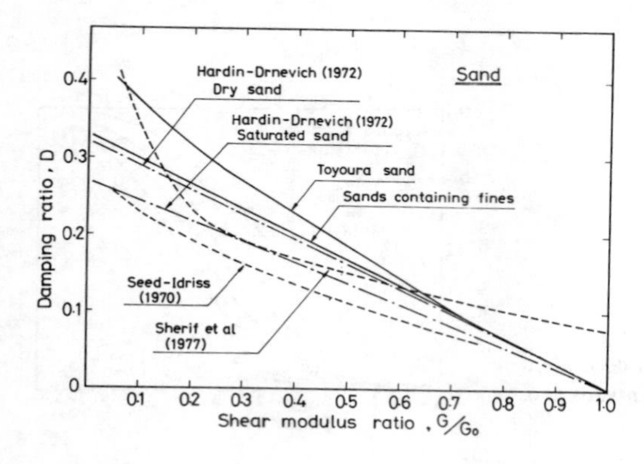

Figure 10.21. Relationship between shear modulus ratio and damping ratio (Tatsuoka & Iwasaki 1978)

10.6 to provide a comparison with the similar relationship obtained from the hyperbolic model. Figure 10.6 appears to indicate that the behaviour of clean sands can be represented with a reasonable degree of accuracy by the use of the hyperbolic model in the range of small to medium shear strains until the shear modulus drops to about one-fourth of the initial modulus. For the larger strain range near failure, the hyperbolic model is incapable of modelling the sand behaviour, giving an intolerably large damping. It is to be noted that the relationship between G/G_0 and D established by the R–O model is linear, as can easily be noted from Equation 10.23. Most of the test data assembled in Figure 10.21 also show a linear relationship in conformity to that deduced from the R–O model. Consequently, it may well be mentioned that the R–O model is suitable for modelling the sand behaviour over practically all ranges of shear strain levels, if parameter r in Equation 10.23 is properly selected.

In order to check the applicability of the proposed models, it would be preferable to refer to the modelling of in-situ behaviour of sands. The test data on undisturbed samples of in-situ sands are limited, but some test results are reported by Kokusho et al. (1982). Using the same sand specimens for which the initial shear moduli were obtained as shown in Figure 10.9, the strain-dependent dynamic behaviour was investigated. The results of the investigation on the shear moduli are shown in Figure 10.22, where the range of obtained test data is indicated along with the average curve. The average curve is roughly consistent with the curve obtained with a confining stress of 100 kN/m^2. The approximate range of the damping data obtained in the same test series is shown in Figure 10.23.

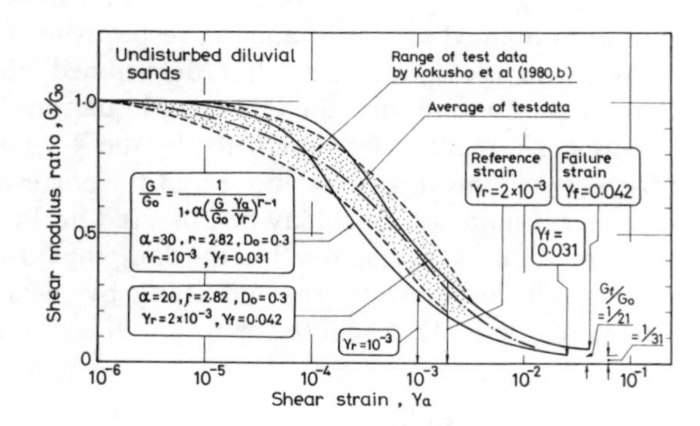

Figure 10.22. Strain-dependent shear modulus of diluvial sands

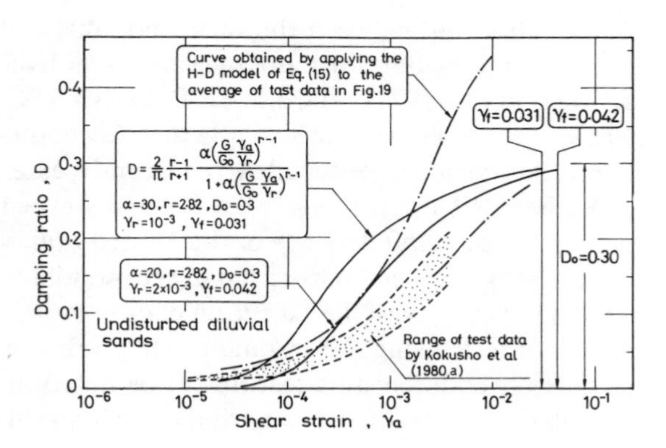

Figure 10.23. Strain-dependent damping ratio of diluvial sands

An attempt is made to model the behaviour of the undisturbed sand based on the test results demonstrated in Figures 10.22 and 10.23. Suppose the behaviour can be represented by the hyperbolic model. Knowing the strain-dependent shear modulus curve shown in Figure 10.22, it is possible to derive the corresponding damping ratio through use of the curve shown in Figure 10.5. The damping ratio versus strain relationship thus established is indicated in Figure 10.23. It may be seen that coincidence with the test data is satisfactory only up to a strain of about 2×10^{-4}. For the range of larger strains, the hyperbolic model tends to give increasingly larger damping ratio than actually observed in the test. The fitting of the above test data to the R–O model can be made by the following procedure. Firstly, it is necessary to determine parameter r in the R–O model. As mentioned in the foregoing section, a proper choice of r can be made from knowledge of damping at the failure state. Although a lack of test data exists for the large strain level near failure, it may well be assumed that the value of D would be approximately equal to 0.3. Hence, the value of r is calculated as $r = 2.82$ from Equation 10.27, assuming the value of G_f/G_0 is approximately equal to zero. The value of the reference strain is estimated to lie between 10^{-3} and 2×10^{-3} on the basis of the appropriately assumed value of shear strength and the initial shear modulus value evaluated from the test data shown in Figure 10.9. Hence, two cases will be considered: one with $\gamma_r = 10^{-3}$ and the other with $\gamma_r = 2 \times 10^{-3}$. For undisturbed medium to dense sands such as those used in the test, the value of failure strain ranges approximately between 2 and 5%. Therefore, to obtain a rounded number for parameter α, the failure strains were assumed to be $\gamma_f = 0.031$ for the case of $\gamma_r = 10^{-3}$, and to be $\gamma_f = 0.042$ for the case of $\gamma_r = 2 \times 10^{-3}$. Then, the values of α are obtained as $\alpha = 30$ and 20, respectively, from Equation 10.24.

With all the parameters thus determined, the strain-dependent modulus and damping ratio were computed by using Equations 10.21 and 10.22. The shear modulus versus strain relationships predicted by the R–O model are indicated in Figure 10.22 for the two cases described above and the corresponding damping relationships are shown in Figure 10.23. It may be observed in these figures that the modelling of the shear modulus is satisfactory, but the damping ratio is not necessarily so. However, for large strain levels where the modelling by means of the hyperbolic model is not adequate, the R–O model seems to give a reasonable value of the damping ratio.

10.8.2 *Cohesive soils*

The relationship between the shear modulus and strain has been investigated in the earlier stage of development by various individuals such as Seed & Idriss (1970), Kovacs et al. (1971), Hardin & Drnevich (1972b) and Taylor & Parton (1973). These authors showed consistently that the modulus of clays decreases with strain once a threshold strain is exceeded. More recently, extensive studies were made by Anderson & Richart (1976) on this subject for undisturbed samples of five clays secured from naturally occurring deposits in the United States. The clays tested had low plasticity indices ranging from 20 to 45 having consolidated undrained shear strengths of about 70 to 85 kN/m^2, with the exception of one clay with a shear strength of 15 kN/m^2. The results of tests using the resonant column device are summarized in Figure 10.24.

One of the interesting features to be noted in the figure is the fact that the shear moduli do not start to drop until the shear strain amplitude is increased to a value of

5×10^{-5}. This is in contrast to the corresponding behaviour in cohesionless soils in which the modulus reduction occurs starting from a smaller strain level of about 5×10^{-6}, as observed in Figures 10.18 and 10.22.

Andreasson (1979, 1981) also investigated the strain-dependency of the shear modulus of plastic clays from three sites in the Gothenburg region of Sweden. The plasticity index of the clays was about 20 to 60. Undisturbed samples were tested in the laboratory using the resonant column device and at the same time in-situ screw plate tests were carried out to determine the modulus values at large levels of strain. The results of the tests are shown in Figure 10.24, where it may be seen that the modulus reduction curve for the Gothenburg clay is similar in shape to that for the US clays reported by Anderson & Richart (1976) (Figure 10.24).

A series of cyclic triaxial tests using a device equipped with highly sensitive displacement sensors was conducted by Kokusho et al. (1982) on undisturbed samples of soft clays secured from an alluvial deposit in Teganuma, Japan. The plasticity index of the clay ranged from 40 to 100 and the natural water content was in a range between 100% and 170%. The initial shear moduli at very small levels of strains were in a range as small as 2,500 to 7,500 kN/m^2. The shear modulus values obtained in the tests as functions of shear strain are shown in Figure 10.25. The damping values measured in the same series of the test are summarized in Figure 10.26. It may be seen that the damping ratio at the failure strain level is approximately 16% and appears to be much less than the corresponding damping ratio of sandy soils.

The reported test data on Teganuma clay contained information on both shear modulus and damping values up to the strain at failure. Therefore, it seems of interest to

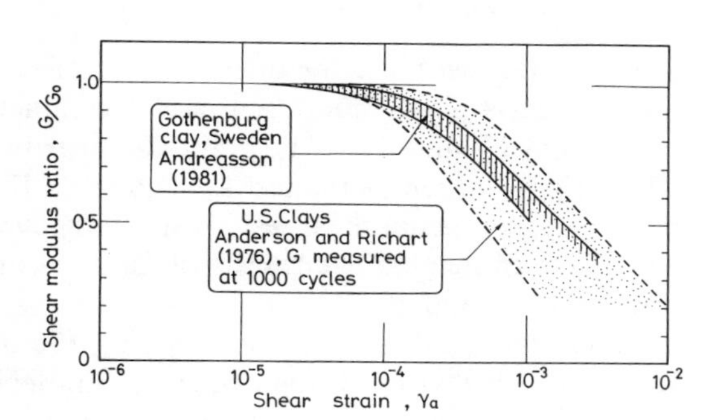

Figure 10.24. Modulus reduction curves for clays

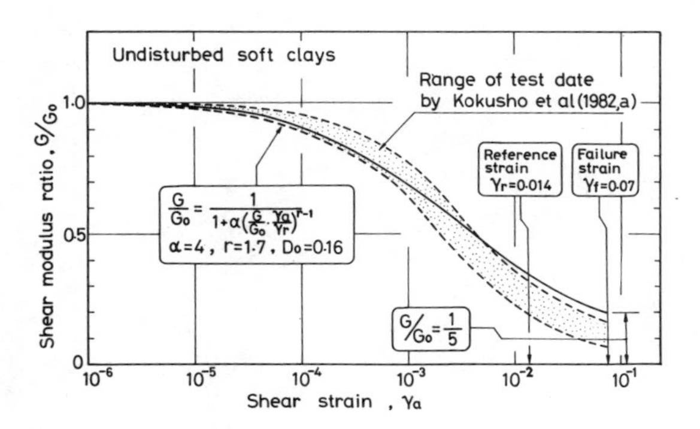

Figure 10.25. Strain-dependent shear modulus of soft clays from Teganuma

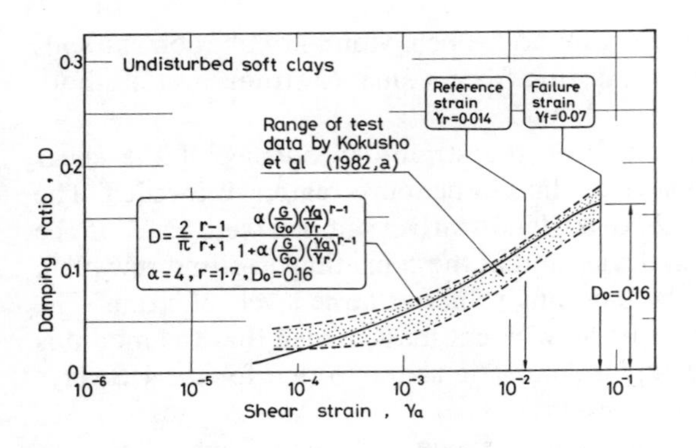

Figure 10.26. Strain-dependent damping ratio of soft clays from Teganuma

see how the observed test data fit the material model. From the knowledge of the damping at failure strain, $D_0 = 0.16$, parameter, r, in the R–O model is computed to be 1.7. Also, the value of reference strain was estimated to be $\gamma_r = 0.014$ based on the strength and initial shear modulus values obtained in the same series of tests. Using a failure strain of $\gamma_f = 0.07$ for the Teganuma clay, parameter α was assessed to be 4. With the use of the parameters thus determined, the modulus reduction curve and damping versus strain curve were established from Equations 10.21 and 10.22, as indicated in Figures 10.25 and 10.26. It may be seen that the R–O model with a properly chosen set of parameters is capable of representing the modulus compatible with the damping characteristics of soft clays.

10.8.3 *Coarse-grained materials*

In connection with the site investigation for the construction of a rockfill dam and a nuclear power plant, Studer et al. (1980) conducted a series of cyclic triaxial shear tests on two gravel materials. The grain-size distribution curves of crushed limestone and alluvial gravel used in the test are shown in Figure 10.11. The tests were performed using a cyclic triaxial shear test apparatus accommodating samples 15 cm in diameter and 30 cm in height. The results of the tests on the alluvial gravel are presented in Figures 10.27 and 10.28.

Along with the tests for determining the initial shear modulus as cited above, Kokusho (1980) also investigated the strain-dependency of the modulus and damping of coarse-grained materials for which grain-size distribution curves are shown in Figure 10.11. The results of the tests are presented in Figures 10.27 and 10.28. It can be

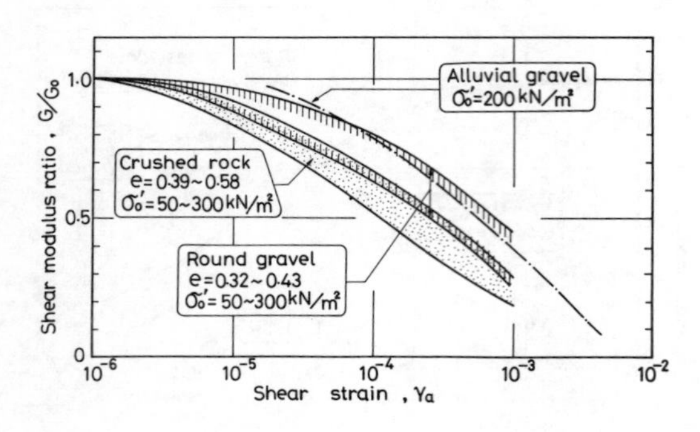

Figure 10.27. Strain-dependent shear modulus of coarse-grained materials

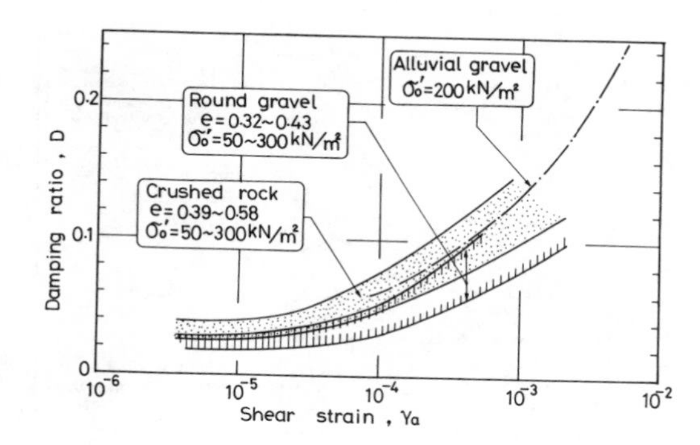

Figure 10.28. Strain-dependent damping ratio of coarse-grained materials

seen in Figure 10.27 that the modulus reduction curves for the crushed stone and round gravel tend to shift towards the left as compared to similar curves for sands, indicating that the nonlinearity of coarse-grained materials starts to appear at a smaller level of strain than fine-grained sands. It can also be noted in Figure 10.27 that results for the crushed rock and round gravel obtained by Kokusho show almost similar modulus reduction characteristics as the test data reported by Studer et al. (1980).

With respect to the damping, the test results from two data sources presented in Figure 10.28 show nearly the same trend.

10.8.4 *Strain-dependent shear moduli and damping by in-situ measurements*

An attempt to measure in-situ shear wave velocity as functions of shear strain was made by Miller et al. (1975) and Troncosco et al. (1977) by means of what is called impulse test. In this test, three or more receiver boreholes are drilled in a linear array together with a source borehole. Then vertically polarized shear wave generated in the source borehole is detected at the same depths in the receiver boreholes, and the velocity of shear wave travelling between two neighbouring boreholes are determined. Since the soil nearer the source borehole is subjected to a larger shear strain than the soil distant from it, the measured shear wave velocity near the source borehole shows lower values of shear wave velocity. During the in-situ measurements, particle velocity in the vertical direction induced by the travelling wave is also measured in each of the receiver boreholes. The particle velocity divided by the velocity of shear wave propagation permits the determination of shear strains involved in the propagation of shear wave during the test. In one of the test series conducted in a compacted dense silty sand fill, the modulus reduction curves obtained by in-situ measurements compared well with the similar curve obtained from the cyclic triaxial tests on compacted specimens of the same soil. However, the modulus reduction curves for natural soil deposits with different geologic histories showed significant difference. It was then considered necessary to perform in-situ measurements for individual deposits of sites in question.

Another kind of attempt to study the strain-dependent modulus and damping characteristics of soils was made by Abdel-Ghaffer & Scott (1979a and b), who analyzed acceleration records obtained at the crest and base of an existing earth dam during recent medium-size earthquakes. The dam was constructed of rolled-fills and the core and shell materials consist basically of clay, sand, gravel and boulder. Frequency analyses of the recorded motions indicated that the earth dam responded primarily in

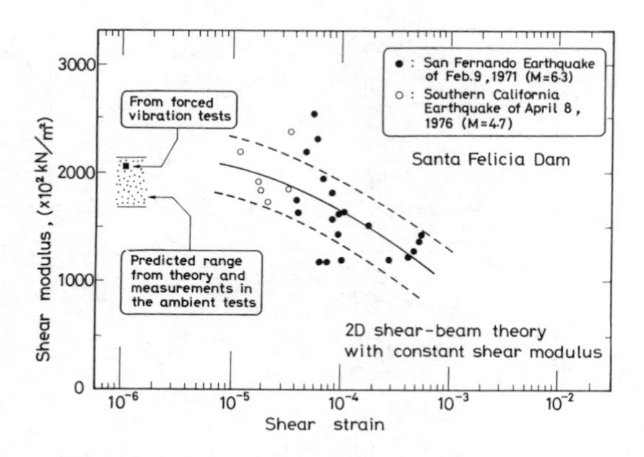

Figure 10.29. Strain-dependent shear moduli of an actual dam material as determined from acceleration records during earthquakes

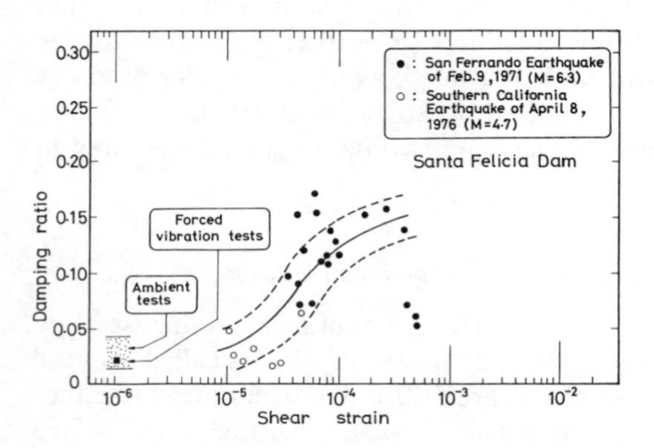

Figure 10.30. Strain-dependent damping ratio of an actual dam material as determined from acceleration records during earthquakes

its fundamental mode of vibration in the upstream-downstream direction. The fundamental mode response in this direction was treated as that of a single-degree-of-freedom hysteretic system. The recorded motions at the crest and base were digitally band-path filtered in order to single out the hysteretic response characteristics in the fundamental mode. Using these filtered records, the relative displacement of the crest with respect to the base was plotted versus the absolute acceleration of the dam. The slope and area of the hysteresis loop thus obtained could be related to the shear modulus and damping factor if a structural model such as a one-dimensional shear beam model or a finite element model is incorporated. On the basis of the procedure is above, overall average values of shear modulus and damping of the materials composing the dam were back-calculated. These values are plotted in Figures 10.29 and 10.30 versus the amplitude of shear strain which is readily determined in the above procedure. Also shown in these figures are the modulus and damping values determined directly from ambient tests and forced virbration tests. It may be seen that the dynamic properties of actual dam materials as back-calculated by the above procedure can yield a reasonable range of values which are in accord with the values determined from other test procedures.

10.9 EVALUATION OF PARAMETERS IN THE RAMBERG-OSGOOD MODEL

10.9.1 *Parameter* α

As mentioned in the foregoing section, parameter α in the R–O model can be determined through Equation 10.24 on the basis of the values of the reference strain and the failure strain of a given soil.

The magnitude of strain at failure, γ_f, depends upon several variables such as soil type, degree of cementation, density, confining pressure, drainage conditions and so forth. However, experiences in the past on statical monotonic loading tests on a variety of soil materials have shown that there is no consistent trend in the failure strain indicating a difference due to soil type. Most soils show failure strains ranging between 3 to 7%. Therefore, it may be assumed that the value of γ_f is from 3 to 7% for all kinds of soils and is indicated accordingly in Figure 10.31.

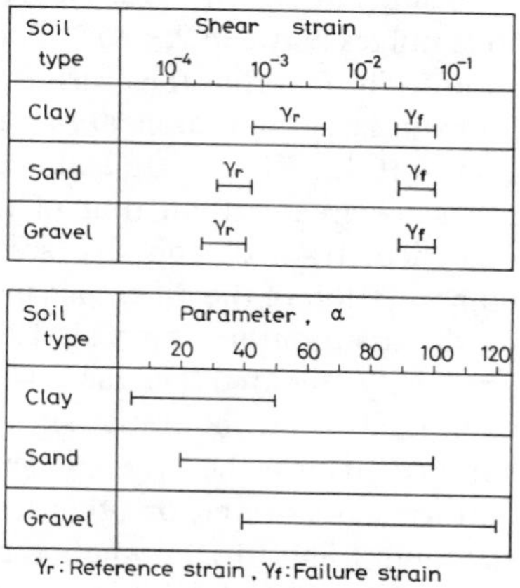

Soil type	Shear strain 10^{-4} 10^{-3} 10^{-2} 10^{-1}
Clay	γ_r γ_f
Sand	γ_r γ_f
Gravel	γ_r γ_f

Soil type	Parameter, α 20 40 60 80 100 120
Clay	
Sand	
Gravel	

γ_r: Reference strain, γ_f: Failure strain

Figure 10.31. Approximate ranges of reference strain, failure strain and parameter α in the Ramberg-Osgood model

Reference strain, γ_r 10^{-4} 10^{-3} 10^{-2}	Soil type	References
	Clay	Wilson-Dietrich (1960)
	Diluvial clay	Takenaka-Nishigaki (1970)
	Clay	D'Appolonia et al (1971)
	Clay	Hardin-Drnevich (1972)
	Clay	Hara et (1974)
	Clay	Anderson (1974)
I (Average)	Clay	Seed-Idriss (1970)
	Clay	Zen et al (1978)
(σ_0'=50~200kN/m²)	Sand	Eq. (37)
(σ_0'=50~200kN/m²)	Round gravel	Eq. (34)
σ_0'=50~200 kN/m²	Crushed rock	Eq. (34)

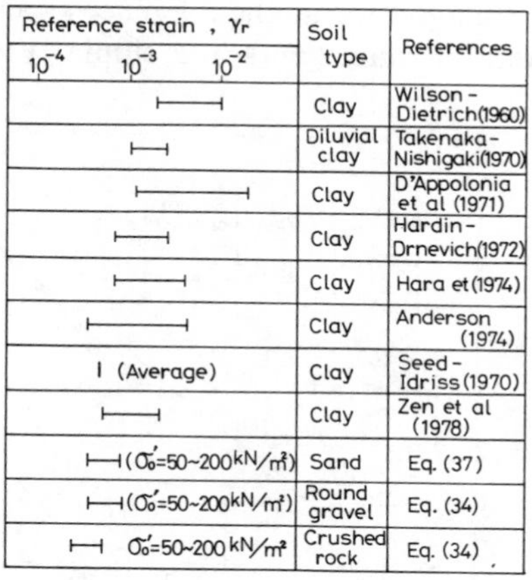

Figure 10.32. Ranges of reference strain based on reports by various investigators

Before the idea of reference strain was proposed by Hardin & Drnevich (1972b), several investigations had been performed to determine the relationship between shear strength and shear modulus at very small levels of strain. The major incentive for such studies appears to have been the development of an empirical procedure to assess in-situ strength of soils based on field measurement of the velocity of shear wave propagation. Various investigators such as Wilson & Dietrich (1960), D'Appolonia & Poulous (1971), Takenaka & Nishigaki (1970) and Hara et al. (1974) were involved in this type of problem and reported results of tests mainly on cohesive soils. The test data by these researchers were mostly presented in terms of the ratio of shear modulus to shear strength, G_0/τ_f. As pointed out by Richart (1981), this ratio is nothing but the reciprocal of the reference strain. The values of the reference strain thus obtained are demonstrated in Figure 10.32. The other test results presented in this figure were obtained apparently from an intent to evaluate the reference strain. For cohesionless materials, an attempt in the above context has seldom been made. Therefore, approximate ranges of reference strain were estimated on the basis of values of initial shear modulus and the strength evaluated from empirical formulae and the Mohr-Coulomb failure criterion. Figure 10.32 shows that the reference strain of cohesionless soils ranges between 2×10^{-4} and 9×10^{-4} with the crushed rock taking the smallest values. The fact that the crushed rock takes the smallest reference strain is accounted for by its largest initial shear modulus, among other cohesionless materials, as shown in Figure 10.12. Figure 10.32 also shows that the reference strain of cohesive soils is generally greater than that of the cohesionless materials. The broad range of the reference strain of cohesive soils as observed in Figure 10.32 is interpreted as a manifestation of the difference in the plasticity of individual soil. To provide evidence for this assumption, Zen et al. (1978) conducted a series of the resonant column tests on artificially prepared specimens with different plasticity indices. The reference strain was computed using the initial shear moduli thus obtained, together with the strength values defined as the stress causing 1% single amplitude strain in the cyclic triaxial tests performed separately on other specimens of the same soils. The results of these tests are shown in Figure 10.33, where it can be seen that the reference strain decreases from a value of 2×10^{-3} down to a value of about 1.5×10^{-4}, as the plasticity index of the soil decreases. The reference strains of non-plastic soils are shown to be on the same order of magnitude as the reference strain of clean sands under low confining stresses. The data arrangements in a similar vein were also made by Kokusho et al. (1982), who

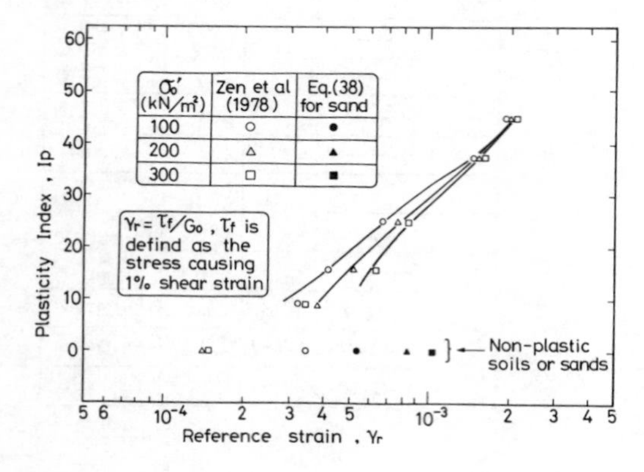

Figure 10.33. Relationship between reference strain and plasticity index of soils (from test data by Zen et al. 1978)

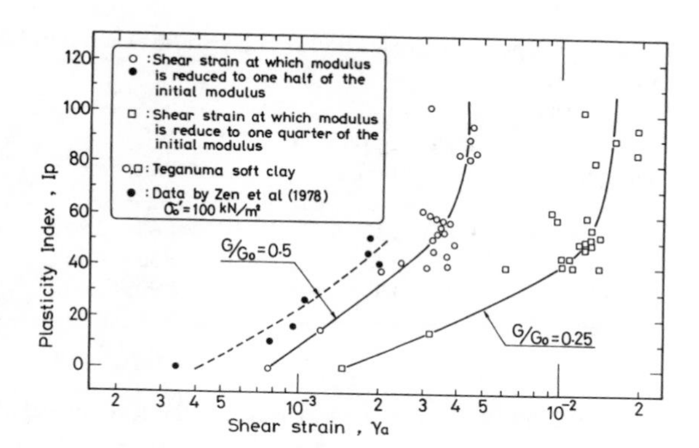

Figure 10.34. Relationship between plasticity index and strain amplitude at which modulus is reduced to one-half and one-quarter of initial modulus (Kokusho et al. 1982)

plotted the shear strains corresponding to one-half and one-quarter reductions of the modulus against the plasticity index for the soft undisturbed Teganuma clays, as shown in Figure 10.34. It is to be remembered here that the reference strain can be determined from modulus reduction curves such as those shown in Figures 10.24 and 10.25 if the type of model to be used is given. For example, if a modulus reduction curve is to be represented by the hyperbolic model, the reference strain can be read off on the curve as the strain at which the modulus is reduced to one-half. When using the R–O model, the strain at which the modulus is reduced to somewhere between one-third and one-quarter may be taken as the reference strain, depending upon the choice of other parameters appearing in the R–O model. With this fact in mind, the plot of the type shown in Figure 10.34 may be deemed as representing a general trend of the reference strain increasing with an increase in the plasticity index of cohesive soils. Also superimposed in Figure 10.34 for comparison are the test data by Zen et al. (1978). Interpreted more quantitatively, the test data in Figure 10.34 show that the reference strain for clays ranges roughly between 5×10^{-4} and 10^{-2}. Such a wide range of variation seems to provide an inferential evidence to the validity of the broad range of reference strain summarized in Figure 10.32.

On the basis of the above considerations, approximate ranges of the reference strain representative of most frequently encountered soils are established and indicated in Figure 10.31 for three kinds of soil materials. Once the reference strain is known, parameter α in the R–O model is determined from Equation 10.24. The ranges in variation of parameter α thus estimated are indicated in Figure 10.31 for three kinds of soil materials.

10.9.2 *Parameter r*

As mentioned in the foregoing section, parameter r is preferably determined from the hysteretic characteristics of soils, as represented by the damping ratio at a large strain near failure. Unfortunately, test data on damping at strains as large as a few percent are very scarse and somewhat inferential estimates must be made at present. From extrapolation of a number of damping versus strain curves such as those shown in Figures 10.26 and 10.30, the damping ratio may be estimated as seen in Table 10.1. The range in variation of the damping ratio for sands was estimated from the data summarized in Figure 10.21. It may be mentioned that the damping ratio for cohesive

Table 10.1 Approximate values of Damping ratio at large strains near failure

Soil type	Damping ratio at large strain D_0	Parameter r
Clay	$0.15 \sim 0.30$	$1.65 \sim 2.80$
Sand	$0.25 \sim 0.40$	$2.30 \sim 4.40$
Gravel	$0.20 \sim 0.35$	$1.90 \sim 3.40$

Table 10.2. Examples of the parameters for the R–O model (clay)

Soil type	α	r	References
Plastic clays	6.25	3.0	Anderson–Richart (1976)
Low-plasticity clays	5.1	2.5	Andersson (1981)
Soft plastic clays	4.0	1.7	Kokusho et al. (1982)

soils is generally smaller than that for cohesionless materials. The values of parameter, r, computed by Equation 10.27 with $G_f/G_0 = 0$ are also listed in Table 10.1 for three kinds of soil materials.

10.9.3 *Examples of the parameters in the model*

The modulus reduction curves shown in Figure 10.24 were fitted to the R–O model. Anderson & Richart (1976) showed that the best fitting parameters for the clays they tested are $r = 3$, $\alpha = 1.0$ and $C_1 = 0.4$ for the model as expressed by Equation 10.19. The test data by Andreasson (1981) were also shown to fit well if the parameters are chosen as $r = 2.5$, $\alpha = 1.8$ and $C_1 = 0.5$ in the model of Equation 10.19. If the test data are to be represented by the model in the form of Equation 10.20, parameter α in Equation 10.20 must be replaced by α/C_1^{r-1} appearing in Equation 10.19. The parameters thus modified to fit the model of Equation 10.20 are shown in Table 10.2. It can be seen that the parameters for cohesive soils are in the range indicated in Figure 10.31.

10.10 CONCLUDING REMARKS

A survey of the reported literatures showed that factors influencing shear moduli at small levels of strain have been almost thoroughly investigated for laboratory prepared specimens of sands and clays. The same can also be said for strain dependent characteristics of the shear modulus and damping ratio. In this respect, several important investigations have been and are being made on soils in naturally occurring deposits by testing undisturbed specimens recovered from these deposits, as well as on coarse-grained materials such as railway ballast and materials used for earth dams. It appears still necessary, however, that test data on dynamic properties be accumulated on a variety of soils encountered in practice.

Most of the tests performed to date have been limited to the nonlinearity of soil properties within the range of small to medium levels of strain, and attempts to represent the soil properties by models have not been, accordingly, complete. In spite of the difficulty in conducting tests and also in summarizing the test results at large strains on the order of a few percent, information in regard to these aspects is very much in need. In view of these circumstances, procedures were suggested in this paper to determine parameters entering in the hyperbolic model and Ramberg-Osgood model so that the model representation can cover a broad range of strains up to a few percent. When these models are used in combination with the Masing rule, it was shown that the hyperbolic model embracing two parameters tends to yield too much damping at large levels of strain and, therefore, it can not be used to properly represent soil behaviour at large strains. To overcome this drawback, the use of the R–O model was shown to be appropriate, if the four parameters of the model are properly chosen in conformity to the soil behaviour at large strains. On the basis of the survey of available test data, approximate ranges of the four parameters are suggested to be incorporated into the R–O model for clays, sands and coarse-grained materials.

ACKNOWLEDGEMENTS

In compiling the draft of this chapter, Dr. T. Kokusho of the Central Research Institute of Electric Power Industry offered some original drawings of the test data and Dr. I. Towhata conducted some of the numerical calculation, pertaining to the models. Dr. K. Mori edited the draft of this chapter. The writer wishes to acknowledge the kindnesses of these persons.

REFERENCES

Abdel-Ghaffar, A.M. & R.F. Scott 1979a. Analysis of an earth dam response to earthquakes. In *Proceedings of ASCE* 105 (GT 12): 1379–1404.

Abdel-Ghaffar, A.M. & R.F. Scott 1979b. Shear moduli and damping factors of an earth dam. In *Proceedings of ASCE* 105 (GT 12): 1405–1426.

Afifi, S.S. & F.E. Richart Jr. 1973. Stress-history effects on shear modulus of soils. *Soils and Foundations* 13 (1): 77–95.

Afifi, S.S. & R.D. Woods 1971. Long-term pressure effects on shear modulus of soils. In *Proceedings of ASCE* 97 (SM 10): 1445–1460.

Anderson, D.G. & F.E. Richart Jr. 1976. Effects of straining on shear modulus of clays. In *Proceedings of ASCE* 102 (GT 9): 975–987.

Anderson, D.G. & R.D. Woods 1975. Comparison of field and laboratory shear moduli. In *Proceedings of ASCE Conference on In-Situ Measurement of Soil Properties* (Raleigh, N.C.), 1: 69–92.

Anderson, D.G. & R.D. Woods 1976. Time-dependent increase in shear modulus of clay. In *Proceedings of ASCE* 102 (GT 5): 525–537.

Andreasson, B. 1979. Deformation characteristics of soft, high-plastic clays under dynamic loading conditions. Department of Geotechnical Engineering, Chalmers University of Technology, Gothenburg, Sweden.

Andreasson, B. 1981. Dynamic deformation characteristics of a soft clay. In *Proceedings of International Conference on Recent Advances in Geotechnical Earthquake Engineering and Soil Dynamics* (St. Louis, Missouri), 1: 65–70.

D'Appolonia, D.J. & H.G. Poulous 1971. Initial settlement of structure on clay. In *Proceedings of ASCE* 97 (SM 10): 1359–1376.
Dobry, R. & W.F. Swiger 1979. The threshold strain and cyclic behavior of cohesionless soils. In *Proceedings of 3rd ASCE/EMDE Specialty Conference* (Austin, Texas): 521–525.
Hara, A. 1980. Dynamic deformation characteristics of soils and seismic response analyses of the ground. Dissertation submitted to the University of Tokyo.
Hara, A., T. Ohta, M. Niwa & S. Tanaka 1974. Shear strength of cohesive soils. *Soils and Foundations* 14 (3): 1–12.
Hardin, B.O. & W.L. Black 1968. Vibration modulus of normally consolidated clay. In *Proceedings of ASCE* 94 (SM 2): 353–369.
Hardin, B.O. & W.L. Black 1969. Closure to vibration modulus of normally consolidated clay. In *Proceedings of ASCE* 95 (SM 6): 1531–1537.
Hardin, B.O. & V.P. Drnevich 1972a. Shear modulus and damping in soils: measurement and parameter effects. In *Proceedings of ASCE* 98 (SM 6): 603–624.
Hardin, B.O. & V.P. Drnevich 1972b. Shear modulus and damping of soils: design equation and curves. In *Proceedings of ASCE* 98 (SM 7): 667–692.
Higuchi, Y., Y. Umehara & H. Ohneda 1981. Evaluation of dynamic properties of the sand deposits under deep seabed. In *Proceedings of 36th Annual Convention of the Japanese Society of Civil Engineering*, 3: 50–51.
Humphries & H.E. Wahls 1968. Stress history effects on dynamic modulus of clay. In *Proceedings of ASCE* 94 (SM 2): 371–389.
Ishihara, K. 1981. Strength of cohesive soils under transient and cyclic loading conditions. State-of-Art report in O. Ergunay & M. Erdik (eds.), *Earthquake engineering*: 154–169. Turkish National Committee on Earthquake Engineering.
Iwasaki, T. & F. Tatsuoka 1977. Effects of grain size and grading on dynamic shear moduli of sands. *Soils and Foundations* 17 (3): 19–35.
Iwasaki, T., F. Tatsuoka & Y. Takagi 1978. Shear moduli of sands under cyclic torsional shear loading. *Soils and Foundations* 18 (1): 39–56.
Kokusho, T. 1980. Cyclic triaxial test on dynamic soil properties for wide strain range. In *Proceedings of 7th World Conference on Earthquake Engineering* (Istanbul), 3: 305–312.
Kokusho, T. & Y. Esashi 1981. Cyclic triaxial test on sands and coarse materials. In *Proceedings of 10th International Conference on Soil Mechanics and Foundation Engineering* (Stockholm), 1: 673–676.
Kokusho, T., Y. Yoshida & Y. Esashi 1982. Dynamic properties of soft clays for wide strain range. *Soils and Foundations* 22 (4): 1–18.
Kondner, R.L. & J.S. Zelasko 1963. A hyperbolic stress-strain formulation of sands. In *Proceedings of 2nd Pan American Conference on Soil Mechanics and Foundation Engineering* (Brazil): 289–324.
Kovacs, W.D., H.B. Seed & C.K. Chan 1971. Dynamic modulus and damping ratio for a soft clay. In *Proceedings of ASCE* 97 (SM 1): 59–75.
Marcuson, W.F. & H.E. Wahls 1972. Time effects on dynamic shear modulus of clays. In *Proceedings of ASCE* 98 (SM 2): 1359–1373.
Miller, R.P., J.H. Troncoso & F.R. Brown 1975. In-situ impulse test for dynamic shear modulus of soils. In *Proceedings of ASCE Conference on In-Situ Measurement of Soil Properties* (Raleigh, N.C.), 1: 319–335.
Prange, B. 1981. Resonant column testing of railroad ballast. In *Proceedings of 10th International Conference on Soil Mechanics and Foundation Engineering* (Stockholm), 3: 273–278.
Richart, F.E. Jr. 1975. Some effects of dynamic soil properties on soil structure interaction. In *Proceedings of ASCE* 101 (GT 12): 1193–1240.
Richart, F.E. Jr. 1981. Influence of soil properties on wave propagation during earthquakes. State-of-the-Art report in *Earthquake engineering*. In *Proceedings of 7th World Conference on Earthquake Engineering* (Istanbul).
Schnabel, P.B., J. Lysmer & H.B. Seed 1972. SHAKE, a computer program for earthquake response analysis of horizontally layered sites. Report No. 72–12, University of California, Berkeley.
Seed, H.B. & I.M. Idriss 1969. Influence of soil conditions on ground motions during earthquakes. In *Proceedings of ASCE* 94 (SM 1): 99–137.
Seed, H.B. & I.M. Idriss 1970. Soil moduli and damping factors for dynamic analysis. Report No. EERC 70–100, University of California, Berkeley.

Sherif, M.A., I. Ishibashi & A.H. Gaddah 1977. Damping ratio for dry sands. In *Proceedings of ASCE* 103 (GT 7): 743–756.
Studer, J., N. Zingg & E.G. Prater 1980. Investigation of cyclic stress-strain characteristics of gravel material. In *Proceedings of 7th World Conference on Earthquake Engineering* (Istanbul), 3: 355–362.
Takenaka, J. & Y. Nishigaki 1970. Dynamic behavior of diluvial soil deposits in Ohsaka. In *Proceedings of 5th Japan National Convention of Geotechnical Engineering*: 89–91 (in Japanese).
Tatsuoka, F. & T. Iwasaki 1978. Hysteretic damping of sands under cyclic loading and its relation to shear modulus. *Soils and Foundations* 18 (2): 25–40.
Taylor, P.W. & J.M. Parton 1973. Dynamic torsion testing of soils. In *Proceedings of 8th International Conference on Soil Mechanics and Foundation Engineering* (Moscow), 1 (2): 425–432.
Troncoso, J.H., F.R. Brown & R.P. Miller 1977. In-situ impulse measurements of shear modulus of soils as a function of strain. In *Proceedings of 6th World Conference on Earthquake Engineering* (New Delhi) 3: 2316–2321.
Wilson, S.D. & R.J. Dietrich 1960. Effect of consolidation pressure on elastic and strength properties of clay. In *Proceedings of ASCE Research Conference on Shear Strength of Cohesive Soils* (Boulder, Colorado): 419–435.
Yokota, K., T. Imai & M. Konno 1981. Dynamic deformation characteristics of soils determined by laboratory tests. Oyo Technical Report No. 3: 13–37.
Yoshimi, Y., F.E. Richart Jr., S. Prakash, D.D. Balkan & Ilyichev 1977. Soil dynamics and its application to foundation engineering. State-of-the-Art report in *Proceedings of 9th International Conference on Soil Mechanics and Foundation Engineering* (Tokyo), 2: 605–650.
Zen, K., Umehara, Y. & K. Hamada 1978. Laboratory tests and in-situ seismic survey on vibratory shear modulus of clayey soils with various plasticities. In *Proceedings of 5th Japanese Earthquake Engineering Symposium*: 721–728.

Liquefaction and densification of cohesionless granular masses in cyclic shearing 11

S. NEMAT-NASSER

11. INTRODUCTION

Cyclic shearing of a sample of cohesionless granules (sand) under fixed confining pressure produces overall densification. The granules move relative to each other in response to the applied shearing, and the sample tends to compact after each cycle in the presence of confinement. If the sample is saturated and is undrained, the densification tendency produces an increase in pore water pressure which in turn reduces the interparticular friction forces. These frictional forces diminish considerably when the pore water pressure attains values close to the confining pressure. The shearing resistance of the sample at such a state is very small, and the sample momentarily behaves like a liquid (hence, the term liquefaction). If on the other hand, the saturated sample is drained, its cyclic shearing at suitably small frequencies increases its density and hence, its shearing resistance.

Large-scale liquefaction has been observed as a result of earthquake ground motion, as well as of explosions. Well-known examples are earthquake-induced liquefaction in Niigata, Japan (1964), and massive soil failure in Anchorage, Alaska (1964). A brief historical account is given by Seed & Idriss (1982), in a recent monograph addressed to practical aspects of this phenomenon. Since the Niigata and Alaska earthquakes, considerable effort has been devoted to examining the mechanism of liquefaction: Seed & Lee (1966), Seed & Idriss (1967), Peacock & Seed (1968), Finn et al. (1970b), De Alba et al. (1975). Ishihara et al. (1975), Martin et al. (1975) and Nemat-Nasser & Shokooh (1977, 1978, 1979). Although a major part of this activity has been experimental, some attempt has been made to quantify the estimate of the liquefaction potential of cohesionless sand by relating the number of cycles required for liquefaction to the normalized shear stress amplitude; see, Martin et al. (1975). A systematic approach based on an energy consideration, for both liquefaction and densification estimates, has been presented by Nemat-Nasser & Shokooh (1978, 1979); see also Nemat-Nasser (1982), where, in addition to the energy approach, liquefaction is analyzed with the aid of dimensional analysis.

Theories of this kind assume isotropic materials and are essentially of a one-parameter description; this parameter usually is the initial void ratio, or equivalently, the relative density. Therefore, they do not take into account any anisotropy that may exist prior to shearing.

It has been shown experimentally by Finn et al. (1970a), and later confirmed by Ishihara et al. (1975), Seed et al. (1977), Ishihara & Okada (1978) and Nemat-Nasser & Tobita (1982), that pre-shear-straining of a drained sample at a relatively large shear-strain amplitude reduces considerably the sample's resistance to liquefaction. Nemat-Nasser & Tobita (1982) have employed a micro-mechanical model of dilatancy, proposed by Nemat-Nasser (1980b), in order to explain this phenomenon.

The micro-mechanical consideration suggests, and laboratory experiments confirm, that the resistance to liquefaction of a pre-shear-strained (drained), or pre-liquefied (undrained) sample diminishes considerably if the pre-straining or the pre-liquefaction in simple shearing under fixed confinement is terminated at zero shear stress, whereas this resistance certainly does not decrease, and perhaps increases, if the pre-straining or the pre-liquefaction is terminated at zero shear strain; see Nemat-Nasser & Tobita (1982) and Nemat-Nasser & Takahashi (1983).

This remarkable result seems to have important practical implications, because it is of considerable concern whether or not a site liquefied during an earthquake possesses any resistance to further liquefaction in aftershocks.

In this chapter, first the unified liquefaction and densification theory of Nemat-Nasser & Shokooh (1978, 1979) is summarized, and it is indicated how the theory has been used to estimate liquefaction potential for random loading; see Pires et al. (1983). The basic equations are reported and the experimental verifications used by Nemat-Nasser & Shokooh are reproduced. Then, the micro-mechanical theory of Nemat-Nasser (1980b) is summarized and its application to the analysis of liquefaction phenomena is reviewed. Finally, some recent experimental results by Nemat-Nasser & Tobita (1982) and Nemat-Nasser & Takahashi (1983) are reported. The experiments have been designed in the light of the micro-mechanical consideration of Nemat-Nasser (1980b), with the specific purpose of bringing into the open the influence of induced, as well as inherent, anisotropy (fabric) on the mechanical response of cohesionless granules subjected to cyclic shearing.

11.2 A SIMPLE LIQUEFACTION AND DENSIFICATION THEORY

First we consider cyclic shearing at constant shear stress amplitude, $\bar{\bar{\tau}}$, and a fixed wave-form under constant pressure, σ_c. Then we examine cyclic shearing with variable amplitude and wave-form.

11.2.1 *Cyclic shearing*

For cyclic shearing, we set:

$$\tau(t) = \tau_0\phi(t), \quad -1 \leq \phi(t) \leq 1, \quad \tau_0 = \bar{\bar{\tau}}/\sigma_c \tag{11.1}$$

where $\phi(t)$ defines the wave-form, and τ_0 is the normalized shear stress amplitude. We also normalize the pore water pressure, p_w, as:

$$p = p_w/\sigma_c \tag{11.2}$$

and note that liquefaction initiation corresponds to the state at which $p \to 1$.

11.2.2 *An energy approach* (*Nemat-Nasser & Shokooh* 1979)

To change the void ratio of a saturated drained granular sample from e to e + de by cyclic shearing, an increment of energy, dW, is required for rearranging the particles. This required energy increases as the void ratio, e, decreases, and becomes rather large when e attains its absolute minimum e_m. If the sample is undrained, then the tendency toward densification results in an increase in the pore water pressure. This in turn decreases the effective forces between contacting granules, and therefore reduces the energy, dW, required for particle rearrangement. Hence one may write:

$$dW = -\tilde{v}\frac{de}{f(1+p)g(e-e_m)} \tag{11.3}$$

where $\tilde{v}$ is a constant with the physical dimensions of dW, and f and g are two non-decreasing material functions with the following properties:

$$f(1) = 1, \quad f' \geq 0, \quad g(0) = 0, \quad g' \geq 0 \tag{11.4}$$

For the undrained sample, we have:

$$de = -\frac{e\sigma_c}{\kappa_w}dp \tag{11.5}$$

which reduces Equation 11.3 to:

$$dW = v\frac{edp}{f(1+p)g(e-e_m)}, \quad v = \tilde{v}\sigma_c/\kappa_w \tag{11.6}$$

The liquefaction initiation is identified with the state where the value of the pore water pressure approaches that of the confining pressure, i.e. when p $(= p_w/\sigma_c) \rightarrow 1$. For problems relating to earthquake-induced liquefaction, the confining pressure σ_c is rather small, say, of the order of 10 to 100 psi; most experimental data are for $\sigma_c < 10$ psi. Since the bulk modulus of water is about 300,000 psi, the volumetric strain and the corresponding work per unit volume per unit confining pressure (dimensionless work) are of the order of 10^{-5} to 10^{-4}, and hence can be ignored without introducing any measurable error. Therefore, the void ratio e in Equation 11.6 can be replaced by its initial value e_0. With this substitution and upon integration, we obtain from Equation 11.6:

$$\Delta W = \frac{ve_0}{g(e_0-e_m)}\int_0^p \frac{dp'}{f(1+p')} \tag{11.7}$$

It should be noted that the representation of Equation 11.3 is somewhat special in the sense that no coupling between p and $(e - e_m)$ is included. A more general form would be to use, instead of $f(1 + p)\,g(e - e_m)$, the function $F(p, e - e_m)$. However, since for the drained case, p = 0, and for the undrained case, $e \sim e_0$, it appears that Equation 11.3 is adequate for our purposes.

Nemat-Nasser & Shokooh (1977, 1978, 1979) consider the following simple approximations for f and g:

$$g(e-e_m) = (e-e_m)^n, \quad n > 1 \tag{11.8}$$

$$f(1 + p) = (1 + p)^r, \quad r > 1 \tag{11.9}$$

and obtain good correlation with some published experimental results for both densification of dry sand and liquefaction of saturated undrained sand. Pires et al. (1983), by relating the theory of Nemat-Nasser & Shokooh to experimentally-based observations of Seed et al. (1976), note an alternative form for the function f:

$$f(1 + p) = \left[\theta\pi \sin^{2\theta - 1}\left(\frac{\theta\pi p}{2}\right)\cos\left(\frac{\pi p}{2}\right)\right]^{-1} \tag{11.10}$$

where $\theta = 0.7$, according to these authors. Being based on experimental observations, Equation 11.10 may indeed produce better results than the simple form of Equation 11.9. In the present review, however, we side with simplicity and thus use Equation 11.9 in the sequel.

Substitution from Equation 11.9 into 11.7 yields:

$$\Delta W = \frac{\bar{\bar{v}} e_0}{g(e_0 - e_m)}[1 - (1 + p)^{1-r}], \quad r > 1 \tag{11.11}$$

where $\bar{\bar{v}} = v/(r - 1)$.

For the densification of the drained sample, on the other hand, $p = 0$, and one obtains, from Equations 11.3 and 11.8, by integration:

$$e = e_m + [(e_0 - e_m)^{1-n} + \bar{v}\Delta W]^{1/(1-n)}, \quad n > 1 \tag{11.12}$$

where $\bar{v} = (n - 1)/\tilde{v}$.

In Equations 11.11 and 11.12 $\bar{\bar{v}}$ and $\bar{v}$ are as yet free parameters, and ΔW is the total energy used to attain the corresponding pore water pressure in Equation 11.11 and densification in Equation 11.12. To complete the formulation, we seek to estimate the energy increment ΔW in Equations 11.11 and 11.12.

Since ΔW is the energy consumed in rearranging the granules, Nemat-Nasser &

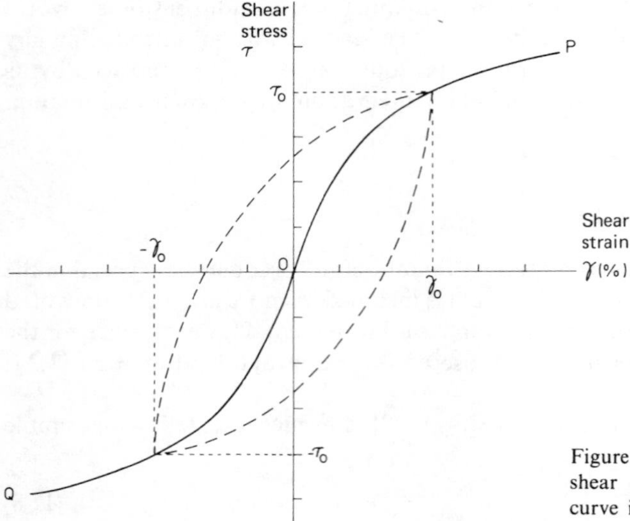

Figure 11.1. A typical maximum shear stress – maximum shear strain curve in cyclic shearing (from Hardin & Drnevich 1972)

Shokooh (1979) suggest that it may be estimated in terms of the corresponding hysteretic loop. A typical loop of this kind is sketched in Figure 11.1. Let A_i be the area enclosed by this loop. It is easy to show (Nemat-Nasser & Shokooh 1979) that, in view of symmetry, one may set $A_i = h_i\tau_0^{1+\alpha}$ with α a positive even integer, and that h_i is an increasing function of the number of preceding cycles, N. If the incremental work for each cycle is taken to be proportional to A_i, one may write:

$$\Delta W = \sum_{i=1}^{N} \lambda_i h_i \tau_0^{1+\alpha} = \bar{h}(N)\tau_0^{1+\alpha} \tag{11.13}$$

Furthermore, for large stress amplitudes, one may assume $\bar{h}$ proportional to N (see Nemat-Nasser & Shokooh for a discussion of this), i.e. $\bar{h} \sim hN$, and obtain:

$$\tau_0^{1+\alpha}N = \overset{*}{v}\,\frac{e_0}{g(e_0 - e_m)}[1-(1+p)^{1-r}], \quad \overset{*}{v} = \bar{v}/h \tag{11.14}$$

Since α is even, it can be fixed immediately by inspection of experimental data. For Monterey No. 0 sand data from De Alba et al. (1975) suggest $\alpha = 4$. This has been verified by experiments at Northwestern University by the author and his former student Mr. Y. Tobita (1982). Indeed, setting $N = N_1$ for $p = 1$ (at liquefaction), Equation 11.14 yields:

$$\tau_0^{1+\alpha} = \frac{\eta e_0}{N_1 g(e_0 - e_m)} \tag{11.15}$$

where η is a material constant. Figure 11.2 shows typical experimental results on

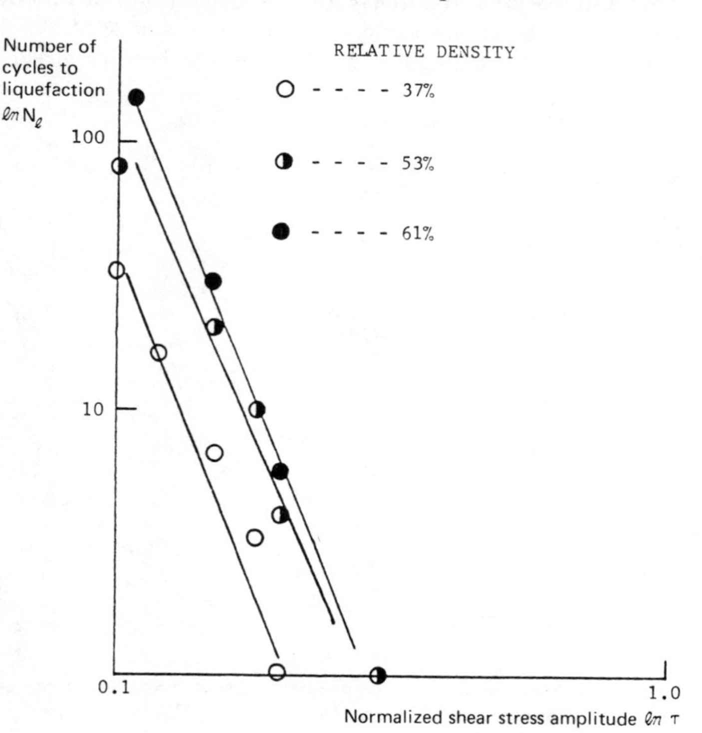

Figure 11.2. Normalized shear stress amplitude versus number of cycles to liquefaction (from Nemat-Nasser 1980a)

Table 11.1. Results obtained from Equation 11.16 on the basis of data of De Alba et al. (1975)

$e_0 - e_m$	σ_c (kPa (psi))	N_1	τ_0	$\eta \times 10^6$
0.132 ($D_r = 54\%$)	55.6 (8.07)	8	0.155	0.859
	55.4 (8.03)	3.25	0.185	0.846
	31.2 (4.53)	12.5	0.144	0.929
	55.3 (8.02)	16	0.135	0.861
	56.1 (8.14)	63	0.104	0.920
0.092 ($D_r = 68\%$)	55.8 (8.09)	15	0.171	0.790
	55.1 (7.99)	4	0.230	0.927
	55.7 (8.08)	53	0.134	0.824
	55.6 (8.06)	6	0.211	0.903
0.052 ($D_r = 82\%$)	55.2 (8.00)	10	0.239	0.406
	55.4 (8.03)	28.5	0.188	0.348
	55.7 (8.08)	15.5	0.211	0.406

circular cylindrical samples of Monterey No. 0 sand, tested in a dynamic simple shear apparatus; see Nemat-Nasser & Tobita (1982) for a description of the test procedure.

Nemat-Nasser & Shokooh use Equation 11.8 and show that the quantity:

$$\eta = N_1 \tau_0^{1+\alpha}(e_0 - e_m)^n / e_0 \tag{11.16}$$

with $\alpha = 4$ and $\eta = 3.5$, is indeed a constant for a number of experiments reported by De Alba et al. (1975) for relatively loose to dense (but not very dense, i.e. for D_r less than, say, 70%) samples, see Table 11.1. Nemat-Nasser & Shokooh suggest that, for dense samples, h in the expression $\bar{h} = hN$ tends to become larger, and, since η is inversely

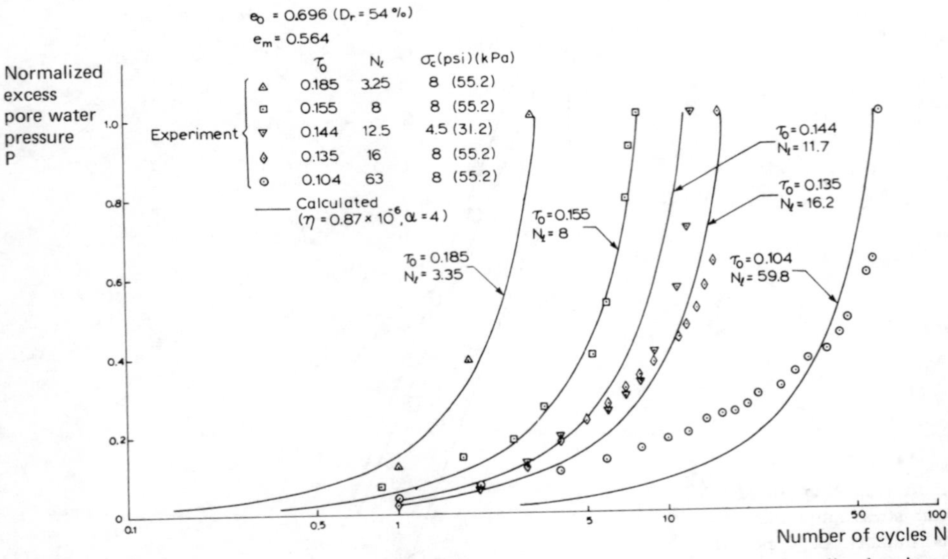

Figure 11.3. Normalized excess pore water pressure p versus number of cycles in cyclic shearing of undrained saturated sand (data from De Alba et al. 1975)

proportional to h (Equation 11.14), it decreases with increasing D_r; see Table 11.1.

With η fixed and g given by Equation 11.8, Equation 11.14 yields the pore water pressure in terms of the number of cycles, for various densities and stress amplitudes; see Figures 11.3 and 11.4. If the number of cycles is normalized with respect to N_1, then we obtain the standard result given by Figure 11.5. The fit here is as good as the simple

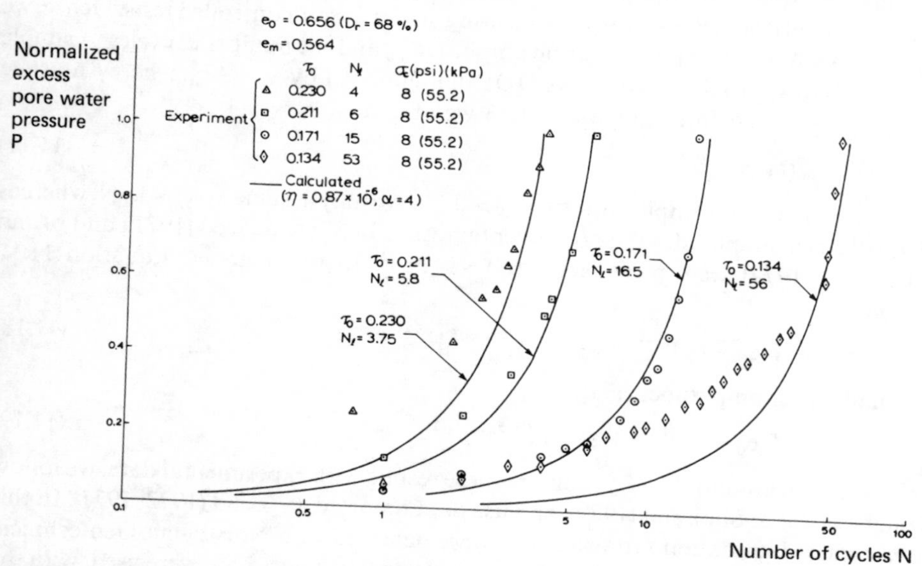

Figure 11.4. Normalized excess pore water pressure p versus number of cycles in cyclic shearing of undrained saturated sand (data from De Alba et al. 1975)

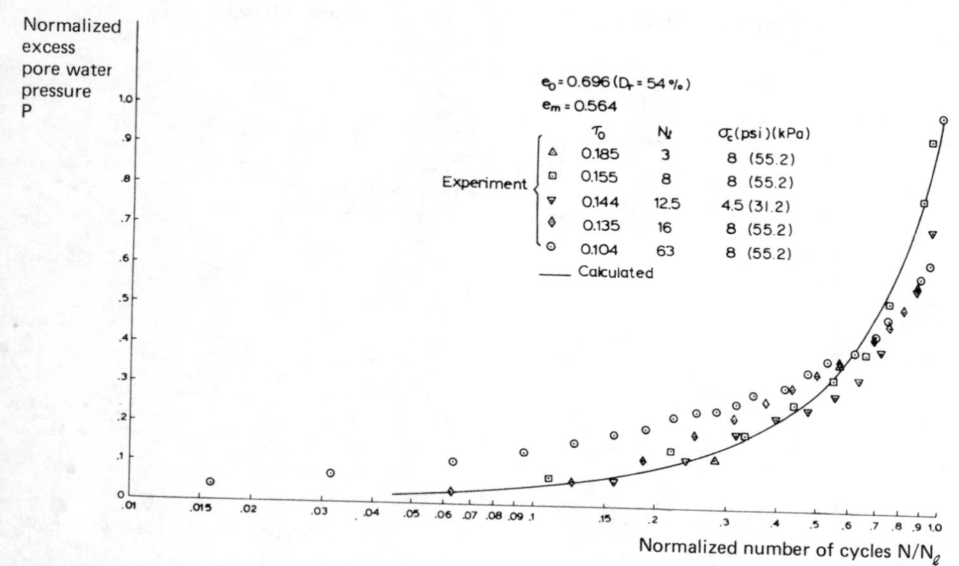

Figure 11.5. Normalized excess pore water pressure p versus normalized number of cycles in cyclic shearing of undrained saturated sand (data from De Alba et al. 1975)

assumption, Equation 11.9, permits. To improve this fit, a more elaborate expression, e.g. Equation 11.10, must be used. Figure 11.6 compares the calculated results of Equation 11.16 with the experimental data; additional comparison with experiments is given by Nemat-Nasser & Shokooh (1979).

To illustrate how the densification estimate of Equation 11.12 relates to experimental data on sand, we observe that most experimental results reported for the densification of drained sand in cyclic shearing are for strain-controlled tests. Hence, we must express ΔW in terms of the strain amplitude γ_0 and the number of cycles. A simple approach is to approximate the curve QOP in Figure 11.1 by $\gamma_0 \sim A\tau_0^\beta$, where β must be an odd integer, and from Equation 11.13 we obtain:

$$\Delta W = \bar{\bar{k}}(N)\gamma_0^{(1+\alpha)/\beta} \tag{11.17}$$

Again, for large strain amplitudes, say, $\gamma_0 > 0.1\%$, we may assume $\bar{\bar{k}}(N) \sim k_1 N$, whereas for small strain amplitudes the experimental results of Silver & Seed (1971a and b) and Youd (1970, 1972) seem to suggest $\bar{\bar{k}}(N) \sim k_2 N^{1/2}$. In this manner, Equation 11.12 becomes:

$$e = e_m + [(e_0 - e_m)^{1-n} + k_2 N^{1/2} \gamma_0^{(1-\alpha)/\beta}]^{1/(1-n)} \tag{11.18}$$

for small strain amplitudes, and:

$$e = e_m + [(e_0 - e_m)^{1-n} + k_1 N \gamma_0^{(1-\alpha)/\beta}]^{1/(1-n)} \tag{11.19}$$

for large strain amplitudes. To compare these results with experimental data, we follow Nemat-Nasser & Shokooh (1979), and use data reported by Youd (1972, 1977). In this case, standard gradation Ottawa sands were densified in a Norwegian Geotechnical Institute type simple shear apparatus. Strains from 0.043% to 8.5% were used, with the number of cycles from 1 to 150,000. Figures 11.7 and 11.8 compare these with the estimates obtained from Equations 11.18 and 11.19, with $\alpha = 4$, $\beta = 5$, $n = 3.5$ and $k_2 = 7000$ for Equation 11.18 and $k_1 = 1000$ for Equation 11.19.

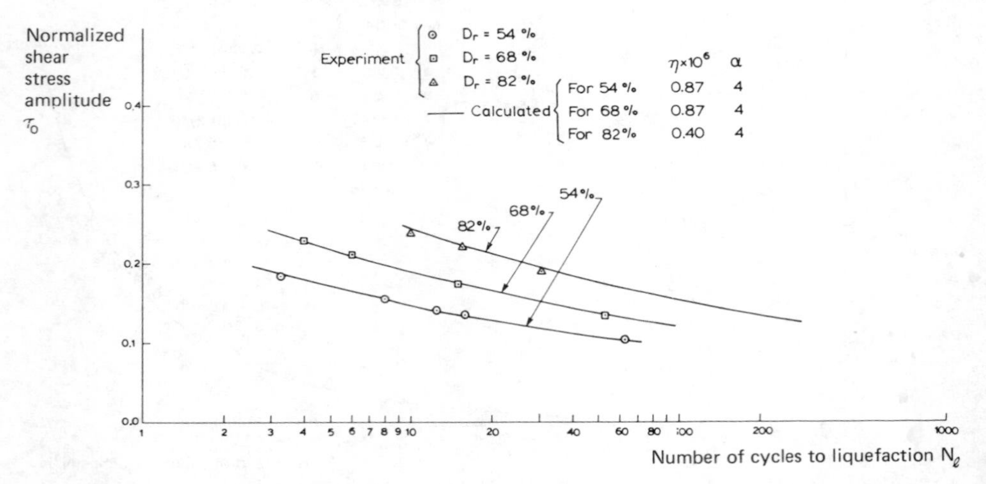

Figure 11.6. Normalized shear stress amplitude τ_0 versus number of cycles to liquefaction (data from De Alba et al. 1975)

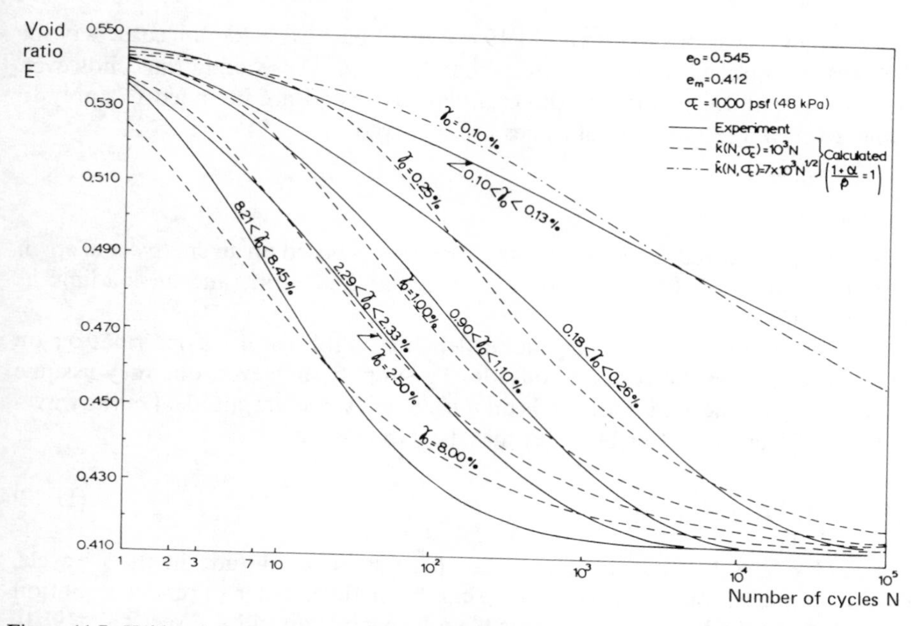

Figure 11.7. Void ratio e versus number of cycles in cyclic shearing of dry sand (data from Youd 1972)

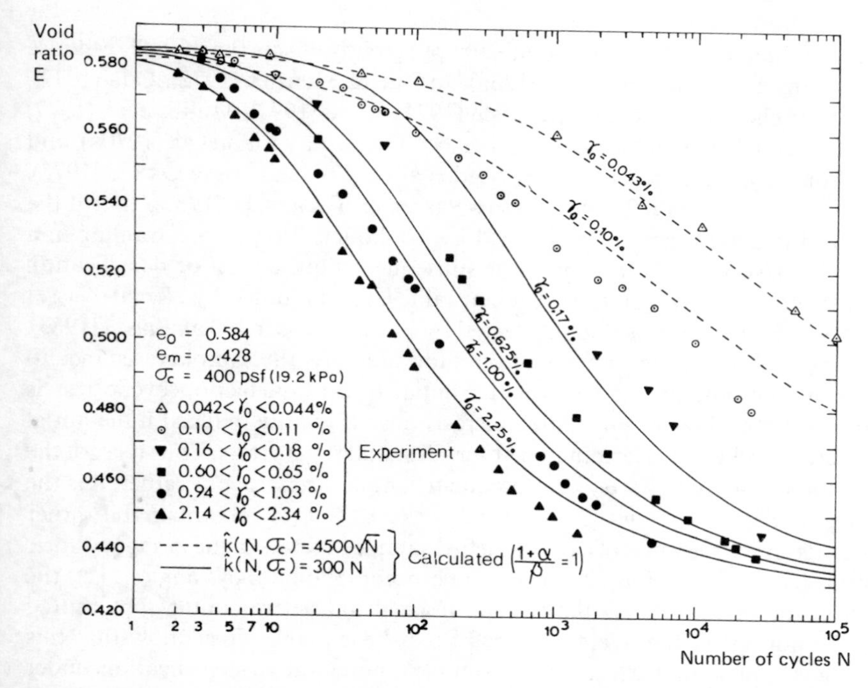

Figure 11.8. Void ratio e versus number of cycles in cyclic shearing of dry sand (data from Youd 1977)

It is clear that Equations 11.18 and 11.19 do encompass the essential features of the densification phenomenon of sand in cyclic shearing. These equations, however, assume no inherent anisotropy for the granular mass, and are intended to apply to cases where very large numbers of cycles are involved.

11.2.3 *Irregular shearing*

Since the theory presented in the preceding subsection is based on an energy concept, it lends itself to extension for application to irregular and even random loading, as discussed by Pires et al. (1983).

The area of the hysteretic loop, in general, depends on the history of deformation, on the wave-form, and on the stress amplitude. To simplify, however, one may assume that, for the ith cycle, the area A_i depends on the current stress amplitude. Let the wave-form be fixed. Then ΔW after N cycles may be expressed as:

$$\Delta W = \sum_{i=1}^{N} h_i(e_i, \tau_i) \tag{11.20}$$

where e_i and τ_i are the void ratio and the shear stress amplitude in the ith cycle, respectively. Some simplifying assumptions can be made in order to render Equation 11.20 useful. A detailed account of a possible approach is given by Pires et al. (1983).

11.3 EFFECT OF FABRIC ON LIQUEFACTION AND DENSIFICATION POTENTIAL OF GRANULAR MASSES

It is well known that granular fabric (or anisotropy) greatly affects the overall response of the granular mass; see, e.g., Lafeber (1966), Arthur & Menzies (1972), Oda (1972), Mahmood & Mitchell (1974), Oda & Konishi (1974), Ladd (1977), Mulilis et al. (1977) and Oda et al. (1980, 1982). In particular, it has been shown by Finn et al. (1970a), and further confirmed by Ishihara et al. (1975), Seed et al. (1977). Ishihara & Okada (1978), Nemat-Nasser & Tobita (1982) and Nemat-Nasser & Takahashi (1983), that if the fabric of a saturated sand sample is changed by pre-liquefaction or prestraining in a stress-controlled test, then the potential for subsequent liquefaction or densification under cyclic loading greatly increases. A remarkable fact, first noted by Nemat-Nasser & Tobita (1982) and recently thoroughly tested by Nemat-Nasser & Takahashi (1983), is that in cyclic shearing under constant confining pressure, the sample resistance to further liquefaction actually increases if the initial (pre-) liquefaction cyclic test is terminated at zero residual strain, while this resistance is greatly reduced if the initial (pre-) liquefaction cyclic test is terminated at zero residual shear stress. Thus it is not the pre-liquefaction per se that affects the subsequent sample strength, but rather, it is the fabric of the granular mass. Indeed, Nemat-Nasser & Tobita (1982) explain this rather dramatic change in the response of a pre-liquefied sample in terms of the granular fabric characterized by the distribution of the microscopic dilatancy angles, i.e. the distribution of the orientation of the contact normals (at active contacts) measured relative to the normal of the overall macroscopic shear plane. They show that this distribution has a profound effect on the sample's potential to densification under drained conditions and, therefore, on its liquefaction potential when saturated and undrained. This result also suggests that the distribution of the dilatancy angles and, therefore, the fabric of a granular material in simple cyclic shearing, is more directly related to the total shear strain rather than to the shear stress.

In this section, we shall first review the micro-mechanical model of Nemat-Nasser (1980b), and then relate this model to the observed fabric-induced changes of the liquefaction and densification potential of sand in cyclic shearing.

11.3.1 *A micro-mechanical model in simple shearing* (*Nemat-Nasser* 1980*b*; *and Nemat-Nasser* & *Tobita* 1982)

Consider uniform shearing of a sample of a granular mass under constant (uniform) normal stress σ, and variable (uniform) shear stress τ; see Figures 11.9(a) and (b). For the sake of modelling, it will be assumed that the sample area; A, is constrained to remain constant, so that volume changes are accompanied by appropriate changes of sample height, h. The macroscopic volume $V = Ah$ is regarded to be statistically representative.

The macroscopic shearing is in the x-direction, but on the micro-scale, grains must override each other and thus, their motion occurs over planes which pass through active contact points, as illustrated in Figure 11.9(c).

Denote by ν the angle which the 'sliding' direction at a typical active contact (C in Figure 11.9(c)) forms with the x-axis. ν is called the dilatancy angle. It is positive if the direction of motion is upward, and negative when it is downward, the first contributing to volume expansion, and the second to contraction; here positive dilatancy refers to volume expansion.

At each instant the macroscopic volume V contains a large number of active contacts, each associated with its own dilatancy angle. Let V contain n families with dilatancy angles ν_i, $i = 1, 2, \ldots, n$. For the sake of modelling, we assign elementary volume V_i, area A_i and height h_i to the ith family, in such a manner that:

$$V_i = A_i h_i, \quad V = \sum_{i=1}^{n} V_i \tag{11.21}$$

and denote by:

$$p_i = \frac{V_i}{V} \tag{11.22}$$

the volume fraction of the ith family.

Consider the forces acting on the ith family which are symbolically shown in Figure 11.9(d) by a granule of dilatancy angle ν_i. The local stresses τ_i and σ_i, in general, are different from the overall macroscopic stresses τ and σ. In Figure 11.9(d):

$$T_i = A_i \tau_i, \quad N_i = A_i \sigma_i \tag{11.23}$$

The actual motion occurs in the x*-direction. Therefore, in the x*, y*-coordinate system the tangential and normal forces, denoted by T_i^* and N_i^*, must relate in accordance with the friction law:

$$T_i^* = N_i^* \tan \phi_\mu \tag{11.24}$$

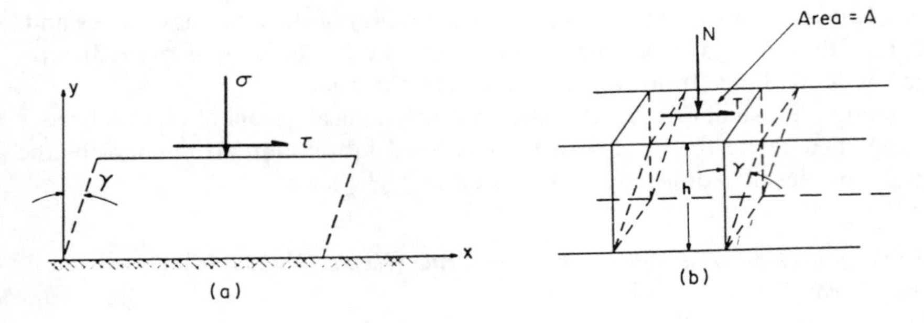

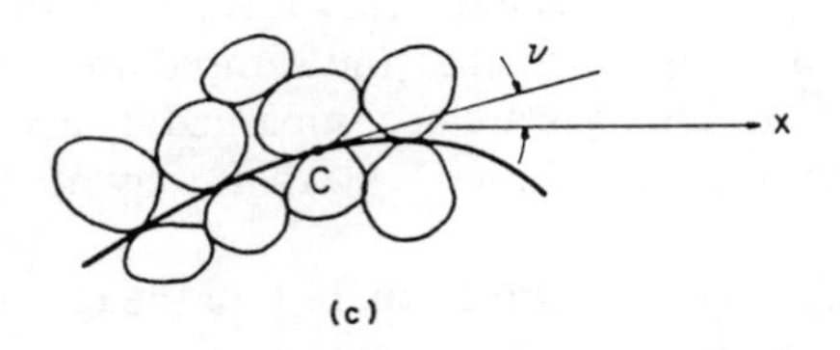

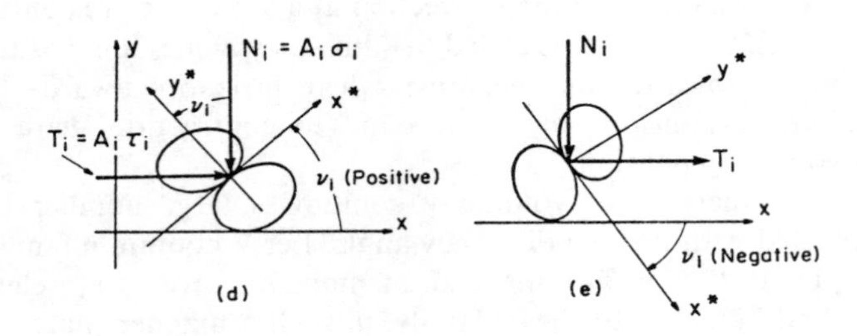

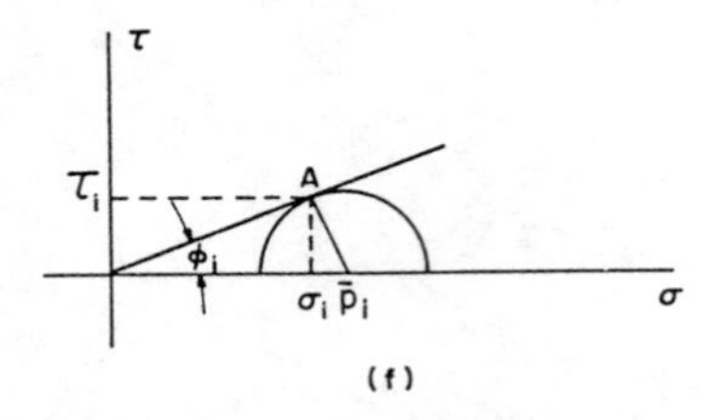

Figure 11.9. (a) Simple shearing (plane strain) under constant normal stress σ; τ is the shear stress and γ is the corresponding shear strain
(b) A statistically representative sample of volume $V = Ah$; $N = \sigma A$ is the total normal force and $T = \tau A$ is the total tangential force acting on the sample
(c) At the active contact C, the sliding direction makes the angle ν with the macroscopic shearing x^--direction; ν is the dilatancy angle
(d) Forces acting on a granule with positive dilatancy angle ν_i; note that normal force N_i hinders sliding for $\nu_i > 0$
(e) An active contact with negative dilatancy angle; note that normal force N_i assists sliding
(f) The Mohr-Coulomb failure criterion is assumed to apply at the local micro-level; σ_i and τ_i are the local normal shear stresses, and $\bar{p}_i$ is the local pressure

where ϕ_μ is regarded as the actual angle of friction. If we define:

$$\tan \phi_i = \frac{\tau_i}{\sigma_i} \tag{11.25}$$

then from the balance of forces we obtain:

$$\tan \phi_\mu = \tan(\phi_i - \nu_i) \quad \text{or} \quad \phi_i - \nu_i = \phi_\mu \tag{11.26}$$

showing that granules with negative dilatancy angles become active first.

Consider now the motion of the ith granule in the x*-direction, and calculate the rate of work per unit volume by:

$$\dot{w}_i = \frac{\dot{W}_i}{V_i} = \frac{1}{V_i} T_i^* \frac{\dot{h}_i}{\sin \nu_i} = \frac{\tau_i \sin \phi_\mu}{\sin(\phi_\mu + \nu_i) \sin \nu_i} \frac{\dot{V}_i}{V_i} \tag{11.27}$$

where $\dot{h}_i / \sin \nu_i$ is the velocity in the x*-direction, and $\dot{V}_i = A_i \dot{h}_i$.

We assume that at the micro-level, the Mohr-Coulomb failure criterion applies. Then the local pressure $\bar{p}_i$ equals $\tau_i / \sin \phi_i \cos \phi_i$ (Figure 11.9(f)), and hence:

$$\dot{w}_i = \tau_i \dot{\gamma}_i - \bar{p}_i \frac{\dot{V}_i}{V_i} = \tau_i \dot{\gamma}_i - \frac{\tau_i}{\sin \phi_i \cos \phi_i} \frac{\dot{V}_i}{V_i} \tag{11.28}$$

From Equations 11.27 and 11.28 we get:

$$\frac{1}{V} \frac{\dot{V}}{\dot{\gamma}} = \sum_{i=1}^{n} \frac{\dot{\gamma}_i}{\dot{\gamma}} p_i \frac{\cos(\phi_\mu + \nu_i) \sin \nu_i}{\cos \phi_\mu} \tag{11.29}$$

where p_i is defined by Equation 11.22.

For the considered macroscopic model problem, Equation 11.29 is exact, subject to the basic assumptions that the Mohr-Coulomb failure criterion applies at the micro-level, and that ϕ_μ is constant. Let us now assume that $\dot{\gamma}_i$ of the micro-motion does not differ substantially from the macro-quantity* γ, so that $\gamma_i / \gamma \sim 1$. Then Equation 11.29 becomes (Nemat-Nasser 1980b):

$$\frac{1}{V} \frac{\dot{V}}{\dot{\gamma}} = \sum_{i=1}^{n} p_i \frac{\cos(\phi_\mu + \nu_i) \sin \nu_i}{\cos \phi_\mu} \tag{11.30}$$

When the sample contains a very large number of active granules, one may consider a continuous distribution of the dilatancy angles, and write:

$$\frac{1}{V} \frac{\dot{V}}{\dot{\gamma}} = \int_{\nu^-}^{\nu^+} p(\nu) \frac{\cos(\phi_\mu + \nu) \sin \nu}{\cos \phi_\mu} d\nu \tag{11.31}$$

Here, $p(\nu)d\nu$ is the volume fraction of active granules whose dilatancy angles are between ν and $\nu + d\nu$, $p(\nu)$ is the corresponding density function, and ν^+ and ν^- are the upper and lower limits of the (active) dilatancy angles.

As has been pointed out by Nemat-Nasser (1980b), granules with negative dilatancy angles are initially activated and, therefore, $p(\nu)$ is initially biased toward negative dilatancy angles. Physically this is because the normal force N_i hinders the motion of an

*Assumptions of this kind are extensively used in calculating the polycrystal response in terms of that of the single crystals, and date back to the work of G.I. Taylor (1938).

active granule with a positive dilatancy angle, whereas it assists if the granule has a negative dilatancy angle; compare Figure 11.9(d) with Figure 11.9(e). This explains the observed initial densification. Note, however, that even with a distribution function p(ν) which is symmetrical with respect to $\nu = 0$ and hence $|\nu^-| = \nu^+$, the right-hand side of Equation 11.31 would be negative; that is, Equation 11.31 incorporates in a natural manner the nonsymmetrical influence of the normal stress σ on 'upgoing' and 'downgoing' granules.

As the shearing proceeds, the distribution function p(ν) becomes more biased toward the positive dilatancy angles. Hence, more weight is given to positive ν's in Equation (11.31), as shearing progresses. Eventually, when a suitable bias toward positive dilatancy angles is attained, the integral in Equation 11.31 vanishes, marking the attainment of a minimum void ratio (or maximum density). After this state, continual shearing in the same direction results in a volume expansion, until the peak stress is reached, which, according to the theoretical considerations of Nemat-Nasser & Shokooh (1980), must correspond to a maximum rate of dilatancy. After that the rate of dilatancy begins to decrease in the post-failure regime, presumably becoming zero asymptotically at the critical state.

The three loading regimes – initial densification (I) leading to dilatancy up to the peak stress (II) and then continuing in the post-failure behaviour to the critical state (III) – are shown schematically in Figure 11.10. Subdivisions of this kind play an important role in the understanding of the effect of history on subsequent behaviour of granular materials in cyclic loading. Before this is examined in connection with drained and undrained experiments, an important point relating to load reversal should be emphasized in the context of Equation 11.31.

Suppose the sample in Figure 11.9(a) is sheared clockwise under constant σ, beginning with $\tau = 0$ and continuing until a state in loading regime II is attained, where further shearing in the same direction results in volume expansion. At this state the distribution function p(ν) is strongly biased toward positive ν's. Suppose now the magnitude of the shear stress is gradually reduced to zero, keeping the normal stress σ constant. It is clear that some active contacts with previously (i.e. during the clockwise loading) large positive dilatancy angles, such as the one shown in Figure 11.9(d), may begin to move down as their corresponding shear stress is reduced; for example in Figure 11.9(d), if ν_i is large enough and T_i is reduced, while N_i remains essentially the same, the particle may move down along the x*-axis under the action of N_i, if there are no other constraints. Because of this, it is expected that the distribution of contact

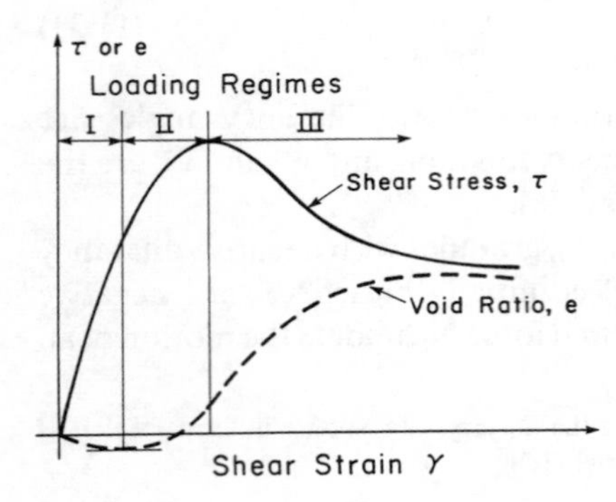

Figure 11.10. Three loading regimes in monotone shearing: regime I corresponds to densification, regime II begins with dilatancy and ends at the peak stress, and regime III pertains to post-failure response

normals as well as the dilatancy angles will change somewhat as the shear stress is reduced to zero. However, it is reasonable to expect that even with these changes, when τ in Figure 11.9(a) is reduced to zero, a strong bias toward positive dilatancy angles for clockwise shearing will still remain. Now, upon load reversal, namely as τ is gradually increased counter-clockwise from zero in Figure 11.9(a), the direction of τ_i in Figure 11.9(d) will change, and a previously positive dilatancy behaves now as a negative one; the direction of the x*-axis is now reversed. Thus, a strong tendency toward large densification is expected upon shear load reversal. This means that, for example, a pre-straining into the loading regime II under undrained conditions can lead to immediate liquefaction, if load reversal is implemented under undrained (saturated) conditions. This is indeed observed by Finn et al. (1970a) and Ishihara & Okada (1978) and is re-established by Nemat-Nasser & Tobita (1982) and Nemat-Nasser & Takahashi (1983). In the sequel we shall describe a number of experiments under both drained and undrained conditions, which were specifically designed to test the implications of Equation 11.31 and, therefore, the effect of fabric.

Figure 11.11, taken from Nemat-Nasser & Tobita (1982), represents the results of a series of experiments on the effects of prestraining and pre-liquefaction on subsequent behaviour under undrained conditions. The saturated drained sample of initial void ratio 0.649 is first stressed over the path OC. At point C the drainage of the saturated sample is discontinued, so that branch CDE represents unloading (from C to D) and load reversal (from D to E) under undrained conditions. Since during the drained

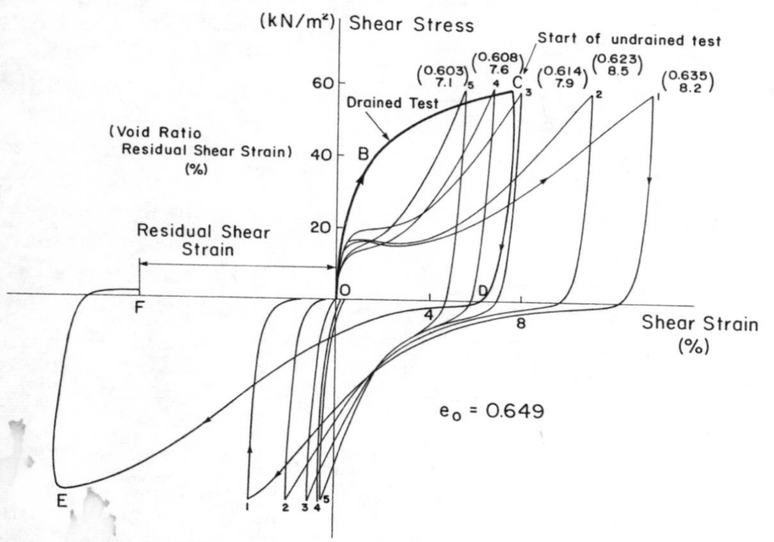

Figure 11.11. Under 100 kN/m^2 confining pressure, a drained sample is quasi-statically sheared over the loading branch OBC. At C the drainage is discontinued, and over (quasi-static) unloading (C to D) and load reversal (D to E) the sample liquefies. At F, residual strain is 8.2%. The sample is drained and reconsolidated to 100 kN/m^2 showing 0.635 void ratio. Then the new liquefaction test is performed (curve marked 1). Other curves are obtained in the same manner. Numbers at upper and lower peaks denote the number of pre-liquefactions. The upper number in parentheses denotes the void ratio, and the lower one, the corresponding residual shear strain (%). Shear stress and shear strain are the horizontal force and displacement divided respectively by the sample area and the sample height (from Nemat-Nasser & Tobita 1982)

loading from O to B to C, the stress state moves into loading regime II with $p(v)$ developing a strong bias toward positive dilatancy angles, a tendency toward liquefaction occurs immediately upon load reversal. The test is terminated at point F with a residual strain of approximately 8.2%, the sample is drained and reconsolidated to the void ratio of 0.635, and a new test under undrained conditions is performed. The result is shown by the curve numbered 1 at the upper and lower peaks in Figure 11.11. The other curves in this figure are obtained by reconsolidation of the sample after liquefaction. The numbers at the upper and lower peaks denote the number of times the sample has been pre-liquefied; the upper number within parentheses associated with each curve gives the corresponding void ratio, and the lower number the corresponding residual shear strain. As expected, the residual shear strain decreases with the number of pre-liquefactions, which may be attributed to the resulting smaller void ratio. The sample, once liquefied, shows a large displacement around an average shear stress of about 15 kN/m^2, regardless of the number of preceding liquefactions. Thus, densification due to reconsolidation after liquefaction does not have much effect on the subsequent undrained behaviour up to liquefaction initiation, although it does reduce the observed strain amplitude to a certain extent.

To further emphasize the directional dependency, namely, the fabric formation due to pre-straining, a virgin sample is subjected to loading, unloading, reverse loading, and then unloading to zero stress, with an overall stress amplitude of 67.4 kN/m^2, under drained conditions. The void ratio and the residual shear strain at the completion of this cycle are 0.649 and 5.1%, respectively. An undrained test is then performed at the overall stress amplitude of 21 kN/m^2. The corresponding stress path is shown in Figure

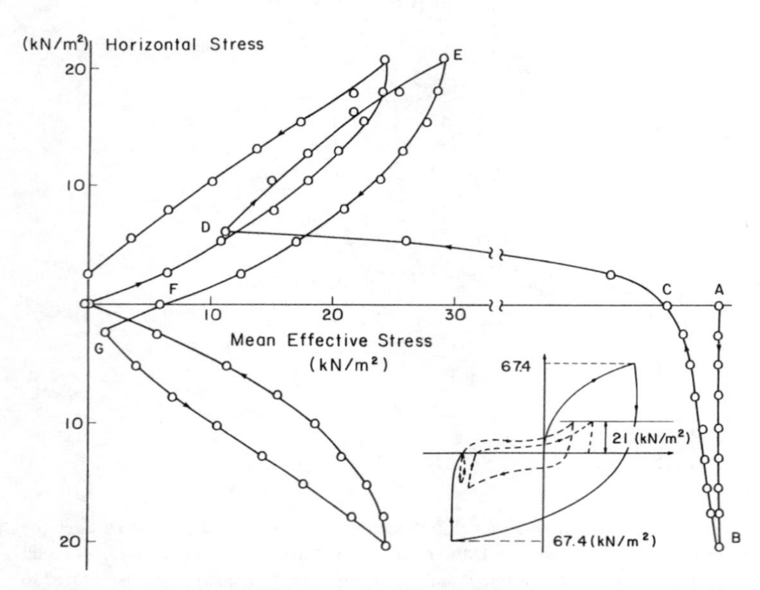

Figure 11.12. A drained sample is subjected quasi-statically to a stress cycle (solid line in the insert figure) of amplitude 67.4 kN/m^2 under 100 kN/m^2 confining pressure. ABCDEF is the load path of the subsequent quasi-static undrained test. AB is the reloading in the direction of the final half-cycle of the drained test; it shows no pore pressure build-up. CD is in the opposite direction; it shows dramatic pore pressure build-up (from Nemat-Nasser & Tobita 1982)

11.12. The loading path AB represents reloading in the same direction as the final half-cycle during the drained test. Hence, essentially no pore pressure build-up occurs. Some pressure is generated over the unloading branch BC, and a dramatic pore pressure is developed upon load reversal over the loading branch CD. This last loading branch is in the direction opposite to the direction of loading and unloading that has been implemented under the drained conditions, and which has resulted in 5.1% residual shear strain. Such strong directional dependency supports the concept of the formation of bias in the distribution of dilatancy angles during pre-straining. Branch DE corresponds to dilatancy and, therefore, pore pressure decreases. Thus, upon unloading from E to F and reverse loading from F to G, extensive pore pressure is generated.

In all the above tests the pre-straining is terminated at zero shear stress. From the micro-mechanical consideration of Nemat-Nasser (1980b), one concludes that, if the fabric as it is described by the distribution of the dilatancy angles, p(ν), is the key ingredient in introducing dramatic changes in sample behaviour because of pre-straining, then a pre-straining over a cycle of relatively large strain amplitude which terminates with zero residual shear strain, should result in a considerably different subsequent response from the one which terminates with zero residual shear stress. This is because, as the shear strain approaches zero, the particles tend to move into a more isotropic distribution of the dilatancy angles than when the shear stress is brought to zero. In fact, as discussed before, unless a very large strain amplitude is involved, unloading to zero shear stress does not completely destroy the fabric, whereas unloading to zero shear strain does to a large extent.

To provide experimental support for the above assertions, two tests are performed on a sample of essentially the same void ratio.

In the first test, shown in Figure 11.13, the sample first is subjected to the stress half-cycle, ABC, under drained conditions, terminating at point C with zero shear stress.

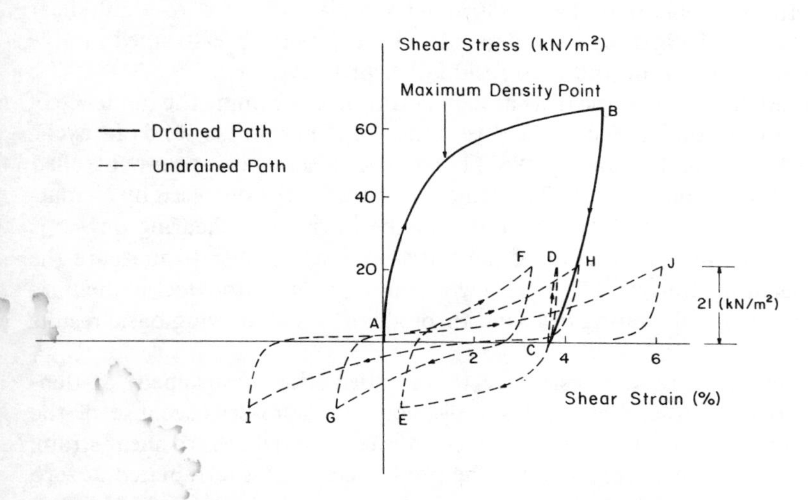

Figure 11.13. Under 100 kN/m^2 confining pressure, a drained sample is subjected quasi-statically to a half-cycle of stressing (ABC) with zero residual shear stress at C. The sample liquefies in the third cycle of 21 kN/m^2 average shear stress amplitude applied quasi-statically under undrained conditions (from Nemat-Nasser & Tobita, 1982)

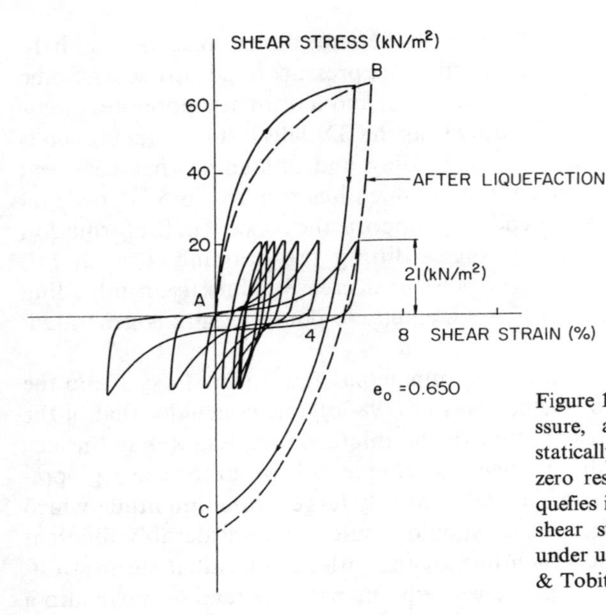

Figure 11.14. Under 100 kN/m^2 confining pressure, a drained sample is subjected quasi-statically to a half-cycle of straining (ABC) with zero residual shear strain at C. The sample liquefies in the seventh cycle of 21 kN/m^2 average shear stress amplitude applied quasi-statically under undrained conditions (from Nemat-Nasser & Tobita 1982)

Then under undrained conditions, cyclic stress of amplitude 21 kN/m^2 is applied, where the sample liquefies between the second and third cycles; note no pore pressure build-up from A to D; note also ratcheting toward the left in Figure 11.13.

In the second test the pre-straining under drained conditions is terminated at zero shear strain, as shown in Figure 11.14 by branches ABC. The sample is then subjected to cyclic loading of stress amplitude 21 kN/m^2, and it is observed that liquefaction occurs during the seventh cycle, which essentially is the same as for the virgin sample. Further examination reveals that the pre-strained sample with zero residual shear strain does not show a strong directional dependency, as does the pre-strained sample with finite residual shear strain (but zero residual shear stress).

In an effort to further verify these facts and, in addition, to examine the influence of the sample preparation on the mechanical response of cohesionless sands in cyclic shearing, Nemat-Nasser & Takahashi (1983) have made a series of strain-controlled tests on Monterey No. 0 sand samples. The same apparatus as the one used by Nemat-Nasser & Tobita (1982) is employed, except that the horizontal shearing device is modified in such a manner as to control the horizontal stroke and to measure the corresponding resulting horizontal force. Two sample preparation techniques are used: moist tamping and pluviating dry sand through air. The following basic results are obtained:

1. In cyclic simple shearing, the resistance to re-liquefaction (undrained) or densification (drained) of a pre-liquefied sample actually increases, because of the concomitant densification, if the pre-liquefaction is terminated at zero shear strain, but this resistance becomes very small, if the pre-liquefaction is terminated at zero residual shear stress.
2. The inherent anisotropy associated with sample preparation techniques affects both the densification and liquefaction potential of the sample.
3. Within each cycle of simple shearing, the induced anisotropy is essentially wiped out

in the neighbourhood of the zero shear strain, and the anisotropy that exists at this state is basically due to the sample preparation techniques (i.e. it is the inherent anisotropy), provided that the sample is not very loose and the strain amplitude is not very large.

4. For simple cyclic shearing, the distribution of the dilatancy angles characterizing the fabric may be related to the shear strain and, in this manner, the densification pattern may be estimated.

ACKNOWLEDGEMENT

This work has been supported by the US Air Force Office of Scientific Research under grant no. AFOSR-80-0017 to Northwestern University, Evanston, Illinois.

REFERENCES

Arthur, J.R.F. & B.K. Menzies 1972. Inherent anisotropy in a sand. *Géotechnique* 22 (1): 115–128.

De Alba, P.S., C.K. Chan & H.B. Seed 1975. Determination of soil liquefaction characteristics by large scale laboratory tests. Report no. EERC 75–14, College of Engineering, University of California, Berkeley.

Finn, W.D.L., P.L. Bransby & D.J. Pickering 1970a. Effect of strain history on liquefaction of sand. *J. Soil Mech. Fdns. Div.* (ASCE) 96 (SM 6): 1917–1934.

Finn, W.D.L., J.J. Emery & Y.P. Gupta 1970b. A shaking table study of the liquefaction of saturated sands during earthquakes. In *Proceedings of 3rd European Symposium on Earthquake Engineering* (Sofia): 253–262.

Hardin, B.O. & V.P. Drnevich 1972. Shear modulus and damping in soils: measurement and parameter effects. *J. Soil Mech. Fdns. Div.* (ASCE) 98 (SM 6): 603–624.

Ishihara, K. & S. Okada 1978. Effects of stress history on cyclic behavior of sand. *Soils and Foundations* 18 (4): 31–45.

Ishihara, K., F. Tatsuoka & S. Yasuda 1975. Undrained deformation and liquefaction of sand under cyclic stresses. *Soils and Foundations* 15 (1): 29–44.

Ladd, R.S. 1977. Specimen preparation and cyclic stability of sands. *J. Geotech. Engng. Div.* (ASCE) 103 (GT 6): 535–547.

Lafeber, D. 1966. Soil structural concepts. *Engng. Geology* 1: 261–290.

Mahmood, A. & J.K. Mitchell 1974. Fabric-property relationships in fine granular materials. *Clays and Clay Minerals* 22 (5/6): 397–408.

Martin, G.R., W.D.L. Finn & H.B. Seed 1975. Fundamentals of liquefaction under cyclic loading. *J. Geotech. Engng. Div.* (ASCE) 101 (GT 5): 423–438.

Mulilis, J.P., H.B. Seed, C.K. Chan, J.K. Mitchell & K. Arulanandan 1977. Effects of sample preparation on sand liquefaction. *J. Geotech. Engng. Div.* (ASCE) 103 (GT 2): 91–108.

Nemat-Nasser, S. 1980a. Liquefaction of soil during earthquakes. In G. Hart (ed.), 1981. *Proceedings of 2nd ASCE Specialty Conference on Dynamic Response of Structures: Experimentation, Observation, Prediction and Control* (Atlanta): 376–385. New York: ASCE.

Nemat-Nasser, S. 1980b. On behavior of granular materials in simple shear. *Soils and Foundations* 20 (3): 59–73.

Nemat-Nasser, S. 1982. On dynamic and static behaviour of granular materials. In G.N. Pande & O.C. Zienkiewicz (eds.), *Soil mechanics–transient and cyclic loads*: 439–458. Chichester: Wiley.

Nemat-Nasser, S. & A. Shokooh 1977. A unified approach to densification and liquefaction of cohesionless sand. Earthquake Research and Engineering Laboratory technical report no. 77–10–3, Dept. of Civil Engineering, Northwestern University, Evanston, Illinois.

Nemat-Nasser S. & A. Shokooh 1978. A new approach for the analysis of liquefaction of sand in cyclic shearing. In *Proceedings of 2nd International Conference on Microzonation for Safer Construction–Research and Application* (San Francisco) 2: 957–969.

Nemat-Nasser, S. & A. Shokooh 1979. A unified approach to densification and liquefaction of cohesionless sand in cyclic shearing. *Can. Geotech. J.* 16 (4): 659–678.

Nemat-Nasser, S. & A. Shokooh 1980. On finite plastic flows of compressible materials with internal friction. *Int. J. Solids and Structs.* 16 (6): 495–514.

Nemat-Nasser S. & K. Takahashi 1983. Does preliquefaction or prestraining reduce sands' resistance to reliquefaction or densification? Earthquake Research and Engineering Laboratory technical report no. 83–7–54, Dept. of Civil Engineering, Northwestern University, Evanston, Illinois.

Nemat-Nasser, S. & Y. Tobita 1982. Influence of fabric on liquefaction and densification potential of cohesionless sand. *Mechanics of mater.* 1 (1): 43–62.

Oda, M. 1972. The mechanism of fabric changes during compressional deformation of sand. *Soils and Foundations* 12 (2): 1–18.

Oda, M. & J. Konishi 1974. Microscopic deformation mechanism of granular material in simple shear. *Soils and Foundations* 14 (4): 25–38.

Oda, M., J. Konishi & S. Nemat-Nasser 1980. Some experimentally based fundamental results on the mechanical behavior of granular materials. *Geotechnique* 30 (4): 479–495.

Oda, M., J. Konishi & S. Nemat-Nasser 1982. Experimental micromechanical evaluation of strength of granular materials: effects of particle rolling. *Mechanics of mater.* 1 (4): 269–283.

Peacock, W.H. & H.B. Seed 1968. Sand liquefaction under cyclic loading simple shear conditions. *J. Soil Mech. Fdns. Div.* (ASCE) 94 (SM 3): 689–708.

Pires, J.E.A., Y.K. Wen & A.H.S. Ang 1983. Stochastic analysis of liquefaction under earthquake loading. Civil Engineering Studies, Structural Research series no. 504, technical report no. UILU-ENG-83-2005, ISSN: 0069–4274, Dept. of Civil Engineering, University of Illinois at Urbana-Champaign, Urbana.

Seed, H.B. & I.M. Idriss 1967. Analysis of soil liquefaction: Niigata earthquake. *J. Soil Mech. Fdns. Div.* (ASCE) 93 (SM 3): 83–108.

Seed, H.B. & I.M. Idriss 1982. *Ground motions and soil liquefaction during earthquakes.* Berkeley, California: Earthquake Engineering Research Institute.

Seed, H.B. & K.L. Lee 1966. Liquefaction of saturated sands during cyclic loading. *J. Soil Mech. Fdns. Div.* (ASCE) 92 (SM 6): 105–134.

Seed, H.B., P.P. Martin & J. Lysmer 1976. Pore-water pressure changes during soil liquefaction. *J. Geotech. Engng. Div.* (ASCE) 102 (GT 4): 323–346.

Seed, H.B., K. Mori & C.K. Chan 1977. Influence of seismic history on liquefaction of sands. *J. Geotech. Engng. Div.* (ASCE) 103 (GT 4): 257–270.

Silver, M.L. & H.B. Seed 1971a. Deformation characteristics of sands under cyclic loading. *J. Soil Mech. Fdns. Div.* (ASCE) 97 (SM 8): 1081–1098.

Silver, M.L. & H.B. Seed 1971b. Volume changes in sands during cyclic loading. *J. Soil Mech. Fdns. Div.* (ASCE) 97 (SM 9): 1171–1182.

Taylor, G.I. 1938. Plastic strain in metals. *J. Inst. Metals* 62 (1): 307–324.

Youd, T.L. 1970. Densification and shear of sand during vibration. *J. Soil Mech. Fdns. Div.* (ASCE) 96 (SM 3): 863–880.

Youd, T.L. 1972. Compaction of sands by repeated shear straining. *J. Soil Mech. Fdns. Div.* (ASCE) 98 (SM 7): 709–725.

Youd, T.L. 1977. Personal communication.

Three-dimensional behaviour and parameter evaluation of an elastoplastic soil model 12

POUL V. LADE

12.1 INTRODUCTION

Successful calculation of deformations and stresses in soil masses due to changes in load using e.g. the finite element method depends to a large extent on a constitutive law for characterization of all aspects of the stress-strain behaviour of the soil. Several stress-strain models have been developed for soils based on concepts from elasticity and plasticity theory (see e.g. Drucker et al. 1957, Roscoe & Burland 1968, Lade & Duncan 1975, Sandler et al. 1976, Lade 1977, Prevost 1978, Dafalias & Herrmann 1980, Mroz et al. 1981).

The stress-strain model which will be presented below is that proposed by Lade (1977). The basic principles involved in this model are outlined, and the ability of the model to account for various aspects of observed three-dimensional soil behaviour is briefly described. The specific procedures employed for determination of material parameters are then presented in connection with derivation of parameters for Antelope Valley sand.

12.2 CONSTITUTIVE MODEL FOR SOILS WITH CURVED YIELD SURFACES

The elastoplastic stress-strain model originally developed for cohesionless soil (Lade 1972, Lade & Duncan 1975) reflects many of the characteristics of sand behaviour observed in laboratory tests. Results of conventional triaxial tests and cubical triaxial tests on cohesionless soil (Lade 1972, Lade & Duncan 1973) and concepts from elasticity and plasticity were employed in formulating the original model. Further developments involve employment of curved yield and failure surfaces, addition of a cap-type yield surface, and use of work-hardening as well as–softening relationships (Lade 1977). The working principles of the model in its present form are briefly reviewed below.

For the purpose of modelling the stress-strain behaviour of soils by an elastoplastic theory, the total strain increments, $\{d\varepsilon_{ij}\}$, are divided into an elastic component, $\{d\varepsilon_{ij}^{e}\}$, a plastic collapse componenet, $\{d\varepsilon_{ij}^{c}\}$, and a plastic expansive component, $\{d\varepsilon_{ij}^{p}\}$, such that:

$$\{d\varepsilon_{ij}\} = \{d\varepsilon_{ij}^{e}\} + \{d\varepsilon_{ij}^{c}\} + \{d\varepsilon_{ij}^{p}\} \tag{12.1}$$

These strain components are calculated separately, the elastic strains by Hooke's law, the plastic collapse strains and plastic expansive strains by a plastic stress-strain theory that involves, respectively, a cap-type yield surface and a conical yield surface with apex at the origin of the stress space.

Figure 12.1(a) shows the parts of the total strain that are considered to be elastic, plastic collapse, and plastic expansive components of strain in a drained triaxial compression test. Typical observed variations of stress difference $(\sigma_1 - \sigma_3)$ and volumetric strain ε_v, with axial strain ε_1, are shown in this figure for a test performed with constant confining pressure σ_3. Both elastic (recoverable) and plastic (irrecoverable) deformations occur from the beginning of loading of a cohesionless soil, the stress-strain relationship is nonlinear, and a decrease in strength may follow peak failure. The volumetric strain is initially compressive and this behaviour may be followed by expansion (as shown in Figure 12.1(a)) or by continued compression. The plastic strains are initially smaller than the elastic strains, but at higher values of stress difference the plastic strains dominate the elastic strains.

The yield surfaces for the plastic strain components are indicated on the triaxial plane in Figure 12.1(b). The conical yield surface may be curved in planes containing the hydrostatic axis, or it may be straight as in the original model (Lade & Duncan 1975). In principal stress space the yield and failure surfaces are shaped like asymmetric bullets with their pointed apices at the origin of the stress space as shown in Figure 12.2(a). Typical cross-sections of these surfaces are shown in Figure 12.2(b). They have been shown to model the experimentally determined three-dimensional strengths of sand as well as normally consolidated clay with good accuracy (Lade & Duncan 1975, Lade & Musante 1977 and 1978).

The yield surface corresponding to the plastic collapse strains forms a cap on the open end of the conical yield surface as shown in Figure 12.1(b). The collapse yield surface is shaped as a sphere with centre in the origin of the principal stress space. It should be noted that yielding resulting from outward movement of the cap does not result in eventual failure. Failure is controlled entirely by the conical yield surface.

The plastic expansive strains are calculated from a nonasssociated flow rule, whereas the plastic collapse strains are calculated from an associated flow rule. The nonassociated flow rule implies that the plastic expansive strain increment vectors superimposed on the stress space form angles with the conical yield surfaces different from 90° as indicated in Figure 12.3(a).

The result of a change in stress is shown in the triaxial plane in Figure 12.3(b). The plastic strain increment vectors are superimposed on the stress space in this diagram. Both plastic collapse and plastic expansive strains are caused by the change in stress from point A to point B, because both yield surfaces are pushed out. The magnitudes of the strain increments are indicated by the lengths of the vectors, and the total plastic strain increment is calculated according to Equation 12.1 as the vector sum of the two components. The elastic strain components are further added (not shown in Figure 12.3) to obtain the total strain increment for the stress change from A to B.

The elastoplastic stress-strain model presented here is applicable to general three-dimensional stress conditions, but the material parameters required to characterize the soil behaviour can be derived entirely from the results of isotropic compression and conventional drained triaxial compression tests. The accuracy of the model has been evaluated by comparing predicted and measured strains for several types of

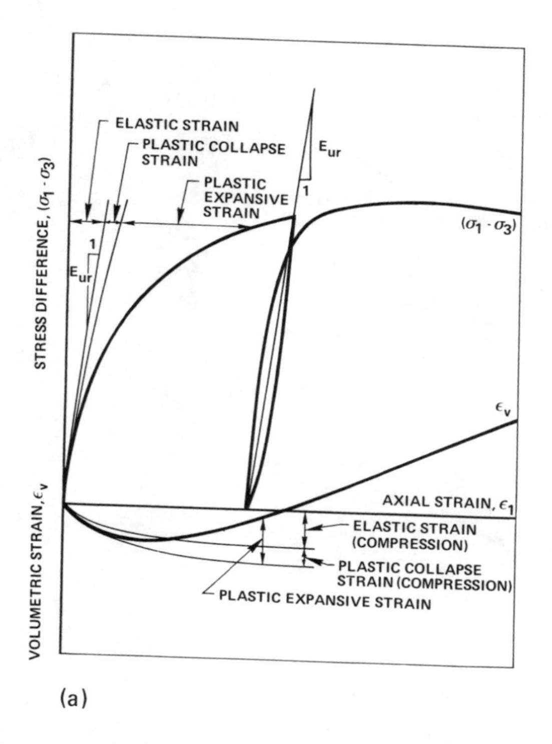

(a)

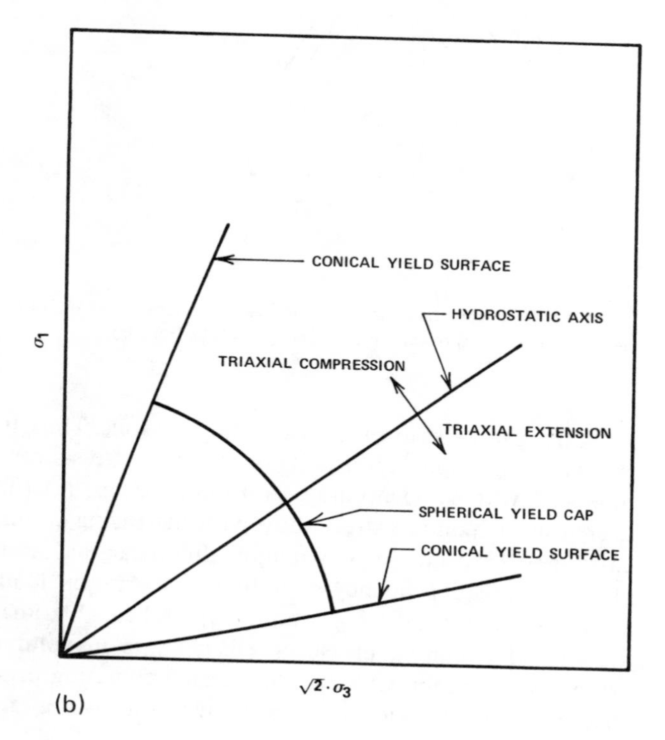

(b)

Figure 12.1. Schematic illustrations of: (a) elastic, plastic collapse, and plastic expansive strain components in drained triaxial compression test, and (b) conical and spherical cap yield surface in triaxial plane

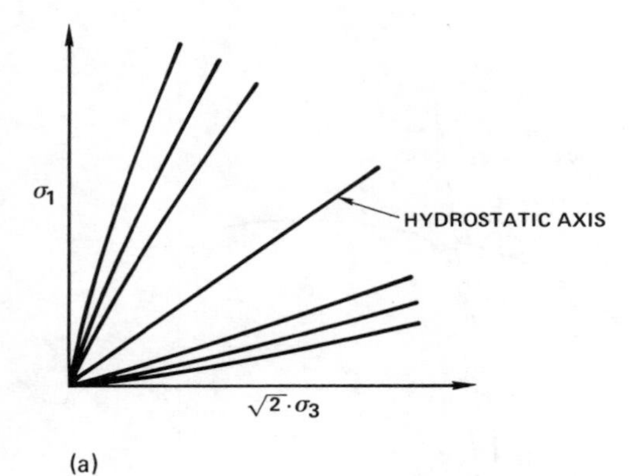

(a)

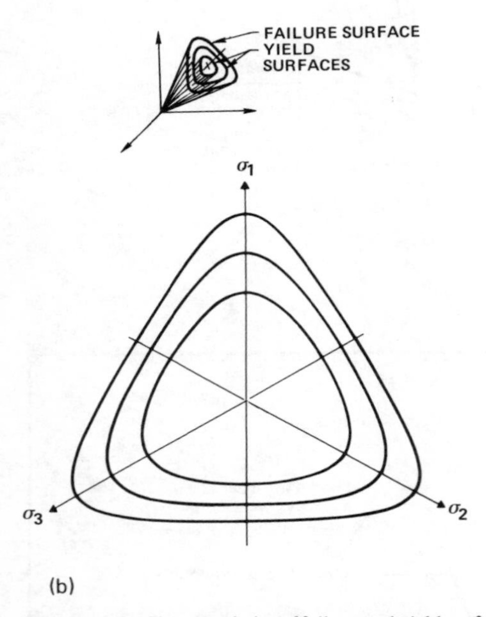

(b)

Figure 12.2. Characteristics of failure and yield surfaces shown in principal stress space. Traces of failure and yield surfaces in: (a) triaxial plane, and (b) octahedral plane

laboratory tests performed on cohesionless soils. Thus, it has been demonstrated that the following important aspects of observed stress-strain and strength behaviour are modelled with good accuracy: (1) nonlinearity; (2) effects of σ_2 and σ_3 including decreasing maximum stress ratio with increasing confining pressure (curved failure envelope); (3) decrease in strength after peak as been reached (strain softening); (4) stress-path dependency including proportional loading and stresses as well as strains occurring under K_0-conditions; (5) shear-dilatancy effect and its variation over a range of confining pressures, i.e. changes in volumetric strain behaviour from expansive to compressive with increasing confining pressure; (6) pore pressures and effective stress-paths for undrained conditions over a range of confining pressures, as

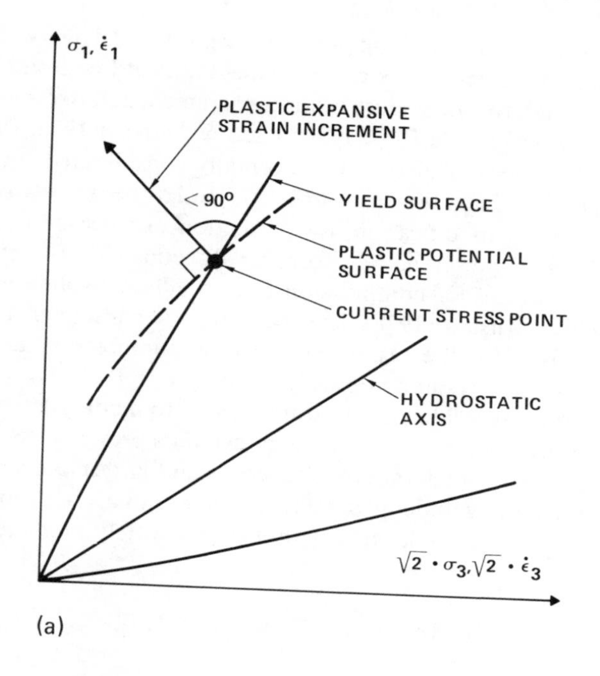

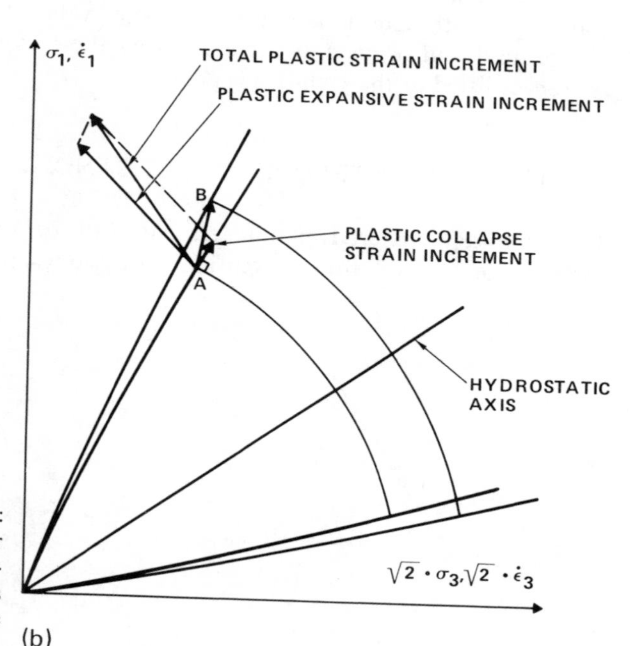

Figure 12.3. Schematic diagrams of: (a) yield and plastic potential surfaces for plastic expansive strains, and (b) yielding process with both yield surfaces activated and combination of plastic strain increments

well as critical confining pressure for given soil density or void ratio; (7) coincidence of strain increment and stress increment axes at low stress levels (elastic behaviour) with transition to coincidence of strain increment and stress axes at high stress levels (plastic behaviour) (Lade 1977, 1978; Lade & Duncan 1975, 1976).

Because the behaviour of normally consolidated clay in many respects resembles that of sand (Lade & Musante 1977), it has been possible to use the same model (with one minor modification) for prediction of stress-strain, pore pressure, and strength behaviour observed in isotropically consolidated, cubical triaxial tests and for prediction of K_0-compression of normally consolidated, remolded clay (Lade 1979, Lade & Musante 1978). The necessary soil parameters for this purpose can be derived entirely from the results of isotropic compression and isotropically consolidated, undrained triaxial compression tests (Lade 1979).

Whereas soil behaviour during small to moderate stress reversals may be predicted with reasonable accuracy, the model in its present form cannot accurately predict soil response during large stress reversals and large changes in stress-involving unloading and reloading under general three-dimensional conditions. Similarly, the behaviour of initially anisotropic soils cannot be accurately predicted by the present model.

12.3 MATHEMATICAL EXPRESSIONS

The mathematical expressions used for modelling the various components of the elastoplastic stress-strain theory are presented without proof or discussion and used in connection with determination of parameter values for Antelope Valley Sand. The background and arguments for employing these expressions are contained in the references listed at the end of this paper.

12.4 TESTS ON ANTELOPE VALLEY SAND

The necessary material parameters may all be determined from an isotropic compression test and three conventional drained triaxial compression tests performed

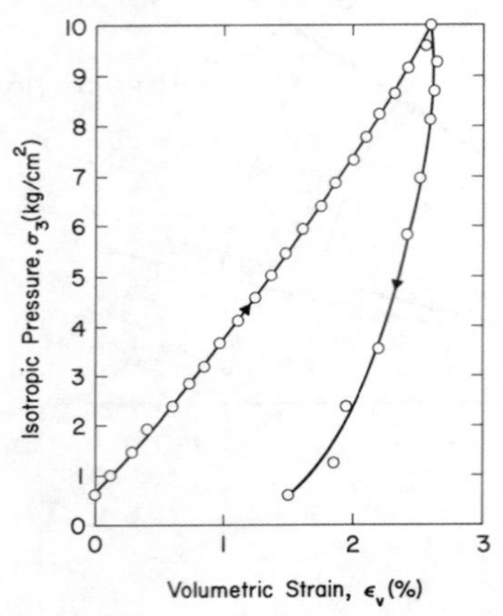

Figure 12.4. Result of isotropic compression test on Antelope Valley sand

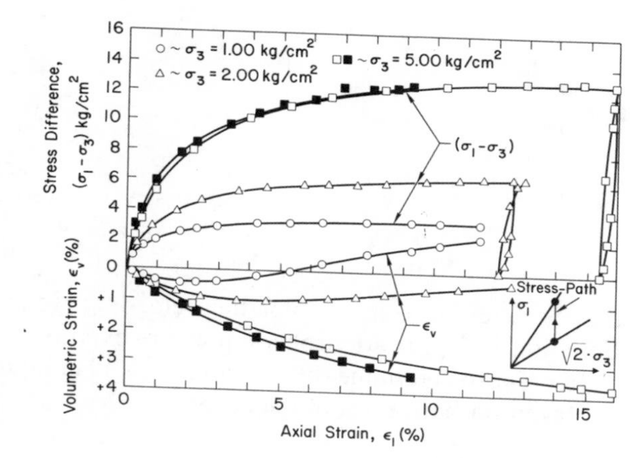

Figure 12.5. Results of drained triaxial compression tests on Antelope Valley sand

at three different, constant confining pressures selected to cover the range of stresses of interest to the field problem to be analyzed. To define the material parameters involved in the elastic modulus, it is necessary to perform unloading-reloading cycles in the triaxial compression tests.

For demonstration of parameter determination for sand, the results of drained tests on Antelope Valley sand # 10– # 20 prepared at a void ratio of 0.83 corresponding to a relative density of 53% will be used. The result of the isotropic compression test is shown in Figure 12.4, and the results of triaxial compression tests are shown in Figure 12.5. In Figure 12.4 the isotropic compression pressure σ_3 is plotted versus the volumetric strain ε_v, which has been corrected for membrane penetration. In Figure 12.5 the vertical stress difference $(\sigma_1 - \sigma_3)$ and the volumetric strain ε_v are both plotted versus the axial strain ε_1. These tests were all performed in a conventional triaxial apparatus, and the type of stress-path used in the triaxial compression tests is shown on the triaxial plane in the insert on Figure 12.5. Unloading-reloading cycles are shown for two tests in the same figure.

The results of two tests performed with a confining pressure of 5.00 kg/cm^2 are shown in Figure 12.5 to demonstrate that some scatter in test results is always observed. Therefore, the precision with which modelling of the stress-strain and volume change relations can be accomplished should be evaluated in the light of the natural scatter of test results.

12.5 DETERMINATION OF MATERIAL PARAMETERS

12.5.1 *Elastic strains*

The elastic strains, which are recoverable upon unloading, are calculated from Hooke's law using the unloading-reloading modulus defined as:

$$E_{ur} = K_{ur} \cdot p_a \cdot \left(\frac{\sigma_3}{p_a}\right)^n \tag{12.2}$$

The dimensionless, constant values of the modulus number K_{ur} and the exponent n

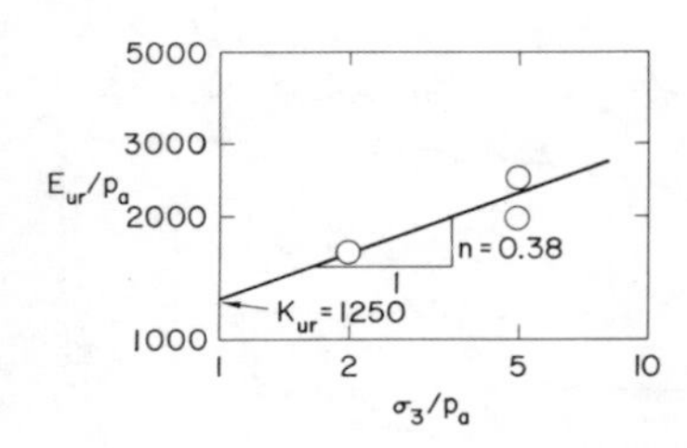

Figure 12.6. Variation of elastic modulus with confining pressure for Antelope Valley sand

may be determined from unloading-reloading cycles in triaxial compression tests. In Equation 12.2 p_a is atmospheric pressure expressed in the same units as E_{ur} and σ_3.

In order to determine the values of K_{ur} and n, Equation 12.2 is rearranged and logs are taken on both sides of the equation:

$$\log\left(\frac{E_{ur}}{p_a}\right) = \log K_{ur} + n \cdot \log\left(\frac{\sigma_3}{p_a}\right) \tag{12.3}$$

Thus, by plotting E_{ur}/p_a versus σ_3/p_a on log–log scales, as shown in Figure 12.6, the value of K_{ur} is determined as the intercept between the best fitting straight line and the vertical line corresponding to $\sigma_3/p_a = 1$. The slope of the straight line corresponds to the exponent n.

A value of Poisson's ratio of 0.2 is used for calculation of elastic strains in sand.

12.5.2 *Plastic collapse strains*

The plastic collapse strains are calculated from a simple plastic stress-strain theory which incorportes (1) a cap-type spherical yield surface with centre at the origin of the principal stress space, (2) an associated flow rule, and (3) a work-hardening relationship which can be determined from an isotropic compression test. The yield criterion, f_c, and the plastic potential function, g_c, have the form:

$$f_c = g_c = I_1^2 + 2 \cdot I_2 \tag{12.4}$$

where I_1 and I_2 are the first and the second stress invariants:

$$I_1 = \sigma_1 + \sigma_2 + \sigma_3 = \sigma_x + \sigma_y + \sigma_z \tag{12.5}$$

$$I_2 = -(\sigma_1 \cdot \sigma_2 + \sigma_2 \cdot \sigma_3 + \sigma_3 \cdot \sigma_1) \tag{12.6}$$

$$= \tau_{xy} \cdot \tau_{yx} + \tau_{yz} \cdot \tau_{zy} + \tau_{zx} \cdot \tau_{xz} - (\sigma_x \cdot \sigma_y + \sigma_y \cdot \sigma_z + \sigma_z \cdot \sigma_x) \tag{12.7}$$

Performing all the necessary derivations, the final form of the plastic collapse stress-strain relationships becomes:

$$\begin{Bmatrix} \Delta\varepsilon_x^c \\ \Delta\varepsilon_y^c \\ \Delta\varepsilon_z^c \\ \Delta\varepsilon_{yz}^c \\ \Delta\varepsilon_{zx}^c \\ \Delta\varepsilon_{xy}^c \end{Bmatrix} = \frac{dW_c}{f_c} \cdot \begin{Bmatrix} \sigma_x \\ \sigma_y \\ \sigma_z \\ \tau_{yz} \\ \tau_{zx} \\ \tau_{xy} \end{Bmatrix} \tag{12.8}$$

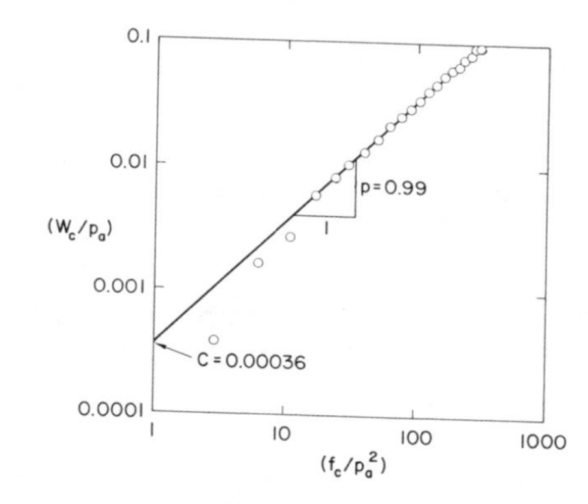

Figure 12.7. Relation between plastic collapse work, W_c, and the value of f_c for Antelope Valley sand

where dW_c is an increment in work per unit volume for a given value of f_c and a given increment of the yield function, df_c. The plastic collapse work is calculated from:

$$W_c = \int \{\sigma_{ij}\}^T \{d\varepsilon_{ij}^c\} \tag{12.9}$$

which for an isotropic compression test reduces to:

$$W_c = \int \sigma_3 \cdot d\varepsilon_v^c \tag{12.10}$$

The plastic collapse strains in Equation 12.10 are calculated from the results of the isotropic compression test by substracting the elastic strains from the measured strains shown in Figure 12.4. W_c then is plotted as a function of f_c which for isotropic compression can be calculated from:

$$f_c = 3 \cdot \sigma_3^2 \tag{12.11}$$

The diagram in Figure 12.7 shows the relationship between W_c and f_c plotted on log–log scales for Antelope Valley sand. This relationship can be modelled by the following expression:

$$W_c = C \cdot p_a \cdot \left(\frac{f_c}{p_a^2}\right)^p \tag{12.12}$$

where the collapse modulus C and the collapse exponent p are determined as shown in Figure 12.7. On this diagram C is the intercept with $(f_c/p_a^2) = 1$ and p is the slope of the straight line. The increment in plastic collapse work is then determined from:

$$dW_c = C \cdot p \cdot p_a \cdot \left(\frac{p_a^2}{f_c}\right)^{1-p} \cdot d(f_c/p_a^2) \tag{12.13}$$

and used in connection with Equation 12.8 for calculation of plastic collapse strains.

12.5.3 *Plastic expansive strains*

The curved yield surfaces which best describe the behaviour of cohesionless soils are expressed in terms of the first and the third stress invariants:

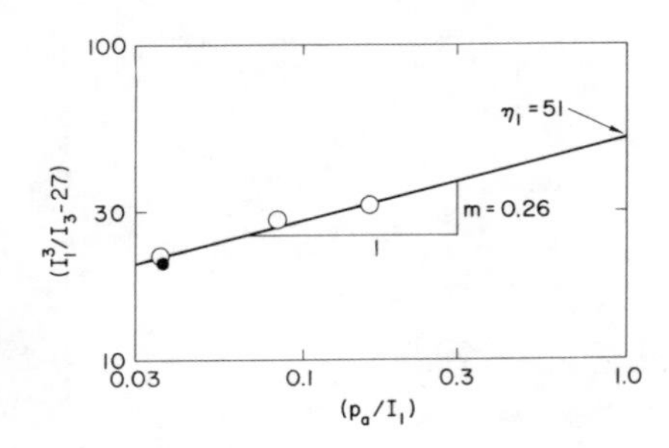

Figure 12.8. Determination of the Values of η_1 and m Involved in Failure Criterion for Antelope Valley sand

$$f_p = (I_1^3/I_3 - 27)\cdot(I_1/p_a)^m \tag{12.14a}$$

$$f_p = \eta_1 \text{ at failure} \tag{12.14b}$$

where I_1 is given by Equation 12.5 and I_3 is given by:

$$\begin{aligned} I_3 &= \sigma_1\cdot\sigma_2\cdot\sigma_3 \\ &= \sigma_x\cdot\sigma_y\cdot\sigma_z + \tau_{xy}\cdot\tau_{yz}\cdot\tau_{zx} + \tau_{yx}\cdot\tau_{zy}\cdot\tau_{xz} \\ &\quad - (\sigma_x\cdot\tau_{yz}\cdot\tau_{zy} + \sigma_y\cdot\tau_{zx}\cdot\tau_{xz} + \sigma_z\cdot\tau_{xy}\cdot\tau_{yx}) \end{aligned} \tag{12.15}$$

In Equations 12.14a and b, the values of η_1 and m are constants to be determined for specific soils at the desired density. For failure conditions the apex angle, indicated on the diagram in Figure 12.2(a), increases with the value of η_1, and the curvature of the failure surface increases with the value of m. The values of η_1 and m can be determined by plotting $(I_1^3/I_3 - 27)$ versus (p_a/I_1) at failure on a log–log diagram as shown in Figure 12.8. On this diagram η_1 is the intercept with $(p_a/I_1) = 1$ and m is the slope of the straight line.

The nonassociated flow rule employed for the plastic expansive strains is characterized by a plastic potential function, g_p, of a form similar to the yield criterion:

$$g_p = I_1^3 - \left(27 + \eta_2\cdot\left(\frac{p_a}{I_1}\right)^m\right)\cdot I_3 \tag{12.16}$$

where η_2 is a constant for given values of f_p and σ_3. The values of η_2 can be determined from the results of triaxial compression tests using the following expression:

$$\eta_2 = \frac{3\cdot(1+\nu^p)\cdot I_1^2 - 27\cdot\sigma_3\cdot(\sigma_1 + \nu^p\cdot\sigma_3)}{\left(\frac{p_a}{I_1}\right)^m\cdot\left[\sigma_3\cdot(\sigma_1+\nu^p\cdot\sigma_3) - \frac{I_3}{I_1}\cdot m\cdot(1+\nu^p)\right]} \tag{12.17}$$

where ν^p is a function of the plastic strain increments determined by substracting the elastic and plastic collapse strains from the total strains:

$$\nu^p = -\frac{\Delta\varepsilon_3^p}{\Delta\varepsilon_1^p} \tag{12.18}$$

The variation of η_2 with f_p and σ_3 is shown in Figure 12.9 for the triaxial compression tests. This variation can be expressed by a simple equation:

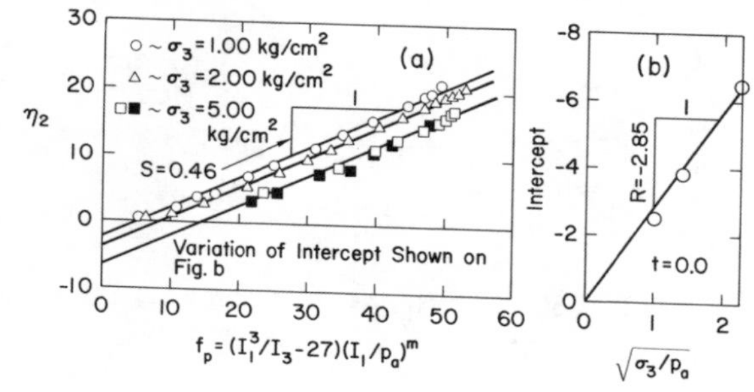

Figure 12.9. Variation of η_2 with f_p for Antelope Valley sand

$$\eta_2 = S \cdot f_p + R \cdot \sqrt{\frac{\sigma_3}{p_a}} + t \tag{12.19}$$

Determination of the values of S, R and t is indicated in Figure 12.9.

The plastic expansive work is calculated from:

$$W_p = \int \{\sigma_{ij}\}^T \{d\varepsilon^p_{ij}\} \tag{12.20}$$

and the variation of W_p with f_p (calculated from Equation 12.14a) is shown in Figure 12.10 for the triaxial compression tests. This variation can be approximated by exponential functions for which the following expresssion is used:

$$f_p = a \cdot e^{-b \cdot W_p} \cdot \left(\frac{W_p}{p_a}\right)^{1/q}, \quad q > 0 \tag{12.21}$$

where the parameters a, b and q are constants for a given value of σ_3. The parameter q can be determined for a constant value of the confining pressure according to:

$$q = \frac{\log\left(\frac{W_{ppeak}}{W_{p60}}\right) - \left(1 - \frac{W_{p60}}{W_{ppeak}}\right) \cdot \log e}{\log\left(\frac{\eta_1}{f_{p60}}\right)} \tag{12.22}$$

where e is the base for natural logarithms, and (W_{ppeak}[4], η_1) and (W_{p60}, f_{p60}) are two sets of corresponding values from the relation between the work input W_p and stress level f_p. These two points correspond to the peak point and the point at 60% of η_1 on the work-hardening part of the $W_p - f_p$ relation, as indicated in Figure 12.10. The variation

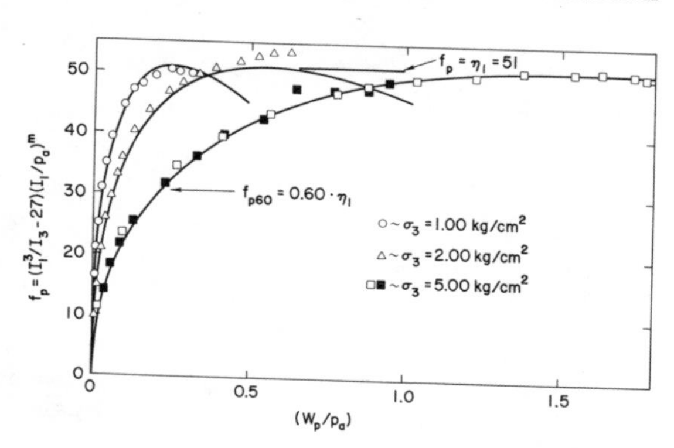

Figure 12.10. Variation of total plastic work with f_p and σ_3 for Antelope Valley sand

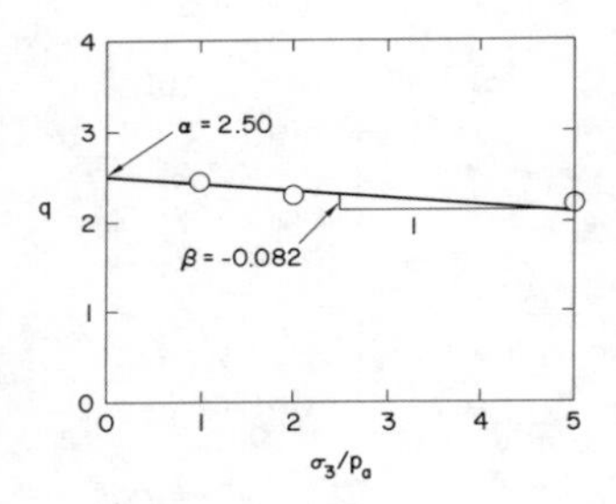

Figure 12.11. Variation of q with confining pressure σ_3 for Antelope Valley sand

of q with confining pressure σ_3 may be modelled by the following simple expression:

$$q = \alpha + \beta \cdot \frac{\sigma_3}{p_a} \tag{12.23}$$

where the values of α and β represent the intercept and the slope of the straight line shown in Figure 12.11.

The parameters a and b in Equation 12.21 are calculated from:

$$a = \eta_1 \cdot \left(\frac{e \cdot p_a}{W_{ppeak}} \right)^{1/q} \tag{12.24}$$

and:

$$b = \frac{1}{q \cdot W_{ppeak}} \tag{12.25}$$

where q is determined from Equation 12.23 and e is the base for natural logarithms. The variation of W_{ppeak} with confining pressure σ_3 can be approximated by the following expression:

$$W_{ppeak} = P \cdot p_a \cdot \left(\frac{\sigma_3}{p_a} \right)^l \tag{12.26}$$

where P and l are constants to be determined as shown in Figure 12.12. In this diagram P is the intercept with $(\sigma_a/p_a) = 1$ and l is the slope of the straight line.

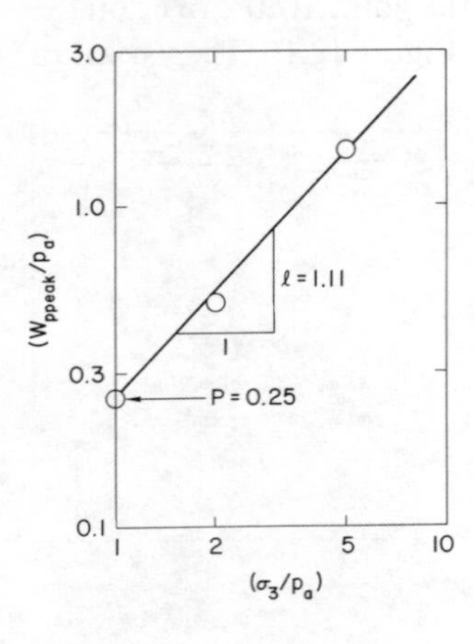

Figure 12.12. Variation of W_{ppeak} with confining pressure σ_3 for Antelope Valley sand

Using the relation between f_p and W_p in Equation 12.21 and the appropriate values of a, b and q as determined above, the solid lines in Figure 12.10 are obtained. These lines are used to represent the experimentally determined work-hardening relationships shown in this figure.

The form of the stress-strain relationships which account for the plastic expansive behaviour of the cohesionless soil can be expressed as:

$$\Delta\varepsilon_{ij}^{p} = \Delta\lambda_p \cdot \frac{\partial g_p}{\partial \sigma_{ij}} \tag{12.27}$$

The derivatives of g_p with regard to the normal stresses become:

$$\frac{\partial g_p}{\partial \sigma_x} = 3 \cdot I_1^2 - \left(27 + \eta_2 \cdot \left(\frac{p_a}{I_1}\right)^m\right) \cdot (\sigma_y \cdot \sigma_z - \tau_{yz}^2) + \frac{I_3}{I_1} \cdot m \cdot \eta_2 \cdot \left(\frac{p_a}{I_1}\right)^m \tag{12.28a}$$

and similar expressions are obtained for the other normal stresses by interchanging the indices on the stresses. The derivatives of g_p with respect to the shear stresses become:

$$\frac{\partial g_p}{\partial \tau_{yz}} = \left(27 + \eta_2 \cdot \left(\frac{p_a}{I_1}\right)^m\right) \cdot (\sigma_x \cdot \tau_{yz} - \tau_{xy} \cdot \tau_{zx}) \tag{12.28b}$$

and similar expressions can be obtained for the other shear stresses by interchanging the indices on the stresses.

The value of the proportionality constant $\Delta\lambda_p$ in Equation 12.27 can be written as:

$$\Delta\lambda_p = \frac{dW_p}{3 \cdot g_p + m \cdot \eta_2 \cdot \left(\frac{p_a}{I_1}\right)^m \cdot I_3} \tag{12.29}$$

where g_p is the plastic potential function and dW_p is the increment in plastic work due to an increase in stress level df_p:

$$dW_p = \frac{df_p}{f_p} \cdot \frac{1}{\left(\frac{1}{q \cdot W_p} - b\right)} \tag{12.30}$$

where f_p is the current value of the stress level.

12.6 SUMMARY OF MATERIAL PARAMETERS

Fourteen soil parameters are incorporated in the elastoplastic stress-strain model for cohesionless soil. The determination of thirteen of these parameters is illustrated for Antelope Valley sand in Figures 12.6–12.12, and Poisson's ratio is 0.2. The parameter

Table 12.1. Summary of parameter values for elastoplastic stress-strain model and Antelope Valley sand #10–20

Parameter	Value	Strain component
Modulus number, K_{ur}	1250	
Modulus exponent, n	0.38	Elastic
Poisson's ratio, ν	0.2	
Collapse modulus, C	0.00036	Plastic collapse
Collapse exponent, p	0.99	
Yield const., η_1	51	
Yield exponent, m	0.26	
Pl. potent. const., R	−2.85	
Pl. potent. const., S	0.46	Plastic
Pl. potent. const., t	0	expansive
Work-hard. const., α	2.50	
Work-hard. const., β	−0.082	
Work-hard. const., P	0.25	
Work-hard. exponent, l	1.11	

values for Antelope Valley sand are summarized in Table 12.1. None of the parameters have dimensions. All dimensions are controlled, where appropriate, by the dimension of the atmospheric pressure, p_a, as e.g. in equation 12.26. The parameters in Table 12.1 may be employed for calculation of strains in Antelope Valley sand for any stress-path involving primary loading, neutral loading, unloading and reloading.

12.7 COMPARISON OF MEASURED AND CALCULATED BEHAVIOUR

Comparison between measured and calculated stress-strain and volume change behaviour for Antelope Valley Sand is shown in Figure 12.13. The points in this figure represent the measured soil behaviour and the solid lines represent the calculations from the model. Most aspects of the soil behaviour measured in the triaxial

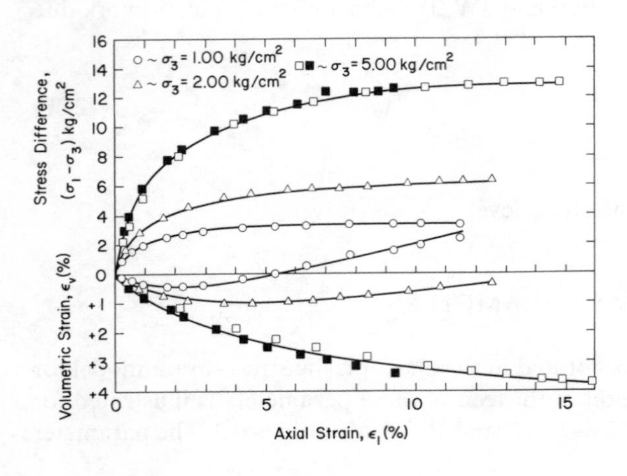

Figure 12.13. Comparison of test data with elastoplastic stress-strain model for Antelope Valley sand

compression tests are reproduced with good accuracy by the constitutive model. Notably, the gradual variation of the volumetric strain behaviour, from being expansive at small confining pressures to being compressive at high confining pressures, is modelled correctly by the elastoplastic stress-strain theory.

Actual predictions of soil behaviour for stress-paths different from those employed in the conventional triaxial compression tests have not been performed. The predictive capabilities were briefly reviewed in Section 12.2.

12.8 SUMMARY AND CONCLUSION

The constitutive model for soils with curved yield surfaces proposed by Lade (1977) has been reviewed, and the procedures used for determination of material parameters have been demonstrated in detail for tests on Antelope Valley sand. Concepts from elasticity and plasticity theory were employed in the development of the stress-strain model which is applicable to general three-dimensional stress-conditions. All material parameters can be determined from an isotropic compression test and three conventional triaxial compression tests. The model in its present form has been shown to be applicable to sand and normally consolidated clay.

REFERENCES

Dafalias, Y.F. & L.R. Herrmann 1980. A bounding surface soil plasticity model. In *Proceedings of International Symposium on Soils under Cyclic and Transient Loading* (Swansea): 335–345. Rotterdam: Balkema.

Drucker, D.C., R.E. Gibson & D.J. Henkel 1957. Soil mechanics and work-hardening theories of plasticity. *ASCE Transactions* 122: 338–346.

Lade, P.V. 1972. The stress-strain and strength characteristics of cohesionless soils. Ph.D. thesis, University of California, Berkeley.

Lade, P.V. 1977. Elasto-plastic stress-strain theory for cohesionless soil with curved yield surfaces. *Int. J. Solids and Structs.* 13: 1019–1035.

Lade, P.V. 1978. Prediction of undrained behavior of sand. *J. Geotech. Engng. Div.* (ASCE) 104 (GT 6): 721–735.

Lade, P.V. 1979. Stress-strain theory for normally consolidated clay. In *Proceedings of 3rd International Conference on Numerical Methods in Geomechanics* (Aachen), 4: 1325–1337.

Lade, P.V. & J.M. Duncan 1973. Cubical triaxial tests on cohesionless soil. *J. Soil Mech. Fdns. Div.* (ASCE) 99 (SM 10): 793–812.

Lade, P.V. & J.M. Duncan 1975. Elasto-plastic stress-strain theory for cohesionless soil. *J. Geotech. Engng. Div.* (ASCE) 101 (GT 10): 1037–1053.

Lade, P.V. & J.M. Duncan 1976. Stress-path dependent behavior of cohesionless soil. *J. Geotech. Engng. Div.* (ASCE) 102 (GT 1): 51–68.

Lade, P.V. & H.M. Musante 1977. Failure conditions in sand and remolded clay. In *Proceedings of 9th International Conference on Soil mechanics and Foundation Engineering* (Tokyo), 1: 181–186.

Lade, P.V. & H.M. Musante 1978. Three-dimensional behavior of remolded clay. *J. Geotech. Engng. Div.* (ASCE) 104 (GT 2): 193–209.

Mroz, Z., V.A. Norris & O.C. Zienkiewicz (1981). An anisotropic, critical state model for soils subject to cyclic loading. *Geotechnique* 31 (4): 451–469.

Prevost, J.-H. 1978. Plasticity theory for soil stress-strain behavior. *J. Engng. Mech. Div.* (ASCE) 104 (EM 5): 1177–1194.

Roscoe, K.H. & J.B. Burland 1968. On the generalized stress-strain behaviour of 'wet' clay. In Heyman & Leckie (eds.), *Engineering plasticity*: 535–609. Cambridge (UK): Cambridge University Press.

Sandler, I.S., F.L. DiMaggio & G.Y. Baladi 1976. Generalized models for geological materials. *J. Geotech. Engng. Div.* (ASCE) 102 (GT 7): 683–699.

On elastoplastic-viscoplastic constitutive modelling of cohesive soils

13

Y.F. DAFALIAS

13.1 INTRODUCTION

The constitutive modelling of engineering materials has been the objective of numerous works during the recent years, primarily because of the increasing complexity of the loading conditions to which modern structures are subjected and the corresponding need for more accurate analysis and response predictions. The parallel development of more powerful and efficient numerical methods of analysis has motivated and allowed the use of sophisticated constitutive models beyond the linear or nonlinear elastic constitutive laws which were utilized in the early stages of continuum mechanics. Particularly in the field of soil mechanics and geotechnical engineering the need of such sophisticated models became a necessity recognizing the highly nonlinear and inelastic character of the soil as a continuum, and prompted a tremendous progress during the last three decades. As a matter of fact it is rather surprising that such progress lagged behind the corresponding development of inelastic constitutive laws for metals, since soils in general exhibit a pronounced inelastic response at much lower stress and strain levels than metals. Apparently one of the reasons is that the traditionally empirical character of the field has delayed the interest of the scientific community towards the area of soil inelasticity, although the very first concept of plasticity has been introduced by the Coulomb's failure criterion for particulate media.

In this chapter we will focus attention on the most recent trends in developing comprehensive constitutive models for cohesive soils within the general framework of elastoplasticity/viscoplasticity. The presentation will be rather descriptive emphasizing the concepts more than the exact analytical formulation, but proper reference to the corresponding works will allow the interested reader to obtain further information. However, this is not intended to be a comprehensive literature review of the subject, a goal which is prohibitive here in view of the enormous number of related works, but it will refer only to some representative papers and ideas closely related in scope and methodology to the author's own work on the subject. In particular the relative merits of different aspects and features of the constitutive models will be discussed within the ever existing dilemma of simplicity versus sophistication which underlines any effort to model physical reality.

Closing the introduction it is pertinent to make an etymological remark. The most representative cohesive soil is the clay and it is interesting to observe that the very origin of the word 'plasticity' is based on the ancient Greek word for clay, 'pelos', and the corresponding derivative word 'plasticotis', which indicates the property of clay to sustain permanent deformations. In the immediately following sections an attempt will be made to qualitatively interpret first the plastic properties of clays from a microstructural point of view.

13.2 GENERAL MACRO- AND MICROSTRUCTURAL CONSIDERATIONS

It is a fact that the macroscopically observed material response described primarily in terms of stress-strain-temperature relations, is the result of microstructural processes. The beauty of continuum mechanics is that it can bypass the differences of the microstructure of different materials and classify their response characteristics within general classes of constitutive laws. For example, it is well known that the microstructure of metals and soils is drastically different, nevertheless, both materials exhibit response features which can be described within the general framework of elastoviscoplasticity. The key word for their common classification is the incremental irreversibility of the deformation for a proper increment and/or range of stress. This does not mean, however, that the particular constitutive models will be identical or even similar. To the contrary, they will exhibit very different properties, such as the dependence or not of the yield criterion on hydrostatic pressure, which will adapt to the particular features of the macroscopic response even if they belong to the same general class of constitutive models. It is there where the microstructural considerations can help us understand and analytically describe better what we observe in macroscopic experimental data. In many occasions the microstructural considerations can even provide us with a quantitative relation between micro- and macromodelling, as for example in the case of sands where the micromechanics of the grain aggregates can be studied quite accurately. Even then though, one has to make a large number of simplifying assumptions which may be equivalent to a more direct and equally realistic macroscopic description.

Such description can be achieved by defining the state of the material in terms of external variables identified here as the stress components σ_{ij}, neglecting for simplicity the explicit inclusion of the temperature which could be considered as a seventh component of a generalized stress-temperature set of external variables (for a strain-space formulation one can consider the strains ε_{ij} instead of stresses), and n scalar or tensor internal variables q_n which embody the past loading history. The n is an identification and not a tensorial index. The q_n include in general proper measures of inelastic deformation, but can also represent other quantities associated with back-stresses, diffusion time-dependent processes etc. Given the state and the first rate $\dot{\sigma}_{ij}$ (a dot indicates material time derivative), the material constitutive relations will be described by proper analytical expressions for the elastic strain rate and the $\dot{q}_n$. One can assume the additive decomposition $\dot{q}_n = \dot{q}_n^v + \dot{q}_n^p$ into a delayed or viscoplastic part $\dot{q}_n^v$ and an instantaneous or plastic part $\dot{q}_n^p$. The $\dot{q}_n^v$ depend only on the state, while the $\dot{q}_n^p$ depend on the state and the rates $\dot{\sigma}_{ij}$, $\dot{q}_n^v$. Although it may seem redundant to write that $\dot{q}_n^p$ depend on both the state and the $\dot{q}_n^v$ since the latter depend on the state, it is done in order to emphasize the coupling that exists in general between plastic and viscoplastic processes in the sense that in addition to $\dot{\sigma}_{ij}$ the $\dot{q}_n^v$ may effect the plastic loading process. Stated equivalently, one may have $\dot{q}_n^p \neq 0$ even if $\dot{\sigma}_{ij} = 0$ due to the existence of $\dot{q}_n^v \neq 0$.

The existence of $\dot{q}_n^v$ renders the response rate- and time-dependent, and if only the $\dot{q}_n^p$ exist and are homogeneous of order one in $\dot{\sigma}_{ij}$, the response is rate-independent.

The choice of using a rate-dependent or rate-independent formulation depends on the material and the loading conditions and can be decided on the basis of the corresponding experimental data. It is very useful however to give at least a qualitative interpretation of these data in terms of the microstructure, because this will help us subsequently to incorporate into the constitutive model meaningful internal variables q_n and will also enhance our confidence on the physical basis of the chosen formulation. For an interpretation of the macroscopic response, one must define first the microelement. The macroscopic constitutive rate equations developed by means of the q_n, refer to the response of the macroelement which by definition consists of many microelements and their ambient space if there exists, but is still small enough for the continuum mechanics field equations approach to apply.

13.3 ON THE MICROMECHANICS OF COHESIVE SOILS

In this section it will be attempted to show qualitatively that on the basis of microstructural considerations the macroscopic response of cohesive soils, clays in particular, must be described within the framework of combined elastoplasticity/viscoplasticity. The clay microstructure is much more complex than that of sand, consisting of particles developing electrophysicochemical bonds among themselves and the surrounding water, without being necessarily always in direct interparticle contact. In addition, natural clays – more than artificial (kaoline) – tend to form clusters of particles which are bonded together in a stronger way than a cluster is bonded to another cluster. In such a case the definition of the microelement presents a dilemma: is it a clay particle or a cluster? The discussion in the remaining of this section will offer a proper answer. The microelement is an aggregate of particles or clusters and the surrounding voids. Upon a stress application the rather pronounced instantaneous elastic deformation of the macroelement is due to the actual elastic deformation and/or relative movement of the clay particles or clusters without slipping at the contact points. Simultaneously, and referring always to a sufficiently large effective stress application, a corresponding instantaneous rearrangement of clusters or particles (if no clusters exist) occurs which can be modelled by rate-independent plastic strains. This instantaneous plastic response is more pronounced for flocculated than dispersed clay structures, since it represents the breaking or disruption of interparticle contacts and the corresponding particle rotations towards positions parallel to the direction of shear. With respect to the word instantaneous, let us observe here that in an actual homogeneous drained experiment the waiting time period for the pore-water pressure to become zero everywhere before deformations are measured, is simply a necessity associated with the fact that drainage occurs from the boundaries of a finite sample. In other words, if it was possible to establish drainage at all the macroelements of the sample, then the above deformation would appear to occur instantaneously in actual time. This deformation process is associated with what is known as primary consolidation (with or without deviatoric stress).

The deformation, however, does not stop here. For definiteness, let us first consider the case where no clusters form and the microelement is the particle. One can distinguish two additional rate- and time-dependent processes. The first consists of a delayed deformation of a viscous nature associated mainly with dispersed structures, as clay particles shear with respect to each other at a rate which depends on the applied stress level and other ambient environmental factors. The second process is not directly related to a viscous shear deformation but it results also in a delayed deformation. More precisely, due to the motion of ions, water molecules and other slow diffusion processes of a physicochemical nature, the partial equilibrium achieved during the first phase of instantaneous deformation among the interparticle contact forces is slowly changing. Thus, further rearrangement of the microelements occurs with time, resulting into a delayed deformation of the macroelement under constant effective stress (drained conditions) since the total stress is kept constant and the pore-water pressure is almost zero. For further details the reader is referred to the comprehensive description by Scott (1963). It is very difficult to separate the contributions of the above two processes and in the present context they can be modelled in a unified way by the evolution of proper q_n within the framework of viscoplastic rate equations. The presence of delayed deformation mechanisms is responsible for the rate dependence of the material response, especially because of the first process of viscous shear, and also for the phenomenon of secondary consolidation. Although both instantaneous (elastoplastic) and delayed (viscoplastic) deformations occur simultaneously, the development rate of the latter is usually slow enough to exhibit measurable effects only after considerable time has elapsed, especially with respect to secondary compression, thus offering the means for the calibration of the parameters entering the rate equations for the corresponding q_n. It must be emphasized, however, that the presence of instantaneous plastic deformations acts as an upper bound on the increase of stress necessary to sustain increasing rates of deformation. That is, if one assumed that no instantaneous plastic deformations occur, it would be necessary to increase indefinitely the stress as the deformation rate increases contrary to experimental evidence. The existence of instantaneous plastic deformation on the other hand will bear almost the full responsibility for the description of very rapid deformations when the delayed phenomena do not have the time to exhibit themselves and bulk portions of material may even rupture with respect to each other. This can be called the phenomenon of rate-dependence saturation, and shows that an accurate and general description of the clay response cannot be realistically described by means of elastoplasticity alone or elastoviscoplasticity alone, but one must work within the general framework of combined elastoplasticity/viscoplasticity.

The delayed deformation becomes much more pronounced if clusters form. In this case it is not only the above viscous and physicochemical processes which effect the intercluster contact forces and cause cluster rearrangement with time, but in addition the clusters themselves deform further with time because of a very simple mechanical process: due to the closer packing and stronger bonding of the particles in a cluster, the intracluster pore water oozes out in the intercluster pore space at a much slower rate than the intercluster water drains during primary consolidation. During this process it is not so much the cluster deformation as such which reflects into the delayed strain of the macroelement consisting of many clusters. This is because the transfer of water from intracluster to intercluster pore space cannot cause alone a volumetric strain of the macroelement; whatever is the reduction of the clusters' volume, is balanced exactly by the increase of the intercluster voids by the above mentioned transfer of incompressible water, recalling that voids are part of the macroelement by definition. It is again the corresponding perturbation of equilibrium of the contact forces, resulting from the relatively small cluster deformation, which facilitates further clusters' rearrangement and, therefore, further and more pronounced delayed deformation of the macroelement. Observe that during this process of secondary consolidation the perturbed intercluster contact forces restore continuously their average value as clusters repack, and pore-water pressure remains zero, thus the effective stress remains constant. In fact one can argue that there are two kinds of effective stresses: the intercluster effective stress obtained by reckoning the intercluster contact forces over the total area, and the intracluster effective stress associated with the particles in the clusters. The former is applied instantaneously among the clusters upon an external effective stress application on the boundary of the macroelement and remains essentially constant with time, during which the latter increases and tends towards a certain value. The dilemma of choosing the proper effective stress is identical to the dilemma of choosing the proper microelement in the presence of cluster formations.

In the previous discussion it appears preferable to choose the cluster as the microelement and the intercluster effective stress as governing the elastic and inelastic response. The intracluster effective stress can be considered indirectly as an internal variable whose evolution due to the mechanical process of water drainage from intra- to inter-cluster voids is related to the delayed viscoplastic deformation for constant intercluster effective stress, in addition to the physicochemical internal processes contributing to it. In this way one can suggest the common framework of combined elastoplasticity/viscoplasticity for the description of the macroscopic constitutive relations of clays, with or without cluster formations.

13.4 RATE-INDEPENDENT ELASTOPLASTICITY FOR COHESIVE SOILS

It is clear from the previous discussion that elastoplastic (instantaneous) and viscoplastic (delayed) deformations occur simultaneously in cohesive soils, but also it is clear that the development rate of the latter is sufficiently slow to permit a realistic first approximation of uncoupling the two processes for relatively rapid loading conditions under which the elastoplastic response dominates. Thus, the general framework of rate-independent elastoplasticity will be utilized subsequently in order to present different aspects of the constitutive modelling of cohesive soils, before the viscoplastic contribution and its coupling with the elastoplastic part are considered towards the end of this chapter. The summation convention will be employed over all repeated tensorial indices, all stresses are effective and compressive strains and stresses are considered positive.

13.4.1 *The elastic response*

Assuming the additive strain rate decomposition $\dot{\varepsilon}_{ij} = \dot{\varepsilon}^{e}_{ij} + \dot{\varepsilon}^{p}_{ij} + \dot{\varepsilon}^{v}_{ij}$ with the superscripts e, p and v denoting the elastic, plastic and viscoplastic parts respectively, the elastic response is given by:

$$\dot{\varepsilon}^{e}_{ij} = C_{ijkl}\dot{\sigma}_{kl}, \quad \dot{\sigma}_{ij} = E_{ijkl}\dot{\varepsilon}^{e}_{kl} \tag{13.1}$$

where the fourth order tensors C_{ijkl} and E_{ijkl} are the elastic compliances and moduli functions of the state in general. Requirements of energy conservation in a purely elastic response impose restrictions on these tensors such that the elastic strain and the stress can be derived from proper elastic potentials. For example, for elastically isotropic soils the elastic constants can be expressed by means of the bulk modulus K and the shear modulus G; it has been established that the former depends on the hydrostatic pressure and often it is assumed that so does G (by assuming a constant Poisson's ratio) with the complete exclusion of dependence on deviatoric stresses. This clearly violates the requirements for the existence of elastic potentials.

13.4.2 *The classical yield surface concept and its shortcomings*

In reference to subsequent notation when two juxtaposed tensors with subscripts n appear in a term which is a scalar, the repetition and summation of the corresponding trace operation over all n are implied. Considering only the plastic part q_n^p of the internal variables (without the viscoplastic part q_n^v), the classical rate-independent plastic response with a smooth yield surface can be expressed by the following set of equations:

Yield surface:
$$f(\sigma_{ij}, q_n^p) = 0 \tag{13.2}$$

Loading index:
$$L = \frac{1}{K_p}\frac{\partial f}{\partial \sigma_{ij}}\dot{\sigma}_{ij} \tag{13.3}$$

Rate equations:
$$\dot{\varepsilon}_{ij}^p = \langle L \rangle R_{ij}^p \tag{13.4a}$$

$$\dot{q}_n^p = \langle L \rangle r_n^p \tag{13.4b}$$

Consistency condition:
$$\dot{f} = 0 \Rightarrow K_p = -\frac{\partial f}{\partial q_n} r_n^p \tag{13.5}$$

where the brackets $\langle\,\rangle$ express the operation $\langle A \rangle = A$ if $A > 0$ and $\langle A \rangle = 0$ if $A \leq 0$, R_{ij}^p and r_n^p are functions of the state only (for normality $R_{ij}^p = (\partial f/\partial \sigma_{ij})$), and the plastic modulus K_p enters the definition of in order to consider both stable ($K_p > 0$) and unstable ($K_p \leq 0$) material response (Dafalias 1979a, 1981a; Dafalias & Herrmann 1980). The q_n^p include in general the ε_{ij}^p and the separate Equation 13.4a for the latter was written only for clarity. Not all the q_n^p must necessarily enter Equations 13.2 and 13.5. Soil experiments are usually performed under triaxial loading conditions and the corresponding yield surfaces can be visualized in the classical triaxial p–q stress space with a direct generalization to the space of two or three stress invariants for isotropic response. Based on the Coulomb's failure criterion Drucker & Prager (1952) extended the formulation to two stress invariants and Drucker et al. (1957) proposed the concept of a cap to control dilation, thus, formally introducing yielding in addition to failure. Cap models have been further refined (DiMaggio & Sandler 1971) and along a parallel but slightly different line the critical state soil mechanics (Roscoe et al. 1963, Schofield & Wroth 1968) captured even better the cohesive soil response and is today a basic frame of reference for soil plasticity, especially for clays. The shapes of the different $f = 0$ vary from spherical to elliptic, logarithmic etc. curves, but the most important common feature is that the only q_n^p entering the corresponding yield surface Equation 13.2 and expressing the hardening, is a measure of the plastic volumetric change either by means

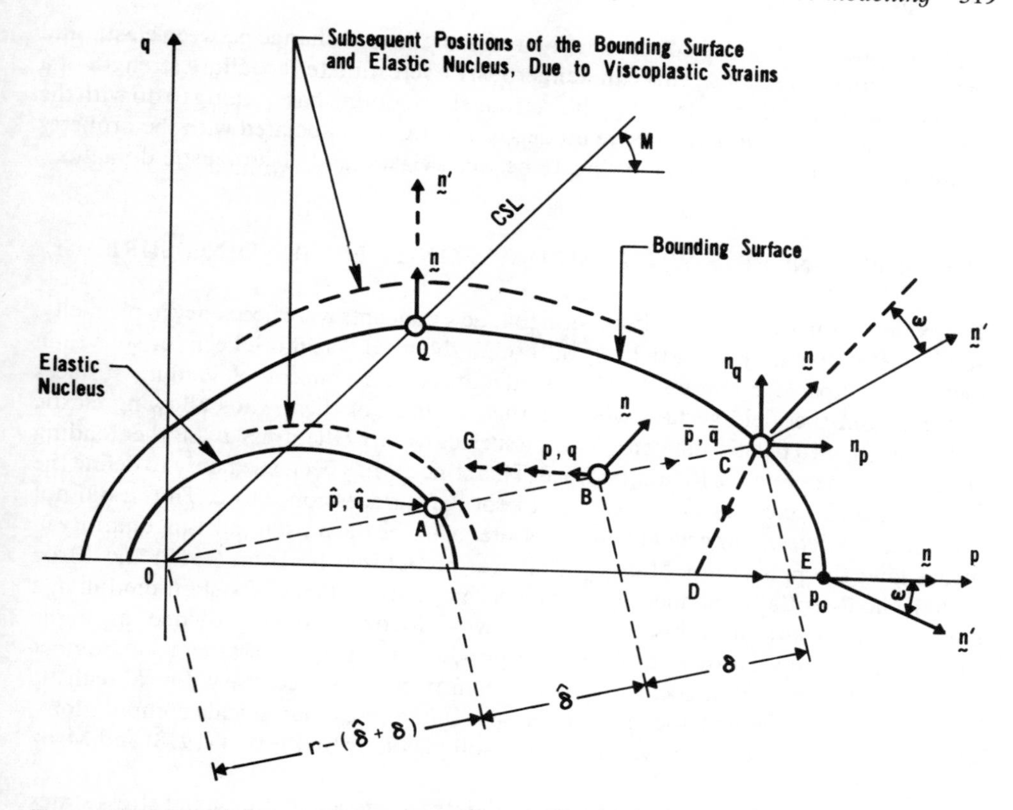

Figure 13.1. Schematic illustration in the triaxial stress space of the concepts of bounding surface, elastic nucleus, overstress, viscoplastic response and anisotropy by a varying non-associated flow rule

of strains or plastic void ratio change. A typical shape is shown in Figure 13.1 for the p–q space, with CSL representing the projection of the critical state line of slope M and where the reader should for the time being consider as yield surface what has been named bounding surface in this figure. The plastic volumetric hardening assumption in conjunction with Equation 13.5 and the normality condition yields a variable K_p on $f=0$ such that $K_p=0$ for point Q (intersection with CSL), and $K_p>0$, $K_p<0$ for the parts to the right and left of Q, respectively. The $K_p=0$ implies critical failure with zero volumetric and infinite deviatoric plastic strain rate while $K_p>0$, $K_p<0$ imply consolidation and dilation, respectively. All the above apply to the generalized representation in the space of two and three isotropic stress invariants.

While the soil response can realistically be described when the stress lies on $f=0$, a serious shortcoming appears for overconsolidated states. The response would be purely elastic until the stress reaches again the yield surface, contrary to experimental evidence that clearly shows a considerable amount of plastic deformation at states within $f=0$. Although this deviation could be approximately neglected for a monotonic loading until failure, it has grave consequences under cyclic loading conditions where most important is the accumulated effect of the repeated loading/unloading/reloading on the progressive pore-water pressure built-up (positive or negative) under undrained conditions as a result of the interchange between elastic and plastic volumetric strains. This can dangerously overestimate the failure strength of a cyclically loaded soil. It is clear that this serious shortcoming has nothing to do with the particular formulation of the above models, but is rather associated with the property of the yield surface to sharply delineate between elastic and elastoplastic domains.

13.5 NEW CONCEPTS WITH EMPHASIS ON THE BOUNDING SURFACE

It is evident from the previous discussion that new concepts were necessary in plasticity. Again metal plasticity leads the way, although the need was more keen for soils than metals. Phillips & Sierakowski (1965) introduced the concept of loading surfaces distinct from the yield surface which is the 'smallest' of them, thus allowing elastic unloading from a plastic state on a loading surface to carry the stress at another loading state within the previous loading surface. This concept has been used only to define the yield points in a more accurate way, and kept the surfaces concentric. Thus it did not offer much in determining what it appears after all to be the most important quantity in a rate law: the plastic modulus K_p. This was recognized by Iwan (1967) and Mroz (1967) who essentially spanned the stress space by surface loci of constant moduli in a continuous (Iwan) or a discrete (Mroz) way. Mroz's model provided a simple kinematic/isotropic hardening of the corresponding surfaces, but a rather large number of such surfaces and associated internal parameters is necessary for a realistic description which renders the model expensive in the case of large scale computations. Mroz's multisurface concept was applied to soil plasticity by Prevost (1978) and Mroz et al. (1978).

Generalizing uniaxial cyclic experimental data for metals to multiaxial stress states and motivated by the concept of hardening saturation from dislocation mechanics, Dafalias & Popov (1974, 1975, 1976) have introduced the concept of the bounding surface in stress space, also proposed independently by Krieg (1975). The initial version, most appropriate for metals, considered simultaneously the bounding surface and an enclosed yield surface, which in the limit could 'shrink' into a point for vanishing elastic range (Dafalias 1979a, Dafalias & Popov 1977), both surfaces hardening in a coupled way. The essence of the bounding surface concept and general formulation was more abstractly presented by Dafalias (1981a) in a way that allows the development of different previous and future versions within the same general framework. Following this reference, for a given state σ_{ij}, q_n^p the bounding surface is defined in stress space analytically by:

$$F(\bar{\sigma}_{ij}, q_n^p) = 0 \tag{13.6}$$

where the functional form of $F=0$ can be similar to the form of $f=0$. Bars over stress quantities indicate points on $F=0$, and the actual stress lies in or on the $F=0$. A properly defined mapping rule associates a unique 'image' stress $\bar{\sigma}_{ij}$ to σ_{ij}, such that the gradient $\underset{\sim}{\nabla} F$ or the corresponding unit normal $\underset{\sim}{n}$ at $\bar{\sigma}_{ij}$ defines the loading/unloading direction at σ_{ij}. This mapping rule must satisfy the integrability and identity conditions (Dafalias 1981a) guaranteeing the existence of a loading surface $f=0$ passing through σ_{ij} and preventing the outward crossing of the bounding surface by the stress. Let us emphasize that the use of a two-surface model (yield and bounding) is not but one way to express the mapping rule by defining a $\bar{\sigma}_{ij}$ with the same $\underset{\sim}{n}$ on $F = 0$ as the $\underset{\sim}{n}$ at σ_{ij} on $f = 0$. Another version is based on the 'radial mapping' rule (Dafalias 1979b, 1981a and b) and illustrated for the triaxial stress space in Figure 13.1. Using a proper stress point inside $F = 0$ as a projection centre, in this case the origin, the actual stress point **B** is projected radially onto the bounding surface at point C, thus, defining the 'image' stress components $\bar{p}, \bar{q}$ for given components p, q. The projection centre does not have to be the origin and following the formulation for metals, Dafalias (1982a) recently introduced a variable projection centre on the p-axis different than the origin for the purpose of describing better the response for large OCR. This essentially corresponds to the concept of a hydrostatic 'back-stress'.

The essence of the bounding surface concept is now the proposition that the plastic modulus K_p at σ_{ij} (p, q in Figure 13.1) is related to a bounding plastic modulus $\bar{K}_p$ at $\bar{\sigma}_{ij}$ ($\bar{p}, \bar{q}$ in Figure 13.1) by means of a proper function of the Euclidean distance δ between σ_{ij} and $\bar{\sigma}_{ij}$ such that $K_p \geq \bar{K}_p$ for $\delta \geq 0$. The $\bar{K}_p$ is defined by the consistency condition $\dot{F} = 0$ in a way analogous to Equation 13.5, which necessitates the expression of the loading index also in terms of $\dot{\bar{\sigma}}_{ij}$ and $\bar{K}_p$ (Dafalias 1981a). The rate Equations 13.3 and 13.4 remain the same with $(\partial F/\partial \bar{\sigma}_{ij})$ substituting $(\partial f/\partial \sigma_{ij})$. Thus the plastic states are not restricted only to those lying on $F = 0$ as in classical yield surface plasticity, and this is the great advantage of the bounding surface concept. As an example of a recently introduced form of the K_p, $\bar{K}_p$, δ relation (Dafalias 1981a and b) we present:

$$K_p = \bar{K}_p + \hat{H}\frac{\delta}{\langle r - s\delta \rangle} \tag{13.7}$$

where $\hat{H}$ is a proper scalar function of the state (often a constant), r is the Euclidean distance between the 'image' stress and the projection centre (Figure 13.1), and $s \geq 1$ is the elastic factor. This name is given to s because for $\delta \geq r/s$ the bracketed quantity in Equation 13.7 becomes negative and according to the definition of the operation $\langle \ \rangle$ Equation 13.7 yields a $K_p = \infty$ and, therefore, defines indirectly a range of purely elastic response called elastic nucleus whose boundary is shown in Figure 13.1. Observe that the boundary of the elastic nucleus is not identical to a yield surface because a stress point moving from the inside of the nucleus can cross its boundary with a smooth elastoplastic transition according to Equation 13.7 and move outside (no consistency condition necessary). For any given state the corresponding $(\delta/r) =$ constant defines a loading surface $f = 0$ homothetic to $F = 0$ with the projection centre acting as the centre of homothecy. A very interesting property of Equation 13.7 and similar forms, is that $\bar{K}_p$ may be negative indicating a contraction of the bounding surface for states on the dry side of the critical state, but K_p can still be positive for sufficiently large value of the ratio δ/r. As this ratio diminishes with the stress point approaching the contracting surface, the K_p will become eventually zero and subsequently negative and again zero on the CSL in a smooth way, describing the important phenomenon of an initially rising deviatoric stress-strain curve with the stress crossing the CSL before it turns and falls back on the CSL (unstable softening response) where critical failure occurs ($K_p = \bar{K}_p = 0$).

The two-surface version of the bounding surface formulation has been used in soil plasticity for the triaxial space by Mroz et al. (1979), the zero-elastic range version for the triaxial space by Dafalias (1979a), and the radial mapping version for the triaxial, two and three stress invariants spaces in a series of papers by Dafalias (1979b) and Dafalias & Herrmann (1980, 1982a and b). In the latter references the details of the material parameters calibration procedures and the successful comparison with experimental data for cyclic and monotonic loadings at different OCR in compression and extension are well established. The calibration and predictions for raw experimental data of natural clays can be found also in Dafalias et al. (1981). Along parallel lines Hashiguchi & Ueno (1977) and Hashiguchi (1980, 1981) have developed the concept of subyield states within the yield surface with a smooth elastoplastic transition. Dafalias (1981c) has shown that in principle their formulation falls within the general framework of the earlier developed bounding surface plasticity. The particular radial mapping version with the stress origin as projection centre, however, appears to have been formulated in a different format first by Hashiguchi & Ueno (1977). The radial mapping version is most appropriate for soils where the concept of subyield states, which essentially is that of the loading surfaces of Phillips & Sierakowski (1965), is well established. The bounding/yield surface version (two-surface formulation) is better fitted to describe metal response (Dafalias & Popov 1976) with a proper updating of discrete memory parameters, because the existence of strongly pronounced subyield states (lowering of yield at reloading after partial unloading) is definitely against experimental evidence contrary to the statements made by Hashiguchi (1981).

13.6 OTHER ASPECTS OF COHESIVE SOIL PLASTICITY

13.6.1 *Undrained loading*

In many of the earlier attempts to present constitutive models for soils the loading under undrained conditions was treated by a separate model than that of drained loading, contrary to the basic principle of a unified constitutive law. The concept of effective stress in conjunction with the additive elastic and plastic strain rate decomposition, offers the means to obtain the corresponding formulation under undrained conditions by merely imposing the internal constraint of zero total volumetric strain rate, assuming that the solid and fluid phases are incompressible for fully saturated samples. Thus, the resulting mechanism of interchange between elastic and plastic volumetric strain rates yields the corresponding equations under undrained conditions (Schofield & Wroth 1968), without referring to empirical parameters describing the pore-water pressure change. Of particular interest for the calibration of the model parameters is the resulting differential equation of the undrained stress path (and its integration) which was given recently by Dafalias & Herrmann (1982b) for the most general case in the space of three stress invariants for isotropic soils. For numerical applications, however, it is convenient to have a common formulation for drained and undrained conditions and this can be accomplished if the slight compressibility of the soil particles and pore water is recognized for undrained conditions and treated by means of a very large bulk modulus Γ in association with the drained formulation and the techniques for the analysis of nearly incompressible solids (Herrmann et al. 1982). For drained conditions and referring to final effective stress application one has $\Gamma = 0$, while partial saturation without global water movement can be treated by a variable Γ as a function of the total void ratio. For actual global flow of water it is necessary to perform a coupled flow-stress analysis for which the plastic constitutive relations refer to effective stresses.

13.6.2 *Flow rules*

While the inelastic deformation process of cohesionless soils is clearly frictional, the corresponding deformation process of cohesive soils is far from being so (Scott 1963). Thus, non-associated flow rules were properly and necessarily applied to the plasticity of sands, but no such clear conclusion applies to clay plasticity. The early critical state models use associated flow rules in general, and the deviation of material response predictions from experimental data, especially for heavy overconsolidation, prompted many authors to introduce non-associated flow rules for clays by means of proper plastic potentials (Banerjee & Stipho 1978). It is possible, however, to obtain very good agreement with experimental data using associated flow rules if one relaxes the requirement of the original critical state soil plasticity versions to have yield surfaces of a specific shape (logarithmic, elliptic with fixed ratio of major axes etc.). Dafalias & Herrmann (1980) for example, have introduced a bounding surface (or yield surface for classical plasticity) consisting of two ellipses and one hyperbola and have obtained excellent agreement with experiments using the associated flow rule for the full range of OCR values. It may be argued that such arbitrary definition of the yield surface shape is heuristic, contrary to the original critical state derivation of the shape on the basis of certain work dissipation assumptions. This last derivation is based on assumptions nevertheless, and it appears rather subjective what one can call more or less fundamental assumption in a continuum mechanics phenomenological approach, especially in view of better description of the material response. This is not to be confused, however, with the rigorousness of the general continuum mechanics formulation irrespective of the motivation to introduce the one or the other assumption.

13.6.3 *Instability and vertex formation*

The instability associated with bifurcation problems and shear band formations occupies a very extensive area of previous and recent research in the mechanics of solids. Setting aside the effect of rate dependence, geometrical effects and material imperfections, instability may arise because of two properties of a constitutive law: a falling stress-strain curve (negative plastic modulus) and the existence of a local vertex or a global corner in the geometry of the loading surface. The first property can be realistically described by means of the bounding surface model (recall comments following Equation 13.7). A classical yield surface will give a rather unrealistic abrupt transition from a rising to a falling stress-strain curve. The most important question, however, is whether the experimentally observed falling stress-strain curve is due to material instability or is an artifact of the experimental set-up due to the inhomogeneity of the sample and the formation of shear bands which may generate a pseudo-softening response of the sample as a whole. Without covering the extensive literature on the subject, it appears that the truth lies somewhere in between. Such softening response can be observed for rather homogeneous samples of clay before shear bands form to an appreciable degree. They can be explained by means of energy release occurring as interparticle bonds break during dilation of heavily overconsolidated samples loaded in triaxial compression or extension. The exact micromechanics of such a phenomenon is very complex involving the electrophysicochemical forces between particles in flocculated systems in particular, and the reader is referred to Scott (1963) for further information. A constitutive model must possess sufficient flexibility in describing the more or less pronounced effects of such unstable behaviour, as the experimental interpretation should dictate, by means of calibrating appropriate parameters. This can be achieved in the bounding surface model by controlling the proximity of the surface to the CSL for the dry side (Dafalias & Herrmann 1980).

With respect to the vertex or corner formations no decisive conclusion exists for cohesive soils. In case of two intersecting yield surfaces such corner formation arises naturally although it largely complicates the corresponding equations in view of the coupled hardening of both surfaces. The introduction of a local vertex along the lines presented by Rudnicki & Rice (1975) appears to be straightforward, although not much has been done to study the subsequent increase or decrease of the sharpness of such vertex, or the process of reappearance of such vertex under random loading involving unloading/reloading at different points on the yield surface. Recently Dafalias (1982b) has introduced a Rudnicki-Rice vertex in the bounding surface formulation in general terms allowing the description of such vertex evolution, but not sufficient investigation has been carried out so far.

13.6.4 *Anisotropy*

Anisotropy of cohesive soils occurs due to the preferred orientation of particles and the development of residual stresses. Inclusion of anisotropy in a constitutive model presents a two-fold difficulty: proper formulation which must not be prohibitively complex, and calibration of the corresponding additional model parameters. One thing that must be emphasized is that clays are soft materials and, therefore, develop anisotropic properties in the course of plastic deformation as easily as they may loose them. Thus, inclusion of initial anisotropy alone without providing the proper formulation for its evolution and possibly its eventual demise, is only slightly more satisfactory than not including it at all. Another point of interest is that preferred microstructural orientations cause both elastic and plastic anisotropy. For small strains, the consideration of elastic anisotropy is important especially in relation to elastic wave-propagation problems. For large strains, plasticity is predominant and elastic anisotropy can be neglected. Plastic anisotropy will be the focus of our subsequent discussion.

The evolution of anisotropy is described by means of the q_n^p entering Equations 13.2–13.5. Assuming that R_{ij}^p, r_n^p are related to f (e.g. normality), anisotropy can be defined by means of the functional form of f. For given initial material symmetries with respect to a reference configuration, the general conclusion is that f must be a function of the direct and joint invariants of σ_{ij}, q_n^p under the group of orthogonal coordinate transformations which are associated with the initial material symmetries. This does not imply necessarily that the material will retain the same symmetries with respect to subsequent strained reference configurations due to the evolution of q_n^p (Dafalias 1979c). Initial anisotropy alone without evolution can be obtained by assuming f a function of proper invariants of σ_{ij} and the direct isotropic invariants of q_n^p. Such is the method of stress pseudo-invariants (Baladi 1978), using instead of isotropic stress invariants equivalent stress quantities which remain invariants under specific rotations of the reference configuration.

The evolution of anisotropy for clays has been modelled by a number of different methods within the above general framework. Residual-stress anisotropy was modelled by kinematic hardening (Prevost 1978, Mroz et al. 1978, Dafalias 1979a), although such hardening was restricted to the inner surface(s) of multi- or two-surface models. In such a case initial anisotropy is defined by means of the initial values of the kinematic hardening parameters. Textural evolution of anisotropy can be modelled by rotation and/or translation of the yield surface (or the outer surface for several surfaces models) as proposed by Mroz et al. (1979) and Hashiguchi (1979). However, it is not clear to which extent such description can adequately control the change of the p-q slope at critical failure since the yield surface depends now also on the deviatoric plastic strains, a property which renders $K_p \neq 0$ at the critical state Q (Figure 13.1), therefore no critical failure occurs there. The simplification to consider the $f=0$ permanently rotated without any dependence on the deviatoric strain, falls within the category of permanent initial anisotropy (Prevost 1978) without evolution.

It was the above difficulty that prompted Dafalias (1981d) to introduce the concept of a varying non-associated flow rule as an alternative simpler but not the most general way to consider anisotropy. The concept was motivated by the experiments of Lewin (1978) and is illustrated Figure 13.1. It was observed that after the loading path BCD, a negative deviatoric strain appeared along the path DE which implies that the plastic strain increment at E for example, is directed along $\underset{\sim}{n}'$, a direction deviating by an angle ω from the normal $\underset{\sim}{n}$ at E along which it should be directed if the soil was isotropic. Thus the degree of deviation from normality, represented by a variable angle ω in triaxial space under the assumption of an isotropic yield (or bounding) surface, can be considered as an indirect measure of anisotropy. Normality is always maintained at point Q, i.e. $\underset{\sim}{n} = \underset{\sim}{n}'$, where due to the plastic volumetric strain hardening of $f=0$, the $K_p=0$ and critical failure occurs with a slope M always. The above concepts are analytically described in a general stress space by means of a properly defined deviatoric second-order anisotropy tensor, which evolves with plastic strains (Dafalias 1981d). It is clear that many assumptions are involved here, as for example the fact that the loading/yield surface itself does not account for anisotropy (the changing deviation ω represents an anisotropic plastic potential) or that the M remains constant while experimentally it has been found to vary, if not very much. Again it becomes a question of simplicity versus sophistication, and it appears that the above idea provides certain features of simplicity which may compensate for the above shortcomings.

13.6.5 *Cyclic response*

The cyclic response has already been discussed in relation to the classical yield surface and the more recent bounding surface formulations, as well as the advantage of the latter over the inability of the former to describe such a response realistically. There are a few more points of interest. The first refers to the stabilization or not of the cyclic loops especially under undrained conditions. The experimental evidence is quite complex, but one can conclude in general that depending on the amplitude of q stabilization may or may not occur before failure (Sangery et al. 1969). Such a response can be described by the bounding surface model in connection with the elastic nucleus. If the amplitude is small, the progressing towards the origin undrained cyclic stress path will enter the elastic nucleus with complete stabilization. For large q amplitudes the undrained stress path may partially enter the elastic nucleus and stabilize when the 'image' stress is at the vicinity of Q (no plastic volumetric strain rate), but still deviatoric plastic strain will continue to accumulate for non-zero mean q value, leading eventually to failure. Caution should be exercised, however, in interpreting the experimental data, especially for natural clays, because the viscoplastic time-independent effects may be as important as the cyclic loading effects for a large number of cycles which require a long period of time to be performed properly.

Another important point is the development of anisotropy for one-sided cyclic deviatoric loading and must be considered according to the discussion of the previous subsection. Equally important is finally the phenomenon of cyclic degradation, which is expected since such kind of loading expedites the breaking and disruption of interparticle bonds. A yield (or bounding) surface which hardens on the basis of volumetric plastic strain cannot describe such degradation, and recently Mroz et al. (1981) introduced a number of different ways to describe the problem; basically the accumulated deviatoric plastic strain is used as a measure of degradation which allows the surface to 'shrink' in the process, even if the simultaneous consolidation may have the opposite effect. In view of Equation 13.7 it appears possible to also render $\hat{H}$ a decreasing function of the accumulated deviatoric strain. In this way, as the undrained stress path progresses towards the CSL, the 'image' stress tends towards Q (Figure 13.1), thus $\bar{K}_p \rightarrow 0$ and for $\hat{H} \rightarrow 0$ one has also $K_p \rightarrow 0$ obtaining degradation. More experimental studies are necessary at this stage, to obtain further systematic information on these processes.

13.6.6 *Numerical implementation and model calibration*

The efficiency of the numerical implementation of a constitutive model depends primarily on the space required to store all the internal model variables (not material constants), and the cost of their updating. The bounding surface model is particularly well suited for such implementation since the only internal model variable is the plastic void ratio change, and the updating is performed by analytical expressions which are given in close form. The inability to consider anisotropy, as it is now formulated, is the price we pay for the corresponding simplicity.

The calibration of a model requires two sets of information: the initial state and the material (or model) constants. The first set presents a problem for any constitutive model because the in-situ determination of initial state (stress, overconsolidation etc.) remains one of the most difficult practical problems today. The second set depends largely on the type of model and the kind of experimental data necessary for calibration. The bounding surface model has the nice property to require a number of constants which either express well defined material properties (compressive and swell indices, M etc.), or are model constants which can be determined by a trial and error procedure, with the advantage that each model constant plays a clearly defined role in the material response. Further elaboration on these points can be found in Herrmann et al. (1982).

13.7 THE VISCOPLASTIC RESPONSE

The extensive discussion of Section 13.3 has clearly shown the necessity to formulate the clay constitutive relations within the framework of combined elastoplasticity/viscoplasticity. Such a formulation employing the concept of the bounding surface in its radial mapping version was presented in general for any material and in particular for cohesive soils by Dafalias (1982a). The reader is referred to the above reference for the details of the analytical formulation, and here we will only describe the basic concepts.

Referring to Figure 13.1 for illustration, the radial mapping rule defines not only the 'image' stress point C on the bounding surface for a given actual stress point B, but also a second 'image' stress point A with coordinates $\hat{p}, \hat{q}$ on the boundary of the elastic nucleus. The distance $\hat{\delta}$ between B and A can be considered as a measure of the classical overstress concept $\Delta\hat{\sigma}$ (Perzyna 1963), which causes viscoplastic flow to occur according to:

$$\dot{\varepsilon}^{v}_{ij} = \langle \Phi\,(\Delta\hat{\sigma}) \rangle R^{v}_{ij} \tag{13.8a}$$

$$\dot{q}^{v}_{n} = \langle \Phi_{n}(\Delta\hat{\sigma}) \rangle r^{v}_{n} \tag{13.8b}$$

where we can define $\Delta\hat{\sigma} = (AB/OA) = [\hat{\delta}/(r - (\hat{\delta} + \delta))]$ and Φ_n (and Φ) is such that $\Phi_n > 0$ when $\Delta\hat{\sigma} > 0$, $\langle \Phi_n \rangle = 0$ when $\Delta\hat{\sigma} \leq 0$, and Φ_n is continuous at $\Delta\hat{\sigma} = +0$. The hardening of the bounding surface depends now on the total inelastic volumetric strain rate (or inelastic void ratio rate) which has both plastic and viscoplastic parts, the plastic occurring when loading takes place. Figure 13.1 illustrates the effect of the viscoplastic response under undrained conditions. As viscoplastic volumetric strain develops at point B due to the existence of $\Delta\hat{\sigma} > 0$, it is balanced off by elastic swelling and effective stress relaxation as shown by the arrows emanating from B and directed towards G. Simultaneously, an expansion of the bounding surface and the elastic nucleus occurs as the viscoplastic void ratio change is negative (viscoplastic consolidation), and shown by the dashed lines in Figure 13.1. This will lead the undrained stress path BG either towards the inside of the elastic nucleus for small values of q where stabilization occurs, or it will reach a position ('image' stress at Q) such that no further volumetric strain occurs but continuous deviatoric strain leads to failure, since $\hat{\delta}$ will remain positive. The above phenomena have been systematically studied experimentally by Arulanandan et al. (1971).

However, this is only half the story. The above concept of the overstress viscoplastic rate (Equations 13.8a and b) cannot be simply superimposed and supplement the plastic equation without coupling. The continuous evolution implied by these equations will necessarily affect the bounding surface and the implied loading surfaces within, or the yield surface for the classical formulation. For such classical formulation the loading index L (Equation 13.3), must be modified in order to account for such coupling. This implies the inclusion of such terms as $\langle \Phi_n \rangle (\partial f/\partial q_n)\, r^{v}_{n}$ in L, where now f is a function of q_n (not only q^{p}_{n}), with $\dot{q}_n = \dot{q}^{p}_{n} + \dot{q}^{v}_{n}$. Such coupling was considered first by Naghdi & Murch (1963) for plastic-viscoelastic response and later by Mandel (1971). The formulation in terms of a bounding surface rather than a yield surface (Dafalias 1982a) presents certain novel features which will not be discussed in detail here. It suffices to say that a modification of the loading index L is necessary not only for points on the bounding surface $F(\bar{\sigma}_{ij}, q_n) = 0$ (observe that F depends on q_n now, not just q_n^p), but also for points within.

Recalling the discussion of Section 13.3, let us observe finally that when the point reaches the $F = 0$, under the particular soil model of the present formulation the rate effect disappears from the total inelastic rate equations. That is, $F = 0$ acts as an upper bound on rate sensitivity, what we have called rate dependence saturation. This is in sharp contrast with the behaviour of plain overstress viscoplastic models which do not have an 'upper ceiling' on the amount of required stress to cause increasing levels of strain rates. Such a representative model for soils has been presented by Oka (1981).

13.8 CONCLUSION

Based on micro- and macroscopic observations of the structure and response of cohesive soils, it is concluded that their constitutive relations can be best obtained within the general framework of combined elastoplasticity/viscoplasticity rather than each one separately. The fact that cohesive soils do not exhibit a sharp delineation between elastic and inelastic domains in stress space, renders the concept of the bounding surface particularly well suited to describe their response. Such concept, initially developed for rate-independent response, can be extended to rate-dependent as well. Different aspects associated with either the classical yield surface or the bounding surface formulation are discussed. One thing becomes clear: a comprehensive constitutive model may become very complex, but the effects of such complexity can be reduced if the model is modularizable. That is, if it can include certain features related to the one or the other particular material response characteristic, by simply assigning non-zero values to model constants related to these features, without altering otherwise the basic framework of the model. From the point of view of practical application a model must posses the feature of easy and inexpensive numerical implementation, and the calibration of the material (model) constants should be possible by fitting the experimental data obtained by more or less standard experimental procedures.

REFERENCES

Arulanandan, K., C.K. Shen & R.B. Young 1971. Undrained creep behavior of a coastal organic silty clay. *Geotechnique* 21: 359–375.

Baladi, G.Y. 1978. An elastic-plastic constitutive relation for transverse-isotropic three-phase earth materials. Miscellaneous paper S-78-14, Final report of US Army Engineer Waterways Experiment Station.

Banerjee, P.K. & A.S. Stipho 1978. Associated and non-associated constitutive relations for undrained behavior of isotropic soft clays. *Int. J. for Numer. and Analyt. Meth. in Geomech.* 2: 35–56.

Dafalias, Y.F. 1979a. A model for soil behavior under monotonic and cyclic loading conditions. In *Proceedings of 5th International SMiRT Conference* (Berlin): Paper K 1/8.

Dafalias, Y.F. 1979b. A bounding surface plasticity mode. In *Proceedings of 7th CANCAM Conference* (Sherbrooke, Canada): 89–90.

Dafalias, Y.F. 1979c. Anisotropic hardening of initially orthotropic materials. *ZAMM* 59: 437–446.

Dafalias, Y.F. 1981a. The concept and application of the bounding surface in plasticity theory. In *Proceedings of IUTAM Symposium on Physical Non-Linearities in Structural Analysis* (Senlis, France, 1980): 56–63. Berlin: Springer Verlag.

Dafalias, Y.F. 1981b. A novel bounding surface constitutive law for the monotonic and cyclic hardening response of metals. In J. Boehler (ed.), *Proceedings of 6th International SMiRT Conference* (Paris): Paper L 3/4.

Dafalias, Y.F. 1981c. Discussion on the Reference of this paper: Hashiguchi, K., 1980. *J. Appl. Mech.* (ASME) 48: 211–212.

Dafalias, Y.F. 1981d. Initial and induced anisotropy of cohesive soils by means of a varying non-associated flow rule. In J. Boehler (ed.), *Proceedings of 'Colloque Inter, nationale du CNRS No. 319: Le Comportement Plastique des Solides Anisotropes* (Villard-de-Lans, France).

Dafalias, Y.F. 1982a. Bounding surface elastoplasticity-viscoplasticity for particulate cohesive media. In *Proceedings of IUTAM Symposium on Deformation and Failure of Granular Materials* (Delft): 97–107. Rotterdam: Balkema.

Dafalias, Y.F. 1982b. A vertex bounding surface plasticity model (abstract). In *Proceedings of 9th U.S. National Congress on Applied mechanics* (Ithaca, USA): 82.

Dafalias, Y.F. & L.R. Herrmann 1980. A bounding surface soil plasticity model. In C.N. Pande & O.C. Zienkiewicz (eds.), *Proceedings of International Symposium on Soils under Cyclic and Transient Loading* (Swansea), 1: 335–345. Rotterdam: Balkema.

Dafalias, Y.F., L.R. Herrmann & J.S. DeNatale 1981. Description of natural clay behavior of a simple bounding surface plasticity formulation. In R.N. Yong & H.Y. Ko (eds.), *Proceedings of North American Workshop on Limit Equilibrium, Plasticity and Generalized Stress-Strain in Geotechnical Engineering* (Montreal), sponsored by NSF/NSERC. ASCE publication.

Dafalias, Y.F. & L.R. Herrmann 1982a. Bounding surface formulation of soil plasticity. In G.N. Pande & O.C. Zienkiewicz (eds.), *Soil Mechanics–transient and cyclic loads*: 253–282. Chichester: Wiley.

Dafalias, Y.F. & L.R. Herrmann 1982b. A generalized bounding surface constitutive model for clays. In R.N. Yong & E.T. Selig (eds.), *Application of plasticity and generalized stress-strain in geotechnical engineering*, ASCE Special Publication Volume: 78–95.

Dafalias, Y.F. & E.P. Popov 1974, 1975. A model of nonlinearly hardening materials for complex loadings. *Proceedings of 7th U.S. National Congress on Applied mechanics* (Boulder, Colorado, 1974); and *Acta Mech.* (1975) 21: 173–192.

Dafalias, Y.F. & E.P. Popov 1976. Plastic internal variables formalism of cyclic plasticity. *J. Appl. Mech.* (ASME) 98: 645–650.

Dafalias, Y.F. & E.P. Popov 1977. Cyclic loading for materials with a vanishing elastic region. *Nuclear Engng. Design* 41: 293–302.

DiMaggio, F.L. & I.S. Sandler 1971. Material models for granular soils. *J. Engng. Mech. Div.* (ASCE) 97 (3): 936–950.

Drucker, D.C., Gibson, R.E. & D.J. Henkel 1957. Soil mechanics and work-hardening theories of plasticity. *Trans. ASCE* 122: 338–346.

Drucker, D.C. & W. Prager 1952. Soil mechanics and plastic analysis of limit design. *Q. Appl. Math.* 10 (2): 157–165.

Hashiguchi, K. 1979. Constitutive equations of granular media with an anisotropic hardening. In *Proceedings of 3rd International Conference on Numerical methods in Geomechanics* (Aachen), 435–439.

Hashiguchi, K. 1980. Constitutive equations of elastoplastic materials with elasticplastic transition. *J. Appl. Mech.* (ASME) 47: 266–272.

Hashiguchi, K. 1981. Constitutive equations of elastoplastic materials with anisotropic hardening and elastic-plastic transition. *J. Appl. Mech.* (ASME) 48: 297–301.

Hashiguchi, K. & M. Ueno 1977. Elastoplastic constitutive laws of granular materials. In *Proceedings of 9th International Conference on Soil mechanics and Foundation Engineering* (Tokyo), Specialty session 9.

Herrmann, L.R., Y.F. Dafalias & J.S. DeNatale 1982. Numerical implementation of a bounding surface soil plasticity model. In *Proceedings of International Symposium on Numerical Models in Geomechanics* (Zurich): 334–343. Rotterdam: Balkema.

Iwan, W.D. 1967. On a class of models for the yielding behavior of continuous and composite systems. *J. Appl. Mech.* (ASCE) 34: 612–617.

Krieg, R.D. 1975. A practical two-surface plasticity theory. *J. Appl. Mech.* (ASME) 97: 641–646.

Lewin, P.I. 1978. The deformation of soft clay under generalized stress conditions. Ph.D. thesis, King's College, University of London.

Mandel, J. 1971. Plasticité classique et viscoplasticité. ICMS, Courses and Lectures No. 97, Udine, Italy.

Mroz, A. 1967. On the description of anisotropic work hardening. *J. Mech. Phys. Solids* 15: 163–175.

Mroz, Z., V.A. Norris & O.C. Zienkiewicz 1978. An anisotropic hardening model for soils and its application to cyclic loading. *Int. J. for Numer. and Analyt. Meth. in Geomech.* 2: 203–221.

Mroz, Z., V.A. Norris & O.C. Zienkiewicz 1979. Application of an anisotropic elastoplastic deformation of soils. *Geotechnique* 29 (1): 1–34.

Mroz, Z., V.A. Norris & O.C. Zienkiewicz 1981. An anisotropic, critical state model for soils subject to cyclic loading. *Geotechnique* 31 (4): 451–469.

Naghdi, P.M. & S.A. Murch 1963. On the mechanical behavior of viscoelastic/plastic solids. *J. Appl. Mech.* (ASME) 30 (3): 321–328.

Oka, F. 1981. Prediction of time-dependent behavior of clay. In *Proceedings of 10th International Conference on Soil Mechanics and Foundation Engineering* (Stockholm):

Perzyna, P. 1963. The study of the dynamic behavior of rate sensitive plastic materials. *Arch. Mech.* 1 (15): 113–128.

Phillips, A. & R.L. Sierakowski 1965. On the concept of the yield surface. *Acta Mech.* 1 (1): 29–35.

Prevost, J.H. 1978. Plasticity theory for soil stress strain behavior. *J. Engng. Mech. Div.* (ASCE) 105: 1177–1196.

Roscoe, K.H., A.N. Schofield & A. Thurairajah 1963. Yielding of clays in state wetter than critical. *Geotechnique* 13 (3): 211–240.

Rudnicki, J.W. & J.R. Rice 1975. Conditions for the localization of deformation in pressure-sensitive dilatant materials. *J. Mech. Phys. Solids* 23: 371–394.

Sangrey, D.A., D.H. Henkel & M.I. Espig 1969. The effective stress response of a saturated clay soil to repeated loading. *Can. Geotech. J.* 6 (3): 241–252.

Schofield, A.N. & C.P. Wroth 1968. *Critical state soil mechanics.* London: McGraw-Hill.

Scott, R.F. 1963. *Principles of soil mechanics.* New York: Addison-Wesley.

Constitutive equations for pressure-sensitive soils: theory, numerical implementation, and examples 14

J.H. PREVOST

14.1 INTRODUCTION

Soil consists of an assemblage of particles with different sizes and shapes which form a skeleton whose voids are filled with various fluids. The stresses carried by the soil skeleton are conventionally termed 'effective stresses' (Terzaghi 1943) in the soil mechanics literature, and those in the fluids are called 'pore-fluid pressures'. It is observed experimentally that the stress-strain behaviour of the soil skeleton is strongly nonlinear, anisotropic and hysteretic. In order to relate the changes in effective stresses carried by the soil skeleton to the skeleton rate of deformation, a general analytical model which describes the nonlinear, anisotropic, elastoplastic, path-dependent, stress-strain-strength properties of the soil skeleton when subjected to complicated three-dimensional, and in particular to cyclic loading paths (Prevost 1977) is presented. A brief summary of the model's basic principles (Prevost 1978b) is included and the consitutive equations are provided. It is shown that the model parameters required to characterize the behaviour of any given soil can be derived entirely from the results of conventional soil tests. The model's accuracy is evaluated by applying it to represent the behaviour of both cohesive and cohesionless soils. Implementation of the proposed formulation in a general finite element program for solution of boundary value problems is discussed.

As for notation, boldface letters denote vectors, second- and forth-order tensors in three dimensions. All stresses effective stresses.

14.2 CONSTITUTIVE EQUATIONS

The constitutive equations for the solid skeleton are written in one of the following forms:

$$\mathbf{C}:\dot{\boldsymbol{\varepsilon}} = \begin{cases} \dot{\boldsymbol{\sigma}}' & \text{small deformations} \\ \overset{\nabla}{\boldsymbol{\sigma}}' + \boldsymbol{\sigma}' \operatorname{div} \mathbf{v} & \text{finite deformation} \end{cases} \tag{14.1}$$

in which $\boldsymbol{\sigma}'$ = effective (Cauchy) stress tensor; $\mathbf{v}$ = (spatial) velocity of solid phase; $\dot{\boldsymbol{\varepsilon}}$ = rate of deformation tensor for the solid phase (= symmetric part of the spatial solid velocity gradient); a dot denotes the material derivative; and $\overset{\nabla}{\boldsymbol{\sigma}}'$ = Jaumann

(1903) derivative, viz.:

$$\overset{\nabla}{\boldsymbol{\sigma}}' = \dot{\boldsymbol{\sigma}}' + \boldsymbol{\sigma}' \cdot \mathbf{w} - \mathbf{w} \cdot \boldsymbol{\sigma}' \tag{14.2}$$

where $\mathbf{w}$ = spin tensor for the solid phase (= skew-symmetric part of the spatial solid velocity gradient). In Equation 14.1, C_{abcd} is an (objective) tensor valued function of, possibly, $\boldsymbol{\sigma}'$ and the solid deformation gradients. Many nonlinear material models of interest can be put in the above form (e.g., *all* nonlinear elastic materials, and many elasto-plastic materials). The finite deformation form of the constitutive equation above was first proposed by Hill (1958) within the context of plasticity theory.

For soil media, the form of the $\mathbf{C}$ tensor is given as follows (Prevost 1978b):

$$\mathbf{C} = \mathbf{E} - \frac{(\mathbf{E}:\mathbf{P})(\tilde{\mathbf{Q}}:\mathbf{E})}{H' + \mathbf{Q}:\mathbf{E}:\mathbf{P}} \tag{14.3}$$

in which H' is the plastic modulus; $\mathbf{P}$ and $\mathbf{Q}$ are symmetric second-order tensors, such that $\mathbf{P}$ gives the direction of plastic deformations and $\mathbf{Q}$ the outer normal to the active yield surface; and $\mathbf{E}$ is the fourth-order tensor of elastic moduli, assumed isotropic, viz.:

$$E_{abcd} = \Lambda \gamma_{ab}\gamma_{bc} + G(\gamma_{ac}\gamma_{bd} + \gamma_{ad}\gamma_{bc}) \tag{14.4}$$

where Λ and G = Lame's constants; γ_{ab} = Kronecker delta. The yield function is selected of the following form:

$$f = \tfrac{3}{2}(\mathbf{S} - \boldsymbol{\alpha}):(\mathbf{S} - \boldsymbol{\alpha}) + C^2(p' - \beta)^2 = k^2 \tag{14.5}$$

where $\mathbf{S} = \boldsymbol{\sigma}' - p'\mathbf{1}$ = deviatoric stress tensor; $p' = \frac{1}{3}\mathrm{tr}\,\boldsymbol{\sigma}'$ = effective mean normal stress; $\boldsymbol{\alpha}$ and β are the coordinates of the centre of the yield surface in the deviatoric stress subspace and along the hydrostatic stress axis, respectively; k = size of the yield surface; C = material parameter called the yield surface axis ratio. From Equation 14.5 it follows:

$$\mathrm{grad}\, f = \frac{\partial f}{\partial \boldsymbol{\sigma}'} = 3(\mathbf{S} - \boldsymbol{\alpha}) + \tfrac{2}{3}C^2\,(p' - \beta)\,\mathbf{1} \tag{14.6}$$

and:

$$|\mathrm{grad}\, f|^2 = 6k^2 + 6C^2(\tfrac{2}{9}C^2 - 1)(p' - \beta)^2 \tag{14.7}$$

It is convenient to decompose $\mathbf{P}$ and $\mathbf{Q}$ into their deviatoric and dilatational components, and in the following:

$$\mathbf{P} = \mathbf{P}' + P''\,\mathbf{1} \quad \text{and} \quad \mathbf{Q} = \mathbf{Q}' + Q''\,\mathbf{1} \tag{14.8}$$

where:

$$P'' = \tfrac{1}{3}\,\mathrm{tr}\,\mathbf{P} \quad \text{and} \quad Q'' = \tfrac{1}{3}\,\mathrm{tr}\,\mathbf{Q} \tag{14.9}$$

and:

$$\mathbf{Q} = \mathrm{grad}\, f / |\mathrm{grad}\, f| \tag{14.10}$$

The plastic potential is selected such that the plastic rate of deformation vector remains normal to the projection of the yield surface into the deviatoric stress subspace, viz.:

$$\mathbf{P}' = \mathbf{Q}' \tag{14.11a}$$

and:

$$3P'' = Q'' + A\,\mathrm{tr}\,(\mathbf{Q}')^3/\mathrm{tr}\,(\mathbf{Q}')^2 \tag{14.11b}$$

where:

$$\begin{aligned} \mathrm{tr}\,(\mathbf{Q}')^2 &= \mathbf{Q}' : \mathbf{Q}' = \mathbf{Q}'_{ab}\mathbf{Q}'_{ab} \\ \mathrm{tr}\,(\mathbf{Q}')^3 &= 3\det(\mathbf{Q}') = Q'_{ab}Q'_{bc}Q'_{ca} \end{aligned} \tag{14.12}$$

and A is a material parameter which measures the departure from an associative plastic flow rule. When $A = 0$, the principal directions of **P** and **Q** coincide and consequently the **C** tensor possesses the major symmetry and leads to a symmetric material tangent stiffness. On the other hand, when $A \neq 0$, the principal directions of **P** and **Q** do not coincide, and **C** does not posses the major symmetry

From Equations 14.4, 14.8 and 14.9:

$$\mathbf{Q}:\mathbf{E}:\mathbf{P} = B\,\mathrm{tr}\,\mathbf{P}\,\mathrm{tr}\,\mathbf{Q} + 2G\mathbf{P}':\mathbf{Q}' \tag{14.13}$$

where $B = \Lambda + 2G/3$ and G are the elastic bulk and shear moduli, respectively.

Remarks

1. Under the assumptions spelled above Equations 14.1 and 14.3 write in expanded form as (small deformation case):

$$\dot{\boldsymbol{\sigma}}' = 2G\,\dot{\boldsymbol{\varepsilon}} + \left(B - \frac{2G}{3}\right)\dot{\varepsilon}_v \mathbf{1} - (2G\,\mathbf{Q}' + B\,3P''\,\mathbf{1})\frac{2G\,\mathbf{Q}':\dot{\boldsymbol{\varepsilon}} + B3Q''\,\dot{\varepsilon}_v}{H' + 2G\,\mathbf{Q}':\mathbf{Q}' + B3Q''3P''} \tag{14.14}$$

where $\dot{\varepsilon}_v = \mathrm{tr}\,\dot{\boldsymbol{\varepsilon}}$. Or equivalently, in terms of deviatoric and dilatational components:

$$\dot{\mathbf{S}} = 2G\,\dot{\mathbf{e}} - 2G\,\mathbf{Q}'\frac{2G\,\mathbf{Q}':\dot{\mathbf{e}} + B\,3Q''\,\dot{\varepsilon}_v}{H' + 2G\,\mathbf{Q}':\mathbf{Q}' + B\,3Q''3P''} \tag{14.15a}$$

$$\dot{p}' = B\,\dot{\varepsilon}_v - B\,3P''\frac{2G\,\mathbf{Q}':\dot{\mathbf{e}} + \mathbf{B}\,3Q''\dot{\varepsilon}_v}{H' + 2G\,\mathbf{Q}':\mathbf{Q}' + B3Q''\,3P''} \tag{14.15b}$$

where $\dot{\mathbf{e}} = \dot{\boldsymbol{\varepsilon}} - \dot{\varepsilon}_v\mathbf{1}$ = deviatoric rate of deformation tensor.

When $\mathbf{Q}' = 0$, $3Q''^2 = 1$ and Equation 14.15b simplifies to:

$$\dot{p} = \left(\frac{1}{B} + \frac{1}{H'}\right)^{-1}\dot{\varepsilon}_v \tag{14.16}$$

and the plastic modulus H′ thus plays the role of a plastic bulk modulus. Similarly, when $Q'' = 0$, $\mathbf{Q}':\mathbf{Q}' = 1$ and Equation 14.15a yields:

$$\mathbf{Q}':\dot{\mathbf{S}} = \left(\frac{1}{2G} + \frac{1}{H'}\right)^{-1}\mathbf{Q}':\dot{\mathbf{e}} \tag{14.17}$$

and the plastic modulus H′ thus plays the role of a plastic shear modulus.

2. When $C = 0$, the yield surface plots in stress space as a cylinder (translated Von Mises yield surface) whose axis is parallel to the space diagonal, and the model reduces to the one used by Prevost (1977 and 1978a). When $C \neq 0$, the yield surface

plots in stress space as an ellipsoidal surface of revolution whose axis is parallel to the space diagonal.

In order to allow for the adjustment of the plastic hardening rule to any kind of experimental data, for example data obtained from axial or simple shear soil tests, a collection of nested yield surfaces is used. Much pros and cons have been said about multi-surface versus two-surface plasticity theories and a critical assessment of their relative merits and shortcomings is reported by Prevost (1982c). The surfaces are all similar, i.e., the axis ratio C (Equation 14.5) is the same for all yield surfaces. The yield surfaces, in general, may translate and change in size, but never rotate. The model therefore combines properties of isotropic and kinematic plasticity. In order to avoid overlappings of the surfaces (which would lead to a non-unique definition of the constitutive theory) the isotropic/kinematic hardening rule couples the simultaneous deformation/translation of all yield surfaces. This is further discussed and explained by Prevost (1978b).

A plastic modulus $H'^{(m)}$ and a nonassociative parameter $A^{(m)}$ are associated with each yield surface. In general, both $A^{(m)}$ and $H'^{(m)}$ are allowed to take different values at different locations on any given yield surface, i.e., both $A^{(m)}$ and $H'^{(m)}$ are functions of position. That much degree of complexity and generality is required in order to be able to accommodate any given soil.

Remarks

Several levels of sophistication (and complexity but versatility) of the model can be achieved by appropriate selection of $A^{(m)}$ and $H'^{(m)}$. Up to this date, the following rules have been used.

Cohesive soils (Prevost 1980, 1982a): $A^{(m)}$ is constant on each surface, and:

$$H'^{(m)} = h'^{(m)} + \frac{\operatorname{tr} \mathbf{Q}^{(m)}}{\sqrt{3}} B'^{(m)} \tag{14.18}$$

where $h'^{(m)}$ plays the role of a plastic shear modulus and $(h'^{(m)} \pm B'^{(m)})$ play the role of plastic bulk moduli.

Cohesionless soils: Both $A^{(m)}$ and $H'^{(m)}$ are allowed to vary on each yield surface:

$$A^{(m)} = A_i^{(m)} \begin{cases} i = 1 \text{ if } \operatorname{tr}(\mathbf{Q}'^{(m)})^3 > 0 \\ i = 2 \text{ if } \operatorname{tr}(\mathbf{Q}'^{(m)})^3 < 0 \end{cases} \tag{14.19}$$

and:

$$H'^{(m)} = h_i'^{(m)} \left| \frac{\operatorname{tr}(\mathbf{Q}^{(m)})^3}{A_i} \right| + B_i'^{(m)} \left| 1 - \frac{\operatorname{tr}(\mathbf{Q}'^{(m)})^3}{A_i} \right| \tag{14.20}$$

$$\begin{cases} i = 1 \text{ if } \operatorname{tr} \mathbf{Q}^{(m)} > 0 \\ i = 2 \text{ if } \operatorname{tr} \mathbf{Q}^{(m)} < 0 \end{cases}$$

where $h_i'^{(m)}$ and $B_i'^{(m)}$ play the role of plastic shear and bulk moduli, respectively. The yield surfaces' initial positions and sizes reflect the past stress-strain history of the soil skeleton, and in particular their initial positions are a direct expression of the material's 'memory' of its past loading history. Because the α's are not necessarily all equal to zero, the yielding of the material is anisotropic. Direction is therefore of importance and the physical reference axes (x, y, z) are fixed with respect to the material element and specified coincide with the reference axes of consolidation. For a soil element whose anisotropy initially exhibits rotational symmetry about the y-axis, $\alpha_x = \alpha_z = -\alpha_y/2$, and Equation 14.5 simplifies to:

$$[(\sigma_y' - \sigma_x') - \alpha]^2 + C^2(p' - \beta)^2 - k^2 = 0 \tag{14.21}$$

in which $\alpha = 3\alpha_y/2$. The yield surfaces then plot as ellipses in the axisymmetric stress plane ($\sigma_x' = \sigma_z'$) as shown in Figure 14.1(a). Points C and E on the outermost yield surface define the critical state conditions (i.e., $H' = 0$) for axial compression and extension loading conditions, respectively (Hughes et al. 1979). It is assumed that the slopes of the critical state lines OC and OE remain constant during yielding.

The yield surfaces are allowed to change in size as well as to be translated by the stress point. Their associated plastic moduli are also allowed to vary and in general both k and H′ are functions of the plastic strain history. They are conveniently taken as functions of invariant measures of the amount of plastic volumetric strains and/or plastic shear distortions, respectively (Prevost 1977, 1978b, 1982a).

Complete specification of the model parameters requires the determination of:

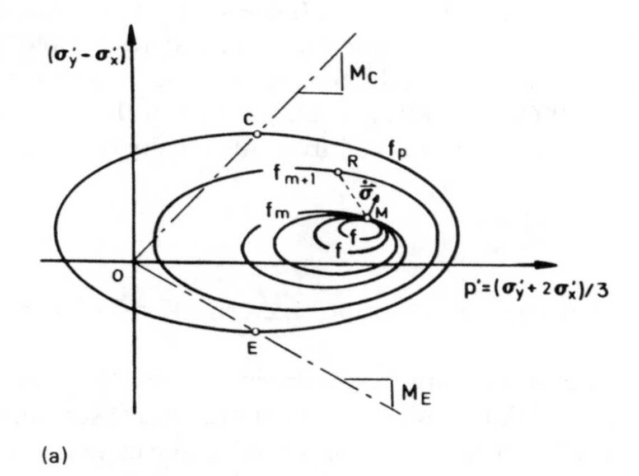

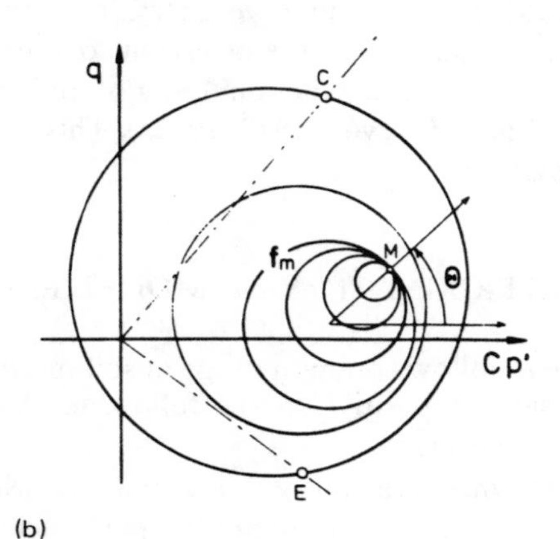

Figure 14.1. Field of yield surfaces: (a) 1 versus p′ plane; (b) q versus cp′ plane

1. the initial positions and sizes of the yield surfaces together with their associated plastic moduli;
2. their size and/or plastic modulus changes as loading proceeds; and finally,
3. the elastic shear G and bulk B ($=\Lambda + 2G/3$) moduli.

Remarks

1. The yield surface $f^{(1)}$ is chosen as a degenerate yield surface of size $k^{(1)} = 0$ which coincides with the stress point. Further, in order to get a smooth transition from the elastic into the plastic regime, $A^{(1)} = 0$ and $H'^{(1)} = \infty$, so that the material behaviour inside $f^{(2)}$ is purely elastic.
2. The dependence of the model parameters upon the effective mean normal stress and volumetric strain assumed is to be of the following form:

$$x = x_1 \left(\frac{p'}{p'_1} \right)^n \quad \text{and} \quad y = y_2 \exp(\lambda \varepsilon_v)$$

respectively, where $x = B$, G and $H'^{(m)}$, and $y = \alpha_{ab}^{(m)}$, $\beta^{(m)}$ and $k^{(m)}$; n is an experimental parameter ($n = 0.5$ for most cohesionless soils (Richard et al. 1970), and $n = 1$ for most cohesive soils); p'_1 = reference effective mean normal stress (i.e., at $\varepsilon_v = 0$ when $p' = p'_1$). It is assumed that when the soil is in a 'normally consolidated' state, the consolidation soil text results plot: (1) as a straight line parallel to the projections of the critical state lines in the Ln p'/p'_1 vs. ε_v diagram (Roscoe & Burland 1968), and (2) as a straight line in the axial stress plane. The parameter λ is then simply determined from the results of K_0-consolidation soil test results, viz.:

$$\lambda = \frac{1}{p'_K} \frac{\dot{p}'_K}{\dot{e}_v^K} \tag{14.22}$$

where the subscript/supercript K refers to K_0-loading conditions.

The soil's anisotropy originally develops during its deposition and subsequent consolidation which, in most practical cases, occurs under no lateral deformations. In the following, the y-axis is vertical and coincides with the direction of consolidation, the horizontal xz-plane is thus a plane of material's isotropy and the material's anisotropy initially exhibits rotational symmetry about the vertical y-axis. The model parameters required to characterize the behaviour of any given soil can then be derived entirely from the results of conventional monotonic axial and cyclic strain-controlled simple shear soil texts (Prevost 1977, 1982a). This is explained and further discussed in the following.

14.3 DETERMINATION OF MODEL PARAMETERS

In order to follow common usage in soil mechanics, compressive stresses and strains are considered positive in the following. All stresses are effective stresses unless otherwise specified.

As explained previously, for a material which initially exhibits cross-anisotropy about the vertical y-axis, the initial positions in stress space of the yield surfaces are defined by the sole determination of the two parameters $\alpha^{(m)}$ and $\beta^{(m)}$ (m = 1,..., p) and Equation 14.5 simplifies to:

$$[q - \alpha^{(m)}]^2 + C^2[p' - \beta^{(m)}]^2 - [k^{(m)}]^2 = 0 \tag{14.23}$$

for axial loading conditions (i.e., $\sigma'_x = \sigma'_z$ and $\tau_{xy} = \tau_{yz} = \tau_{zx} = 0$), where $q = (\sigma'_y - \sigma'_x)$. The yield surfaces then plot as circles in the q versus Cp′ plane (referred to as the axial stress plane hereafter) as shown in Figure 14.1(b). When the stress point reaches the yield surface $f^{(m)}$, then:

$$q = \alpha^{(m)} + k^{(m)} \sin\theta \tag{14.24a}$$

$$p' = \beta^{(m)} + \frac{k^{(m)}}{C}\cos\theta \tag{14.24b}$$

where θ is defined in Figure 14.16, and Equation 14.15 simplifies to:

$$\frac{\dot{\bar{\varepsilon}}}{\dot{q}} = \frac{1}{2G} + \frac{1}{H'^{(m)}}\frac{\sin\theta(\sin\theta + C\gamma\cos\theta)}{\sin^2\theta + \frac{2}{9}C^2\cos^2\theta} \tag{14.25a}$$

$$\frac{\dot{\varepsilon}_v}{\dot{p}'} = \frac{1}{B} + \frac{1}{H'^{(m)}}\frac{\left(\frac{2C}{3}\cos\theta + A^{(m)}\sin\theta\right)\frac{1}{3\gamma}(\sin\theta + C\gamma\cos\theta)}{\sin^2\theta + \frac{2}{9}C^2\cos^2\theta} \tag{14.25b}$$

in which:

$$\dot{\varepsilon}_v = \dot{\varepsilon}_y + 2\dot{\varepsilon}_x \tag{14.26}$$

$$\bar{\varepsilon} = (\dot{\varepsilon}_y - \dot{\varepsilon}_x) \tag{14.27}$$

$$\gamma = \frac{\dot{p}'}{q} \tag{14.28}$$

14.3.1 *Interpretation of monotonic drained axial compression and extension soil test results*

Let θ_C and θ_E denote the values of θ when the stress point reaches the yield surface $f^{(m)}$ in axial compression and extension loading conditions, respectively. Combining Equations 14.24–14.28 one finds that:

$$\tan(\theta_C + \theta_E) = \frac{-2R}{1 - R^2} \tag{14.29}$$

$$\frac{1}{\tan\theta_C} = \frac{3}{2C}\left[3\gamma_C\frac{x_C}{y_C} - A_C^{(m)}\right] \tag{14.30a}$$

$$\frac{1}{\tan\theta_E} = \frac{3}{2C}\left[3\gamma_E\frac{x_E}{y_E} - A_E^{(m)}\right] \tag{14.30b}$$

in which:

$$R = C\frac{p'_C - p'_E\exp[\lambda(\varepsilon_v^C - \varepsilon_v^E)]}{q_C - q_E\exp[\lambda(\varepsilon_v^C - \varepsilon_v^E)]} \tag{14.31}$$

and:

$$\frac{1}{x_C} = \left(\frac{p'_C}{p'_1}\right)^n \frac{\dot{\bar{\varepsilon}}}{\dot{q}} - \frac{1}{2G_1} \tag{14.32}$$

$$\frac{1}{y_C} = \left(\frac{p'_C}{p'_1}\right)^n \frac{\dot{\varepsilon}_v}{\dot{p}'} - \frac{1}{B_1} \tag{14.33}$$

and similarly for x_E and y_E, where the subscripts C and E refer to axial compression and extension loading conditions, respectively. Further:

$$H_C^{\prime(m)} = x_C \sin\theta_C \frac{\sin\theta_C + C\gamma_C \cos\theta_C}{\sin^2\theta_C + \frac{2}{9}C^2\cos^2\theta_C} \tag{14.34}$$

and similarly for $H_E^{\prime(m)}$.

The smooth experimental stress-strain curves obtained in axial tests are approximated by linear segments along which the tangent (or secant) modulus is constant. Evidently, the degree of accuracy achieved by such a representation of the experimental curves is directly dependent upon the number of linear segments used. The model parameters associated with the yield surface $f^{(m)}$ are determined by the condition that the slopes $\dot{q}/\dot{\bar{\varepsilon}}$ are to be the same in axial compression and extension tests when the stress point has reached the yield surface $f^{(m)}$ (Prevost 1977, 1978a and b, 1982a). The corresponding values of θ_C and θ_E are determined by combining Equations 14.29 and 14.30 once a rule has been adopted for $A^{(m)}$ (see previous discussion). The case $A^{(m)} = \text{constant}$ is discussed in detail by Prevost (1982a). Once θ_C and θ_E have been determined, the model parameters associated with $f^{(m)}$ are simply obtained from Equations 14.24 and 14.25 (for example, Prevost 1982a).

14.3.2 *Interpretation of monotonic drained axial compression and extension soil test results*

In undrained tests, $\dot{\varepsilon}_v = 0$, and (from Equation 14.33) $y_C = y_E = -B_1$ in that case. The model parameters associated with the yield surface $f^{(m)}$ are again determined by the condition that the slopes $\dot{q}/\dot{\bar{\varepsilon}}$ are to be the same in axial compression and extension tests when the stress point has reached the yield surface $f^{(m)}$. As previously, the corresponding values θ_C and θ_E are determined from Equations 14.29 and 14.30, in which:

$$R = C\frac{p'_C - p'_E}{q_C - q_E} \tag{14.35}$$

Knowing θ_C and θ_E, the model parameters associated with $f^{(m)}$ are computed from Equations 14.24 and 14.25, in which $\varepsilon_v^C = \varepsilon_v^E = 0$.

14.3.3 *Interpretation of simple shear soil test results*

In simple shear soil tests, $\dot{\varepsilon}_x = \dot{\varepsilon}_y = \dot{\varepsilon}_z = 0$. The necessary algebra for the determination of the model parameters is considerably simplified in that case if the elastic contribution to the normal strains is neglected. Equation 14.15 then yields: $\sigma'_x = \sigma'_z$.

$$\frac{\dot{\gamma}_{xy}}{\dot{\tau}_{xy}} = \frac{1}{G} + \frac{2}{h'_m} \tag{14.36}$$

$$\alpha^{(m)} = (\sigma'_y - \sigma'_x) \tag{14.37}$$

$$\beta^{(m)} = \sigma'_y - \tfrac{2}{3}(\sigma'_y - \sigma'_x) \tag{14.38}$$

$$k^{(m)} = \sqrt{3}\,\tau_{xy} \tag{14.39}$$

The model parameters associated with the yield surface $f^{(m)}$ are then simply determined from the above equations and a piecewise linear representation of the shear stress-strain curves obtained in a simple shear test. Note that the sole use of simple shear test results does not allow the determination of the parameters $B'^{(m)}$ and $A^{(m)}$. On the other hand, it is apparent from Equations 14.36 and 14.39 that the degradation of the mechanical properties of the material under cyclic shear loading conditions, i.e.:

$$k^{(m)}(\bar{e}) \quad \text{and} \quad h'^{(m)}(\bar{e}) \tag{14.40}$$

with:

$$\bar{e} = \int \{\tfrac{2}{3}\dot{\mathbf{e}}' : \dot{\mathbf{e}}'\} \qquad \dot{\mathbf{e}} = \dot{\boldsymbol{\varepsilon}} - \tfrac{1}{3}(\mathrm{tr}\,\dot{\boldsymbol{\varepsilon}})\mathbf{1} \tag{14.41}$$

where the integration is carried along the strain path, is most conveniently determined from the results of cyclic strain-controlled simple shear tests ($\bar{e} = \int 1/\sqrt{3}\ |\dot{\gamma}|$ in that case). This is explained and further discussed by Prevost (1977).

14.4 MODEL EVALUATION

14.4.1 *Laboratory prepared kaolinite clay*

The soil data to be used in this section are part of the ones which were collected by the organizing committee of the NSF/NSERC North American Workshop on Plasticity and Generalized Stress-Strain Applications in Soil Engineering held May 28–30, 1980 at McGill University, Montreal, Canada. Laboratory axial test data on a laboratory-prepared kaolinite clay had then been transmitted to the author. Predictions about the constitutive behaviour of the soil subjected to loading stress paths not identified in the data had been requested by the organizing committee. This section describes the test results, their analysis, and compares the model predictions (Prevost 1980) with observed behaviour in the tests.

The experimental tests had been conducted on cylindrical samples in a torsional shear testing device. All samples had first been K_0-consolidated with a cell pressure of 58 psi and a back-pressure of 18 psi, and then left to rebound to an equal-all-around cell pressure of 58 psi, and a back-pressure of 18 psi (in other words, the excess axial load necessary for K_0-consolidation was then released). All the tests were stress-controlled and performed under constant volume conditions (i.e., undrained).

Figure 14.2 shows in dashed lines the experimental results obtained in conventional

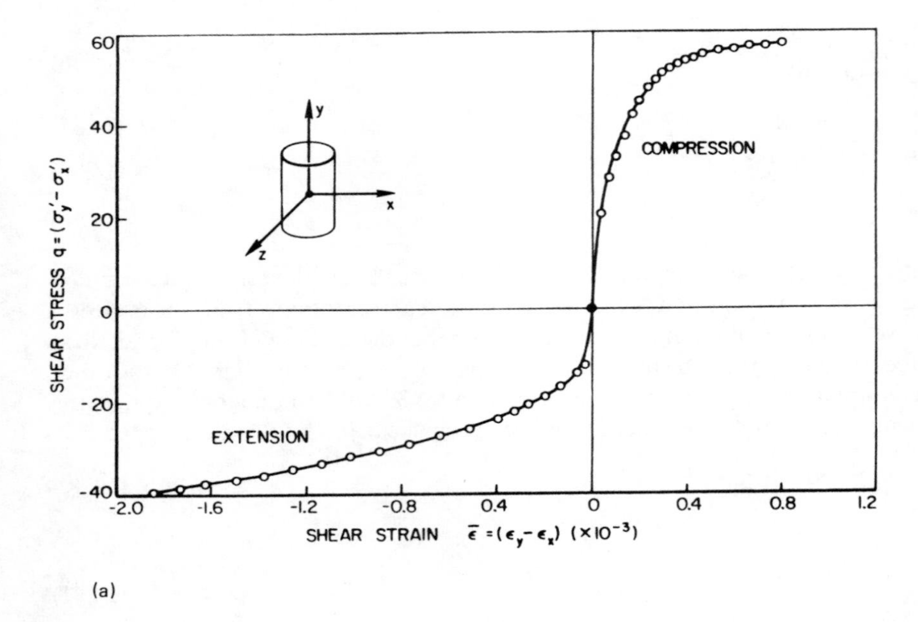

(a)

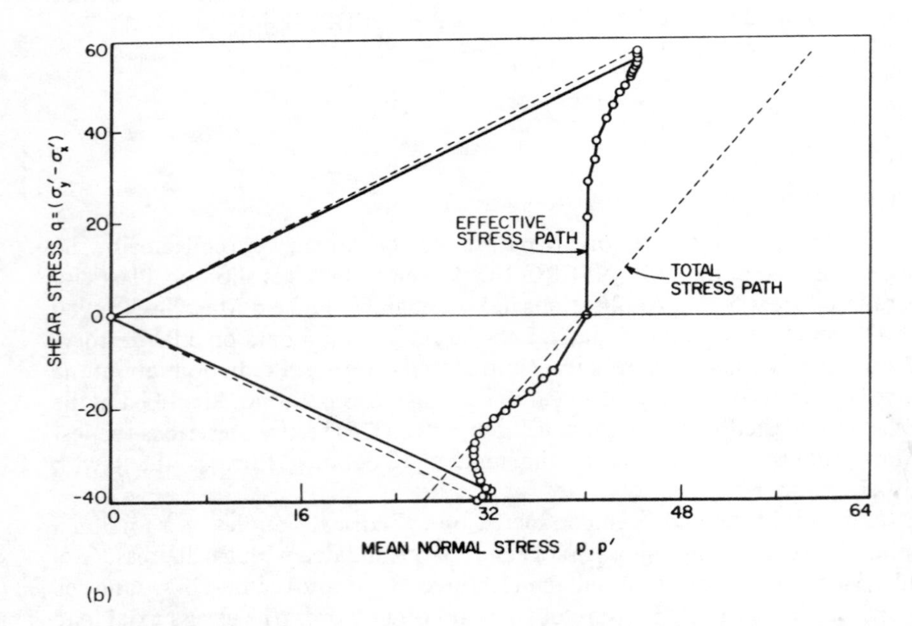

(b)

Figure 14.2. Kaolinite clay–undrained axial soil test results: (a) shear stress vs. shear strain; (b) shear stress vs. effective mean normal stress.

undrained monotonic axial compression/extension soil tests, and in solid lines the design curves used to determine the model parameters for that clay. Note that some data points close to failure have been ignored when selecting the design curves because they are not consistent with the rest of the data. This inconsistency may be due to experimental difficulties in capturing failure states in stress-controlled testing devices.

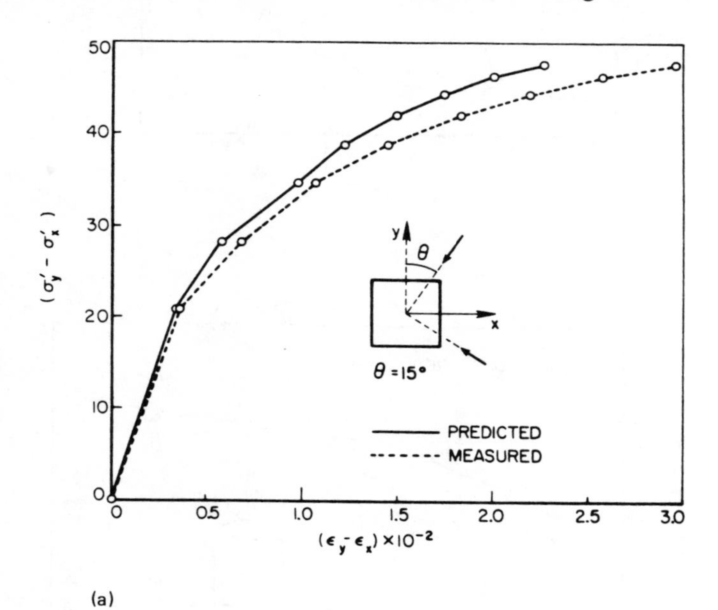

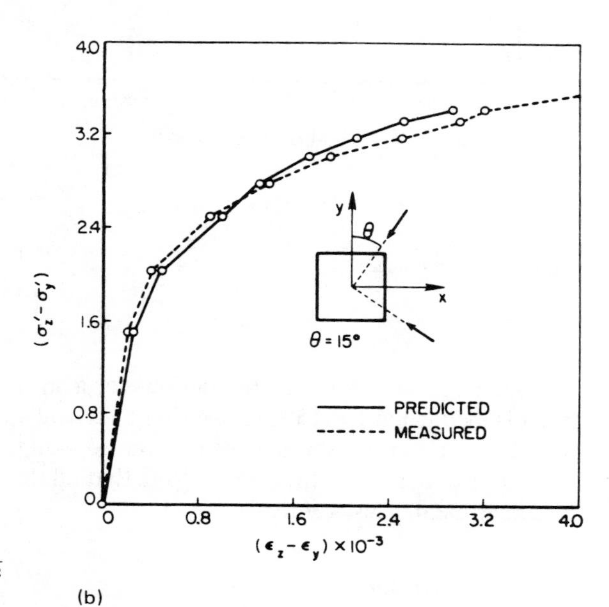

Figure 14.3. Kaolinite clay–undrained shear test with $\theta = 15°$:
(a) $(\sigma_x - \sigma_y)$ vs. $(\varepsilon_x - \varepsilon_y)$
(b) $(\sigma_y - \sigma_z)$ vs. $(\varepsilon_y - \varepsilon_z)$
(c) $(\sigma_z - \sigma_x)$ vs. $(\varepsilon_z - \varepsilon_x)$
(d) Pore-fluid pressure vs.

$$\frac{1}{\sqrt{2}}\sqrt{(\varepsilon_x - \varepsilon_y)^2 + (\varepsilon_y - \varepsilon_z)^2 + (\varepsilon_z - \varepsilon_x)^2}$$

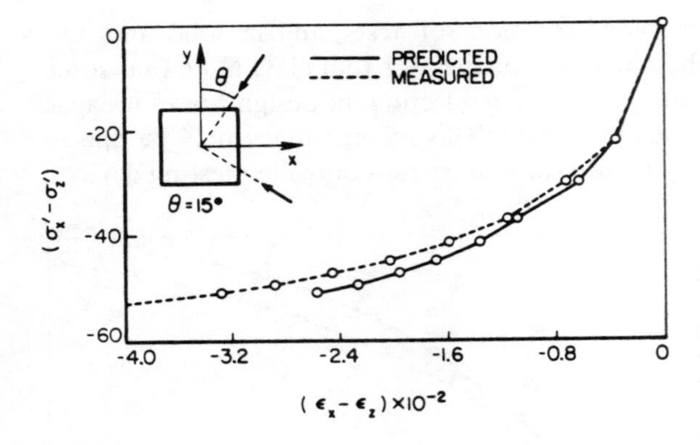

(c)

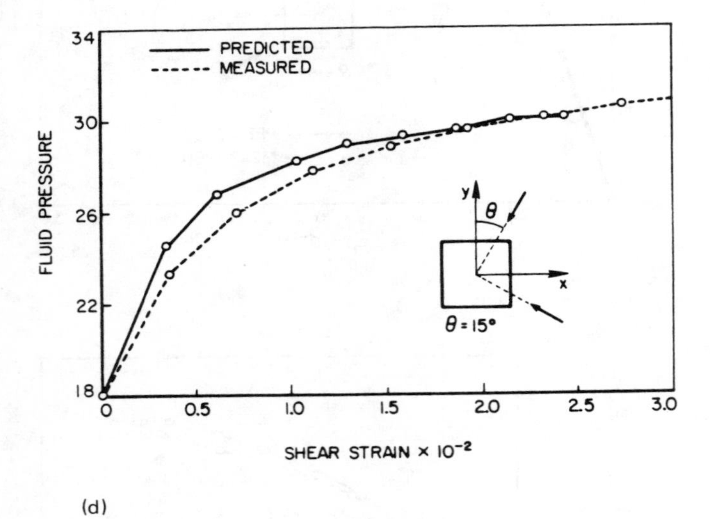

(d)

Figure 14.3. (Continued)

Figure 14.3 shows model predictions for a shear test in which the major principal stress is inclined at $\theta = 15°$ relative to the vertical axis of the soil specimen. This figure also shows a comparison between predicted (solid lines) and observed (dashed lines) behaviour of the soil in these tests. Note that all the model predictions agree well with the experimental test results.

14.4.2 *Dilating sand*

'Dilating sand' is a synthesized data set generated from real sand data. Figure 14.4 shows the corresponding conventional drained axial compression/extension test data.

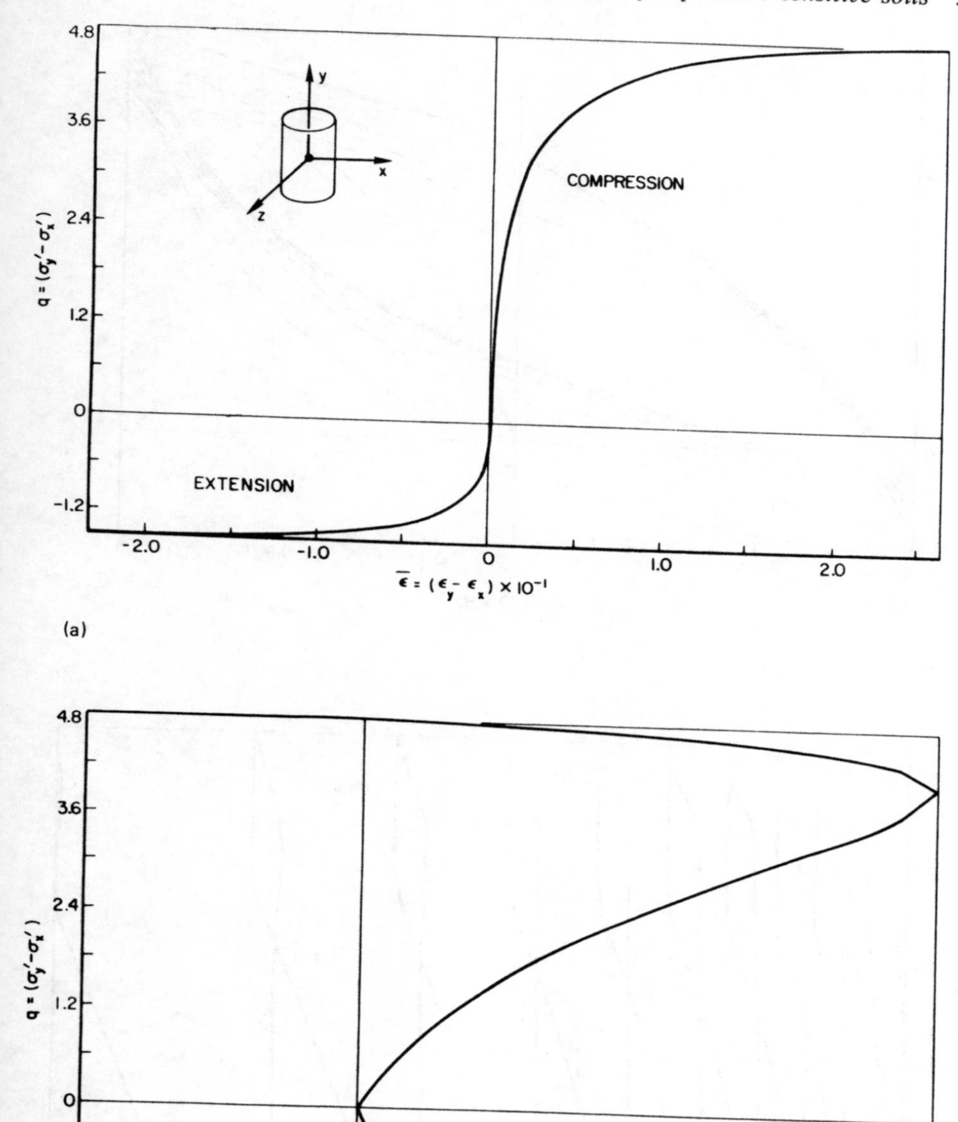

Figure 14.4. Dilating sand – drained axial soil test data: (a) shear stress vs. shear strain; (b) shear stress vs. volumetric strain

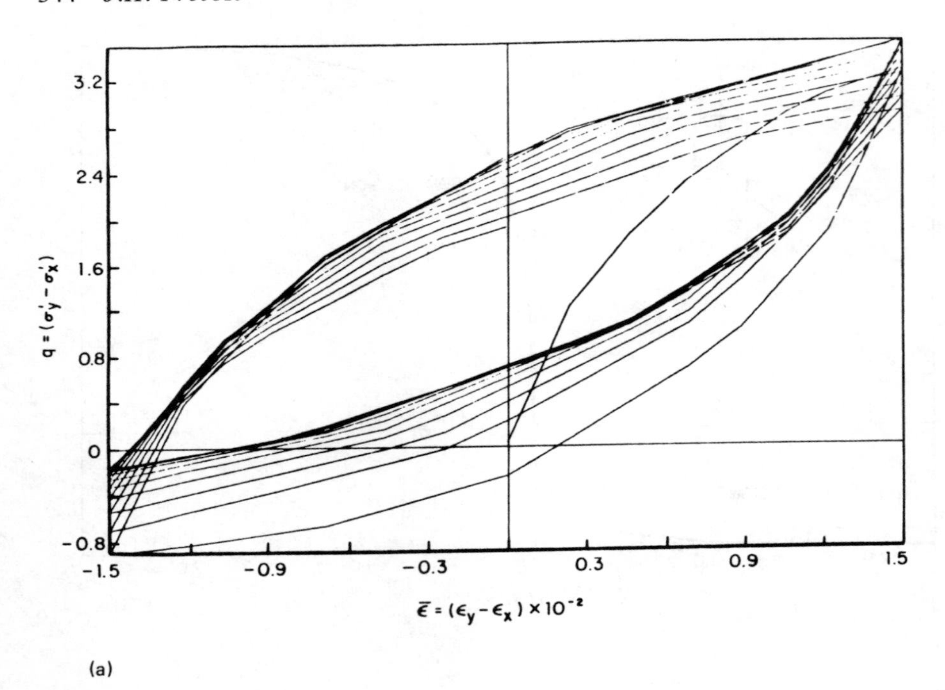

(a)

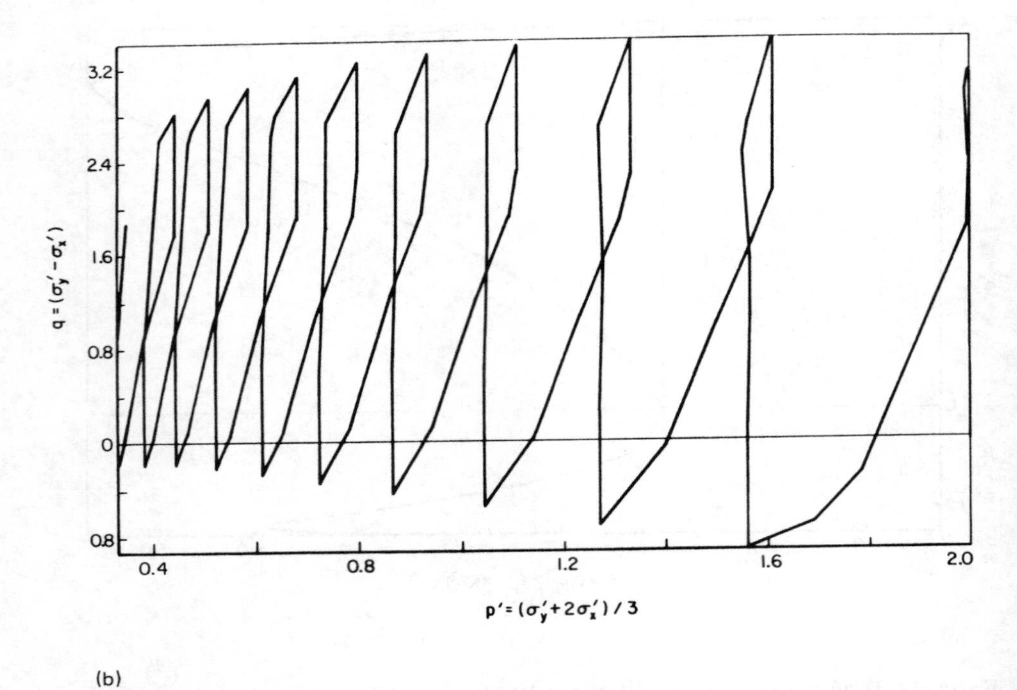

(b)

Figure 14.5. Dilating sand – undrained cyclic axial strain-controlled test (axial strain = 1%): (a) shear stress vs. shear strain; (b) shear stress vs. effective mean normal stress

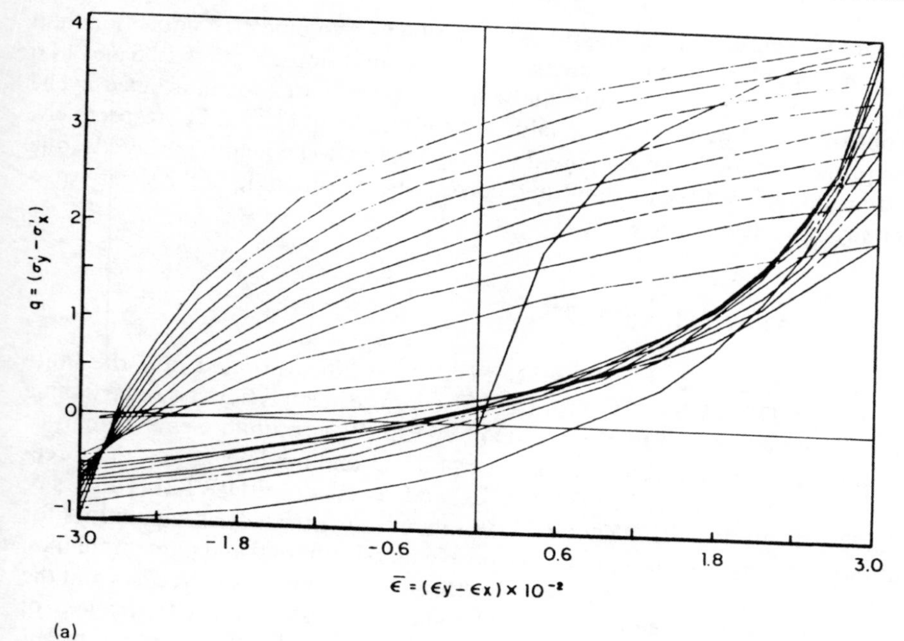

(a)

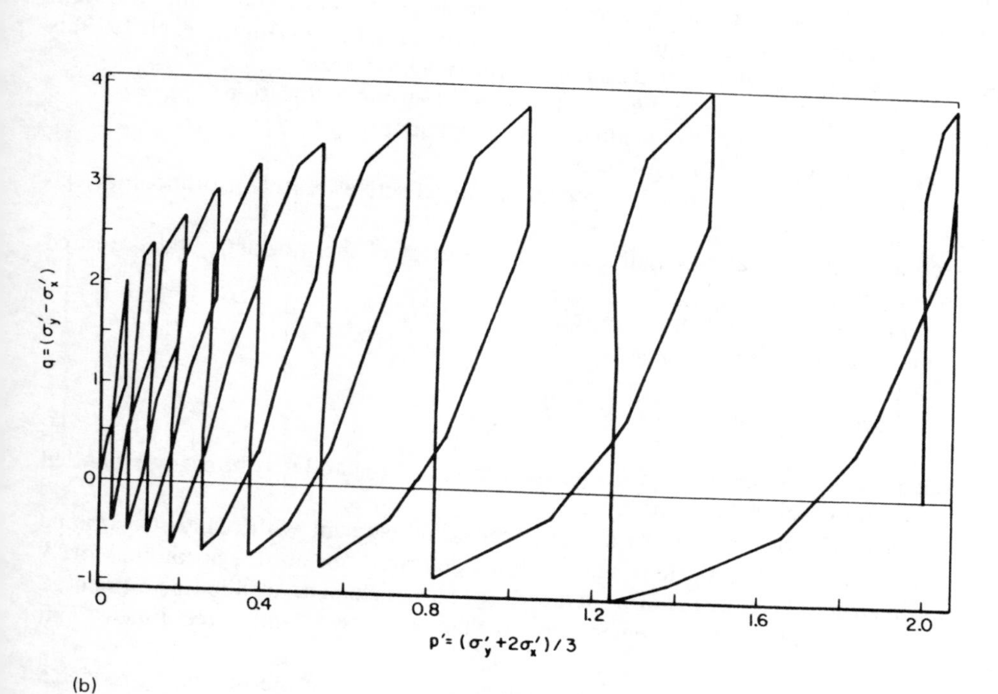

(b)

Figure 14.6. Dilating sand – undrained cyclic axial strain-controlled test (axial strain = 2%): (a) shear stress vs. shear strain; (b) shear stress vs. effective mean normal stress

As shown in this figure, the sand is assumed to exhibit first volumetric compaction than dilation in both compression and extension loading conditions. Figures 14.5 and 14.6 show the computed hysteresis loops and effective stress paths for undrained cyclic strain-controlled tests performed at axial strain amplitudes of 1% and 2%, respectively. Note the progressive build-ups of pore-fluid pressures as cyclic loading proceeds (due to the extension phase of the loading), and corresponding softening of the shear-stress vs. shear-strain hysteresis loops.

14.5 COMPUTATIONAL ASPECTS

The soil model described above has been coded and incorporated into the finite element program DYNA-FLOW (Prevost 1981a) for use in analysis of boundary value problems of interest in soil mechanics. DYNA-FLOW is a finite element analysis program for the static and transient response of linear and nonlinear two- and three-dimensional systems. DYNA-FLOW is an expanded version of DIRT II (Hughes & Prevost 1979). In particular, DYNA-FLOW offers transient analysis capabilities for both parabolic and hyperbolic initial value problems in both solid and fluid mechanics. There are no restrictions on the number of elements, the number of load cases and the number or bandwidth of the equations. Despite large system capacity, no loss of efficiency is encountered in solving small problems. In both static and transient analyses, an implicit-explicit predictor-multicorrector scheme (Hughes et al. 1979) is used. Some features which are available in the program area:

– selective specification of high- and low-speed storage allocations;

– both symmetric and non-symmetric matrix equation solvers;

– eigenvalue/vector solution solver;

– reduced/selective integration procedures, for effective treatment of incompressibility constraints;

– coupled field equation capabilities for treatment of thermoelastic and saturated porous media;

– isoparametric data generation schemes;

– mesh optimization options;

– plotting options;

– interactive options.

The element and material model libraries are modularized and may be easily expanded without alteration of the main code.

The element library contains a two-dimensional element with plane stress/plane strain and axisymmetric options, and full finite deformations may be accounted for. A three-dimensional element is also included. A contact element, a slide-line element, a truss element and a beam element are available for two- and three-dimensional analysis.

The material library contains a linear elastic model, a linear thermo-elastic model, a Newtonian fluid model, and a family of elasto-plastic models developed by the author.

Accuracy and versatility of the computer code DYNA-FLOW in applications of interest in geotechnical engineering have been demonstrated in examples reported in a

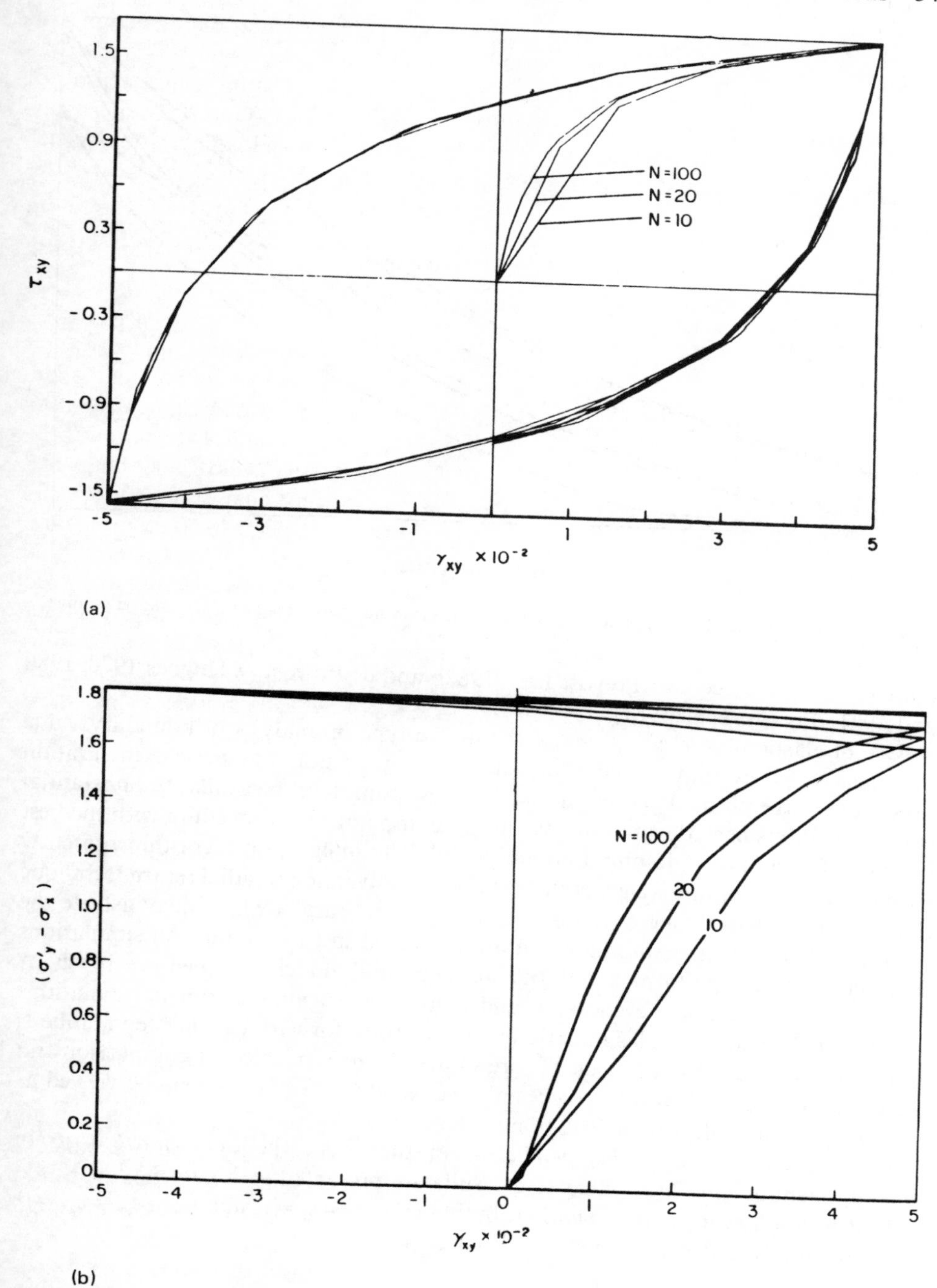

Figure 14.7. Dilating sand – sample shear test simulation: (a) shear stress τ_{xy} vs. shear strain γ_{xy}; (b) shear stress $(\sigma_y - \sigma_x)$ vs. shear strain γ_{xy}

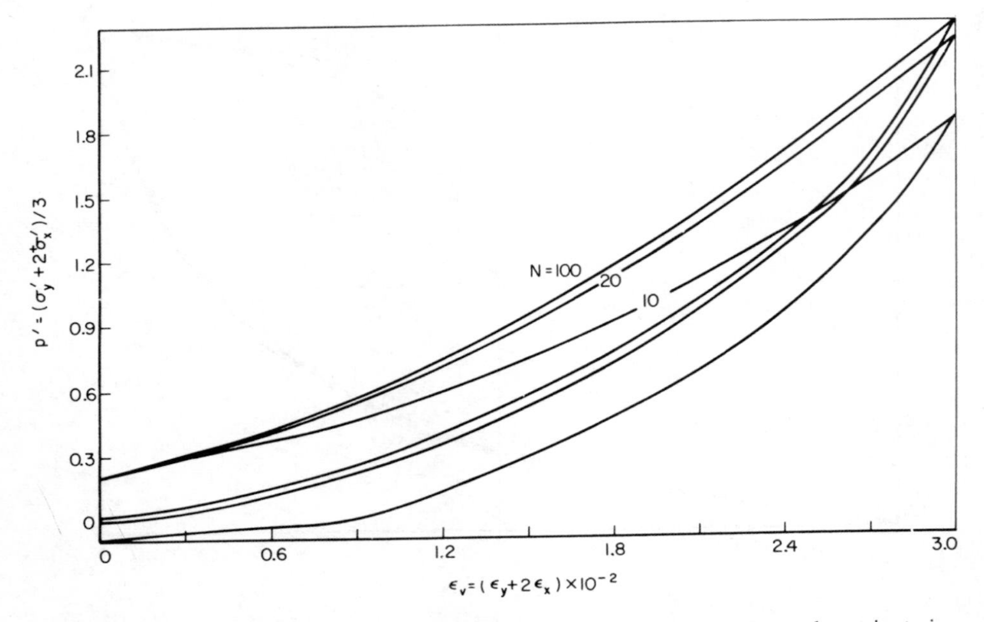

Figure 14.8. Dilating sand – hydrostatic test simulation: mean normal stress p′ vs. volumetric strain ε_v

number of papers (see also Prevost 1981/1982b and d; Prevost & Hughes 1978, 1980 and 1981; Prevost et al. 1981).

Use of elastic-plastic equations of the above type in analysis of boundary value problems requires that an efficient, 'sturdy' and accurate numerical integration procedure of the plasticity equations at the stress point level be available. Substantial efforts have thus been devoted to designing a computational procedure with the best balance of accuracy and computational speed. The integration algorithm presently used in the stress routine is a generalization of the conventional radial return technique (Krieg & Key 1976, Krieg & Krieg 1977). Numerical results which demonstrate the accuracy of the stress-point algorithm are presented in this section. All simulations reported here were performed with the 'dilating sand' model described previously. A number of monotonic and cyclic, axial and simple shear strain-experiment simulations are reported hereafter. Each simulation was performed for various load step numbers, and the computed stress-strain curves are shown on the same plot for comparison and evaluation of the stress routine accuracy. The results for 100 steps may be viewed as converged in the following calculations:

1. *Simple shear strain loading condition*: In that case, the only nonzero strain component is $\gamma_{xy} = 2\varepsilon_{xy}$. Calculated results are presented in Figure 14.7.
2. *Hydrostatic strain loading condition*: In that case, $\varepsilon_x = \varepsilon_y = \varepsilon_z$ and $\varepsilon_{xy} = \varepsilon_{yz} = \varepsilon_{zx} = 0$. Calculated results are presented in Figure 14.8.

14.6 SUMMARY AND CONCLUSIONS

A general analytical model which describes the nonlinear, anisotropic, elastoplastic, stress-strain-strength properties of the soil skeleton when subjected to complicated three-dimensional loading paths is proposed. A brief summary of the model's basic principle is included and the constitutive equations are provided. It is shown that the model parameters required to characterize the behaviour of any given soil can be derived entirely from the results of conventional soil tests. The model's accuracy is evaluated by applying it to represent the behaviour of both cohesive and cohesionless soils. Implementation of the proposed model in a general finite element program for solution of boundary value problems is discussed.

ACKNOWLEDGEMENTS

This work was supported in part by grants from the Naval Civil Engineering Laboratory, Port Heuneme, California, and the National Science Foundation (CEE-8120757). These supports are most gratefully acknowledged. Computer time was provided by Princeton University Computer Center.

REFERENCES

Hill, R. 1958. A general theory of uniqueness and stability in elastic plastic solids. *J. Meth. Phys. Solids* 6: 236–249.

Hughes, T.J.R., K.S. Pister & R.L. Taylor 1979. Implicit-explicit finite elements in nonlinear transient analysis. *Comp. Meth. Appl. Mech. Engng.* 17/18: 159–182.

Hughes, T.J.R. & J.H. Prevost 1979. DIRT II: a nonlinear quasi-static finite element analysis program. California Institute of Technology, Pasadena.

Hvorslev, M.J. 1937. Ueber die Festigkeitseigenschaften gestorler bindiger Boden. Ingeniørvidenskabelige Skrifter, A. no. 45, Copenhagen.

Jaumann, G. 1903. *Grundlagen der Bewegungslehre.* Leipzig.

Krieg, R.D. & S.W. Key 1976. Implementation of a time-independent plasticity theory into structural computer programs. In *Constitutive Equations in viscoplasticity theory: computational and eigineering aspects,* AMD vol. 20. New York: ASME.

Krieg, R.D. & D.B. Krieg 1977. Accuracies of numerical solution methods for the elastic-perfectly plastic model. *J. of Pressure Vessel Technology* (ASME) 99 (4): 510–515.

Prevost, J.H. 1977. Mathematical modeling of monotonic and cyclic undrained clay behavior. *Int. J. for Numer. and Analyt. Meth. in Geomech.* 1 (2): 196–216.

Prevost, J.H. 1978a. Anisotropic undrained stress-strain behavior of clays. *J. Geotech. Engng. Div.* (ASCE) 104 (GT 8): 1075–1090.

Prevost, J.H. 1978b. Plasticity theory for soil stress-strain behavior. *J. Engng. Mech. Div.* (ASCE) 104 (EM 5): 1177–1194.

Prevost, J.H. 1980. Constitutive theory for soil. In *Proceedings of the NSF/NSERC North American Workship on Plasticity and Generalized Stress-Strain Applications in Soil Engineering* (Montreal).

Prevost, J.H. 1981a. DYNA-FLOW: a nonlinear transient finite element analysis program. Report 81-SM-1, Department of Civil Engineering, Princeton University.

Prevost, J.H. 1981b. Consolidation of anelastic porous media. In *Proceedings of ASCE* 107 (EM 1): 169–186.

Prevost, J.H. 1982a. Nonlinear anisotropic stress-strength behaviour of soils. In *ASTM STP* 740 *on Shear Strength of Soils*: 431–455.

Prevost, J.H. 1982b. Nonlinear transient phenomena in saturated porous media. *Comp. Meth. Appl. Mech.* 30 (1): 3–18.

Prevost, J.H. 1982c. Two-surface vs. multi-surface plasticity theories. *Int. J. for Numer. and Analyt. Meth. in Geomech.*

Prevost, J.H. 1982d. Nonlinear transient phenomena in elastic-plastic solids. *J. Engng. Mech. Div.* (ASCE)

Prevost, J.H., B. Cuny, T.J.R. Hughes, R.F. Scott 1981. Foundations of offshore gravity structures: analysis. *J. Geotech. Engng. Div.* (SSCE) 107 (GT 2): 143–165.

Prevost, J.H. & T.J.R. Hughes 1978. Analysis of gravity offshore structure foundations subjected to cyclic wave forces. In *Proceedings of Offshore Technology Conference* (Houston, Texas), 1: 1809–1818.

Prevost, J.H. & T.J.R. Hughes 1980. Finite element solution of boundary value problems in soil mechanics. In *Proceedings of International Symposium on Soils under Cyclic and Transient Loading* (Swansea): 263–276. Rotterdam: Balkema.

Prevost, J.H. & T.J.R. Hughes 1981. Finite element solution of elasticplastic BVP. *J. Appl. Mech.* (ASME) 48 (1): 69–34.

Richard, R.E., R.D. Woods & J.R. Hall 1970. *Vibrations of soils and foundations*. Englewood Cliffs, New Jersey: Prentice-Hall.

Roscoe, K.H. & J.B. Burland 1968. On the generalized stress-strain behavior of 'wet' clay. In Heymand & Leckis (eds.), *Engineering plasticity*: 535–609. Cambridge (UK): Cambridge University Press.

Terzaghi, K. 1943. *Theoretical soil mechanics*. New York: Wiley.

Numerical modelling of soil response to cyclic loading using 'stress-reversal surfaces' 15

V.A. NORRIS

15.1 INTRODUCTION

The response of soils to cyclic loading may be modelled to various levels of sophistication, depending on the task in hand. For the development of practical models it is normal to fit the model to experimental data at only one point in each cycle (Carter et al. 1982, Van Eekelen 1982). Models based on complete stress-strain histories may be too complex for use in practice and too sophisticated for fast development, but are expected to provide a sound basis for simpler, practical models in the long term. Amongst such models are those which use stress-reversal surfaces to incorporate the dependence of material behaviour on stress path history.

The concept of a stress-reversal surface may be introduced most simply by using a yield surface as an example. Stress-reversal in this case occurs when a stress point initially moving in contact with the yield surface moves away from it, the constitutive law changing from plastic to elastic. When the stress point recontacts the yield surface the constitutive law changes back to plastic. In general a stress-reversal surface may be defined as a surface dividing stress space into zones which have different strain-hardening rules. The need for such surfaces in the detailed simulation of soil response to cyclic loading will be made apparent as their historical basis is described.

15.2 HISTORICAL PERSPECTIVE

The concept of a stress-reversal surface may be traced back to the work of both Mroz (1967) and Iwan (1967). Mroz uses a single yield surface together with a set of hypersurfaces to define the variation of the plastic modulus whereas Iwan uses a set of yield surfaces, but essentially their models serve the same purpose.

Iwan's model is more directly related to classical plasticity theory and will for this

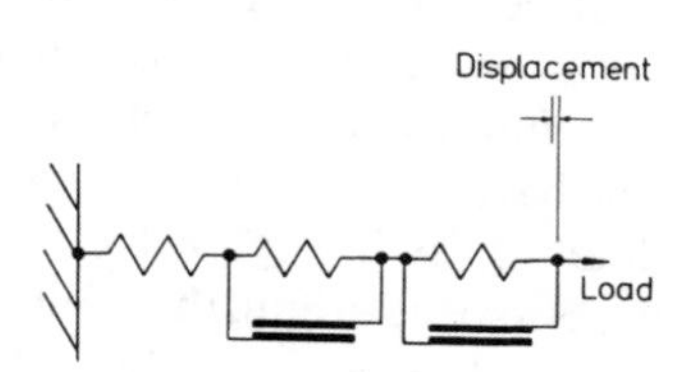

Figure 15.1. Series-parallel, spring-slider model

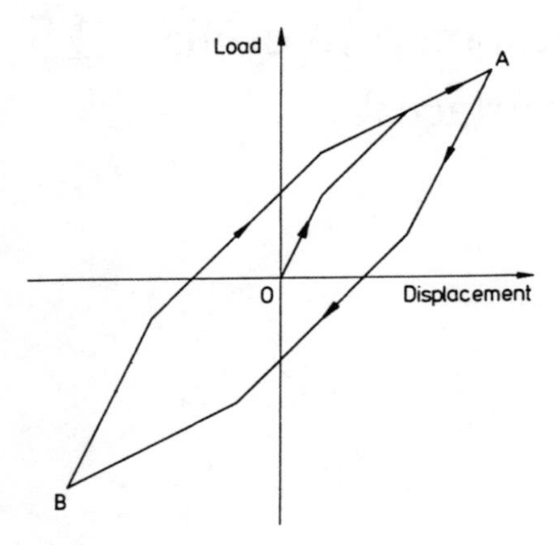

Figure 15.2. Load-displacement response to symmetric, two-way loading

reason be described first. He assumes that material behaviour may be described by the combined action of a set of elastoplastic elements, such as those shown in Figure 15.1 for uniaxial loading. The load-displacement response of such a system is shown in Figure 15.2 for symmetric, two-way loading. Here the unloading and reloading paths are exact copies of the initial loading path enlarged by a factor of two, (Bauschinger effect of the Massing type), in approximate agreement with real material behaviour. The same behaviour may be described in terms of average stress by the combined action (see Koiter 1953) of a set of yield surfaces (Figure 15.3) each of which exhibits linear kinematic hardening – the surface translates with the stress point at a rate dependent on the plastic strain rate. For uniaxial conditions each yield surface contacted by the stress point must move with it until stress-reversal occurs. After stress-reversal the stress point has twice as far to move as during initial loading before it recontacts a yield surface (see Figures 15.3 (b) and (c)). The resulting stress-strain path thus has the same form as the load-displacement path of Figure 15.2.

In order to apply the multiple yield surface technique to more than one dimension it is necessary to adopt a rule defining the direction of yield surface translation. Iwan assumes in an example that each surface in contact with the stress point moves in the direction of its corresponding plastic strain rate, but other rules may be used.

As described so far the multiple yield surface model is capable of approximating nonlinear behaviour in a piecewise, linear manner. Continuously nonlinear simulation can, however, be achieved by the use of a continuous distribution of yield surfaces. A model formed in this manner will be used to illustrate the function of a stress-reversal surface.

For simplicity, the example of uniaxial cyclic loading handled by a finite number of yield surfaces (Figure 15.3) will be reexamined using a continuous distribution of surfaces. Instead of defining individual surfaces by means of individual geometries, locations, and hardening moduli it is now necessary to use functions to describe the continuous variation of these properties. As an example, when all surfaces are circles the hardening modulus may be assumed to be a function of the surface radius. During loading, as in Figure 15.3(a), for the case of circular surfaces initially centred at the stress-space origin, the radius of the yield surface just contacted is equal to the uniaxial stress. Hence the radius of the stress-reversal surface is equal to the maximum

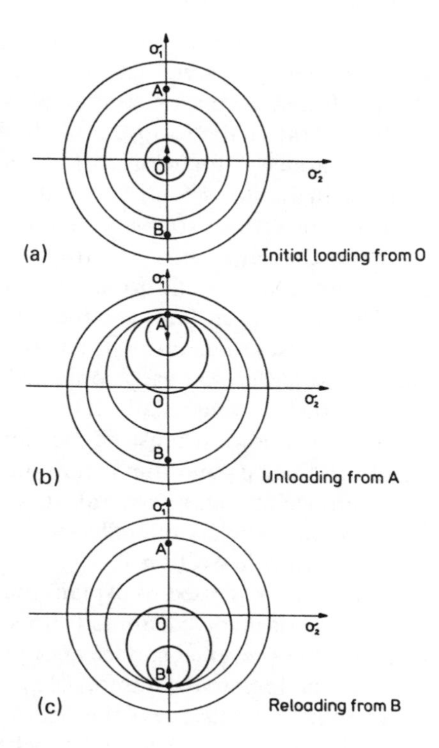

Figure 15.3. Multiple yield surface model

compressive stress. During the early stages of unloading, however, any surface just contacted by the stress point passes through the stress-reversal point, so that its radius is equal to half the distance between the stress point and the stress-reversal point. During the later stages of unloading, if the maximum stress in tension is greater than the maximum stress in compression, the stress point goes beyond the stress-reversal surface and the newly contacted yield surfaces are now centred on the stress-space origin, as during initial loading. Thus the stress-reversal surface serves to divide the stress-space into zones with different yield surface distributions and hence with different hardening rules.

In the case of Mroz's hardening surface model it is the overall plastic modulus, K_p, that is specified by the surfaces, so that there is no need to sum the contributions from a set of surfaces.

$$d\boldsymbol{\varepsilon}^p = \frac{\bar{\mathbf{n}}\mathbf{n}^T d\boldsymbol{\sigma}}{K_p}, \quad \mathbf{n}^T d\boldsymbol{\sigma} \geq 0 \tag{15.1}$$

where:

$d\boldsymbol{\varepsilon}^p$ = plastic strain increment vector;
$d\boldsymbol{\sigma}$ = stress increment vector;
$\mathbf{n}$ = unit vector normal to the yield surface and the hardening surface in an outward direction;
$\bar{\mathbf{n}}$ = unit vector defining the direction of the plastic strain increment.

For associative plastic flow $\bar{\mathbf{n}} = \mathbf{n}$. For non-associative plastic flow a rule is used to define $\bar{\mathbf{n}}$. It may be seen that when the stress path is tangential to the hardening surface $\mathbf{n}^T d\boldsymbol{\sigma} = 0$ and no plastic straining takes place (unless, of course, $K_p = 0$).

In diagrammatic form the model has the same appearance as the multiple yield surface model of Figure 15.3. In order to model the Bauschinger effect the surfaces move with the stress point, just as do the yield surfaces of Iwan's model. By appropriate selection of surface sizes and hardening moduli it is possible to make both models simulate the same uniaxial stress-strain behaviour. Whether they can be made to simulate the same multi-axial behaviour depends upon whether the translation rules used for the yield surfaces and the hardening surfaces are compatible. In his 1967 paper Mroz proposes that, as the stress point moves from one point of surface contact to the next, those surfaces already contacted should move with the stress point so as to touch the next surface tangentially at the stress point. Prevost (1978a), in fact, insists that this tangential condition must be met for the sake of 'uniqueness of the plastic loading function'. Such a condition makes the hardening surface model incompatible with the yield surface model, in general. It is considered by the author to be an unnecessary restraint on the behaviour of hardening surface models and an argument for its removal is put forward in Section 15.4.

In the example used to explain the function of the stress-reversal surface only one such surface is involved. Should stress-reversal occur whilst the stress point is inside an existing stress-reversal surface another such surface is formed (see Figure 15.4). There is, in fact, no limit to the number of possible stress-reversal surfaces. However, when the stress point contacts one of these surfaces and pushes it along the latter ceases to serve as a stress-reversal surface, and may be forgotten. Its function is taken over by a larger stress-reversal surface with which it was in contact before being pushed by the stress point. Nevertheless, there is a need to avoid the possible computational expense and complexity of an unlimited number of stress-reversal surfaces. This may be done by retaining only the n most recent stress-reversal surfaces and assuming some rule for the distribution of the hardening surfaces between the outermost stress-reversal surface and the boundary surface (Mroz et al. 1981).

The simplest example of such reduced models is that in which the only stress-reversal surface is the yield surface. Effectively, such models were first proposed by

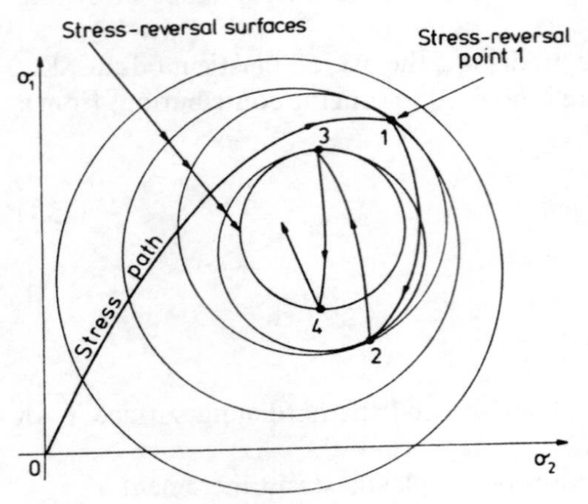

Figure 15.4. Multiple stress-reversal surface model

Dafalias & Popov (1975) and Krieg (1975), defining the plastic modulus in terms of the distance between the stress point and a point on the boundary surface rather than using a continuum of plastic hardening surfaces. Such a model was applied to soils by Mroz et al. (1979), employing a full critical state ellipse as the boundary surface. Realising that there were certain disadvantages in using such a simplified model, Mroz et al. (1981) added another stress-reversal surface to their model and defined the plastic modulus by means of a continuous distribution of hardening surfaces, but simultaneously reduced the yield surface to a point to retain computational simplicity.

A comparison of the results obtained using the various forms of stress-reversal surface model will now be given, reference being made to those of other models of similar design.

15.3 APPLICATION OF THE STRESS-REVERSAL SURFACE TECHNIQUE

Application of the full stress-reversal surface model, with no limitation on the number of reversal surfaces, has been published only by Mroz et al. (1978). The disadvantage of using such a model is illustrated in Figure 15.5. For constant stress, two-way, cyclic, undrained, triaxial loading of a normally consolidated soil each stress-reversal occurs inside the latest stress-reversal surface so that all such surfaces are retained. Only the latest reversal surface is used during the cyclic loading but all the surfaces would be used during subsequent monotonic loading to failure. An important feature to note is that the stress path stabilises to the right of the boundary surface centre, a physically real phenomenon which some models cannot match.

Little work has been carried out using this multiple stress-reversal surface model, but much has been done by Prevost using models of the form of Mroz's original

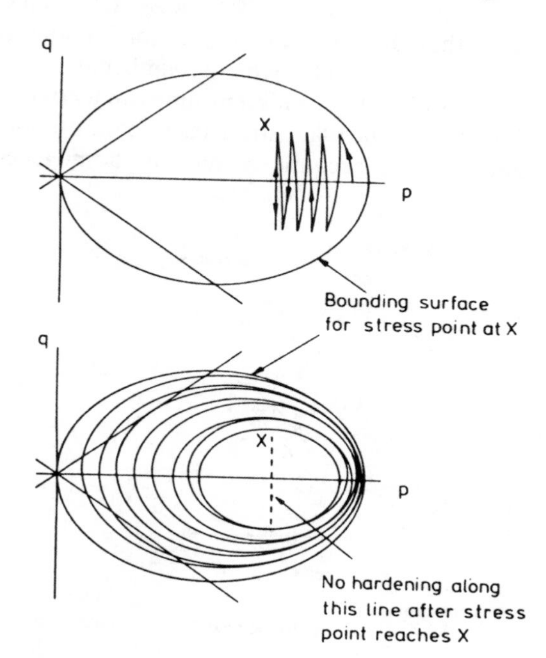

Figure 15.5. Multiple stress-reversal surface model – constant stress amplitude, two-way, cyclic, undrained, triaxial loading of a normally consolidated soil

multiple hardening surface model. Such models have the disadvantages that they require the use of a number of surfaces and are piecewise linear so that they record the occurrence of stress-reversal only approximately. They do, however, have the advantage that the number of surfaces required is independent of the stress path and may be specified to suit the degree of accuracy required. Prevost (1978b) points out that the technique of using a number of distinct surfaces is completely general, whereas using a particular formula to predict stress-strain behaviour is not. The strength of this argument is, however, reduced by the fact that the formulae required to describe the size (and possibly shape) variation and also the translation of these distinct surfaces, as well as their associated plastic moduli, are themselves not completely general. Prevost (1977) initially developed a multiple surface model for undrained analysis based on Von Mises yield criterion. Using a set of circles in deviatoric stress space he was able to simulate deviatoric stress-strain histories but not pore pressures. He later developed (Prevost 1978b) a general model using a full critical state ellipse for the boundary surface. He states that the hardening surfaces (which he calls yield surfaces) must, for uniqueness, always touch tangentially when moving with the stress point. As a result, a restraint is placed on the direction of the normal to the yield surface which, though perhaps not important in deviatoric stress space, is very important where a hydrostatic stress component is involved. To deal with this restraint he employs a non-associative flow rule for predicting the direction of plastic strain. (Use of the latter is to be avoided, if possible, for computational reasons, and an alternative is considered in Section 15.4.)

In applying his model to soil behaviour Prevost first evaluates the model's parameters by means of standard test data. In his 1977 paper, where he employs his deviatoric stress model for undrained conditions, he uses parameters obtained from monotonic and cyclic strain-controlled simple shear tests to predict the results of (1) cyclic stress-controlled simple shear tests, (2) monotonic triaxial compression and extension tests, and (3) cyclic stress- and strain-controlled triaxial tests. The results are good, though in the case of cyclic loading the results are plotted versus the number of cycles, i.e. the detailed stress-strain behaviour within each cycle is not given. Later he (Prevost 1978a) bases his undrained analyses on triaxial tests and predicts the results of simple shear and plane strain tests. In the case of simple shear a finite element analysis is used to take account of the non-homogeneity of the stress distribution. These

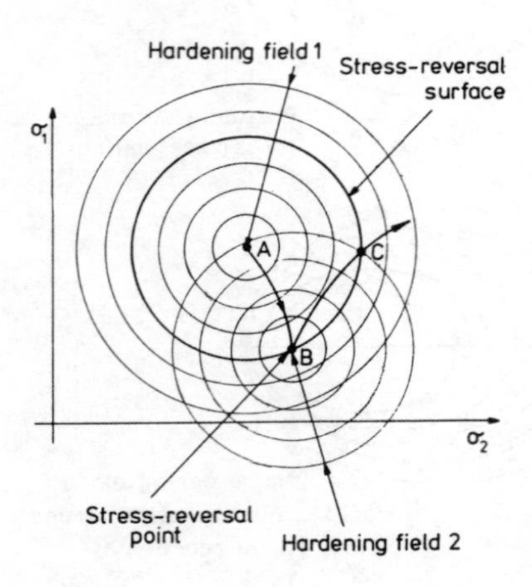

Figure 15.6. Hueckel & Nova multiple, stress-reversal surface model

undrained analyses do not predict pore pressures as no account is taken of volumetric plastic strains. With his more general model, using a full effective stress tensor and a boundary surface based on a full critical state ellipse, he shows (Prevost 1979) how the model parameters are derived from a consolidation-swelling test together with a triaxial test, but does not compare its predictions with real soil behaviour.

Petersson & Popov (1977) also have developed a variation on Mroz's multiple hardening surface model, but have not applied it to soils.

Another model of similar function is that proposed by Nova & Hueckel (1980). They do not use plasticity theory inside their boundary surface, but they do use continuously distributed sets of hardening surfaces to define 'paraelastic' constitutive laws. At stress-reversal the new hardening surfaces are centred at the stress-reversal point instead of being made to contact it. The effect is illustrated in Figure 15.6. As the stress point moves from point A to B hardening is controlled by hardening field 1. After stress-reversal at B hardening is controlled by hardening field 2 until the stress point goes outside the stress-reversal surface at point C. Thereupon hardening is once again controlled by hardening field 1. The functioning of this model is thus similar to that of the multiple stress-reversal surface model of Mroz et al. (1978, 1981), the main difference being that the paraelastic behaviour determined by a particular continuum of hardening surfaces centred at the stress reversal point is piecewise independent of the stress path prior to stress-reversal.

Of the elasto-plastic, stress-reversal surface models the two-surface model has been most widely used. Dafalias & Popov (1975) were the first to develop such a model followed by Krieg (1975), but without special reference to soil phenomena. Mroz et al. (1978, 1979, 1981) developed two-surface models for soils, using a full, effective stress vector and a boundary surface founded on the complete critical state ellipse. Confining their initial investigations to triaxial plane stress paths they were able to use a two-dimensional model (see Figure 15.7). Rather than trying to match stress-strain plots accurately they tried first of all to obtain all-round, qualitative agreement with real soil behaviour, particularly with regard to effective stress paths. Figure 15.8 shows analytical and experimental (Henkel 1960) effective stress paths for undrained triaxial tests on soil at various overconsolidation ratios. In general agreement with real soil behaviour stress paths to the right of the boundary surface centre curve to the left, and vice versa. Figure 15.9 compares predicted K_o consolidation and swelling stress paths with those of a remoulded clay. It may be seen that the general form is the same, the main difference being due to the difference in the slope of the consolidation paths. This can be attributed to the form of the boundary surface, i.e. it is a defect of the critical state ellipse rather than the kinematic hardening model. A correction could be made by allowing the boundary surface to translate and/or rotate (Mroz et al. 1979) but the additional complexity involved discourages further investigation at present. (For an

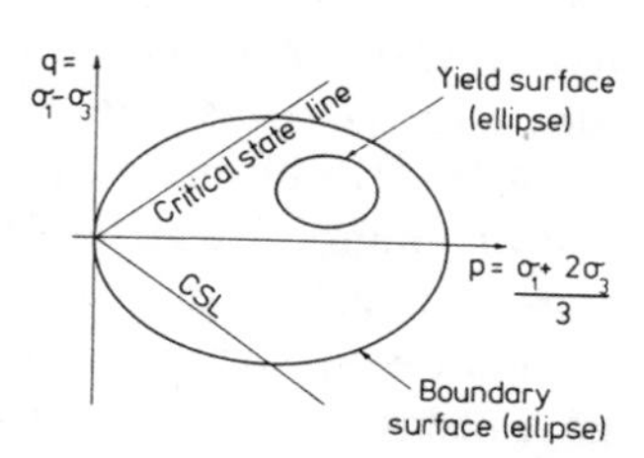

Figure 15.7. Two-surface model in p – q stress plane

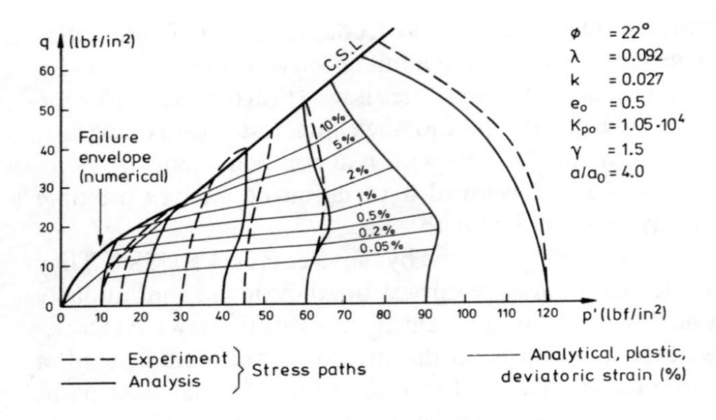

Figure 15.8. Undrained stress paths for the two surface model in p′, q space in comparison with experimental results for weald clay (after Hankel 1960)

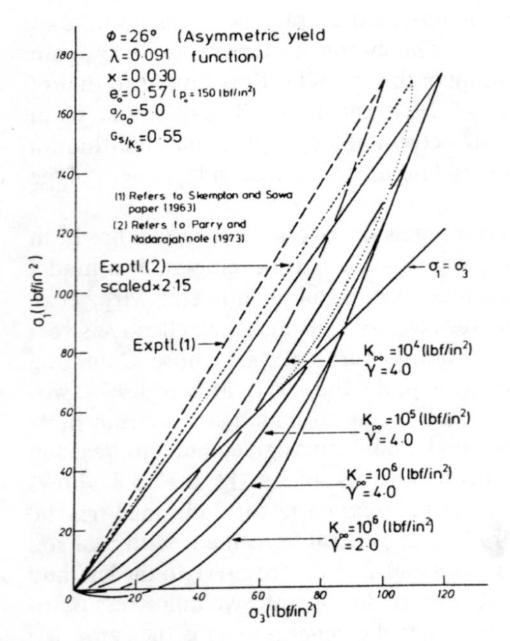

Figure 15.9. Two-surface model – stress paths for K_o consolidation and swelling

experimental investigation of this matter see Lewin 1973). Perhaps the most important finding concerning the two-surface model is that for cyclic, undrained, triaxial loading the stress path always stabilises beneath the apex of the boundary surface when an associative plastic flow rule is used. Thus for normally or lightly overconsolidated soil the stress path travels to the left but for OCR greater than about 2 the stress path travels to the right. Such behaviour is not true of real soils, especially overconsolidated soils, for which the stress path generally travels to the left. Also, for stress cycles of sufficient amplitude the ultimate maximum stress state is normally close to the critical state line. Use of a non-associative plastic flow rule would change the location of ultimate stress states and might improve the performance of the two-surface model to some extent.

In order to overcome this defect of the two-surface model two alternative techniques have been considered by Mroz et al. (1981): (1) reduction of the boundary surface size as a function of accumulative deviatoric plastic strain, so that eventually the cyclic stress path always reaches the critical state line at the apex of the critical state ellipse, and (2) the introduction of a new surface to undertake this same function whilst the boundary surface remains unaffected. The first technique is simpler but has the disadvantage that after cyclic loading has caused the stress path to approach the critical state line the strength predicted for monotonic loading is unrealistically low. Neither technique is capable of removing two other defects of the two-surface model. Firstly, stress-strain hysteresis loops can be modelled realistically only when the stress cycle is approximately centred on the hydrostatic axis. The variation of the plastic modulus, which is a function of distance from the boundary surface, will be quite different for loading and unloading if the cyclic loading is significantly asymmetric. Secondly, none of the two-surface models can predict realistically the location of stable states which arise for stress cycles of relatively small amplitude. Sangrey et al. (1969) found that for a particular OCR this location is linearly related to the stress cycle amplitude (see Figure 15.10).

A two-surface model has also been developed by Ghaboussi & Momen (1982). As for Prevost's original model (1977) they use the yield surface to predict only deviatoric strains, and hence avoid the problems involved in predicting effective stress paths using a full effective stress tensor. They do, however, predict the volumetric strains using a separate semi-empirical technique. Their model is intended for drained as well as undrained analyses so that full three-dimensional surfaces are required. By restricting the model to sands, and not trying to model softening or post-peak behaviour, they are able to use a conical failure surface of the Mohr-Coulomb type. The yield surface

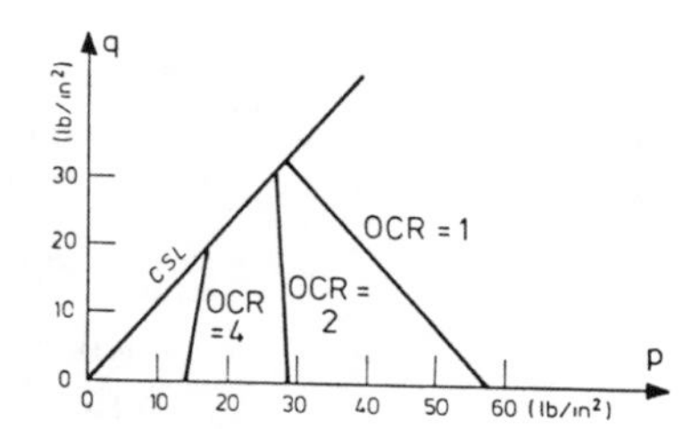

Figure 15.10. Loci of steady, cyclic states (Sangrey et al. 1969)

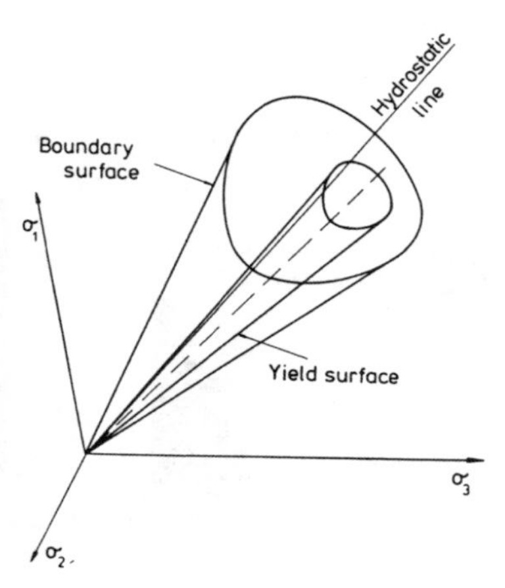

Figure 15.11. Two-surface, Mohr–Coulomb type model of Ghaboussi & Momen

rotates about the point at which it meets the apex of the boundary surface at the stress space origin (see Figure 15.11). An interesting feature of this model is that the value of the plastic modulus after stress-reversal is dependent on the angle of stress-reversal, being a maximum when this angle is equal to π (cf. Nova & Hueckel's (1980) model). Both stress-strain plots and effective stress paths are given for monotonic and cyclic triaxial loading and the results are quite accurate, membrane penetration effects being taken into account in the case of undrained loading. However, most cyclic test results are for two-way symmetric loading and no stress-strain results are shown for asymmetric loading. It seems likely that the hysteresis defect of other two-surface models in this respect would apply to this model also.

Mroz et al. (1981), having realised the defects of their two-surface model, made a partial return to the multiple, stress-reversal surface model by introducing one extra surface. At the same time they reduced the yield surface to a point to retain the model's relative simplicity. The direction of plastic flow was now defined, as for the complete multiple stress-reversal surface model, by the direction of the normal to the current hardening surface (see Figure 15.12). All hardening surfaces inside the stress-reversal surface contact this surface tangentially at the stress-reversal point. All surfaces outside are centred on the line joining the centre of the stress-reversal surface to that of the boundary surface as a linear function of surface size.

Figure 15.13 shows the extent of the new model's ability to predict the location of stability states for various amplitudes of cyclic loading. The right-hand set of stability states, for a normally consolidated soil, has an almost linear upper bound passing through the apex of the boundary surface. Thus the stress path can never reach the

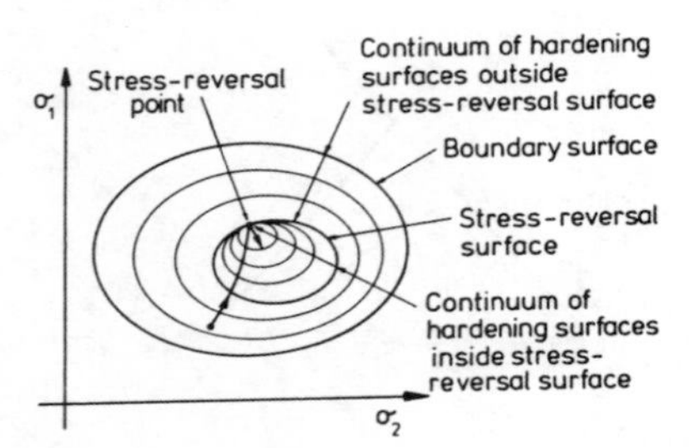

Figure 15.12. Model with point yield surface and one stress-reversal surface

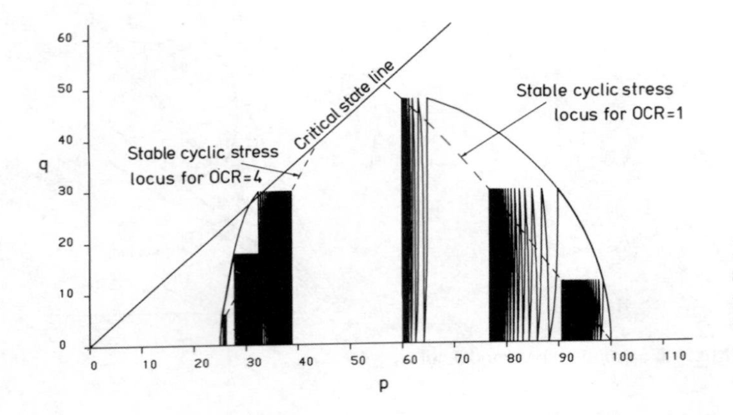

Figure 15.13. Point yield surface model – effective stress paths for various, constant stress amplitude, one-way, cyclic, undrained, triaxial loading

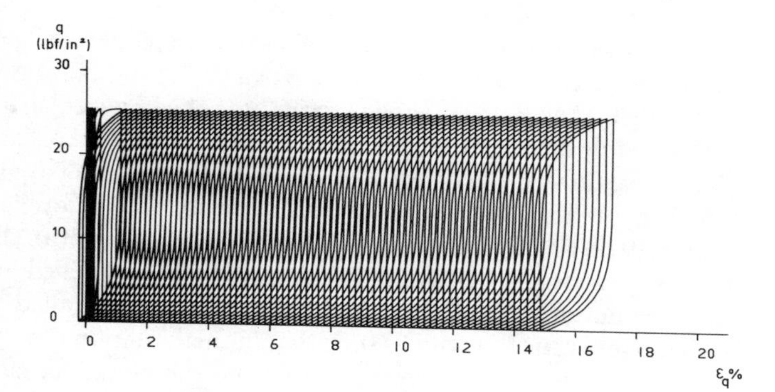

Figure 15.14. Point yield surface model – stress-strain plot for constant stress amplitude, one-way, cyclic, undrained, triaxial loading

critical state line during constant stress cyclic loading. The form of the maximum stress locus is, however, similar to that obtained experimentally by Sangrey et al. (1969) (Figure 15.10), and all that is required is some modification of the model to rotate the locus in an anti-clockwise direction. In the case of an overconsolidated soil the locus of maximum ultimate stress states is again of the correct form, but the stress path should normally travel to the right only during initial loading and thereafter travel to the left. Again the ultimate state locus needs to be rotated in an anti-clockwise direction. In their 1981 paper Mroz et al. dealt with this problem by proposing an additional shift of the stress-reversal surface towards the origin as a function of accumulated deviatoric strain. Because this technique is not very satisfactory for overconsolidated soil, analysis of the latter is considered using the model proposed in Section 15.4.

Another advantage of the new model is its ability to predict hysteresis realistically, even for asymmetric loading (see Figure 15.14).

15.4 FUTURE DEVELOPMENTS

Of the soil models compared in this chapter it is the author's opinion that only those which 'memorise' at least the latest two stress-reversal events are able to simulate cyclic soil phenomena realistically, i.e. the nonlinear, multiple, stress-reversal surface model of Mroz et al. (1978) and its simplified version with a zero size yield surface and only one stress-reversal surface (1981), the corresponding piecewise linear hardening surface

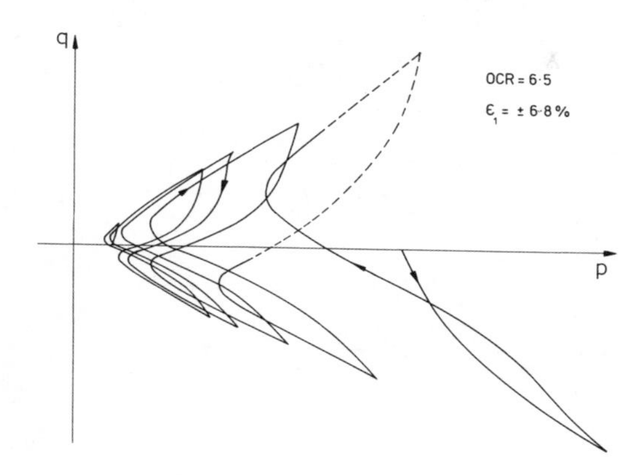

Figure 15.15. Effective stress path for two-way, constant strain amplitude, cyclic, undrained, triaxial loading of slate dust (Lewin 1979)

models of Prevost (1977, 1978a, b, 1979) and the paraelastic, nonlinear, multiple stress-reversal surface model of Hueckel & Nova (1979) and Nova & Hueckel (1980). None of these models has yet been shown, to the author's knowledge, to simulate effective stress paths reliably in the case of undrained cyclic loading. Prevost's technique of using a non-associative flow rule may be satisfactory for some stress paths, but difficult to apply to others, e.g. that shown for undrained, two-way, triaxial cyclic loading in Figure 15.15 (Lewin 1979) or similar results obtained by Takahashi et al. (1980). In his 1978b paper Prevost proposed a hardening rule for which, in undrained axial tests, the stress path joins the deviatoric apices of the hardening surfaces (called yield surfaces). His non-associative plastic flow rule is such that in this situation the hydrostatic component of plastic strain is always directed away from the boundary surface centre so that one would expect the effective stress path to be directed towards it. This result is much the same as that of the two-surface model of Mroz et al. (1979), i.e. a final stress state beneath the apex of the boundary surface, in disagreement with experimental observations. Nova & Hueckel's model has the advantages of a plasticity model with a non-associative flow rule, since the direction of the strain vector is not controlled by the direction of normality to the current hardening surface used to define stress-reversal. What is different is that the loading surface is centred at the previous stress-reversal point instead of being tangential to the previous stress-reversal surface at this point; hence its location is dependent only on the location of the stress-reversal point, not the preceding stress path. No attempt will be made here to discuss the relative merits of the two types of model; such a task should, perhaps, await the further development of their

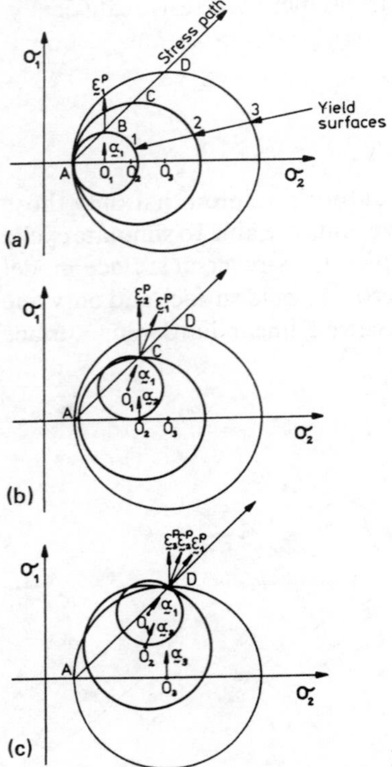

Figure 15.16. Application of Iwan's multiple yield surface model

constitutive laws as more detailed sets of experimental data become available. What will be attempted is the development of a form of hardening surface model which retains the advantages of an associative plastic flow rule without placing unnecessary restrictions on the direction of the plastic strain vector.

In order to provide a physical basis for this proposed model reference will again be made to Iwan's multiple yield surface model. Iwan assumed in an example that each yield surface translates in the direction of the plastic strain increment associated with that yield surface. The application of this rule is illustrated in Figure 15.16. After elastic behaviour, as the stress point moves from A to B (Figure 15.16(a)), yield surface 1 predicts plastic straining in the vertical direction for an associative plastic flow rule. Hence surface 1 initially translates vertically. As both the stress point and the yield surface move in different directions the direction of plastic strain rotates towards the stress path. The direction of translation of the yield surface consequently changes likewise so that its centre approaches the stress path asymptotically. When the stress path reaches point C (Figure 15.16(b)) yield surface 2 begins to translate in a similar manner, though out of phase with yield surface 1. The same applies to yield surface 3 when the stress point reaches D (Figure 15.16(c)). Thus this example illustrates an important fact not pointed out in Iwan's 1967 paper: yield surfaces in contact with the stress point are not, in general, tangential to each other. It follows that when stress-reversal occurs for one yield surface it does not necessarily occur for all other yield surfaces in contact with the stress point. Which surfaces are affected depends on the direction of the stress increment.

If such behaviour is to be modelled by means of hardening surfaces these also will not, in general, be tangential to each other at the stress point. Such a situation is depicted in Figure 15.17, using a finite number of surfaces for illustration purposes. Surface 4 is the stress-reversal surface and surface 1 the yield surface. If at stress-reversal the stress path enters surface 1 elastic behaviour is predicted. If, however, it is directed outside surface 1 elasto-plastic behaviour occurs, the initial plastic modulus being dependent on the surfaces between which the stress path passes, i.e. it is dependent on the direction of the stress-path. Should the stress path remain on the stress-reversal surface the behaviour is the same whether defined by the elastic zone within the yield surface or the stress-reversal surface, since the latter predicts zero plastic strain in this case. Thus the condition of continuity of the constitutive law at the stress-reversal surface is preserved. After stress-reversal the condition of uniqueness of the stress-strain law may be maintained by using the largest surface in contact with the stress point to define the plastic hardening modulus and direction of plastic strain. The use of this rule will be considered for stress path OAB in Figure 15.17. If the stress path initially enters

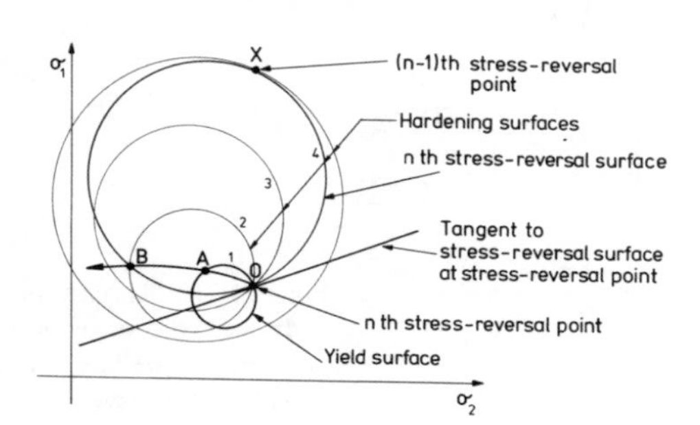

Figure 15.17. Proposed model – hardening surfaces at stress-reversal

the yield surface, then contacts surface 1 at A and surfaces 2 and 4 at B, the plastic hardening modulus changes from a relatively high value inside surface 2 to a relatively low value at B defined by hardening surface 4, the stress-reversal surface. Outside surface 4 the plastic hardening modulus will be defined in the same way as before stress-reversal occurred, i.e. by a set of surfaces passing through another stress-reversal point, X. This example shows that the model as described so far has two defects: (1) the plastic modulus may change suddenly as the stress point contacts the stress-reversal surface, and (2) the condition that the yield surface is tangential to the current hardening surface (see Equation 15.1) requires that the yield surface suddenly rotates when the stress point reaches B. Such discontinuous behaviour could be avoided by allowing the hardening surfaces inside the stress-reversal surface, but not yet contacted by the stress point, to gradually rotate towards a state of tangential contact with the stress-reversal surface; the final state would be reached as the stress point contacted the stress-reversal surface. It has, however, been found that even with a discontinuity the model is capable of handling undrained cyclic loading of some soils quite well.

The application of the proposed model will be illustrated by means of the experimental stress path shown in Figure 15.15. Here an isotropically overconsolidated slate dust (Lewin 1979) has been subjected to two-way, constant strain, cyclic loading under undrained conditions. It is assumed that for isotropic stress history and current loading the translation of the hardening surfaces is along the hydrostatic axis, as for older models. The state of the model's surfaces after (a) isotropic consolidation and (b) isotropic swelling is shown in Figures 15.18(a) and (b). The new type of behaviour begins with triaxial unloading (Figure 15.18(c)). The model's predictions, determined by the hardening surface most recently contacted, are the same as for older models. For an associative plastic flow rule the hydrostatic component of the plastic

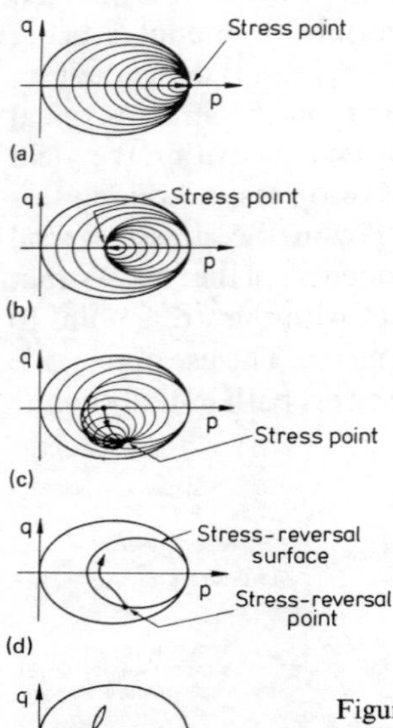

Figure 15.18. Application of proposed model – simulation of two-way, cyclic, undrained triaxial loading of overconsolidated soil: (a) isotropic consolidation; (b) isotropic swelling; (c) triaxial, undrained unloading; (d) triaxial, undrained reloading; (e) stress path after second unloading

strain increment is directed towards the origin, i.e. the volumetric plastic strain increment, $d\varepsilon_v^p$, is negative (dilatant). For undrained conditions the total volumetric strain is zero (if the compressibility of the pore fluid and soil granules is ignored). Therefore:

$$d\varepsilon_v = d\varepsilon_v^e + d\varepsilon_v^p = 0 \tag{15.2}$$

where:

$d\varepsilon_v$ = total volumetric strain increment;
$d\varepsilon_v^e$ = elastic volumetric strain increment.

It follows that:

$$dp\ (= K_e d\varepsilon_v^e) = -K_e d\varepsilon_v^p \tag{15.3}$$

where K_e = elastic bulk modulus; i.e. the mean effective stress increases. The translation of the contacted surfaces, however, is quite different from that of other models so that upon stress-reversal the model's predictions also are quite different. To prevent the stress path from travelling to the right, as takes place with other models, the hardening surfaces may now be arranged so that a compressive, volumetric, plastic strain is predicted and hence the stress path travels to the left (see Figure 15.18(d)). Eventually, as the stress path approaches the stress-reversal surface the distribution of hardening surfaces approaches compatibility with this surfaces and the stress path turns back to the right. Once the stress path passes outside the stress-reversal surface the model's predictions are similar to those during the initial triaxial unloading and the stress point continues travelling to the right. Upon stress-reversal the same behaviour is repeated inversely so that the stress path takes the form shown in Figure 15.18(e). In this way, for several cycles of loading, it is possible for the model to predict a stress path similar to that of the real soil shown in Figure 15.15.

15.5 CONCLUSIONS

In modelling inelastic behaviour of soil subjected to cyclic loading by means of strain-hardening surfaces it is important to take the following two factors into account:

1. The strain-hardening rule changes when the stress path returns inside the current hardening surface, an event referred to as 'stress-reversal'.
2. The strain-hardening rule reverts to an earlier hardening rule when the stress path reaches a certain locus in stress space. This locus depends on the point of stress-reversal and is therefore referred to as a stress-reversal surface. Its size and location and the hardening rule which it invokes are dependent on a previous stress-reversal event.

In order to take account of these two factors realistically it is necessary for a model to remember at least the two most recent stress-reversal events. Models consisting of only a boundary surface and a yield surface can remember only the latest stress-reversal event and are, therefore, incapable of modelling detailed soil behaviour. Of models known to the author only those of Mroz et al. (1978, 1981), Nova & Hueckel (1980), Prevost (1977, 1978) and Petersson & Popov (1977) are designed to memorise more than one stress-reversal event. Of these authors none has yet published a generally reliable method for predicting volumetric strains for soils. A short-term solution to this problem would be to use the multiple hardening surface model only for deviatoric strains and use a separate semi-empirical technique for volumetric strains, as did Ghaboussi & Momen (1982) with their two-surface model. For a more general solution, without the requirement of a non-associative flow rule, the author proposes that the multiple hardening surface models be freed of the restraint that hardening surfaces must touch tangentially at the stress point.

ACKNOWLEDGEMENT

The author wishes to acknowledge his indebtedness to the SERC for financial assistance in all his research associated with this chapter.

REFERENCES

Carter, J.P., J.R. Booker & C.P. Wroth 1982. A critical state soil model for cyclic loading. In G.N. Pande & O.C. Zienkiewicz (eds.), *Soil mechanics–transient and cyclic loads*: 219–252. Chichester: Wiley.

Dafalias, Y.F. & E.P. Popov 1975. A model of nonlinearly hardening materials for complex loading. *Acta Mech.* 21: 173–192.

Eekelen, H.A.M. van 1982. Fatigue models for cyclic degradation of soils. In G.N. Pande & O.C. Zienkiewicz (eds.), *Soil mechanics–transient and cyclic loads*: 459–468. Chichester: Wiley.

Ghaboussi, J. & H. Momen 1982. Modeling and analysis of cyclic behaviour of sands. In G.N. Pande & O.C. Zienkiewicz (eds.), *Soils mechanics–transient and cyclic loads*: 313–342. Chichester: Wiley.

Henkel, D.J. 1960. The relationships between the effective stresses and water content in saturated clays. *Geotechnique* 10: 41–54.

Hueckel, T. & R. Nova 1979. Some hysteresis effects of the behaviour of geologic media. *Int. J. Solids and Structs.* 15: 625–642.

Iwan, W.D. 1967. On a class of models for the yielding behaviour of continuous and composite systems. *J. Appl. Mech.* 34: 612–617.

Koiter, W.T. 1953. Stress-strain relations, uniqueness and variational theorems for elastic-plastic materials with a singular yield surface. *Q. Appl. Math.* 11: 350–354.

Krieg, R.D. 1975. A practical two-surface plasticity theory. *J. Appl. Mech.* 42: 641–646.

Lewin, P.I. 1973. The influence of stress history on the plastic potential. In *Proceedings of Symposium on the Role of Plasticity in Soil Mechanics* (Cambridge): 96–106.

Lewin, P.I. 1979. Private communication.

Mroz, Z. 1967. On the description of anisotropic workhardening. *J. Mech. Phys. Solids* 15: 163–175.

Mroz, Z., V.A. Norris & O.C. Zienkiewicz 1978. An anisotropic hardening model for soils and its application to cyclic loading. *Int. J. for Numer. and Analyt. Mech. in Geomech.* 2: 203–221.

Mroz, Z., V.A. Norris & O.C. Zienkiewicz 1979. Application of an anisotropic hardening model in the analysis of elasto-plastic deformation of soils. Geotechnique 29: 1–34.

Mroz, Z., V.A. Norris & O.C. Zienkiewicz 1981. An anisotropic critical state model for soils subject to cyclic loading. *Geotechnique* 31: 451–469.

Nova, R. & T. Hueckel 1980. A 'geotechnical' stress variables approach to cyclic behaviour of soils. In *Proceedings of International Symposium on Soils under Cyclic and Transient Loading* (Swansea): 301–314. Rotterdam: Balkema.

Parry, R.H.G. & V. Nadarajah 1973. A volumetric yield locus for lightly overconsolidated clay. *Geotechnique* 23: 450–454.

Petersson, H. & E.P. Popov 1977. Constitutive relations for generalised loadings. *J. Engng. Mech. Div.* (ASCE) 103: 611–627.

Prevost, J.H. 1977. Mathematical modelling of monotonic and cyclic undrained clay behaviour. *Int. J. for Numer. and Analyst. Meth. in Geomech.* 1: 195–216.
Prevost, J.H. 1978a. Anisotropic undrained stress-strain behaviour of clays. *J. Geotech. Engng. Div.* (ASCE) 104: 1075–1090.
Prevost, J.H. 1978b. Plasticity theory for soil stress-strain behaviour. *J. Engng. Mech. Div.* (ASCE) 104: 1177–1194.
Prevost, J.H. 1979. Mathematical modelling of soil stress-strain-strength behaviour. In *Proceedings of 3rd International Conference on Numerical Methods in Geomechanics* (Aachen): 347–361.
Sangrey, D.A., D.J. Henkel & M.I. Esrig 1969. The effective stress response of a saturated clay soil to repeated loading. *Can. Geotech. J.* 6: 241–252.
Skempton, A.W. & V.A. Sowa 1963. The behaviour of saturated clays during sampling and testing. *Geotechnique* 13: 269–290.
Takahashi, M., D.W. Hight & P.R. Vaughan 1980. Effective stress changes observed during undrained cyclic triaxial tests on clay. In *Proceedings of International Symposium on Soils under Cyclic and Transient Loading* (Swansea): 201–209. Rotterdam: Balkema.

A critical look at constitutive models for soils

16

G.N. PANDE & S. PIETRUSZCZAK

16.1 INTRODUCTION

Many numerical techniques for the solution of load deformation problems of soil masses satisfy the equations of equilibrium, the equations of compatibility of strains and the boundary conditions in a routine manner. It is the choice of an appropriate constitutive law for the behaviour of soils which ultimately determines whether the solution obtained is realistic and meaningful. No wonder that in recent years, much research effort has been devoted to the development of constitutive laws for soils. A rough count shows that there are nearly 40–50 models published in the literature. Two schools of thought on the philosophy of constitutive modelling seem to have emerged. The first advocates the development of many simple models with relatively few parameters, each for a specific application and for specific types of soils like sands, soft clays and overconsolidated clays. However, the second school advocates use of an all embracing model with a relatively large number of parameters, with the possibility that the general model reduces to specific models by suppressing certain aspects or assigning constant values to a certain set of parameters. This approach, though academically elegant, is unlikely to be favoured by the practising engineer.

The main aim of this chapter is to make an appraisal of various constitutive models for soils. It is impossible to cover all the existing models. The authors, therefore, have chosen only those models which are fairly well known. Models are continually being modified and updated. The versions as appearing in Pande & Zienkiewicz (1982) have been adopted as final at this moment in time (September 1982). Some models not appearing in the above reference have been included in the appraisal due to the author's familiarity with them.

Models for monotonic as well as cyclic behaviour of soils are considered. After general preliminaries in Section 16.2, Sections 16.3 and 16.4 present the models reviewed and criteria used for appraisal of their response in monotonic loading respectively. Section 16.5 presents the appraisal of models. Section 16.6 discusses the criteria adopted for appraisal of models of soil behaviour for cyclic/transient loading and Section 16.7 presents the appraisal itself.

16.2 GENERAL PRELIMINARIES

16.2.1 *Classification of models*

It is useful to classify the various constitutive models in different categories depending on the fundamental theories on which they are based. This classification is meant to help the practising engineer and researchers to identify those models which are of relevance in a particular situation, and readily correlate new models with existing theories or models.

Nonlinear elasticity models

This category of models is the simplest of all. Elastic moduli (Young's modulus, E, and Poisson's ratio, ν) are assumed to be nonlinear functions of stress. The models in this category usually involve many constants to which little physical significance can be attached but all constants can be derived from triaxial compression tests. The most popular model in this category appears to be due to Duncan & Chang (1970) and has been used in many practical situations. Some designers have gained sufficient experience in the use of these models and appear to get very good correlation with the field performance of structures.

There are, however, serious objections to the use of nonlinear elasticity models. The main one is that several formulations (Duncan & Chang's, for instance) are not conservative since energy may, under certain circumstances, be continuously extracted from the soil sample by subjecting it to a simple stress cycle.

Also the hypoelastic approach, although consistent in a mechanical sense, may predict an unrealistic material response for some loading paths, in particular for cyclic loading conditions. This is the result of loading – unloading criteria used in hypoelasticity which are defined by the sign of the stress work rate.

Therefore, the plasticity formulation seems to be of advantage and it also proves to be much more convenient in incorporating memory rules of particular loading events.

Elasto-plasticity models

The mathematical theory of elasto-plasticity is well established and has been used extensively in many branches of engineering. The framework of this theory has been a fertile ground for the development of soil models. Various permutations and combinations of yield functions, flow rules and hardening rules give rise to different models and with judicious choice, the theory has considerable flexibility to enable modelling the complex behaviour of soils in both monotonic and cyclic loading.

Isotropic hardening, kinematic hardening and mixed hardening rules with hardening due to plastic volumetric strains and plastic shear strains give rise to models of varying complexity. The Critical State Model (CSM), the Infinite Number of Surfaces (INS) model due to Mroz et al. (1979, 1982), and models due to Lade (1977, 1978) and Nova (1982) are a few examples in this category.

The authors believe that at the present stage of development elasto-plasticity based models are the ones most likely to be able to represent soil behaviour satisfactorily.

Elasto-visco-plasticity models

In certain problems of soil mechanics, the time effects, other than those due to consolidation, have to be included. Here elasto-visco-plastic models with time-dependent plastic strains are most logical.

Endochronic and endochronic-plasticity models

These models have been developed in the past decade and are based on plasticity without yield surfaces. The original work of Valanis (1971) on metals has been developed by Bazant & Bhat (1976) for concrete and Bazant & Krizek (1975) for soils. Valanis & Read (1982) have recently proposed a new endrochronic-plasticity theory. There are a number of arguments and counterarguments for the validity of this class of models. The models involve a large number of parameters and their general lack of transparency is likely to deter the practising engineer in using them, at least for some time to come.

16.3 MODELS REVIEWED

Table 16.1 lists various models reviewed in this paper. The range of applicability of these models is also indicated in the table. The well-established CSM (modified Cam Clay Model) is the basis of many models proposed for cyclic loading and has, therefore, been placed in the first position in Table 16.1. Endochronic models for soils are of recent origin and, being less transparent to many engineers at this stage, have been placed at the end of the table. Other models have been arranged in alphabetical order. If a particular name has been assigned by the proposer of a model, the same has been used, while if no name has been assigned the model is called after the name of the author.

Table 16.1 also gives details of the type of formulation adopted in the models. In the models based on elasto-plastic theory, the type of flow rule adopted, i.e. associated or non-associated, has been indicated as this has in important bearing on the computer implementation of the models. Similarly, the type of hardening rule, i.e. isotropic or kinematic has also been indicated. Thus a reader conversant with soil plasticity should be able to obtain a brief outline of the model from Table 16.1.

Table 16.1. Various constitutive models reviewed and their range of applicability

Sl No.	Model	Applicability		Type of formulation
		Type of soil	Type of loading	
1.	Critical state model (modified Cam Clay Model)	Normally consolidated and lightly overconsolidated clays	Monotonic	Elasto/plastic, isotropic hardening, associated flow rule
2.	Bounding surface plasticity model	Clays[1]	Monotonic, cyclic, transient	Elasto/plastic, kinematic hardening, associated flow rule
3.	Carter et al.	Clays[1]	Monotonic, cyclic, transient	Elasto/plastic + empirical isotropic hardening, associated flow rule
4.	Densification model	Sands	Cyclic, transient	Elasto/plastic + empirical ideal plasticity, non-associated flow rule
5.	Ghaboussi & Momen	Sands	Monotonic, cyclic, transient	Elasto/plastic, kinematic hardening, non-associated flow rule

Table 16.1 (*Contd.*)

Sl No.	Model	Applicability: Type of soil	Type of loading	Type of formulation
6.	Infinite number of surface (INS) model	All soils[2]	Monotonic, cyclic, transient	Elasto/plastic, kinematic hardening, associated flow rule
7.	Lade	Sands	Monotonic	Elasto/plastic, isotropic hardening, non-associated flow rule
8.	Multi-laminate model	All soils[2]	Monotonic	Elasto/plastic, isotropic/kinematic hardening, associated/non-associated flow rule
9.	Nova	All soils	Monotonic, cyclic, transient	Elasto/plastic, isotropic hardening, associated flow rule
10.	Pender	Over-consolidated clays	Cyclic, transient	Elasto/plastic + empirical isotropic hardening, associated flow rule
11.	Prevost	All soils	Monotonic, cyclic, transient	Elasto/plastic, kinematic hardening, non-associated flow rule
12.	RS model	All soils[2]	Monotonic, cyclic, transient	Elasto/plastic, isotropic hardening, non-associated flow rule
13.	Endochronic model (Bazant &)	All soils[3]	Monotonic, cyclic, transient	Endochronic
14.	Endochronic model (Valanis & Read)	All soils[3]	Monotonic, cyclic, transient	Endochronic + plasticity

1. Although models were proposed for clays, application to sands has been made by Zienkiewicz et al. (1982).
2. Detailed testing of the model has been done for clays only.
3. Detailed testing of the model has been done for sands only.

16.4 CRITERIA FOR APPRAISAL OF MODELS OF SOIL BEHAVIOUR FOR MONOTONIC LOADING

16.4.1 *Background*

Many models such as Sl No. 2, 3, 6, 8, 9, 10 and 12 in Table 16.1 are modifications of the CSM.

CSM, developed at Cambridge (UK) over the past 10–15 years, was originally intended to explain the behaviour of normally consolidated and lightly overconsolidated clay (overconsolidation ratio less than 2). It was first implemented in a finite

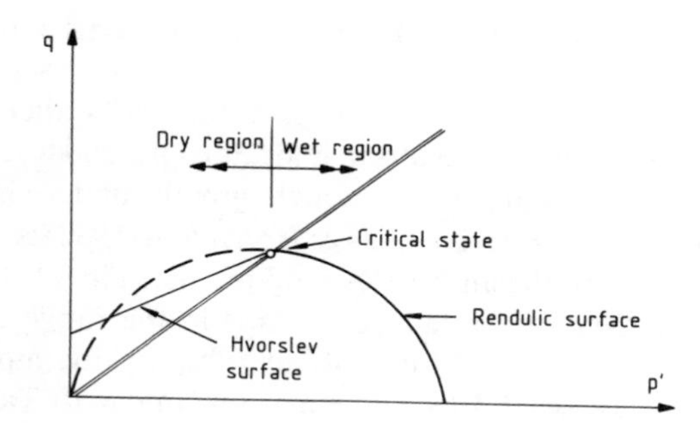

Figure 16.1. Rendulic and Hvorslev surfaces in q – p space

element code by Humpheson (1976), who also assumed a hexagonal section in the Π plane to correspond to observed independence of failure on the intermediate principal stress. In boundary value problems on loading, certain points in the soil mass may unload and their stress paths migrate into the 'dry' region. The description of a model, therefore, from the standpoint of the solution of boundary value problems is incomplete unless the model also defines the constitutive relationships in the overconsolidated region as well.

Many authors have extended the Rendulic surface in the overconsolidated region while many others have used Hvorslev's surface in this region (Figure 16.1). Irrespective of which of the above options is adopted, the CSM is known to have the following features:

– For normally consolidated clays in drained triaxial loading, the CSM predicts shear stress – shear strain response close to the observed experimental results. The volumetric strains and failure stress are also reasonably well predicted.

– For normally consolidated clays in undrained traixial loading, although failure stress, stress path and pore pressures are reasonably well predicted, the shear stress – shear strain curves are overstiff, as compared to the experimental curves, i.e. for a given shear stress, the predicted shear strain is much lower than the observed experimental shear strain.

– The value of K_0 (the ratio of horizontal stress to vertical stress) is overpredicted in all cases.

Prediction of the correct level of shear strain in monotonic loading is at least as important as the prediction of the failure stress if the model is to be extended for application to cyclic and transient loads. Further, the importance of initial stress cannot be over emphasized in any geotechnical analysis as it begins with the assumption of system of initial stresses and the soil models have an important role in predicting the initial stresses (through the prediction of K_0).

In view of the above, the following two criteria were adopted for comparison of models:

1. Octahedral shear strain ($\varepsilon = \frac{2}{3}(\varepsilon_a - \varepsilon_r)$) at 90% of failure stress ($q_f = \sigma_r$) in an undrained compression test. In the above σ_a, σ_r are the axial and radial stress respectively at failure and ε_a, ε_r are the corresponding strains.
2. Value of K_0.

16.5 AN APPRAISAL OF MODELS FOR MONOTONIC LOADING

Table 16.2 shows the % error in predicted axial strain* (ε_a) with reference to experimental results in undrained triaxial tests for various models. It is noted that most critical state based models greatly understimate the strains. The bounding surface model due to Dafalias & Herrmann (1982) is an exception. The reasons for this are not clear to the authors. All models listed in Table 16.2 show that the axial strain attained at 90% of the failure stress is in the range of 1–3% while for the bounding surface plasticity model it is about 10%, a value more in line with experimental evidence.

Table 16.3 lists information similar to Table 16.2 for overconsolidated clays. The picture here is also similar to that of normally consolidated clays. The INS model seriously under-predicts the axial strains. Pender's model is not general and has so far only been developed for triaxial space. It assumes that the undrained stress path in triaxial space is parabolic, based on experimental observations. Although failure

Table 16.2. Performance of various models for normally consolidated clays and loose sands

Model	% error in predicted ε at $0.9q_f$	Predicted ε at $0.9q_f$	Prediction by	Experiments by
Critical	−200%	1%[2]	Carter et al. (1982) [Fig. 9.23]	Taylor & Bacchus (1969)
Critical state (RS model)	−166%	1.25%	Pande & Pietruszczak (1982)	Banerjee & Stipho (1978)
Critical state (INS model) $\zeta = 1$: $\zeta = 0.9$;	Results above apply to this model as well −72%	2%	Mroz SZ & Pietruszczak	Grenoble Int. Workshop Comittee (1982)
Critical state (bounding surface plasticity	0% (?)	10%	Dafalias[3] Hermann (1982) [Fig. 10.6(b)]	Banerjee & Stipho (1978)
Ghaboussi & Momen	≃0%	1.5%	Ghaboussi & Momen (1982) [Fig. 12.10]	Ishihara, Tatsuoka & Yasuda (1975) (Fuji River sand)
Nova	−85%[1]	3%	Nova (1982) [Fig. 13.6(b)]	Tatsuoka (1972) (Fuji River sand)
Prevost	≃0%		Prevost (1977)[4]	Not known-perhaps Prevost's own

1. Without accounting for membrane penetration.
2. Most favourable prediction.
3. Possible anomalies in Figures 10.6(a)–(c). Stress paths plotted in Figure 10.6(a) do not correspond to pore pressures plotted in Figure 10.6(c).
4. Total stress model.

*There is a direct relationship between ε and ε_a for undrained triaxial conditions.

Table 16.3. Performance of various models for overconsolidated clays (undrained triaxial tests)

Model	% error in prediction of peak shear strength	% error in prediction of strain at 90% of peak shear strength	% error in prediction of pore pressure at failure	Prediction by	Experiment by
CSM					
INS	+2%	−360%	0%	Pietruszczak Zienkiewicz (1982)	El-hamrawy (1978), silty clay [OCR = 4]
INS	+6%	−300%	0%	Norris (1979)	Lewin (1973), Llyn Brianne slate dust [OCR = 8]
Bounding surface plasticity	<1%	0%	0%	Dafalias & Herrmann (1982)	Banerjee & Stipho (1978), Kaolin [OCR = 8]
RS model	+8%	?	−8%	Pande & Pietruszczak (1982)	Banerjee & Stipho (1978), Kaolin [OCR = 2]
Prevost	0%	0%	?	Prevost (1979)	Andersen (1976), Drammen clay [OCR = 4]
Pender	0%	0%	120%	Pender (1982)	Anderson (1976), Drammen clay [OCR = 4]

stresses and strains are correctly predicted by this model, prediction of excess pore pressures is far from satisfacoty. Prevost's initial model (1977, 1978) was a total stress model and thus was not capable of predicting pore pressures. Prevost has later proposed an effective stress model (Prevost 1979) but no results of excess pore pressure during undrained triaxial compression are given.

Except the INS model and Lade's model none of the models listed in Table 16.1 appear to predict realistic values of K_0. In the INS model Mroz & Norris (1982) adopt the shape of the consolidation surface represented by an ellipse whose axes do not coincide with $p'(=(\sigma'_1+\sigma'_2+\sigma'_3)/3)$ and $q=(\sigma'_1-\sigma'_3)$ axes; σ'_1 σ'_2 and σ'_3 being effective prinicipal stresses. This inclined shape of the ellipse is defined by an additional parameter, ζ (Figure 16.2). Figure 16.3 shows the variation of K_0 with effective angle of shearing resistance ϕ for various values of ζ. It is seen that experimental values reported for a wide range of soils can be predicted by an appropriate choice of ζ. The quality of K_0 predictions for the models of Ghaboussi & Momen, Nova, Pender and Prevost is not known.

In concluding this section of the chapter it can be stated that the models which are extensions of the CSM suffer from the drawbacks that their respose is too stiff and the K_0-value is overpredicted. The first drawback implies that these models cannot be successfully used for situations of cyclic and transient loading in general. For example,

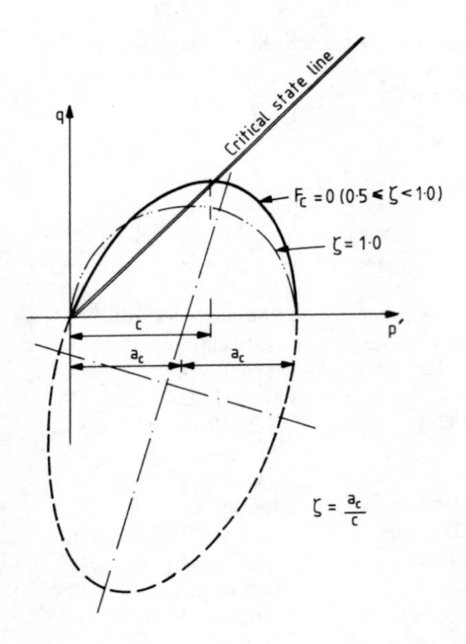

Figure 16.2. Definition of ζ in INS model

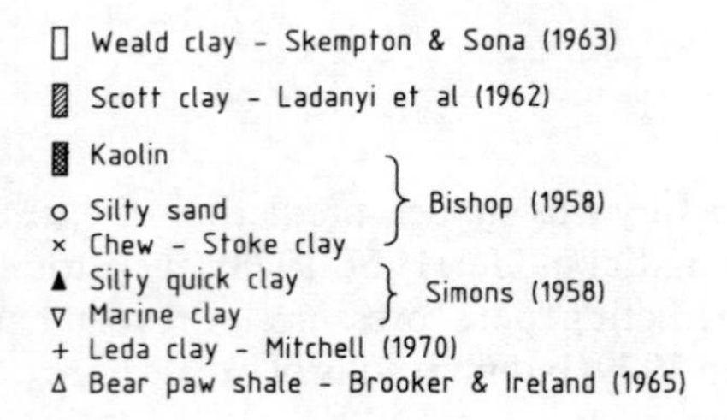

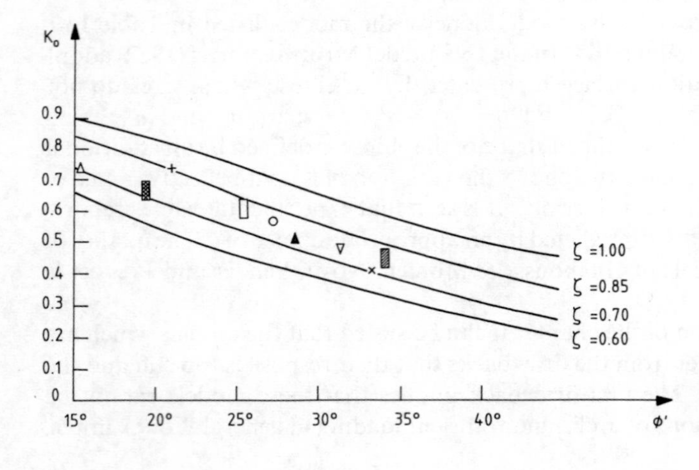

Figure 16.3. Variation of K_0 with ϕ' for different values of ζ (after Mroz & Norris 1982)

Table 16.4. Performance of various models, based on the chosen criteria for evaluation

Criterion	Model											
	CSM	Bounding surface	Carter et al.	Densi-fication	Ghaboussi & Momen	INS	Lade	Multi-laminate soil	Nova	Pender	Prevost	RS model
Normally consolidated soils: Prediction of strains in undrained triaxial conditions	P	G	P	P	G	P	G	P	P	?	G	P
Over-consolidated soils: Prediction of strains in undrained triaxial conditions	P	G	P	P	G	P	G	P	?	G	G	P
K_0 prediction	P	P	P	—	?	G	G	P	?	?	?	P

P=poor, G=good.

if the model predicts that failure stress is reached at 1% shear strain, strain controlled cyclic shear tests with an amplitude of 5% are bound to provide anomalous results as compared to experimental observations. This aspect is discussed further in the next section. The second drawback prohibits the use of the model to predict the initial stress conditions, which is an important role of the model in geotechnical engineering practice.

Table 16.4 gives, in a qualitative sense, the performance of various models based on the chosen criteria for evaluation.

16.6 CRITERIA FOR EVALUATION OF MODELS FOR CYCLIC AND TRANSIENT LOADING

Before we adopt any criteria for comparison of models, we shall look at the typical response of soils in cyclic tests.

16.6.1 *Generation of pore pressures during cyclic tests*

Figure 16.4 shows normalized excess pore pressure against number of cycles in constant total mean stress two-way cyclic triaxial tests reported by Matsui et al. (1980) on Senri clay. The curves are plotted for various levels of stress amplitudes represented

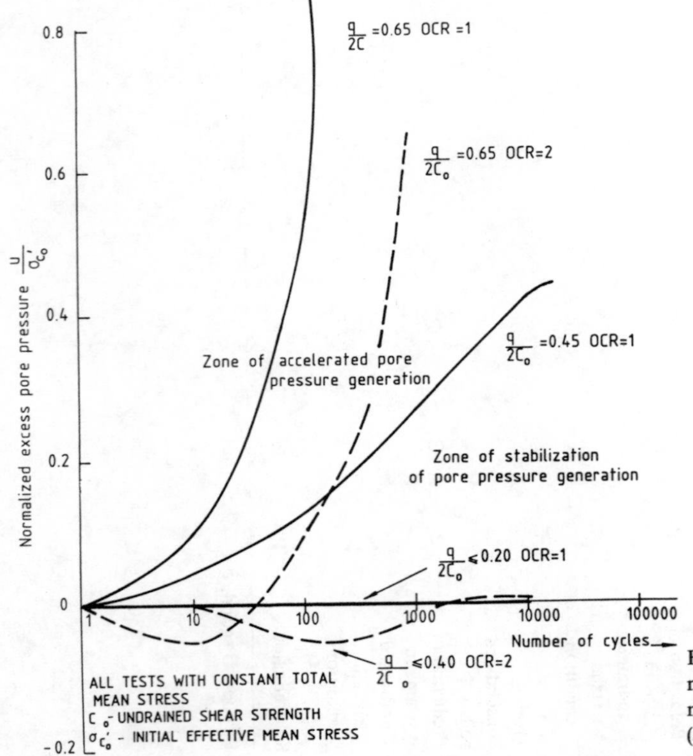

Figure 16.4. Pore pressure generation characteristics of normally consolidated Senri clay (after Matsui et al. 1980)

by $q/2c_{uo}$, where c_{uo} is the initial undrained shear strength in triaxial compression, with initial overconsolidation ratios of 1 and 2. The important observation to be made is that for $q/2c_{uo} = 0.40$, pore pressure generation stabilizes after a finite number of cycles while for $q/2c_{uo} > 0.40$ the pore pressure generation does not stabilize and eventually leads to the liquefaction of the specimen. Similar remarks can be made from the experimental results for Drammen clay in simple shear conducted by Anderson (1976) and many others. Thus it would appear that there is a threshold limit of shear stress amplitude for any soil ($q/2c_{uo} = 0.40$ for the above mentioned Senri and Drammen clays) below which the pore pressures would stabilize and above which the pore pressure generation continues unabated and eventually leads to liquefaction. We adopt this phenomenon as one of the criteria for the evaluation of numerical models. Figure 16.5 shows two typical curves of pore pressure generation – types A and B will be used to describe the characteristics of numerical models in the next section.

Experimental evidence is presently inadequate to identify during which part of the loading cycle the pore pressure generation actually takes place. Difficulties of experimental observation and associated inaccuracies could make this information unreliable. Finn et al. (1982) claim that pore pressure generation takes place during the unloading phase of the cyclic loading. From the point of view of numerical models, it is easy to identify the phase in which the generation of pore pressures takes place. Four different characteristics of pore pressure generation can be identified in the triaxial tests. These are shown in Figure 16.6 and can be summarized as follows:

– Type J: Pore pressure generation during loading as well as unloading phases.

– Type K: Pore pressure generation during loading only – pore pressure reduction on unloading.

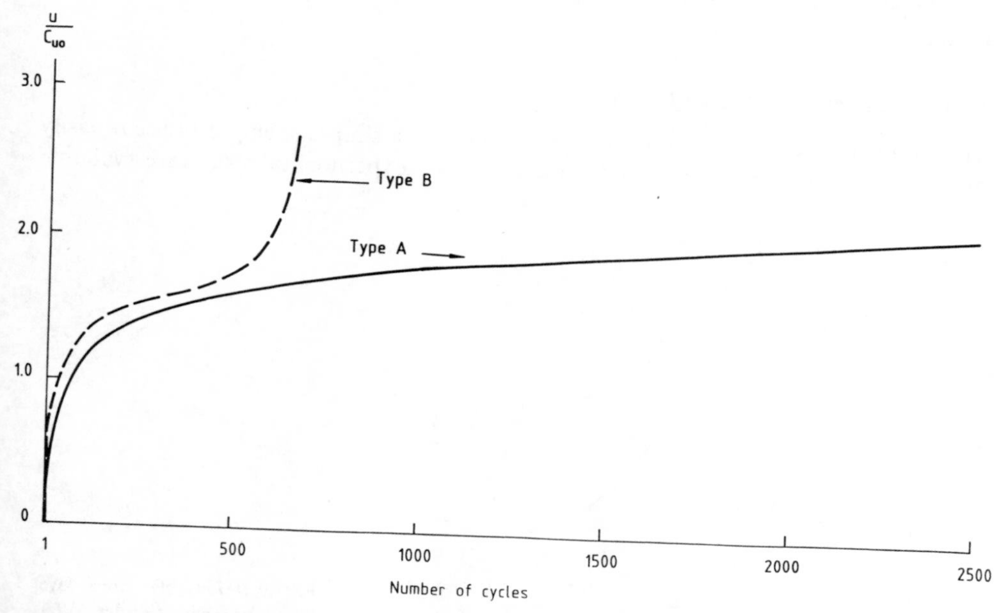

Figure 16.5. Typical generation of excess pore pressure with number of cycles

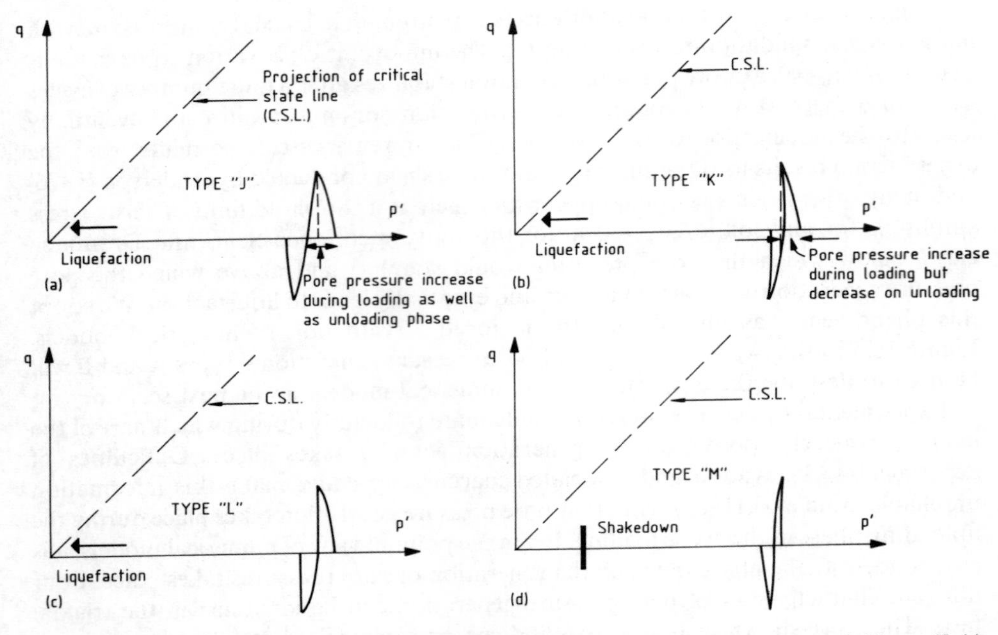

Figure 16.6 Typical characteristics of generation of pore pressure within a cycle e: (a) type J: (b) type K; (c) type L; (d) type M

– Type L: Pore pressure generation during loading, no pore pressure generation during unloading.

– Type M: Pore pressure reduction on loading, pore pressure generation on unloading.

16.6.2 *Number of cycles to failure*

The number of cycles to failure in cyclic triaxial or simple shear tests is an easily monitored observed parameter and numerous sets of experimental results are available

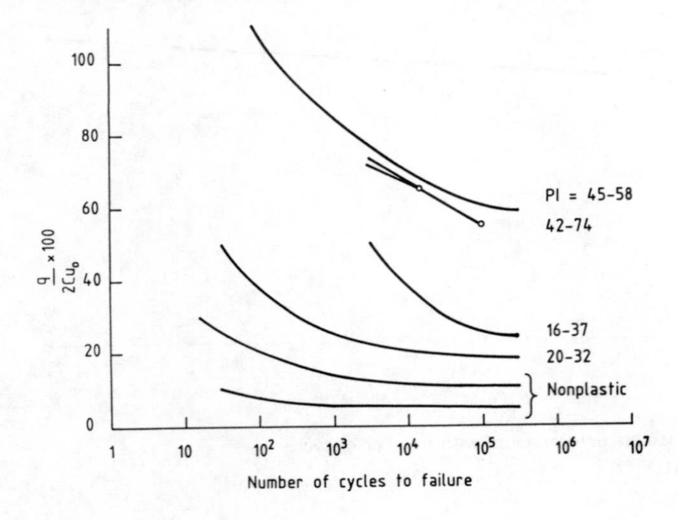

Figure 16.7. Cyclic stress level versus number of cycles to failure of six marine soils (after Houston & Herrmann 1980)

in the literature. Figure 16.7 shows the results of Houston & Herrmann (1980) for six marine soils of varying plasticity indices. The ratio of cyclic shear stress amplitude to initial monotonic failure stress (c_u) is plotted against number of cycles to failure. It is seen that for $q/2c_u = 0.3$, four soils survive up to 700,000 cycles. Results of one soil with plasticity index PI = 45–48% are rather puzzling. For this soil it is possible to have $q/2c_{uo} > 1$ and it takes nearly 100 cycles to cause failure.

We have adopted the capability of a numerical model to predict (even very approximately) the observed number of cycles to failure as one of the criteria for our appraisal of models.

16.6.3 *$G - \gamma$ plot*

From the early days of development in soil dynamics, experiments have been conducted to assess the shear modulus of soils after some cycles of pre-straining (Seed & Idriss 1970, Hardin & Drnevich 1972, Seed 1976, 1979). Tests have been conducted in the simple shear apparatus, on shaking tables and in the triaxial apparatus. The idea is to find the apparent shear modulus (G) of the soil after a finite number (adopted as 10) of shear strain cycles of constant amplitude have been applied to the soil. The tests are repeated at various shear stress amplitudes and results are expressed in the form of a $G/G_0 - \gamma$ plot where normalized apparent shear modulus (G/G_0) is plotted against the logarithm of strain amplitude (γ) (Figure 16.8), G_0 being the initial shear modulus. Experiments show that the apparent shear modulus decreases as the amplitude of shear straining is increased. For most soils, at strain amplitudes of 1%, the apparent shear modulus reduces to some 20–10% of the initial monotonic value (G_0). The test data of the type shown in Figure 16.8 form the basis for the current design practice of many structures and are also used for the evaluation of the liquefaction potential of sand deposits.

The authors believe that any constitutive model should at least qualitatively predict a $G/G_0 - \gamma$ plot of the type shown in Figure 16.8 as designers are unlikely to have confidence in a model which does not.

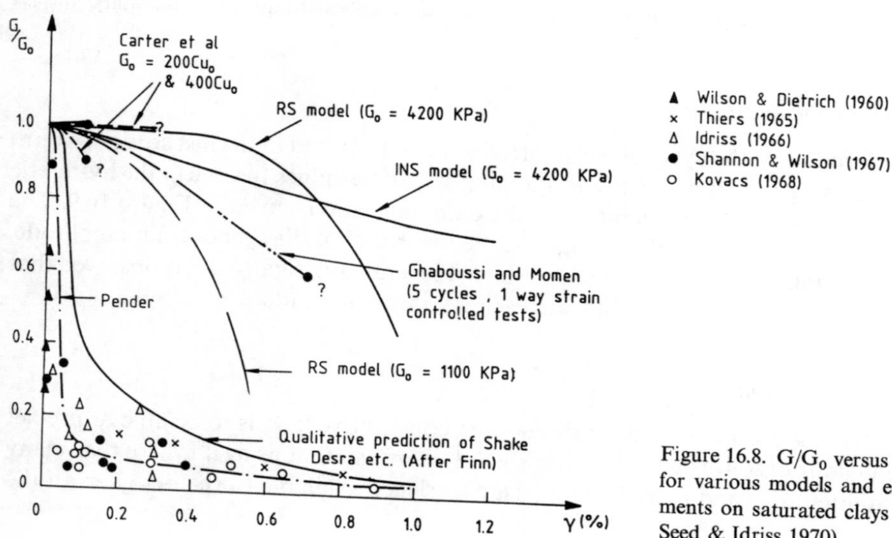

Figure 16.8. G/G_0 versus γ plot for various models and experiments on saturated clays (after Seed & Idriss 1970)

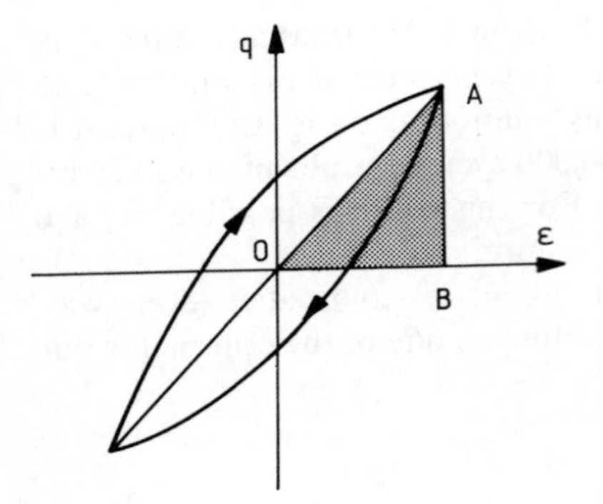

Figure 16.9. Definition of damping ratio

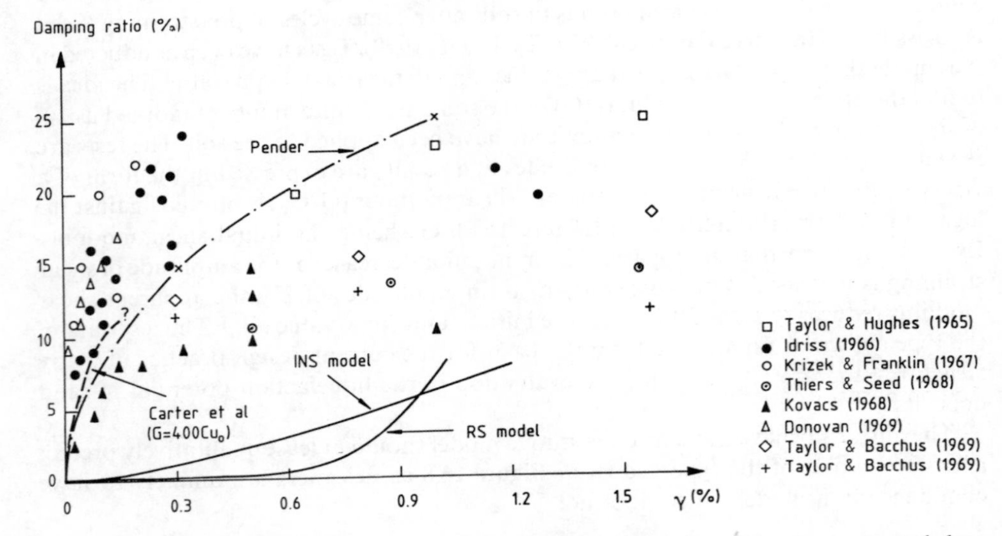

Figure 16.10. Damping ratio versus γ plot for various models and experiments on saturated clays

16.6.4 *Damping ratio – γ plot*

As a by-product of the cyclic strain controlled tests described in the last section, a term called damping ratio is defined. The damping ratio determines the size of the hysteretic loop after ten cycles. The definition of the damping ratio is given in Figure 16.9. The percentage of damping is plotted against the logarithm of the cyclic strain amplitude. Figure 16.10 shows the experimental results of several investigators. It is observed that the damping ratio decreases with increasing strain amplitude.

16.6.5 *Liquefaction*

One of the important reasons for developing constitutive laws is to be able to predict the behaviour of soil structures in transient dynamic situations as it is here that many uncertainties arise and there is little past experience available to the designer. In the

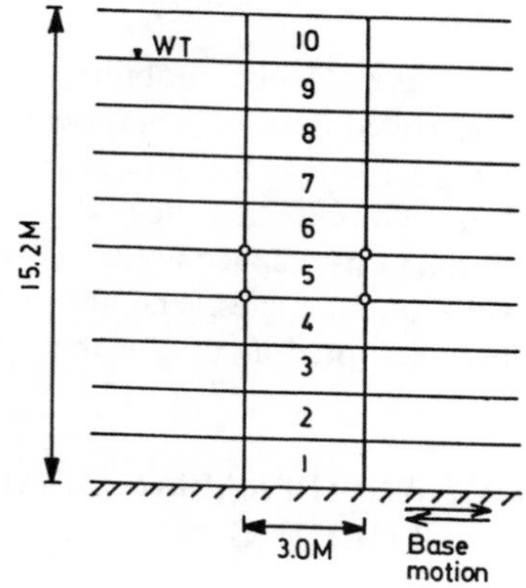

Figure 16.11. Details of typical problem of liquefaction of a sand layer

past few years procedures for dynamic analysis and prediction of liquefaction have been developed and applied to many practical problems (Seed et al. 1975, Seed 1979). Both total stress and effective stress procedures have been proposed (Finn et al. 1977). A number of simplifying assumptions are made and soil non-linearity is accounted for in a pseudo-elastic manner.

The key to the successful solution of dynamic problems of soil structures lies in the constitutive law which should be formulated based on the proper understanding of the underlying complex behaviour of soils. Admittedly, simplifying assumptions have to be made in the constitutive law for it to be of any practical use. The authors of this chapter, while going through the maze of constitutive laws proposed by various authors, decided to compare the capability of some models to predict the liquefaction phenomenon in a horizontal uniform layer of sand subjected to the El Centro earthquake with accelerations scaled to one-tenth of their original intensities. It is a standard problem used by many authors in the past to demonstrate the effectiveness of their methods. The details of this problem are given in Figure 16.11.

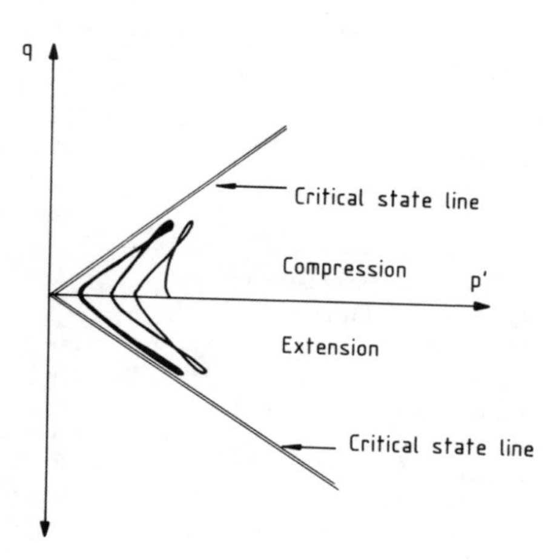

Figure 16.12. Cyclic mobility in two-way stress-controlled triaxial tests on overconsolidated materials

16.6.6 *Cyclic mobility*

The term 'cyclic mobility' has been given to a phenomenon observed in stress-controlled triaxial experiments on overconsolidated clays and dense sands, which may be described as follows: after a certain number of cycles, the material appears to have liquefied but when shear stress is applied, failure does not take place. The typical stress paths in p–q space are as shown in Figure 16.12. The stress path does not change in subsequent cycles. The ability of the constitutive model to model cyclic mobility has been adopted as one of the criteria for the evaluation of models in this chapter.

16.7 AN APPRAISAL OF MODELS FOR CYCLIC AND TRANSIENT LOADING

In this section, comments based on the various criteria stated in the previous section are offered on the various models listed in Table 16.1.

16.7.1 *Pore pressure generation characteristics*

The pore pressure generation characteristics of various models are shown in Table 16.5. It is noted that some models continue to predict generation of pore pressures with number of cycles irrespective of the stress amplitude and will finally liquefy (type B response). On the other hand, some models always predict stabilization of pore pressures after a finite number of cycles (type A response). A threshold value below which pore pressure generation should stabilize and above which pore pressure generation should continue is not demonstrated by any model unless further refinements are made. Knowing a-priori the type of pore pressure generation characteristics required, some models can be tuned to give the desired result. In the tests on Senri clay (Figure 16.4) and many other investigations cyclic stress or strain is applied up to a few thousand cycles and sometimes up to a million cycles. Even with modern computers, it is not trivial to simulate a million cycles. The authors are of the opinion that numerical errors build up rapidly even at a few thousand cycles.

Table 16.5. Summary of pore pressure: generation characteristics of various models

	Characteristic type	
Model	Overall response (see Figure 16.5)	Load/unload response (see Figure 16.6)
Bounding surface plasticity	A	L
Ghaboussi & Momen	?	J, M
INS model	A	J
Nova	A	J
Prevost	—	—
RS model	A or B	J, K, M
Carter et al.	B	L
Densification	B	L

16.7.2 *Number of cycles to failure/liquefaction*

The number of cycles to failure can be predicted reasonably successfully in low-amplitude two-way strain-controlled tests. However, for high-amplitude tests, the number of cycles to liquefaction is not predicted correctly as liquefaction is predicted in very much fewer cycles than observed. Thus while it is easy (?) to reproduce the test results of Taylor & Bacchus (1969) for cyclic amplitudes of 0.3% and 0.8% using many models, attempts to predict results for cyclic amplitude of 1.67% do not appear to have been successful.

16.7.3 *$G/G_0 - \gamma$ plot*

Figure 16.8 shows $G/G_0 - \gamma$ plots for various models. Experimental results due to many investigators are also shown in the figure. Most of the models produce $G/G_0 - \gamma$ plots of quite the wrong type. What it indicates is that most models do not produce enough plastic strains at low amplitudes of shaking. The shape of the $G/G_0 - \gamma$ plot appears to be quite sensitive to initial value of G. A lower value of G gives a sharper drop in G/G_0 for the same amplitude of shaking. Results of Ghaboussi & Momen's model for sands are based on five cycles only instead of ten. They are likely to improve if ten cycles are considered. Pender's model appears to give an excellent fit to the observed experimental data. It is hardly surprising as an equation for G/G_0 as a function of γ is built into the model. The model is not generally applicable to arbitrary stress-strain paths. Qualitative predictions of computer programs 'Shake' and 'Desra' are also plotted on Figure 16.8 and show that they are in agreement with the experimental observations. However, these programs make use only of pseudo-elastic models. On the whole the picture is not very encouraging. Even very sophisticated models like INS do not produce a reasonable $G/G_0 - \gamma$ plot. As a matter of fact, it is

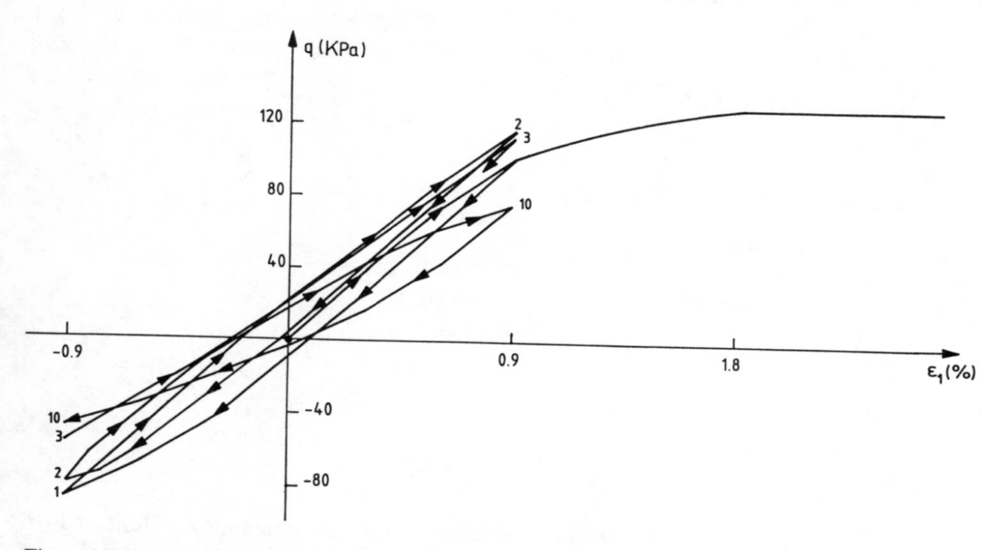

Figure 16.13. Typical stress-strain loops predicted by INS, bounding surface RS model and Carter et al's model

noticed that the INS model and many other models show an increase in G/G_0 in the first few cycles and then a drop (Figure 16.13).

16.7.4 *Damping ratio plot*

Figure 16.10 shows the damping ratio versus shear strain (γ) amplitude for four models. Experimental results of various investigators are also shown. The damping ratio, as indicated by experiments, should increase with larger amplitude of shear straining. Here the INS and RS models show a very slow rise in the damping ratio while Carter et al.'s model shows results consistent with experiments. Pender's model also shows correct response. The remarks on this model made in the last subsection apply here also.

16.7.5 *Prediction of liquefaction*

It has been possible to compare the liquefaction prediction capabilities of only three models, viz. densification model (Zienkiewicz et al. 1978, 1980, 1982), Carter et al.'s model (1982) and the RS model (Pande & Pietruszczak 1982). The problem description is given in Figure 16.11. The material parameters used in the comparison are given below:

For all models:	
Saturated unit weight and sand	$= 18$ kN/m^3
Bulk unit weight of sand above water table	$= 16$ kN/m^3
Effective friction angle, ϕ'	$= 40°$
Initial void ratio, e_0	$= 0.6$
Coefficient of earth pressure at rest, K_0	$= 1.0$
Coefficient of permeability, k	$= 10$ m/s
Young's modulus, E	$= 98\,000$ kN/m^2
Poisson's ratio	$= 0.15$
For Carter, Booker and Wroth's model:	
Compression index, λ	$= 0.0825$
Swelling index, κ	$= 0.0025$
Degradation parameter, θ	$= 0.4$
For the densification model:	
Effective cohesion, c′	$= 10$ kN/m^2
Densification strain parameters,	$= 0.04$
A B	$= 55.5$
ζ	$= 17.2$
For the RS model:	
(λ and κ – as above) and	
β	$= 0.75$

Some of the above data is rather questionable for a study of liquefaction potential. For example, a ϕ' of 40° would signify very dense sand and one would not expect liquefaction in such a layer. A value of c′ equal to 10 kN/m^2 is also unusual. However since our aim is simply to compare the results of various constitutive models the above has been adopted. In a separate paper parametric studies showing the influence of coefficient of permeability, coefficient of earth pressure at rest and initial shear stresses in the ground with more realistic data for the properties of sand have been presented (Shiomi et al. 1982).

The numerical results for the three models are presented in Figures 16.14–16.17.

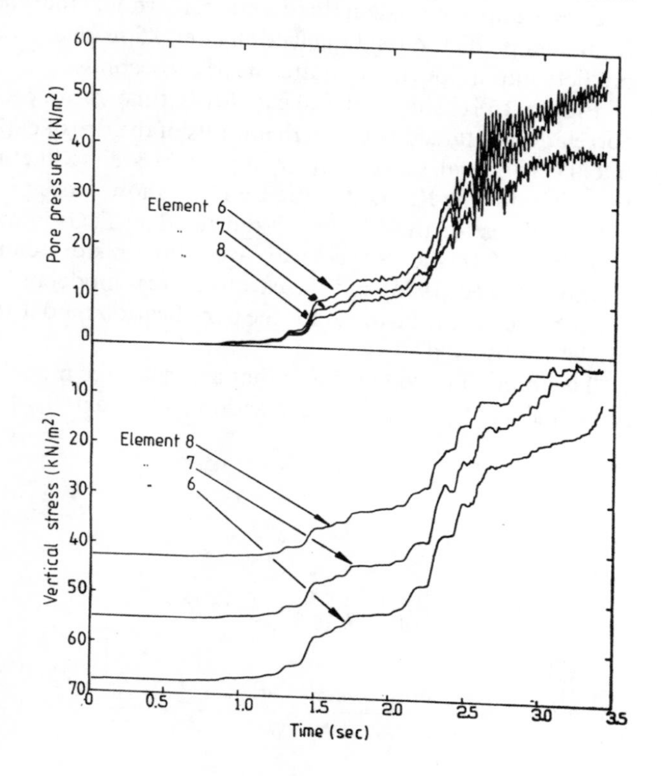

Figure 16.14. Time history of excess pore pressure and vertical stress predicted by the RS model

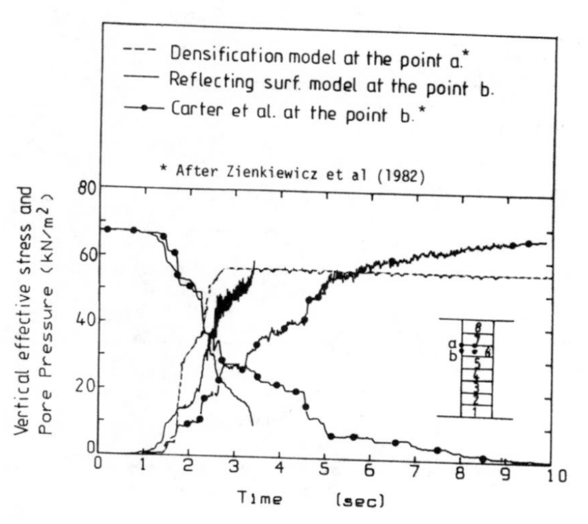

Figure 16.15. Pore pressure time history predicted by different models

Figure 16.14 shows the time history of excess pore pressure and vertical stress predicted by the RS model for elements 6, 7 and 8. It is seen that initial liquefaction (effective vertical stresses reducing to zero) takes place after 3.5 seconds in elements 8 which is 3.8 meters below the surface. Figure 16.15 compares the excess pore pressure generated during the earthquake shock predicted by the different models at the top and bottom locations of element 6 (designated a and b on the inset of the diagram). Vertical effective stresses are also plotted on this diagram. It is noted that the densification model and the RS model predict initial liquefaction after 3.5 seconds, whereas the Carter et al. model predicts initial liquefaction after nearly 9 seconds.

Figure 16.16 shows the shear stress time history of element 8. Again a good correlation obtained between the results of the densification model and the RS model, but in Carter et al.'s model shear stresses are seen to fluctuate for a considerably longer time than for the RS model. Figure 16.17 shows pore pressure profiles with depths at various times for the RS model and the densification model.

Figure 16.18 shows the shear stress time history computed using the total stress method of Seed (1975) and the effective stress model of Finn et al. (1977). A very close resemblance of the histories for the densification model, the RS model and Finn et al.'s model is observed.

The foregoing merely shows that at least three models discussed in this subsection are capable of predicting liquefaction. As a matter of fact all models having pore

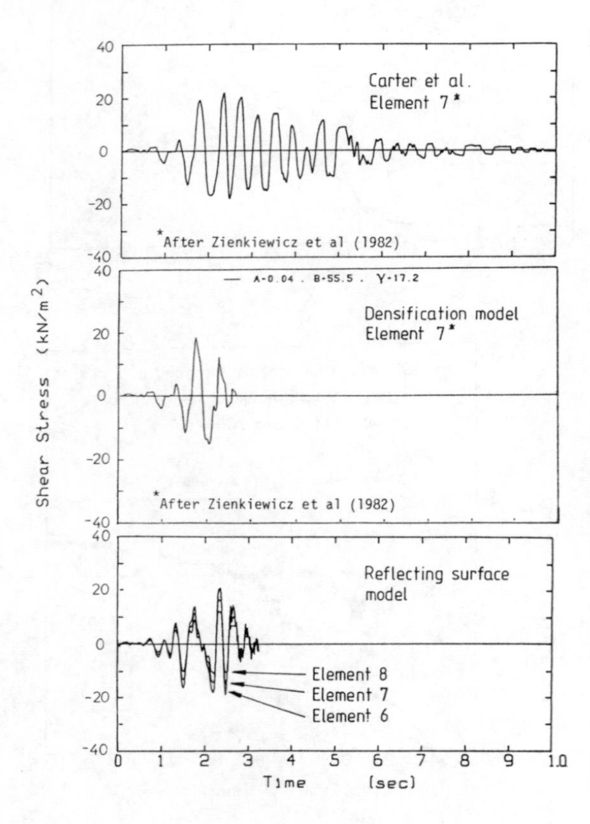

Figure 16.16. Shear stress time history of element 8 predicted by various models

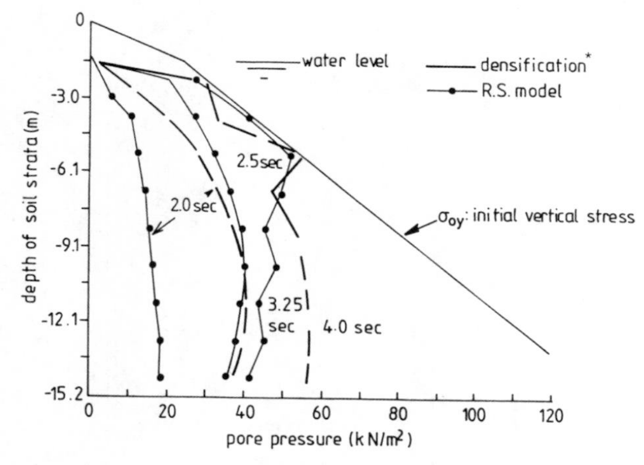

Figure 16.17. Pore pressure profiles predicted by the RS model and the densification model

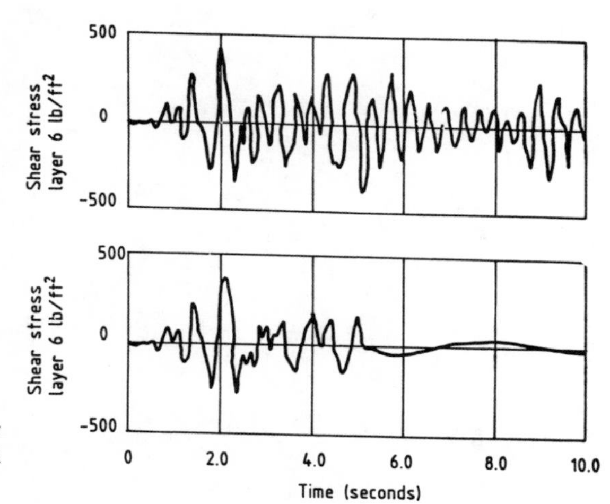

Figure 16.18. Shear stress time history computed using total stress method of Seed and effective stress model of Finn et al.

pressure generation characteristics of type **B** (Figure 16.5) would predict liquefaction while those having characteristics of type A do not predict liquefaction for any shaking.

16.7.6 *Cyclic mobility*

The phenomenon of cyclic mobility was discussed in Subsection 16.6.6. Except for the model of Pender which is specially designed for cyclic mobility, no model is likely to predict this phenomenon. Again Pender's model is rather empirical and cannot be applied in general boundary value problem situations.

Table 16.6 has been drawn up summarizing the evaluation of various models based on the criteria listed in Section 6.6. It is hoped that readers will find it a useful guide.

Table 16.6. Qualitative evaluation of various models for cyclic and transient loading

Criterion	Model											
	CSM	Bounding surface	Carter et al.	Densi-fication	Ghaboussi & Momen	INS	Lade	Multi-laminate soil	Nova	Pender	Prevost	RS model
Generation of pore pressure		A, L	A, B, L	B, J	?, M	A, J			A, J	?	?	A, B, J, K, M
Stable hysteretic		Possible	No	No	No	Yes			?	Yes	Yes	Yes
G/G-plot		P	P	P	G	P			?	G	?	P
Damping ratio plot		?	P	P	?	P			?	G	?	P
Liquefaction		No	Yes	Yes	Yes	No			?	?	Yes	Yes
Cyclic mobility		No	No	No	?	No			?	Yes	No	No

A, B, J, K, L, M = pore pressure generation characteristics; P = poor; G = good.

16.7.7 *Some additional considerations and comments*

Rotation of principal stress axes

Most of the phenomenological models discussed in this chapter follow the overall framework of the mathematical theory of plasticity and attempt to match the experimental behaviour of soils under the standard test configurations such as triaxial compression or extension. In these test configurations the directions of principal stresses remain fixed in contrast to the real boundary value problems where they rotate even under monotonic loading. In cyclic simple shear tests and transient dynamic problems the principal stress directions will rotate in each cycle. The influence of rotation of principal stress axes on the plastic strains developed is ignored in all isotropic hardening models except the multi-laminate model. The critical state model is an isotropic hardening model and by virtue of the definition of isotropic hardening does not account for the rotation of principal stress axes and consequent plastic flow induced anisotropy of strength. The kinematic hardening models like INS, and models due to Prevost and Dafalias & Herrmann do account for the rotation of principal stresses. This is indirectly achieved through the description of a 'translational rule' for the 'yield surface'. These rules are purely speculative, arbitrary and have no physical meaning. Since the parameters for these models are derived from tests in which there is no rotation of principal stress axes they are also arbitrary and without any physical basis. Pande & Sharma (1981, 1983) using the multi-laminate model have shown that increase in plastic volumetric strains due to rotation only of principal stress axes without any change in their magnitudes can be as much as 100%. Much experimental work is believed to be in progress to quantify the influence of rotation of principal stress axes (Arthur et al. 1977, 1980) and it may turn out to be an important factor which cannot be ignored in constitutive modelling.

Number of parameters

The number of parameters required to define a model is an important aspect to be borne in mind for practical applications. Another related aspect is whether parameters involved have physical meaning and whether they can be derived from standard laboratory experiments. The number of parameters involved in various models varies from 5 for CSM to about 20 for Prevost's model (Table 16.7). These figures are meant to be only indicative. It is difficult to be precise here as many parameters can sometimes be claimed (by authors) to be constants. Endochronic models appear to have a relatively large number of parameters with little or no physical meaning. At least soil engineers are likely to have problems in getting familiarized with them.

Implementation in f.e. codes

Problems of coding a soil model in a finite element system are not trivial. To the author's knowledge all models except those of Ghaboussi & Momen, Nova and Pender have been implemented in finite element programs. Of the models in Table 16.1, the CSM and multi-laminate models are the simplest to implement for monotonic loading problems and the densification model for cyclic and transient loading problems.

It is remarkable to note that although most models have been implemented in finite element programs, no boundary value problems of any practical significance appear to have been solved. It will perhaps take a long time before one sees the application of these models in practical civil engineering problems.

Table 16.7. Influence of rotation of principal stress axes, number of parameters and implementation in f.e. code for various models

	Model												Endochronic models	
Criterion	CSM	Bounding surface	Carter et al.	Densi-fication	Ghaboussi & Momen	INS	Lade	Multi-laminate soil	Nova	Pender	Prevost	RS model	Bazant et al.	Valanis & Read
Influence of rotation of principal stress axes	No	No	No	No	Ar.	Ar.	—	SPB	1	No	Ar.	No	?	?
No. of parameters	5	9	6	6	13	8–10	16	5	12	8	20	6	15	18
Whether implemented in f.e. code	Yes	Yes	Yes	Yes	?	Yes	Yes	Yes	?	No	Yes	Yes	Yes	Yes

Ar. = arbitrary; SPB = sound physical basis.

16.8 CONCLUSIONS

The authors have attempted to assess critically various soil models proposed in recent years. We do not claim that we have understood all the complexities of every model. The authors of respective models know their models best. However, an attempt has been made in this chapter to bring out possible weaknesses and strong points of various models.

Certain aspects which are extremely important from the point of view of practical acceptability by the engineering profession have been emphasized and it is hoped that engineers and scientists working in the area of constitutive modelling will find these of interest.

The authors refrain from making any specific conclusions in relation to any models as most of the models are under various stages of development and will, no doubt, be revised in due course. The readers, therefore, should make their own conclusions. All comments are summarized in Tables 16.1–16.7 and should serve as a useful guide. Practising engineers also should find these helpful in making their own judgements in specific situations.

ACKNOWLEDGEMENTS

The authors would like to thank friends and colleagues both within the Swansea group and outside with whom they had many frank and useful discussions on the subject of soil models. We are specially thankful to Dr. D.J. Naylor, Dr. V. Norris and Professor O.C. Zienkiewicz whose criticism was specially helpful.

REFERENCES

Anderson, K.H. 1976. Behaviour of clay subjected to undrained cyclic loading. In *Proceedings of Boss Conference* (Trondheim), 1: 392–403.

Arthur, J.R.F. 1980. Principal stress rotation: a missing parameter. *J. Geotech. Div.* (ASCE) (GT 4): 419–433.

Arthur, J.R.F. et al. 1977. Plastic deformation and failure in granular media. *Geotechnique* 27 (1): 53–74.

Banerjee, P.K. & A.S. Stipho 1978. Associated and non-associated constitutive relations for undrained behaviour of isotropic clays. *Int. J. for Numer. and Analyt. meth. in Geomech.* 2: 35–36.

Bazant, Z.P. & P.D. Bhat 1976. Endochronic theory of inelasticity and failure of concrete. *J. Engng. Mech. Div.* (ASCE) 102 (EM 4): 701–722.

Bazant, Z.P. & R.J. Krizek 1975. Saturated sand as an inelastic two phase medium. *J. Engng. Mech. Div.* (ASCE) 101 (EM 4): 317–332.

Carter, J.P., J.R. Booker & C.P. Wroth 1982. A critical state soil model for cyclic loading. In G.N. Pande & O.C. Zienkiewicz (eds.), *Soil mechanics–transient and cyclic loads*: 219–252. Chichester: Wiley.

Dafalias, Y.F. & L.R. Herrmann 1982. Bounding surface formulation of soil plasticity. In G.N. Pande & O.C. Zienkiewicz (eds.), *Soil mechanics – transient and cyclic loads*: 253–282. Chichester: Wiley.

Duncan, J.M. & C.Y. Chang 1970. Non-linear analysis of stress and strain in soils. *J. Soil Mech. Fdns. Div.* (ASCE) 96 (SM 5): 1629–1651.

El-Ghamrawy, M.K. 1978. A sandy clay till-some properties measured during consolidation and shear. Ph.D. thesis, University of London.

Finn, W.D.L., S.K. Bhatia & D.J. Pickering 1982. The cyclic simple shear test. In G.N. Pande & O.C. Zienkiewicz (eds.), *Soil mechanics – transient and cyclic loads*: 583–607. Chichester: Wiley.

Finn, W.D.L., K.W. Lee & G.R. Martin 1977. An effective stress model for liquefaction. *J. Geotech. Engng. Div.* (ASCE) 101 (GT 6): 517–533.

Ghaboussi, J. & H. Momen 1982. Modelling and analysis of cyclic behaviour of sands. In G.N. Pande & O.C. Zienkiewicz (eds.), *Soils mechanics – transient and cyclic loads*: 313–342. Chichester: Wiley.

Hardin, B.O. & V.P. Drnevich 1972a. Shear modulus and damping in soils: measurement and parameter effects. In *Proceedings of ASCE* 98 (SM 6): 603–624.

Hardin, B.O. & V.P. Drnevich 1972b. Shear modulus and damping of soils: design equation and curves. In *Proceedings of ASCE* 98 (SM 7): 667–692.

Houston, W.N. & H.G. Herrmann 1980. Undrained cyclic strength of marine soils. *J. Geotech. Div.* (ASCE) GT 6: 691–712.

Humpheson, C. 1976. Finite element analysis of elasto/visco-plastic soils. Ph.D. thesis, University of Wales.

International Workshop on Constitutive Models for Soils. Grenoble, September 2–4, 1982.

Ishihara, K., F. Tatsuoka & S. Yasuda 1972. Undrained deformation and liquefaction of sand under cyclic stress. *Soils and Foundations* 15 (1): 29–44.

Lade, P.V. 1977. Elasto-plastic stress-strain theory for cohesionless soils with curved yield surfaces. *Int. J. Solids and Structs.* 913: 1019–1035.

Lade, P.V. 1978. Prediction of undrained behaviour of sand. *J. Geotech. Div.* (ASCE) 104 (GT 6): 721–735.

Lewin, P.I. 1973. Test results of Llyn Brianne slate dust. Private communication.

Matsui, T., H. Ohara & T. Ito 1980. Cyclic stress-strain history and shear characteristics of clay. *J. of Geotech. Div.* (ASCE) GT 10: 1101–1120.

Mroz, Z., V.A. Norris & O.C. Zienkiewicz 1978. An anisotropic hardening model for soil and its application to cyclic loading. *Int. J. for Numer. and Analyt. Meth. in Geomech.* 2: 203–221.

Mroz, Z., V.A. Norris & O.C. Zienkiewicz 1979. Application of an anisotropic hardening model in the analysis of the elasto-plastic deformation of soils. *Geotechnique* 29: 1–34.

Mroz, Z. & V.A. Norris 1982. Elasto-plastic constitutive models for soils with application to cyclic loading. In G.N. Pande & O.C. Zienkiewicz (eds.), *Soil mechanics – transient and cyclic loads*: 173–217. Chichester: Wiley.

Norris, V.A. 1979. Simulation of triaxial tests on Llyn Brianne slate dust by means of an anisotropic plasticity model. Report C/R/355/79, Dept. of Civil Engineering, University College of Swansea.

Nova, R. 1982. Constitutive model of soil under monotonic and cyclic loading. In G.N. Pande & O.C. Zienkiewicz (eds.), *Soil mechanics – transient and cyclic loads*: 343–373. Chichester: Wiley.

Pande, G.N. & K.G. Sharma 1981. Time-dependent multi-laminate model for clays – a numerical study of the influence of rotation of principal stress axes. In *Proceedings of Symposium on Implementation of Computer Procedures and Stress-Strain Laws in Geotechnical Engineering* (Chicago), 2: 570–590. Durham: Acorn Press.

Pande, G.N. & K.G. Sharma 1983. Multi-laminate model of clays – a numerical evaluation of the influence of rotation of principal stress axes, *Int. J. for Numer. and Analyt. Meth. in Geomech.*

Pande, G.N. & S. Pietruszczak 1982. Reflecting surface model for soils. In *Proceedings of International Symposium on Numerical Models in Geomechanics* (Zurich): 50–64. Rotterdam: Balkema.

Pande, G.N. & O.C. Zienkiewicz (eds.), 1982. *Soil mechanics – transient and cyclic loads – constitutive relations and numerical treatment*. Chichester: Wiley.

Pender, M.J. 1982. A model for cyclic loading of overconsolidated soil. In G.N. Pande & O.C. Zienkiewicz (eds.), *Soil mechanics – transient and cyclic loads*: 283–311. Chichester: Wiley.

Pietruszczak, S. & O.C. Zienkiewicz 1981. Application of combined isotropic-kinematic hardening rule in the F.E. analysis of a scale model footing. INME Swansea Report, C/R/388/81.

Prevost, J.H. 1977. Mathematical modelling of monotonic and cyclic undrained clay behaviour. *Int. J. for Numer. and Analyt. Meth. in Geomech.* 1 (2): 195–216.

Prevost, J.H. 1978. Plasticity theory for soil stress-strain behaviour. *J. Engng. Mech. Div.* (ASCE) 104 (EM 5): 1177–1194.

Seed, H.B. 1976. Evaluation of soil liquefaction effects on level ground during earthquakes. ASCE Annual Meeting, October 1976, Philadelphia. Preprint 2752.

Seed, H.B. 1979. Considerations in the earthquake-resistant design of earth and rockfill dams. 19th Rankine Lecture of the British Geotechnical Society. *Geotechnique* 29 (3): 215–263.

Seed, H.B. & I.M. Idriss 1970. Soil moduli and damping factors for dynamic response analyses. Report No. EERC 70–10, Earthquake Engineering Research Centre, University of California, Berkeley.

Seed, H.B., I.M. Idriss, K.L. Lee & F.I. Makdisi 1975. Dynamic analysis of the slide in the Lower San Fernando Dam during the earthquake of February 9, 1971. *J. Geotech. Engng. Div.* 101 (GT 9): 889–911.

Shiomi, T., S. Pietruszczak & G.N. Pande 1982. A liquefaction study of sand layers using the reflecting surface model. In *Proceedings of International Symposium on Numerical Models in Geomechanics* (Zurich): 411–418. Rotterdam: Balkema.

Tatsuoka, F. 1972. Shear tests in a triaxial apparatus – a fundamental study of deformation of sand. Ph.D. thesis, University of Tokyo.

Taylor, P.W. & D.R. Bacchus 1969. Dynamic cyclic strain tests on a clay. In *Proceedings of 7th International Conference on Soil mechanics and Foundation Engineering* (Mexico City), 1: 401–409.

Valanis, K.C. 1971. A theory of visco-plasticity without yield surface (Parts I and II). *Arch. of Mech.* 23: 517–555.

Valanis, K.C. & H.E. Read 1982. A new endochronic plasticity model for soils. In G.N. Pande & O.C. Zienkiewicz (eds.), *Soil mechanics – transient and cyclic loads*: 357–417. Chichester, Wiley.

Zienkiewicz, O.C., C.T. Chang & E. Hinton 1978. Nonlinear seismic response and liquefaction. *Int. J. for Numer. and Analyt. Meth. in Geomech.* 2 (4): 381–404.

Zienkiewicz, O.C., K.H. Leung, E. Hinton & C.T. Chang 1980. Earthdam analysis for earthquakes: numerical solution and constitutive relations for non-linear (damage) analysis. In *Proceedings of Conference on Desing of Dams to Resist Earthquakes*. London: ICE.

Zienkiewicz, O.C., K.H. Leung, E. Hinton & C.T. Chang 1982. Liquefaction and permanent deformation under dynamic conditions. In G.N. Pande & O.C. Zienkiewicz (eds.), *Soil mechanics – transient and cyclic loads*: 71–103. Chichester, Wiley.

Subject index

Italics = main reference